**IET SECURITY SERIES 05**

# Iris and Periocular Biometric Recognition

**IET Book Series on Advances in Biometrics – Call for authors –**
*Book Series Editor: Michael Fairhurst, University of Kent, UK*

This Book Series provides the foundation on which to build a valuable library of reference volumes on the topic of Biometrics. *Iris and Periocular Biometric Recognition, Mobile Biometrics, User-centric Privacy and Security in Biometrics*, and *Hand-based Biometrics* are the first volumes in preparation, with further titles currently being commissioned. Proposals for coherently-integrated, multi-author edited contributions are welcome for consideration. Please email your proposal to the Book Series Editor, Professor Michael Fairhurst, at: m.c.fairhurst@kent.ac.uk, or to the IET at: author_support@theiet.org.

**Forthcoming Titles in this Series include:**

*Iris and Periocular Biometric Recognition* (Christian Rathgeb and Christoph Busch, Eds.): Iris recognition is already widely deployed in large-scale applications, achieving impressive performance. More recently, periocular recognition has been used to augment biometric performance of iris in unconstrained environments where only the ocular region is present in the image. This book addresses the state-of-the-art in this important emerging field of research.

*Mobile Biometrics* (Guodong Guo and Harry Wechsler, Eds.): Mobile biometrics aim to achieve conventional functionality and robustness while also supporting portability and mobility, bringing greater convenience and opportunity for deployment in a wide range of operational environments. However, achieving these aims brings new challenges, stimulating a new body of research in recent years, and this is the focus of this timely book.

*User-centric Privacy and Security in Biometrics* (Claus Vielhauer, Ed.): The rapid emergence of reliable biometric technologies has brought a new dimension to this area of research, allowing the development of new approaches to embedding security into systems and processes, and providing opportunities for integrating new elements into an overall typical security chain. This book provides a comprehensive overview of leading edge research in the area.

*Hand-based Biometric Methods and Technologies* (Martin Drahanský, Ed.): This book provides a unique integrated analysis of current issues related to a wide range of hand phenomena relevant to biometrics. Generally treated separately, this book brings together the latest insights into 2D/3D hand shape, fingerprints, palmprints, and vein patterns, offering a new perspective on these important biometric modalities.

# Iris and Periocular Biometric Recognition

Edited by
Christian Rathgeb and Christoph Busch

The Institution of Engineering and Technology

Published by The Institution of Engineering and Technology, London, United Kingdom

The Institution of Engineering and Technology is registered as a Charity in England & Wales (no. 211014) and Scotland (no. SC038698).

© The Institution of Engineering and Technology 2017

First published 2017

The Institution of Engineering and Technology
Michael Faraday House
Six Hills Way, Stevenage
Herts SG1 2AY, United Kingdom

www.theiet.org

**British Library Cataloguing in Publication Data**
A catalogue record for this product is available from the British Library


**ISBN 978-1-78561-168-1 (hardback)**
**ISBN 978-1-78561-169-8 (PDF)**


Typeset in India by MPS Limited

# Contents

            schemes                                                        303
        13.4.1    Stolen key-inversion attacks                            304
        13.4.2    Attacks via record multiplicity                         305
    13.5    Countermeasures to software attacks                           306
    13.6    Conclusions                                                   308
    References                                                            308

PART V    **Privacy protection and forensics**                           **317**

**14    Iris biometric template protection**                             **319**
    *Christian Rathgeb, Johannes Wagner, and Christoph Busch*

    14.1    Introduction                                                  319
    14.2    Iris template protection schemes                              321
        14.2.1    Iris-biometric cryptosystems                           322
        14.2.2    Cancelable iris biometrics                             324
    14.3    Implementation of iris template protection schemes            326
        14.3.1    Iris fuzzy vault                                       326
        14.3.2    Bin-combo for iris-codes                               329
    14.4    Experimental evaluations                                      329
        14.4.1    Performance evaluation                                 330
        14.4.2    Discussion                                             332
    14.5    Summary and research directions                              334
    References                                                            336

**15    Privacy-preserving distance computation for IrisCodes**          **341**
    *Julien Bringer, Hervé Chabanne, and Constance Morel*

    15.1    Introduction                                                  341
    15.2    Secure distance computation in the semi-honest model          343
        15.2.1    Oblivious transfer                                     343
        15.2.2    Yao's garbled circuits                                 344
        15.2.3    GSHADE in the semi-honest model                       345
        15.2.4    Privacy-preserving distance computation for
                  IrisCodes in the semi-honest model                    346
    15.3    Secure distance computation in the malicious model            349
        15.3.1    Yao's garbled circuits in the malicious setting        349
        15.3.2    GSHADE in the malicious setting                       350
        15.3.3    Privacy-preserving distance computation for
                  IrisCodes in the malicious model                      351
    15.4    Application to other iris representations                     355
    15.5    Conclusion                                                    355
    Acknowledgments                                                       356
    References                                                            356

# Preface

Nowadays, biometric recognition represents an integral component of identity management systems, providing a robust and reliable way of recognizing individuals. Iris recognition has received significant attention in the recent past and is already widely deployed in several large-scale nation-wide projects. Due to its intricate structure the iris texture constitutes one of the most powerful biometric characteristics. Under active participation of captured subjects, existing approaches to iris recognition achieve well-documented resistance against false matches and, hence, auspicious recognition accuracy. Periocular biometric recognition systems, which process the externally visible skin region of the face that surrounds the eye socket, have been introduced to improve biometric performance in unconstrained environments. More recently, the use of periocular biometric recognition in the fields of surveillance as well as mobile applications has become of particular interest.

This book provides a comprehensive collection of current research topics in the field of iris and periocular biometric recognition by a wide variety of experts from academia and industry. The book is intended to complement existing literature in the area of iris and periocular biometric recognition. In-depth investigations, accompanied by comprehensive experimental evaluations, provide the reader with theoretical and empirical explanations of fundamental and current research topics. Moreover, future directions and issues still to be solved in the fields of iris and periocular biometric recognition are discussed.

## Target audiences

While being of primary interest to researchers and practitioners in the field of biometric recognition, this book will appeal to a broad readership. In particular, introductory chapters provide a comprehensive overview to the covered topics, which address readers wishing to gain a brief overview of the current state of the art. Subsequent chapters, which will delve deeper into various research challenges, are oriented towards advanced readers, in particular, graduate students. Moreover, the book provides a good starting point for young researchers as well as a reference guide pointing at further literature. Hence, the book could also be used as a recommended text for courses in the area of biometric recognition and the more general security area.

# Structure and plan

The book is divided into six different parts comprising a total of 20 chapters. Parts, as well as distinct groups of chapters, are intended to be fairly independent and the reader is therefore able to focus on only relevant parts or chapters where this is helpful.

*Part I – Introduction to Iris and Periocular Recognition:* in the first part of this book two introductory chapters provide the reader with an overview of fundamentals in iris recognition (Chapter 1) and periocular recognition (Chapter 2). These chapters should serve as points of entry for both of the main topics of this book.

*Part II – Issues and Challenges:* selective issues and challenges in the field of iris and periocular biometric recognition are elaborated in the second part of the book. Topics include robust iris segmentation (Chapter 3), iris image quality (Chapter 4), indexing and retrieval of iris biometric databases (Chapter 5), an analysis of the discrimination capability provided by different periocular sub-regions (Chapter 6), as well as investigations on novel sensing devices (Chapter 7).

*Part III – Soft Biometric Classification:* the third part of this book provides the reader with a collection of recent achievements in the area of soft biometrics. The first two chapters of this part investigate gender classification based on iris (Chapter 8) and the periocular region (Chapter 9). The third chapter provides insights on iris-based age prediction (Chapter 10).

*Part IV – Security Aspects:* in the fourth part of the book focus is put on different security issues relevant to iris recognition systems. The first two chapters present summaries of iris biometric presentation attack detection in general (Chapter 11) and, more specifically, the detection of textured contact lenses (Chapter 12). Furthermore, an overview of software attacks on iris recognition systems is given (Chapter 13).

*Part V – Privacy Protection and Forensics:* privacy-related topics, in particular iris biometric template protection (Chapter 14) and privacy-preserving comparison score estimation (Chapter 15), are investigated in the fifth part of this book. In addition, forensic topics including the identification of iris sensors from iris images (Chapter 16) and the matching of periocular images to good-quality iris images (Chapter 17).

*Part VI – Future Trends:* the last part of the book discusses future trends including iris recognition for embedded systems (Chapter 18) and mobile iris recognition (Chapter 19). Finally, a concluding chapter provides insights to future trends from an industry perspective (Chapter 20).

Christian Rathgeb and Christoph Busch, Hochschule Darmstadt, Germany
Editors

# Acknowledgements

We would like to express our special thanks to the editors at the Institution of Engineering and Technology (IET) and the editor of this book series Prof. Michael Fairhurst for their patience and helpful advice during the development of this book. We also would like to thank all of the authors for the smooth cooperation and their excellent contributions to this project. Our work on this project was supported by the German Federal Ministry of Education and Research (BMBF) as well as by the Hessen State Ministry for Higher Education, Research and the Arts (HMWK) within the Center for Research in Security and Privacy (CRISP). Finally, we would like to thank our families and friends for their support and encouragement while we worked on this book.

# Foreword

In late 2013 the IET published a book which I edited on *Age factors in Biometric Processing* recognizing the importance, in a world where the practical deployment of biometric systems is becoming increasingly widespread, of understanding how the changes induced by natural ageing are critical in developing more reliable and acceptable future systems. The diversity of the contributions in the book also demonstrated how bringing together different perspectives on a topic of such fundamental importance can provide new insights to illuminate and integrate issues of relevance and concern. This approach led to the IET launching a further Book Series to address other similarly important topics. The present volume is therefore effectively the second book in the Series overall, but the first in what has now become this extended series (the *IET Book Series on Advances in Biometrics*), building on the successful formula established in the initial book.

The landscape of biometrics is changing and developing rapidly. Although the human iris is now well established as a source of information for particularly reliable and robust identity analysis, as systems develop and, especially, as we seek to design systems which can more flexibly capture and analyse data, providing a more effective or convenient interaction experience for users, it has become clear that we need to extend our thinking on how to utilize available data. Periocular biometric processing provides a means of extending the power of iris-based processing and the range of information which is utilized, especially in scenarios where it is not possible to control explicitly the operational environment in which a system is embedded. The gradual appearance of databases containing images which extend the region of interest from the iris alone to a wider area around the eye demonstrates the increasing importance of periocular processing. As will be seen, this book provides a comprehensive survey of current state-of-the-art work both in relation to the iris itself as a biometric data source and to the diversity of factors which can be manipulated in order to extend and enhance how we deal with human identification based on the whole periocular region.

The contributors come from a variety of backgrounds, and the volume overall represents an integration of views from across the spectrum of stakeholders, including, of course, academia and industry. We hope that the reader will find this a stimulating and informative approach, and that this book will take its place in the emerging Book Series as a valuable and important resource which will support the development of influential work in this area for some time to come.

Other books in the Book Series are in production, and we look forward to adding regularly new titles to inform and guide the biometrics community as we continue to grapple with fundamental technical issues and continue to support the transfer of the best ideas from the research laboratory to practical application. It is hoped that this Book Series will prove to be an on-going primary reference source for researchers, system users and students, and for anyone who has an interest in the fascinating world of biometrics where innovation is able to shine a light on topics where new work can promote better understanding and stimulate practical improvements. To achieve real progress in any field requires that we understand where we have come from, where we are now, and where we are heading. This is exactly what this book and, indeed, all the volumes in this Book Series aim to provide.

Michael Fairhurst, University of Kent, UK
Series Editor, Advances in Biometrics Book Series

*Part I*

**Introduction to Iris and Periocular Recognition**

*Chapter 1*

# Fundamentals in iris recognition

*Christian Rathgeb and Christoph Busch*

## 1.1  Introduction

To confirm an individual's identity accurately and reliably iris recognition systems analyse the texture that is visible in the iris of the eye. The iris is the circular shaped muscle between cornea and lens, which regulates the light transmitted to the retina by adapting the width of the pupil. It is a well protected, internal organ of the eye, which is externally visible from distance up to some meters. The rich random pattern of the iris, which consists of arching ligaments, furrows, ridges, crypts, corona, freckles and a zigzag collarette, constitutes a powerful biometric characteristic. Except for eye colour, iris patterns are epigenetic (not genetically determined) characteristics randomly emerging during gestation. The iris's structural pattern has been found to be substantially diverse across a population, where left and right iris patterns of a single subject and even the irides of monozygotic twins are uncorrelated. Moreover, apart from pigmentation changes, iris patterns have been found to be apparently stable over time. Examples of iris images captured under near-infrared (NIR) light and in the visible band of light are shown in Figure 1.1. Anatomical structures of the eye and the iris pattern are schematically illustrated in Figure 1.2.

More than 20 years after the first iris recognition algorithm has been developed by Daugman [1], iris biometrics remains an active and rapidly growing field of research. Iris recognition has legendary resistance to false matches revealing auspicious recognition accuracy enabling reliable biometric identification in large-scale systems. Nationwide deployments of iris recognition technologies underline its tremendous inroads. Nonetheless, to improve iris recognition and to facilitate its use, researchers have to solve several open issues, such as robust iris processing in non-ideal (mobile) environments, cross-sensor and cross-spectral recognition, extraction of soft-biometric information or privacy-related issues.

This introductory chapter is intended to provide a rough overview of fundamentals in iris recognition, while a deeper insight into selective topics will be given in subsequent chapters. The remainder of this chapter is organized as follows: Section 1.2 summarizes key components of generic iris recognition systems. The current landscape of research and industry in the field of iris recognition is briefly summarized in Section 1.3. Further, open issues and challenges are discussed in Section 1.4. Finally, a summary is given in Section 1.5.

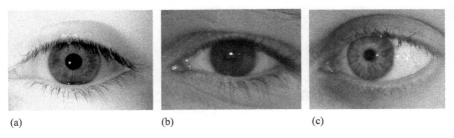

(a)                              (b)                              (c)

*Figure 1.1    Iris images captured under (a) NIR light, (b) visible light (brown) and
(c) visible light (blue) (images taken from BioSecure [2] and UBIRISv2
iris database [3])*

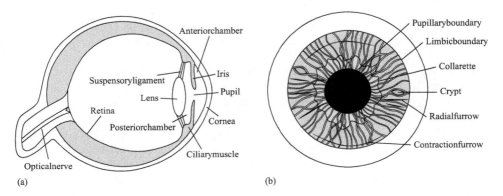

*Figure 1.2    Anatomical structure of (a) the eye and (b) features within the iris
pattern*

## 1.2    Iris recognition processing chain

The ever-increasing demand on biometric systems has entailed continuous proposals
of different iris recognition techniques. Still, the processing chain of traditional iris
recognition systems, as illustrated in Figure 1.3, has remained almost unaltered and
can be divided into four major modules: (1) image acquisition, (2) pre-processing
including segmentation, (3) feature extraction and (4) comparison; the description of
key modules will be primarily based on Daugman's approach [4], which is the core
of most operational deployments of iris recognition.

### 1.2.1    Image acquisition

Image acquisition plays a critical role in iris recognition, since poor imaging conditions negatively affect the recognition accuracy of the system. The certification of

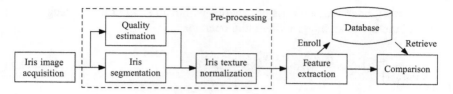

*Figure 1.3    Generic iris biometric processing chain*

inferior quality cameras has led to unpractical biometric performance in the past [5]. The majority of commercial iris acquisition cameras captures iris images in the NIR band between 700 and 900 nm wavelengths, which reveals structural patterns even for strongly pigmented irises, in contrast to images acquired under visible light, cf. Figure 1.1(b). Different types of iris acquisition cameras include mounted access control cameras, portable hand-held cameras, as well as stand-off portal devices [6]. Conventional iris cameras emit NIR light, which is reflected by the iris tissue and collected by the sensor. During exposure time, the iris must be within a sensor-dependent spatial volume and oriented towards the camera within a narrow angular cone to produce a sharp in-focus image. A certain illumination level is required to compensate for motion blur using a fast shutter as well as to maximize the iris area (and to avoid dilated pupils). While early iris recognition systems required subjects to present their iris within a rather close range below 0.5 m, today's stationary systems are capable of capturing iris images at a distance of 2 m within a capture volume of approximately 1 m$^3$ (on the move). Lately, the increasing popularity of smart phones has led to the development of mobile devices with NIR sensing capabilities for iris recognition [7]. Moreover, the usefulness of light-field cameras for iris recognition has recently been explored [8].

## 1.2.2   Pre-processing

The aim of the pre-processing step is to precisely locate the iris in a captured image. The segmentation of the iris involves a detection of inner and outer iris boundaries, a detection of eyelids, an exclusion of eyelashes as well as contact lense rings and a scrubbing of specular reflections [4]. Simultaneously, diverse quality factors of the captured iris image are estimated according to established metrics [9]. Finally, the iris is mapped to dimensionless coordinates, i.e., a normalized rectangular texture, based on which the feature extraction is performed.

Iris segmentation usually starts with a coarse detection of the pupil based on analysis of the (ideally three-valued) horizontal gray-level profile through the pupil. Due to the hollow structure inside the cornea, the gray-level of the pupil is usually at the lowest level and thus, may be easily located. Localization of the sclera can be performed similarly and turns out to be more robust in case pupillary boundaries are less pronounced, in particular in visible wavelength iris images [10].

To detect inner and outer iris boundaries within an image $I(x,y)$, Daugman [1] proposed the use of an integrodifferential operator,

$$\max_{(r,x_o,y_0)} \left| G_\sigma(r) * \frac{\partial}{\partial r} \oint_{r,x_0,y_0} \frac{I(x,y)}{2\pi r} \, ds \right|, \tag{1.1}$$

where the center $(x_0, y_0)$ and radius $r$ describe the circular boundary. The operator applies a Gaussian smoothing function $G_\sigma$ searching for the optimal circular boundary yielding a maximum difference in intensity values. When discretized in implementations, the change in pixel values is estimated and smoothing is progressively reduced by iterative application of the operator. In past years, numerous improvements to this initial concept have been proposed, e.g., [11–13], in order to achieve more robustness in the segmentation process, especially for non-ideal iris images. Alternatively, an iris boundary localization approach based on a circular Hough transform was proposed by Wildes [14]. Representing another popular technique for iris boundary localization the application of circular Hough transform has also been developed further by several researchers, e.g., [15–17].

Based on the above annular iris models, which extract concentric circular boundaries, Daugman [18] suggested to encode boundary shapes accurately using active contours. Initially, a Fourier expansion of $N$ angular data points of radial gradient edge data $\{r_\theta\}$ for $\theta = 0$ to $N - 1$ spanning $[0, 2\pi]$ is computed. Then a set of $M$ discrete Fourier coefficients $\{C_k\}$ is derived from the data sequence $\{r_\theta\}$,

$$C_k = \sum_{\theta=0}^{N-1} r_\theta e^{-2\pi i k\theta/N}. \tag{1.2}$$

From these $M$ discrete Fourier coefficients, an approximation to the inner or outer iris boundary is obtained by the Fourier series $\{R_\theta\}$,

$$R_\theta = \frac{1}{N} \sum_{k=0}^{M-1} r_\theta e^{-2\pi i k\theta/N}, \tag{1.3}$$

where the trade-off between fidelity to the true boundary and the stiffness of the active contour is set by $M$. Due to the fact that iris boundaries are rarely true circles this approach significantly enhances the biometric performance of an iris recognition system [18]. The use of active contours has been further improved by different researchers, e.g., [19,20]. Subsequently, noise removal is performed by employing methods specifically designed to detect occluded iris regions resulting from the previously stated noise sources. Iris segmentation for visible wavelength iris images requires further adaptations and might differ a lot from the above de-facto standard methods for NIR iris images [21], see Section 1.4.1.

Before and after the segmentation process the quality of the iris image is examined in order to minimize time and resources spent on processing non-standard images. Quality metrics such as motion or defocus blur can be assessed prior to the segmentation process. Requirements for relevant iris image quality factors, such as usable iris area, iris–sclera contrast, iris–pupil contrast or pupil boundary circularity, are defined

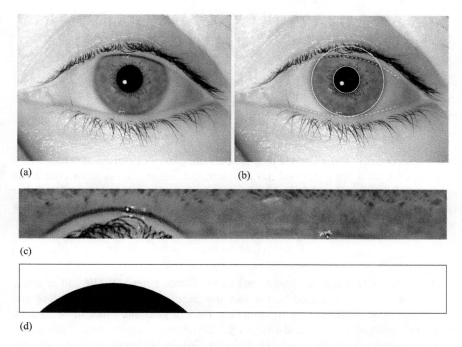

*Figure 1.4    Generic iris biometric pre-processing steps (image taken from BioSecure iris database [2]): (a) iris image, (b) iris segmentation, (c) normalized texture and (d) noise mask*

in [9]. An overview of existing works on iris quality estimation can be found in [22]. Automated feedback on iris image quality improves usability and value of remote or offline enrolments. Iris image quality metrics can serve as a predictor of biometric performance [9], in particular, a fusion of segmentation quality and final comparison scores has been found to improve the recognition accuracy of iris recognition systems [23]. Kalka *et al.* [24] report defocus blur, motion blur and off-angle to be the quality factors with highest impact on recognition accuracy.

Iris texture normalization refers to mapping of the iris image from raw coordinates to a doubly dimensionless non-concentric coordinate system, using the found iris boundaries. Similarly, a binary noise mask is generated based on previously performed noise detection. In the rubbersheet model introduced by Daugman [1], a rendering map is established as a linear combination from inner and outer iris boundary curves. While this approach has been widely applied by researchers, more recently it has been found that nonlinear deformation effects can occur in the case of great variation in pupil dilation. Tomeo-Reyes *et al.* [25] proposed a nonlinear normalization based a biomechanical model of the iris to compensate for such effects. Finally, the contrast of the resulting iris texture might be enhanced by applying histogram stretching methods. Figure 1.4 depicts the processing chain of generic iris pre-processing for a sample iris image.

*Figure 1.5    Encoding of real filter responses of a 2D Gabor wavelet applied to the iris texture of Figure 1.4(c)*

### 1.2.3    Feature extraction

In his pioneer work, Daugman [1] proposed the use of two-dimensional Gabor filters with zero DC to a pre-processed iris image in the dimensionless polar coordinate system, $I(\rho, \phi)$. Local regions of the iris are projected onto quadrature 2D Gabor filters at different scales generating a complex-valued response. Each complex Gabor response is encoded into two bits by the signs of its real and imaginary value

$$h_{\{Re,Im\}} = sgn_{\{Re,Im\}} \int_\rho \int_\phi I(\rho, \phi)e^{-i\omega(\theta_0 - \phi)}e^{-(r_0 - \rho)^2/\alpha^2}e^{-(\theta_0 - \phi)^2/\beta^2} \rho d\rho d\phi \quad (1.4)$$

where $r_0$, $\theta_0$, $\omega$, $\alpha$ and $\beta$ are parameters of the Gabor filter [4]. The use of phase information only is motivated by the fact that amplitude information is less discriminative as it is affected by illumination, camera gain and other noise factors. Due to the correlation of real and imaginary filter responses operational deployments of iris recognition have used only the real part [26]. In [1] resulting iris-codes consist of 2,048 bits. A pre-processed iris texture might also be interpreted as a set of one-dimensional signals averaged from adjacent rows which can be processed using one-dimensional Gabor filters, as proposed by Masek [27]. An example of an iris-code is depicted in Figure 1.5.

Numerous alternative methods have been suggested for the purpose of iris feature extraction. It has been shown that the vast majority of proposed schemes resemble that of [4], in the sense that these quantize responses of other linear transforms or filters (replacing Gabor filters), which are applied to iris textures in a similar manner [28]. Most notably, Krichen *et al.* [29] suggested the use of packets of Gabor wavelets instead of applying the classical wavelet approach at different scales. Monro *et al.* [30] presented an iris feature extraction method based on differences of discrete cosine transform (DCT) coefficients. Tan *et al.* [31,32] proposed different types of custom-built filters, including circular symmetric filters and tripole filters. Further, Tan *et al.* [33] suggested an iris feature extractor based on a characterization of key local variations, i.e., an encoding of minima and maxima obtained from one-dimensional dyadic wavelet transform responses.

Despite the above de-facto standard approaches, some entirely different schemes have been proposed. Belcher and Du [34] suggested the use of scale invariant feature transform (SIFT) to extract feature vectors consisting of SIFT-keypoints and their descriptors. SIFT might also be employed to detect and match macro-features, i.e., structures within the iris texture, such as furrows or crypts, as suggested by Sunder and Ross [35]. More recently, Shen and Flynn [36] proposed a detection and a shape-based comparison technique for iris crypts. While the latter approaches might not outperform traditional systems, neither in terms of recognition accuracy nor in comparison speed, they potentially complement their feature vectors yielding potential improvement in a fusion scenario [35,37].

## 1.2.4 Comparison

In generic iris recognition systems the binary data representation of iris-codes enables a rapid comparison. Typically, comparisons between iris-codes are implemented by the simple Boolean exclusive-OR operator (XOR) applied to a pair of iris-codes, masked (ANDed) by both of their corresponding noise masks to prevent occlusions caused by eyelids or eyelashes from influencing comparisons. The XOR operator ($\oplus$) detects disagreement between any corresponding pair of bits, while the AND ($\cap$) operator ensures that the compared bits are both deemed to have been uncorrupted by noise. The population count ($\|\cdot\|$), i.e., number of ones, of the resulting bit vector and of the ANDed mask template are then measured in order to compute a fractional Hamming distance (*HD*) as a measure of the dissimilarity between a pair of iris-codes, *codeA*, *codeB* and their according noise masks, *maskA*, *maskB* [4],

$$HD = \frac{\|(codeA \oplus codeB) \cap maskA \cap maskB\|}{\|maskA \cap maskB\|}. \tag{1.5}$$

The estimation of *HD* scores is based on logical operations and intrinsic functions, achieving millions of iris-code comparisons per second per CPU core [4]. Non-genuine *HD* scores yield a binomial distribution with mean $p$ and variance $\sigma^2$. Due to a certain correlation of bits in iris-codes *HD* scores are binomially distributed with a reduction in the number of compared bits. The provided degrees of freedom $d = p(1 - p)/\sigma^2$ is an indicator for the entropy of extracted iris-codes, e.g., $d = 249$ in [4]. To prevent false matches resulting from the comparison of only a small number of unmasked bits, Daugman suggests to employ a normalized score [18],

$$HD_{norm} = 0.5 - (0.5 - HD)\sqrt{\frac{n}{960}}, \tag{1.6}$$

where $n$ is the number of unmasked compared bits and 960 is the average number of unmasked compared bits in the implementation of [18]. Hence, all deviations from a 0.5 *HD* score are rescaled in proportion to the square root of the number of bits that were compared when obtaining that score.

In the comparison stage circular bit shifts are applied to iris-codes and *HD* scores are estimated at $K$ different shifting positions, i.e., relative tilt angles. The minimal obtained *HD*, which corresponds to an optimal alignment, represents the final score. Hence, score distributions are skewed towards lower *HD* scores, which (for a given threshold) increases the probability of a false match by the factor $K$ [26]. It is important to note that the number of shifting positions employed to determine an appropriate alignment between pairs of iris-codes may vary depending on the application scenario. Some public deployments of iris recognition go as far as $K = 21$ shifting positions when hand-held cameras are used for which it is obviously more difficult to ensure an upright capture orientation [26].

Different other iris-biometric comparators have been proposed in order to improve the biometric performance of an iris recognition system. In case several iris-codes are extracted during enrolment these can be combined to a single iris-code using a majority decoder [38]. Alternatively, unstable or 'fragile' bits, i.e., bits which exhibit a higher probability than others to flip their value during a genuine comparison, can be masked out, as suggested by Hollingsworth *et al.* [39]. Dong *et al.* [40] proposed the use of personalized weight maps, which is shown to significantly improve recognition accuracy. Moreover, the progression of comparison scores across considered shifting positions can be leveraged in order to improve biometric performance [41].

## 1.3    Status quo in research and industry

After the initial proposal of Daugman [1], first iris recognition systems have been deployed relatively fast compared to other biometric modalities. As previously mentioned, the continuous improvement of iris biometric technologies further enhances usability and, hence, acceptability. Numerous challenges evolve from real-world deployments of iris recognition giving rise to database collections and academic competition, where one driving factor behind rapid progress of research in computer science and its reproducibility is the provision of open source software.

### 1.3.1    Databases, competitions and software

The constant progress in iris biometric research is probably best reflected by the number of open iris datasets, being part of challenges or made available for research purposes. Due to the fact that experimental results obtained from closed or proprietary datasets are difficult to reproduce, public datasets represent a valuable means to compare existing approaches. Public databases comprise standard NIR and visible wavelength iris images as well as high resolution images, synthetic images or images captured at a distance and in unconstrained conditions. Figure 1.6 shows iris image samples of commonly used databases, a comprehensive list of publicly available iris image databases can be found in [42]. General requirements on biometric databases are summarized in [43]. Occasionally, iris image databases are released together with according iris recognition competitions.

Iris recognition technologies reveal auspicious biometric performance in case processed iris images meet standardized quality requirements [9]. This has been confirmed by the Iris Challenge Evaluation (ICE) in 2005 [44] and 2006 [45], which represent the first open benchmarks for iris recognition algorithms. In order to tackle unsolved challenges, further competitions have been organized in past years. Most notably, the Noisy Iris Challenge Evaluation (NICE) [46] part I and part II investigated the challenge of segmentation as well as feature extraction and comparison on unconstrained visible spectrum iris images. More recently, diverse covariates of iris recognition led to further competitions pushing forward research challenges, such as

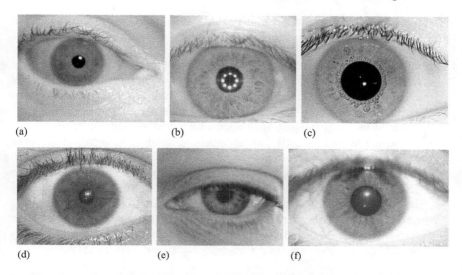

*Figure 1.6*   *Example images of (a)–(c) near-infrared and (d)–(f) visible wavelength iris databases: (a) ICE'06 [45], (b) CASIAv4-Interval [47], (c) IITDv1 [48], (d) UBIRISv1 [49], (e)MobBIO [50] and (f) UTIRIS [51] (visible wavelength)*

cross-sensor interoperability [52], cross-spectral iris recognition [53] or liveness, i.e., presentation attack, detection [54] or mobile environments [55].

Open source iris recognition software development kits (SDKs) should provide an adequate starting point for researchers working on improvements of different key components of an iris recognition system. Unfortunately, the current availability of open iris recognition SDKs turns out to be rather disappointing. Most existing open source solutions reveal non-competitive biometric performance and are hardly maintained [56]. Hence, the research program in most laboratories either relies on in-house software solutions or even commercial products. The Open Source Iris Recognition Software [57] (OSIRIS) and the University of Salzburg Iris Toolkit [56] (USIT) represent commendable exceptions to this. Furthermore, a single point of reference, which summarized available open source iris recognition software, has remained non-existent.

## 1.3.2   Standardization and deployments

Standardization in the field of information technology is pursued by a Joint Technical Committee (JTC1) formed by the International Organization for Standardization (ISO) and the International Electrotechnical Commission (IEC). The international committee for biometric standardization, Subcommittee 37 (SC37), was founded in 2002. Apart from general biometrics-related standards, there are two substantial standards focusing on iris recognition:

1.  ISO/IEC 19794-6 [58] defines biometric data interchange formats for iris image data. Standard iris image formats, e.g., uncropped, cropped or cropped and masked (based on region of interest), are specified. Based on the findings of Daugman and Downing [5] and the IREX-1 report [59] a JPEG 2000 compression of the latter image format allows for a compact storage of iris images requiring approximately 2 kB. Further, an encoding of the iris image data record is established.
2.  ISO/IEC 29794-6 [9] defines terms and quantitative methodologies relevant to characterizing the quality of iris images. A set of requirements for the quality of iris images is specified for which the fulfilment should yield high confidence iris biometric decisions. The required iris image quality metrics are computed from a single image. In addition, guidance to acquisition device manufacturers is provided and an encoding of the iris image quality data record is defined.

Up until now, more than a billion subjects worldwide have had their iris patterns mathematically encoded using the Daugman algorithms [4] for enrollment in national ID or entitlements programmes. Storing iris images in the standardized format has proven to be a success factor for these deployments, in order to maintain interoperability of various vendors and to avoid a vendor-lock-in for the operator. Major goals of these programmes are an enhancement of security as well as improved access to social benefits, financial aids or other entitlements with reduced fraud.

Within the Indian Aadhar project, which has been launched in 2009, the Unique IDentification Authority of India (UIDAI) is biometrically enrolling the entire population. This project is frequently referred to as flagship deployment of iris (and fingerprint) recognition technology, manifesting the extreme resistance of iris recognition against false matches. A similar national ID program has been launched in Indonesia and Mexico. Other programmes employ iris recognition to track and deliver aid to refugees in conflict areas in the Middle East, e.g., in Syria, Jordan or Afghanistan. Frequent traveller border-crossing programmes have successfully employed iris recognition technology for fast automated border control in the UK, the Netherlands, Canada, USA and the United Arab Emirates [60].

## 1.4    Challenges in iris recognition

Progress in iris recognition systems has resulted in several challenges and new opportunities, which have been the focus of recent research efforts. Apart from studying iris recognition in the ideal scenario, the biometric community is actively pursuing open problems, such as heterogeneous recognition for addressing interoperability, the development of presentation attack detection methods or the iris template ageing phenomenon. Hence, open research problems evolved numerous covariates of iris recognition.

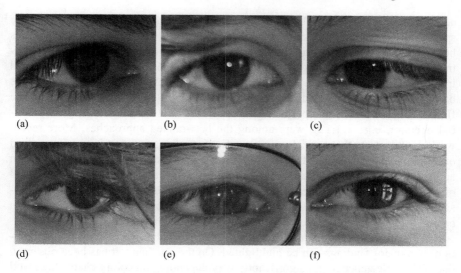

(a)                    (b)                    (c)

(d)                    (e)                    (f)

*Figure 1.7    Visible wavelength iris images in unconstrained environments (images taken from UBIRISv2 iris database [3]): (a) bad lightning, (b) bad focus, (c) off-angle, (d) occlusion, (e) glasses and (f) reflection*

## 1.4.1    Visible wavelength and unconstrained conditions

Among others, iris recognition based on visible wavelength images represents one major challenge. Independent tests [44,61] have confirmed remarkable recognition accuracy for iris recognition in case good quality image data captured under NIR light is available. If captured properly visible wavelength iris images might exhibit a similar level of detail within the iris texture, while in general the amount of available information is expected to be less [62]. However, within visible wavelength iris images possible artefacts, such as specular reflections or shadows, are more pronounced, which generally lead to an increased intra-class variation. In past years, numerous researchers developed specialized recognition strategies in order to enable reliable visible wavelength iris recognition. In addition, the feasibility of iris recognition in unconstrained environments, e.g., on-the-move or at-a-distance, has been subject to recent research efforts. Without assuming subjects' willingness to be recognized, non-cooperative iris recognition would be of particular interest for security purposes [62]. A specific security application is the recognition of individuals in video surveillance footage, where faces are almost entirely occluded or masked but eye regions remain visible. Figure 1.7 shows different examples of iris images captured under visible wavelength and unconstrained conditions.

Accurate segmentation of the iris region in visible wavelength images represents one of the most critical tasks, since errors in the pre-processing stage significantly impact recognition accuracy [63]. Quality requirements on NIR iris images [9], e.g., differences in contrast between pupil, iris and sclera, are not transferable to iris images acquired under visible wavelength. In particular, the majority of methods that were specifically designed to process visible wavelength iris images detect the outer iris boundaries first, since the iris and the sclera exhibit much higher difference in contrast compared to the iris and the pupil, especially in the case of strongly pigmented iris textures. After a coarse localization of the iris a refinement of iris boundaries based on active contour models, e.g., in [20], has revealed promising results. In the first part of the NICE competition [46] optimized versions of the integrodifferential operator, e.g., [11,13], were among the best-ranked approaches. More recent works demonstrate that a pixel-wise classification of iris and non-iris regions based on machine-learning concepts, e.g., [64], further improves segmentation accuracy. In addition to the application of specifically designed noise removal techniques, potential off-axis gaze requires an adequate transformation to compensate differences in perspective [12,18]. A comprehensive overview of most relevant works on segmentation of iris images in visible wavelength images is provided in [62].

With respect to the feature extraction stage in visible wavelength iris recognition, two major findings can be highlighted. On the one hand, it has been observed that iridal reflectance might significantly vary depending on colour channels as well as eye colour [65], which motivates an adaptive use of colour channels in visible wavelength iris recognition [66]. On the other hand, despite the traditional analysis of the extracted iris texture, the use of further sources of (soft-biometric) information available in the iris region and the surrounding ocular region has been found to significantly improve recognition accuracy. In the second part of the NICE competition such strongly multi-biometric approaches, which utilize additional features, e.g., eye colour or skin texture, performed best, e.g., [67,68]. Still, obtained biometric performance is still far from that reported for good quality NIR iris images. The development of reliable visible wavelength iris recognition systems is further motivated by the increasing use of smart-phone, which might as well be used for (iris) biometric recognition [55].

## 1.4.2   Security and privacy protection

The false match rate (FMR) of a biometric system provides an intuitive measure for security. Iris recognition systems can be operated at extremely low FMRs, i.e., 100,000 times lower than generic face recognition systems [61], which motivate the application of iris recognition in high security scenarios. However, there are several possible attack points in a biometric system [69], where presentation (or spoofing) attacks might be considered most relevant. Presentation attacks refer to a presentation to the biometric data capture subsystem with the goal of interfering with the operation of the biometric system [70]. Such attacks can be launched by using artificial attack presentation instruments, e.g., print outs, electronic displays, or even textured contact lenses [71], examples are illustrated in Figure 1.8. In recent years, diverse approaches have been proposed in order to reliably detect presentation attacks. General purpose texture descriptors, e.g., binarized statistical image features [72], in combination

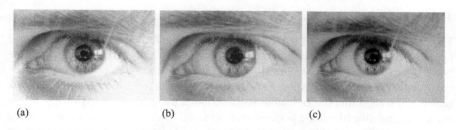

(a)                    (b)                    (c)

*Figure 1.8*   *Example of a (a) real iris image and (b) and (c) attack artefacts*
*(images taken from the VSIA database [73]): (a) real, (b) printout and*
*(c) screen*

with machine-learning techniques trained to make a binary decision (real or fake iris sample) have been shown to reveal promising results for the detection of presentation attacks [73–75].

Concerns against the common use of biometric technologies arise from the storage and misuse of biometric reference data [76], e.g., iris-codes. In order to safeguard individuals' privacy, protection of biometric reference data is of utmost importance. With respect to iris recognition researchers already showed against common belief that, in case an attacker has full knowledge of the employed feature extraction, iris-codes can be utilized in order to reconstruct images of subjects' iris textures [77,78]. Unfortunately, the natural intra-class variance of iris images, which can be caused by several factors, e.g., partial closure of eyelids, prevents from the application of conventional cryptographic techniques. On the one hand, a permanent protection of iris-codes is not feasible using standard encryption, since a comparison of encrypted iris-codes will most likely fail since single bit flips cause drastic changes in the encrypted domain. On the other hand, if decryption is performed at authentication an attacker could glean an iris-code by simply launching an authentication attempt. Technologies of biometric template protection [76,79], which are commonly categorized as biometric cryptosystems [80] and cancelable biometrics [69], offer solutions to privacy preserving biometric authentication. Template protection schemes are required to fulfil two major requirements, irreversibility and unlinkability [81]. Apart from satisfying these properties, ideal biometric template protection schemes shall not cause a decrease in recognition accuracy with respect to the corresponding unprotected system. Focusing on iris recognition, numerous approaches to biometric cryptosystems, e.g., [82–84], and cancelable biometrics, e.g., [85–87], have been proposed in past years. While some proposed techniques are theoretically sound, the majority of these schemes does not guarantee the desired irreversibility and unlinkability properties without significantly degrading biometric performance [88]. That is, up until now template protection schemes yield a clear trade-off between recognition accuracy and privacy protection, such that a provable secure and accuracy-preserving protection of iris biometric reference data remains an open research issue.

### 1.4.3   Cross-sensor and cross-spectral recognition

The increasing popularity of iris biometrics leads to a continuous development and upgradation of new and existing iris sensors, respectively. Differences among multiple types of iris sensors such as optical lens and illumination wavelength yield certain cross-sensor variations of iris texture patterns. Recent studies demonstrated that cross-sensor iris recognition might lead to reduced biometric performance [89,90]. It was shown that relative biometric performance obtained from single-sensor comparisons is not necessarily a reliable predictor of the relative biometric performance obtained cross-sensor comparisons using the same sensors. In addition, relative performance of cross-sensor experiments has been found to be dependent on both, the sensors as well as the employed iris recognition software modules. To assure interoperability of iris recognition systems it is necessary to achieve a reliable matching of heterogeneous iris images captured by different types of iris sensors.

A re-enrolment of data subjects every time a new sensor is deployed is expensive and time-consuming, especially in large-scale applications. This fact motivated the proposal of diverse approaches which aim at improving the biometric performance of cross-sensor iris recognition, e.g., [89,91,92]. Mitigation of cross-sensor variations in iris recognition can be based on a reliable classification of iris sensors and the application of an adequate pre-processing [89]. Alternatively, machine-learning techniques have been employed to select a common feature set which is effective across different types of iris sensors [91] or to learn transformations which constrain samples from different sensors to behave in a similar manner in the transformed domain [92].

Multispectral images contain information across multiple wavelengths or spectral channels of the electromagnetic spectrum. Researchers have explored the use of such images in iris recognition, motivated by the idea that iridal composition is biologically diverse, such that different ranges of the electromagnetic spectrum might better capture certain physical characteristics of the iris pattern [93]. It has been shown that a fusion of iris images acquired under NIR and visible wavelength can improve the biometric performance of an iris recognition system [94,95]. As previously mentioned, it has been found that the nature of iris information presented in different spectral channels varies according to eye colour [65]. An automatic selection of wavelength bands, which reveal rich textural features, has been proposed in [66]. Further, an adaptive method to derive 'pseudo-NIR' images from visible wavelength iris images has been presented in [96] to enable a cross-spectrum matching of iris images to strengthen the interoperability of iris recognition. As previously mentioned, cross-spectral iris recognition has also been subject of a recent iris biometric competition [53].

### 1.4.4   Soft biometrics and template ageing

While not necessarily unique to an individual, soft biometric attributes, e.g., gender or ethnicity, can be employed in biometric systems in conjunction with primary biometric characteristics in order to improve or expedite recognition accuracy. Such attributes are typically obtained from primary biometric characteristics, are classifiable in predefined human-understandable categories and can be extracted in an automated manner [97]. Intuitively, eye colour is the only soft biometric attribute, which can be naturally gleaned from an iris pattern acquired in visible band of light. Eye colour has been utilized as an additional source of information in diverse approaches [68,98], especially in challenging unconstrained environments. However, the existence of further soft biometric attributes in the iris pattern, such as gender, has been alluded in medical literature [99,100]. In past years, it has been shown that soft biometric attributes, including gender and ethnicity [101] as well as age [102], can be predicted from iris patterns using general purpose texture descriptors in combination with machine-learning techniques. Moreover, it has been found that raw iris images, which comprise parts of the periocular region, might be even more useful to extract soft biometric attributes, such as gender [103].

Physiological biometric characteristics like the iris represent dynamic biological systems and, thus, are potentially affected by ageing. Age alterations of biometric features might cause a degradation of biometric performance of the overall system. Lately, several researchers have proposed diverse investigations in order to analyse the effects of ageing on iris recognition, an overview of existing works can be found in [104,105]. The term 'template ageing' has been established in order to describe an increase in intra-class variability and, hence, a degradation in recognition accuracy, with increased time since enrolment [105]. In one of the first studies on the permanence of iris patterns, Baker and Flynn [106] reported that, at a false accept rate of 0.01%, the false reject rate increases by 75% for a rather short time-lapse of four years. However, in an independent analysis conducted by NIST [107] it was concluded that there should be no concern about iris recognition accuracy degrading over time, contradicting previous findings. Further, it is important to note that up until now a photographic evidence of iris pattern changes over time has remained non-existent.

Nonetheless, as mentioned earlier there is good evidence for nonlinear deformation with dilation and constriction of the iris. That is, template ageing effects may also be caused by changes in environmental conditions. Such effects might be avoided by employing segmentation methods, which compensate for nonlinear effects of pupil dilation, e.g., [25]. Furthermore, template ageing might result from sensor ageing, too [108]. As iris (template) ageing has become a topic of controversy, it appears that the debate over temporal stability of iris recognition accuracy is going to continue [109].

## 1.4.5 *Large-scale identification and iris biometric fusion*

On large-scale databases biometric identification attempts or de-duplication checks require an exhaustive one-to-many comparison. Hence, the response time of the system is expected to increase significantly with large number of enrollees, especially in case computational power is limited. In order to circumvent the bottleneck of an exhaustive $1 : N$ comparison different concepts have been proposed in order to reduce the workload in an iris biometric (identification) systems, including coarse classification or 'binning', a serial combination of a computationally efficient and a conventional system, indexing schemes and hardware-based acceleration.

By binning an iris biometric database into $c$ classes, the overall workload for an identification attempt can be reduced by factor $c$, given that irises of registered subjects are equally distributed among these classes. Natural features to be utilized include eye position (left or right) [68] or eye colour [110,111]. Recent advances in the field of soft biometrics suggest further possible classification. Instead of creating tangible, human-understandable classes, it is also possible to rely on distinct iris texture features [112]. Binning is equivalent to the combination of biometric systems. Hence, classification of errors might significantly increase the false non-match rate (FNMR) of the overall system. Moreover, the potential benefit of binning is limited by the number of bins, which determines the factor by which the database size can be reduced.

Within serial combinations computationally efficient biometric systems are used to extract a shortlist of most likely candidates, which is frequently referred to as pre-selection. While generic iris recognition systems already provide a rapid comparison, such more efficient biometric comparators can be obtained by employing compressed versions of original iris-codes during pre-screening [113,114]. Further, a rotation-invariant iris recognition scheme can be applied in the pre-selection step [115]. Similar to binning approaches, a serial combination of a computationally efficient and an accurate (but more complex) scheme might increase the FNMR of the overall system. However, a serial combination enables a more accurate operation of the resulting trade-off between computational effort and accuracy by choosing an adequate size of the returned shortlist.

Indexing schemes aim at constructing hierarchical search structures for iris bio-metric data, which tolerate a distinct amount of biometric variance. Such schemes substantially reduce the overall workload of a biometric identification, e.g., to $\log N$ in the case of a binary search tree. Such search structures might be designed for iris-codes [116,117] as well as iris images [118,119]. While the majority of works report hit/ penetration rates on distinct datasets, required computational efforts for indexing and retrieval are frequently omitted. The application of complex search structures on rather small datasets may as well cloud the picture about actual gains in terms of speed and leaves the scalability of some approaches questionable.

Adapting comparison procedures to adequate hardware, e.g., multiple cores within a CPU, allows for parallelization [120]. By simultaneously executing a number of $t$ threads the overall workload for an identification can be reduced by the factor $t$, since a $1 : N$ comparison can be performed in parallel on $t$ subsets of equal size $N/t$. Moreover, iris-code comparisons can be efficiently performed on the GPU using GPGPU or CUDA [121] or FPGA [122].

Apart from hardware-based acceleration, most of presented schemes either fail to provide a significant acceleration or they suffer from a significant decrease in recognition accuracy. Hence, existing approaches often obtain a trade-off between biometric performance (recognition accuracy) and speed-up, compared to a traditional iris recognition system.

Especially on large-scale databases the observation of multiple biometric char-acteristics has been found to significantly improve the accuracy and reliability, in particular in challenging identification scenarios [20], while recognition systems based on a single biometric indicator often have to contend with unacceptable error rates [123]. However, improvement in biometric performance as a result of biometric fusion should be weighed against the associated overhead involved [124], e.g., additional sensing cost.

Information fusion can take place at various stages of an iris recognition system. Modern iris recognition sensors are capable of acquiring the left and right iris of a data subject simultaneously. Fusing the information extracted from both irises of an individual further improves the robustness of an iris recognition system, where a fusion of normalized scores has been shown to be competitive [124]. Besides this obvious type of information fusion, additional iris biometric fusion techniques have been proposed in past years, a comprehensive overview is given in [125]. Note that a fusion of comparison and quality scores and multispectral iris biometric fusion has already been mentioned in Sections 1.2.2 and 1.4.3, respectively. Different types of feature extraction methods might obtain complementary features, which improves the overall accuracy of the system without requiring any additional sensing [48,126]. The acquisition of multiple instances of a single iris allows for a feature type fusion, which has been found to improve the robustness of an iris recognition system [38,39]. Performance gains have also been reported for a fusion of iris segmentation results obtained by different segmentation algorithms [127]. Finally, raw iris images comprise additional information such as vascular patterns within the sclera, the periocular region or soft-biometric features, which can be utilized for information fusion to improve the biometric performance of an iris recognition system in particular, in unconstrained environments [68].

## 1.5   Summary

The rich random pattern visible in the iris of the eye represents one of the strongest biometric characteristics and considerable research efforts are being invested to push forward iris recognition technologies. Numerous covariates of iris recognition are the focus of current research programmes, which aim at improving different key factors of iris biometric system, e.g., usability, acceptability or interoperability.

This chapter provides a very brief overview of the processing chain of generic iris recognition systems as well as a the current state-of-the-art. Furthermore, current challenges in the field of iris recognition were touched upon. The ever-increasing relevance of iris recognition is probably best perceived by the number and diversity of current challenges, where some of the following chapters will provide a deeper insight into selective topics.

## References

[1]   J. Daugman. "High confidence visual recognition of persons by a test of statistical independence," *IEEE Transactions on Pattern Analysis and Machine Intelligence*, 15(11):1148–1161, 1993.

[2] J. Ortega-Garcia, J. Fierrez, F. Alonso-Fernandez, *et al.* "The multiscenario multienvironment biosecure multimodal database (BMDB)," *IEEE Transactions on Pattern Analysis and Machine Intelligence*, 32(6):1097–1111, 2010.

[3] H. Proença, S. Filipe, R. Santos, J. Oliveira, and L. A. Alexandre. "The UBIRIS.v2: a database of visible wavelength iris images captured on-the-move and at-a-distance," *IEEE Transactions on Pattern Analysis and Machine Intelligence*, 32(8):1529–1535, 2010.

[4] J. Daugman. "How iris recognition works," *IEEE Transactions on Circuits and Systems for Video Technology*, 14(1):21–30, 2004.

[5] J. Daugman and C. Downing. "Effect of severe image compression on iris recognition performance," *IEEE Transactions on Information Forensics and Security*, 3:52–61, 2008.

[6] J. Wayman, N. Orlans, Q. Hu, F. Goodman, A. Ulrich, and V. Valencia. "Technology assessment for the state of the art biometrics excellence roadmap, vol. 2 ver. 1.3," Technical report, The MITRE Corporation, 2009. US Gov Contr. J-FBI-07-164.

[7] S. Thavalengal, I. Andorko, A. Drimbarean, P. Bigioi, and P. Corcoran. "Proof-of-concept and evaluation of a dual function visible/NIR camera for iris authentication in smartphones," *IEEE Transactions on Consumer Electronics*, 61(2):137–143, 2015.

[8] R. Raghavendra, K. B. Raja, and C. Busch. "Exploring the usefulness of light field cameras for biometrics: an empirical study on face and iris recognition," *IEEE Transactions on Information Forensics and Security*, 11(5):922–936, 2016.

[9] ISO/IEC JTC1 SC27 Security Techniques. *ISO/IEC 29794:2015. Information Technology – Biometric Sample Quality – Part 6: Iris Image Data*. International Organization for Standardization, 2015.

[10] H. Proença. "Iris recognition: on the segmentation of degraded images acquired in the visible wavelength," *IEEE Transactions on Pattern Analysis and Machine Intelligence*, 32(8):1502–1516, 2010.

[11] W. Sankowski, K. Grabowski, M. Napieralska, M. Zubert, and A. Napieralski. "Reliable algorithm for iris segmentation in eye image," *Image and Vision Computing*, 28(2):231–237, 2010.

[12] S. Schuckers, N. Schmid, A. Abhyankar, V. Dorairaj, C. Boyce, and L. Hornak. "On techniques for angle compensation in nonideal iris recognition," *IEEE Transactions on Systems, Man, and Cybenetics, Part B: Cybernetics*, 37(5):1176–1190, 2007.

[13] T. Tan, Z. He, and Z. Sun. "Efficient and robust segmentation of noisy iris images for non-cooperative iris recognition," *Image and Vision Computing*, 28(2):223–230, 2010.

[14] R. Wildes. "Iris recognition: an emerging biometric technology," *Proceedings of the IEEE*, 85(9):1348–1363, 1997.

[15] X. He and P. Shi. "A new segmentation approach for iris recognition based on hand-held capture device," *Pattern Recognition*, 40:1326–1333, 2007.

[16] Z. Xu and P. Shi. "A robust and accurate method for pupil features extraction," in *Proceedings of the International Conference on Pattern Recognition (ICPR '06)*, pp. 437–440. IEEE, Piscataway, NJ, 2006.

[17] J. Zuo and N. Schmid. "On a methodology for robust segmentation of nonideal iris images,". *IEEE Transactions on Systems, Man, and Cybernetics, Part B: Cybernetics*, 40(3):703–718, 2010.

[18] J. Daugman. "New methods in iris recognition," *IEEE Transactions on Systems, Man, and Cybernetics, Part B: Cybernetics*, 37(5):1167–1175, 2007.

[19] R. Chen, X. Lin, and T. Ding. "Iris segmentation for non-cooperative recognition systems," *IET Image Processing*, 5(5):448–456, 2011.

[20] S. Shah and A. Ross. "Iris segmentation using geodesic active contours," *IEEE Transactions on Information Forensics and Security*, 4(4):824–836, 2009.

[21] H. Proença and L. Alexandre. "Iris segmentation methodology for non-cooperative recognition," *IEE Proceedings of Vision, Image and Signal Processing*, 153(2):199–205, 2006.

[22] N. A. Schmid, J. Zuo, F. Nicolo, and H. Wechsler. "Iris quality metrics for adaptive authentication," in J. M. Burge and W. K. Bowyer, editors, *Handbook of Iris Recognition*, pp. 67–84. Springer, London, 2013.

[23] Z. Zhou, Y. Du, and C. Belcher. "Transforming traditional iris recognition systems to work in nonideal situations," *IEEE Transactions on Industrial Electronics*, 56(8):3203–3213, 2009.

[24] N. D. Kalka, J. Zuo, N. A. Schmid, and B. Cukic. "Image quality assessment for iris biometric," in *Proceedings of SPIE*, vol. 6202, pp. 61020D.1–61020D.11. SPIE, Bellingham, WA, 2006.

[25] I. Tomeo-Reyes, A. Ross, A. D. Clark, and V. Chandran. "A biomechanical approach to iris normalization," in *International Conference on Biometrics (ICB '15)*, pp. 9–16, 2015.

[26] J. Daugman. "Information theory and the iriscode," *Transactions on Information Forensics and Security*, 11(2):400–409, Feb. 2016.

[27] L. Masek. "Recognition of human iris patterns for biometric identification," Master's thesis, University of Western Australia, 2003.

[28] A. W. K. Kong, D. Zhang, and M. S. Kamel. "An analysis of iriscode," *IEEE Transactions on Image Processing*, 19(2):522–532, 2010.

[29] E. Krichen, M. A. Mellakh, S. Garcia-Salicetti, and B. Dorizzi. "Iris identification using wavelet packets," in *Proceedings of the 17th International Conference on Pattern Recognition (ICPR '04)*, vol. 4, pp. 335–338, 2004.

[30] D. M. Monro, S. Rakshit, and D. Zhang. "DCT-based iris recognition," *IEEE Transactions on Pattern Analysis and Machine Intelligence*, 29(4):586–595, 2007.

[31] L. Ma, T. Tan, Y. Wang, and D. Zhang. "Personal identification based on iris texture analysis," *IEEE Transactions on Pattern Analysis and Machine Intelligence*, 25(12):1519–1533, 2003.

[32]  Z. Sun and T. Tan. "Ordinal measures for iris recognition," *IEEE Transactions on Pattern Analysis and Machine Intelligence*, 31(12):2211–2226, 2009.

[33]  L. Ma, T. Tan, Y. Wang, and D. Zhang. "Efficient iris recognition by characterizing key local variations," *IEEE Transactions on Image Processing*, 13(6):739–750, 2004.

[34]  C. Belcher and Y. Du. "Region-based SIFT approach to iris recognition," *Optics and Lasers in Engineering*, 47(1):139–147, 2009.

[35]  M. S. Sunder and A. Ross. "Iris image retrieval based on macro-features," in *Proceedings of the International Conference on Pattern Recognition (ICPR'10)*, pp. 1318–1321, 2010.

[36]  J. Chen, F. Shen, D. Chen, and P. Flynn. "Iris recognition based on human-interpretable features," *IEEE Transactions on Information Forensics and Security*, 11(7):1476–1485, 2016.

[37]  F. Alonso-Fernandez, P. Tome-Gonzalez, V. Ruiz-Albacete, and J. Ortega-Garcia. "Iris recognition based on SIFT features," in *International Conference on Biometrics, Identity and Security (BIdS'09)*, pp. 1–8, 2009.

[38]  S. Ziauddin and M. N. Dailey. "Iris recognition performance enhancement using weighted majority voting," in *15th IEEE International Conference on Image Processing*, pp. 277–280, 2008.

[39]  K. P. Hollingsworth, K. W. Bowyer, and P. J. Flynn. "The best bits in an iris code," *Transactions on Pattern Analysis and Machine Intelligence*, 31(6):964–973, 2009.

[40]  W. Dong, Z. Sun, and T. Tan. "Iris matching based on personalized weight map," *IEEE Transactions on Pattern Analysis and Machine Intelligence*, 33(9):1744–1757, 2011.

[41]  C. Rathgeb, A. Uhl, and P. Wild. "Iris-biometric comparators: exploiting comparison scores towards an optimal alignment under Gaussian assumption," in *Proceedings of the International Conference on Biometrics (ICB'12)*, pp. 1–6, 2012.

[42]  C. Rathgeb, A. Uhl, and P. Wild. *Iris Biometrics: From Segmentation to Template Security*. Number 59 in Advances in Information Security. Springer, Berlin, 2012.

[43]  P. J. Flynn. "Biometrics databases," in A. K. Jain, P. Flynn, and A. A. Ross, editors, *Handbook of Biometrics*, pp. 529–548. Springer, Berlin, 2008.

[44]  P. Phillips, K. Bowyer, P. Flynn, X. Liu, and W. Scruggs. "The iris challenge evaluation 2005," in *Proceedings of the International Conference on Biometrics: Theory, Applications, and Systems (BTAS'08)*, pp. 1–8. IEEE, Piscataway, NJ, 2008.

[45]  P. J. Phillips, W. T. Scruggs, A. J. O'Toole, *et al.* "FRVT 2006 and ICE 2006 large-scale results," NISTIR 7408 Report, NIST, 2007.

[46]  H. Proença and L. Alexandre. "Toward covert iris biometric recognition: experimental results from the nice contests," *IEEE Transactions on Information Forensics and Security*, 7(2):798–808, 2012.

[47]   Chinese Academy of Sciences, Institute of Automation. CASIA Iris Image Database: http://biometrics.idealtest.org. Retrieved August 2016.
[48]   A. Kumar and A. Passi. "Comparison and combination of iris matchers for reliable personal authentication," *Pattern Recognition*, 43(3):1016–1026, 2010.
[49]   H. Proença and L. A. Alexandre. "UBIRIS: a noisy iris image database," in *Proceedings of the International Conference on Image Analysis and Processing (ICIAP'05)*, pp. 970–977, 2005.
[50]   A. F. Sequeira, J. C. Monteiro, A. Rebelo, and H. P. Oliveira. "MobBIO: a multimodal database captured with a portable handheld device," in *Proceedings of the International Conference on Computer Vision Theory and Applications (VISAPP'14)*, vol. 3, pp. 133–139, Jan. 2014.
[51]   M. Hosseini, B. Araabi, and H. Soltanian-Zadeh. "Pigment melanin: pattern for iris recognition," *IEEE Transactions on Instrumentation and Measurement*, 59(4):792–804, Apr. 2010.
[52]   A. Sgroi, K. Bowyer, and P. Flynn. "Cross sensor iris recognition competition," in *Proceedings of the International Conference on Biometrics: Theory, Applications and Systems (BTAS'12)*, 2012.
[53]   A. F. Sequeira, L. Chen, P. Wild, *et al.*, "Cross-eyed – cross-spectral iris/periocular recognition database and competition," in *Proceedings of the International Conference of the Biometrics Special Interest Group (BIOSIG'16)*, pp. 1–8, 2016.
[54]   D. Yambay, J. S. Doyle, K. W. Bowyer, A. Czajka, and S. Schuckers. "Livdet-iris 2013 – iris liveness detection competition 2013," in *International Joint Conference on Biometrics (IJCB'14)*, pp. 1–8, 2014.
[55]   M. D. Marsico, M. Nappi, D. Riccio, and H. Wechsler. "Mobile Iris Challenge Evaluation (MICHE)-I, biometric iris dataset and protocols," *Pattern Recognition Letters*, 57:17–23, 2015.
[56]   C. Rathgeb, A. Uhl, P. Wild, and H. Hofbauer. "Design decisions for an iris recognition SDK," in W. K. Bowyer and J. M. Burge, editors, *Handbook of Iris Recognition (2nd Edition)*, pp. 359–396. Springer, London, 2016.
[57]   N. Othman, B. Dorizzi, and S. Garcia-Salicetti. "OSIRIS: an open source iris recognition software," *Pattern Recognition Letters*, 11:124–131, 2015.
[58]   ISO/IEC JTC1 SC27 Security Techniques. *ISO/IEC 19794-6:2011. Information Technology – Biometric Data Interchange Formats – Part 6: Iris Image Data*. International Organization for Standardization, 2011.
[59]   P. Grother, E. Tabasi, G. W. Quinn, and W. Salamon. "IREX I – performance of iris recognition algorithms on standard images," NIST interagency report 7629, National Institute of Standards and Technology (NIST), 2009.
[60]   J. Daugman. "Iris recognition at airports and border-crossings," in S. Z. Li and A. Jain, editors, *Encyclopedia of Biometrics*, pp. 819–825. Springer, Boston, MA, 2009.
[61]   P. J. Grother, G. W. Quinn, J. R. Matey, *et al.* "IREX III – performance of iris identification algorithms," NIST interagency report 7836, National Institute of Standards and Technology (NIST), 2012.

[62]    H. Proença. "Unconstrained iris recognition in visible wavelengths," in W. K. Bowyer and J. M. Burge, editors, *Handbook of Iris Recognition (2nd Edition)*, pp. 321–358. Springer, London, 2016.

[63]    H. Proença and L. A. Alexandre. Iris recognition: analysis of the error rates regarding the accuracy of the segmentation stage," *Image and Vision Computing*, 28(1):202–206, 2010.

[64]    N. Liu, H. Li, M. Zhang, J. Liu, Z. Sun, and T. Tan. "Accurate iris segmentation in non-cooperative environments using fully convolutional networks," in *Proceedings of the International Conference on Biometrics (ICB'16)*, pp. 1–8, 2016.

[65]    C. Boyce, A. Ross, M. Monaco, L. Hornak, and X. Li. "Multispectral iris analysis: a preliminary study," in *Proceedings of the International Conference on Computer Vision and Pattern Recognition Workshop (CVPRW'06)*, pp. 31–37, 2006.

[66]    Y. Gong, D. Zhang, P. Shi, and J. Yan. "An optimized wavelength band selection for heavily pigmented iris recognition," *IEEE Transactions on Information Forensics and Security*, 8(1):64–75, 2013.

[67]    G. Santos and E. Hoyle. "A fusion approach to unconstrained iris recognition," *Pattern Recognition Letters*, 33(8):984–990, 2012.

[68]    T. Tan, X. Zhang, Z. Sun, and H. Zhang. "Noisy iris image matching by using multiple cues," *Pattern Recognition Letters*, 33(8):970–977, 2012.

[69]    N. Ratha, J. Connell, and R. Bolle. "Enhancing security and privacy in biometrics-based authentication systems," *IBM Systems Journal*, 40(3): 614–634, 2001.

[70]    ISO/IEC TC JTC1 SC37 Biometrics. *ISO/IEC IS 30107-1. Information Technology – Biometrics Presentation Attack Detection – Part 1: Framework*. International Organization for Standardization, Mar. 2016.

[71]    S. Marcel, M. Nixon, and S. Z. Li. *Handbook of Biometric Anti-Spoofing*. Springer-Verlag Inc., New York, 2014.

[72]    J. Kannala and E. Rahtu. "BSIF: binarized statistical image features," in *21st International Conference on Pattern Recognition (ICPR'12)*, pp. 1363–1366, 2012.

[73]    R. Raghavendra and C. Busch. "Robust scheme for iris presentation attack detection using multiscale binarized statistical image features," *IEEE Transactions on Information Forensics and Security*, 10(4):703–715, 2015.

[74]    J. S. Doyle and K. W. Bowyer. "Robust detection of textured contact lenses in iris recognition using BSIF," *IEEE Access*, 3:1672–1683, 2015.

[75]    K. B. Raja, R. Raghavendra, and C. Busch. "Video presentation attack detection in visible spectrum iris recognition using magnified phase information," *IEEE Transactions on Information Forensics and Security*, 10(10):2048–2056, 2015.

[76]    C. Rathgeb and A. Uhl. "A survey on biometric cryptosystems and cancelable biometrics," *EURASIP Journal on Information Security*, 2011(3), 2011.

[77]    J. Galbally, A. Ross, M. Gomez-Barrero, J. Fiérrez, and J. Ortega-Garcia. "Iris image reconstruction from binary templates: an efficient probabilistic

approach based on genetic algorithms," *Computer Vision and Image Understanding*, 117(10):1512–1525, 2013.

[78] S. Venugopalan and M. Savvides. "How to generate spoofed irises from an iris code template," *IEEE Transactions on Information Forensics and Security*, 6(2):385–395, 2011.

[79] A. K. Jain, K. Nandakumar, and A. Nagar. "Biometric template security," *EURASIP Journal of Advanced Signal Processing*, 2008:1–17, 2008.

[80] U. Uludag, S. Pankanti, S. Prabhakar, and A. K. Jain. "Biometric cryptosystems: issues and challenges," *Proceedings of the IEEE*, 92(6):948–960, 2004.

[81] ISO/IEC JTC1 SC27 Security Techniques. *ISO/IEC 24745:2011. Information Technology – Security Techniques – Biometric Information Protection.* International Organization for Standardization, 2011.

[82] M. Braithwaite, U. C. V. Seelen, and J. Cambier, *et al.*, "Applications-specific biometric template," in *Proceedings of the Workshop on Automatic Identification Advanced Technologies*, pp. 167–171, 2002.

[83] J. Bringer, H. Chabanne, G. Cohen, B. Kindarji, and G. Zemor. "Optimal iris fuzzy sketches," in *Proceedings of the IEEE First International Conference on Biometrics: Theory, Applications, and Systems*, pp. 1–6. IEEE, Piscataway, NJ, 2007.

[84] F. Hao, R. Anderson, and J. Daugman. "Combining cryptography with biometrics effectively," *IEEE Transactions on Computers*, 55(9):1081–1088, 2006.

[85] J. K. Pillai, V. M. Patel, R. Chellappa, and N. K. Ratha. "Secure and robust iris recognition using random projections and sparse representations," *IEEE Transactions on Pattern Analysis and Machine Intellegence*, 33(9):1877–1893, 2011.

[86] C. Rathgeb, F. Breitinger, and C. Busch. "Alignment-free cancelable iris biometric templates based on adaptive bloom filters," in *Proceedings of the Sixth IAPR International Conference on Biometrics (ICB'13)*, pp. 1–8, 2013.

[87] J. Zuo, N. K. Ratha, and J. H. Connel. "Cancelable iris biometric," *Proceedings of the International Conference on Pattern Recognition (ICPR'08)*, pp. 1–4, 2008.

[88] K. Nandakumar and A. K. Jain. "Biometric template protection: bridging the performance gap between theory and practice," *IEEE Signal Processing Magazine – Special Issue on Biometric Security and Privacy*, pp. 1–12, 2015.

[89] S. S. Arora, M. Vatsa, R. Singh, and A. Jain. "On iris camera interoperability," in *International Conference on Biometrics: Theory, Applications and Systems (BTAS'12)*, pp. 346–352, 2012.

[90] R. Connaughton, A. Sgroi, K. Bowyer, and P. J. Flynn. "A multialgorithm analysis of three iris biometric sensors," *IEEE Transactions on Information Forensics and Security*, 7(3):919–931, 2012.

[91] L. Xiao, Z. Sun, R. He, and T. Tan. "Coupled feature selection for cross-sensor iris recognition," in *International Conference on Biometrics: Theory, Applications and Systems (BTAS'13)*, pp. 1–6, 2013.

[92]    J. K. Pillai, M. Puertas, and R. Chellappa. "Cross-sensor iris recognition through kernel learning," *IEEE Transactions on Pattern Analysis and Machine Intelligence*, 36(1):73–85, 2014.

[93]    A. Ross. "Iris recognition: the path forward," *Computer*, 43:30–35, 2010.

[94]    M. J. Burge and M. K. Monaco. "Multispectral iris fusion for enhancement, interoperability, and cross wavelength matching," in *Proceedings SPIE*, vol. 7334, pp. 73341D–73341D-8, 2009.

[95]    P. Wild, P. Radu, and J. Ferryman. "On fusion for multispectral iris recognition," in *International Conference on Biometrics (ICB'15)*, pp. 31–37, 2015.

[96]    J. Zuo, F. Nicolo, and N. A. Schmid. "Cross spectral iris matching based on predictive image mapping," in *2010 Fourth IEEE International Conference on Biometrics: Theory Applications and Systems (BTAS)*, pp. 1–5, 2010.

[97]    A. Dantcheva, P. Elia, and A. Ross. "What else does your biometric data reveal? A survey on soft biometrics," *Transactions on Information Forensics and Security*, 11(3):441–467, 2016.

[98]    A. Dantcheva, N. Erdogmus, and J. L. Dugelay. "On the reliability of eye color as a soft biometric trait," in *IEEE Workshop on Applications of Computer Vision (WACV'11)*, pp. 227–231, 2011.

[99]    M. Larsson, N. L. Pedersen, and H. Stattin. "Importance of genetic effects for characteristics of the human iris," *Twin Research*, 6(3):192–200, 2003.

[100]   R. A. Sturm and M. Larsson. "Genetics of human iris colour and patterns," *Pigment Cell and Melanoma Research*, 22(5):544–562, 2009.

[101]   S. Lagree and K. W. Bowyer. "Predicting ethnicity and gender from iris texture," in *Proceedings of the International Conference on Technologies for Homeland Security (HST'11)*, pp. 440–445, 2011.

[102]   M. D. Costa-Abreu, M. Fairhurst, and M. Erbilek. "Exploring gender prediction from iris biometrics," in *International Conference of the Biometrics Special Interest Group (BIOSIG'15)*, pp. 1–11, 2015.

[103]   D. Bobeldyk and A. Ross. "Iris or periocular? Exploring sex prediction from near infrared ocular images," in *Proceedings of the International Conference of the Biometrics Special Interest Group (BIOSIG'16)*, pp. 1–12, 2016.

[104]   K. Bowyer and E. Ortiz. "Iris recognition: does template ageing really exist?" *Biometric Technology Today*, 2015(10):5–8, 2015.

[105]   S. P. Fenker, E. Ortiz, and K. W. Bowyer. "Template aging phenomenon in iris recognition," *IEEE Access*, 1:266–274, 2013.

[106]   S. E. Baker and K. W. B. P. J. Flynn. "Empirical evidence for correct iris match score degradation with increased time-lapse between gallery and probe matches," in *Proceedings of the International Conference on Biometrics (ICB'09)*, pp. 1170–1179, 2009.

[107]   P. Grother, J. Matey, E. Tabassi, G. Quinn, and M. Chumakov. "IREX VI – temporal stability of iris recognition accuracy," Technical Report NIST Interagency Report 7948, National Institute of Standards and Technology (NIST), 2013.

[108]   T. Bergmüller, L. Debiasi, A. Uhl, and Z. Sun. "Impact of sensor ageing on iris recognition," in *Proceedings of the International Joint Conference on Biometrics (IJCB'14)*, pp. 1–8, 2014.

[109]   J. L. Wayman. "Editorial: two views on iris ageing," *IET Biometrics*, 4(4):191, 2015.

[110]   J. Fu, H. J. Caulfield, S.-M. Yooc, and V. Atluri. "Use of artificial color filtering to improve iris recognition and searching," *Pattern Recognition Letters*, 26(14):2244–2251, 2005.

[111]   N. B. Puhan and N. Sudha. "Coarse indexing of iris database based on iris colour," *International Journal on Biometrics*, 3(4):353–375, 2011.

[112]   H. Zhang, Z. Sun, T. Tan, and J. Wang. "Iris image classification based on hierarchical visual codebook," *IEEE Transactions on Pattern Analysis and Machine Intelligence*, 36(6):1120–1133, 2014.

[113]   J. E. Gentile, N. Ratha, and J. Connell. "An efficient, two-stage iris recognition system," in *Proceedings of the International Conference on Biometrics: Theory, Applications, and Systems (BTAS'09)*, pp. 1–5. IEEE, Piscataway, NJ, 2009.

[114]   J. E. Gentile, N. Ratha, and J. Connell. "SLIC: short-length iris codes," in *Proceedings of the International Conference on Biometrics: Theory, Applications, and Systems (BTAS'09)*, pp. 1–5. IEEE, Piscataway, NJ, 2009.

[115]   M. Konrad, H. Stögner, A. Uhl, and P. Wild. "Computationally efficient serial combination of rotation-invariant and rotation compensating iris recognition algorithms," in *Proceedings of the International Conference on Computer Vision Theory and Applications (VISAPP'10)*, vol. 1, pp. 85–90, 2010.

[116]   F. Hao, J. Daugman, and P. Zielinski. "A fast search algorithm for a large fuzzy database," *IEEE Transactions on Information Forensics and Security*, 3(2):203–212, 2008.

[117]   C. Rathgeb, F. Breitinger, H. Baier, and C. Busch. "Towards bloom filter-based indexing of iris biometric data," in *Proceedings of the International Conference on Biometrics (ICB'15)*, 2015.

[118]   R. Gadde, D. Adjeroh, and A. Ross. "Indexing iris images using the burrows-wheeler transform," in *Proceedings of the International Workshop on Information Forensics and Security (WIFS)*, 2010.

[119]   H. Proença. "Iris biometrics: indexing and retrieving heavily degraded data," *IEEE Transactions on Information Forensics and Security*, 8(12), 2013.

[120]   R. Rakvic, B. Ulis, R. Broussard, R. Ives, and N. Steiner. "Parallelizing iris recognition," *IEEE Transactions on Information Forensics and Security*, 4(4):812–823, 2009.

[121]   N. Vandal and M. Savvides. "CUDA accelerated iris template matching on graphics processing units (GPUs)," in *Proceedings of the Fourth International Conference on Biometrics: Theory Applications and Systems (BTAS)*, 2010.

[122]   M. López, J. Daugman, and E. Cantó. "Hardware-software co-design of an iris recognition algorithm," *IET Information Security*, 5(1):60–68, 2011.

[123]   A. Ross and A. K. Jain. "Information fusion in biometrics," *Pattern Recognition Letters*, 24(13):2115–2125, 2003.

[124]   A. K. Jain, B. Klare, and A. A. Ross. "Guidelines for best practices in biometrics research," in *Proceedings of the International Conference on Biometrics (ICB'15)*, pp. 1–5, 2015.

[125]   P. Radu, K. Sirlantzis, G. Howells, F. Deravi, and S. Hoque. "A review of information fusion techniques employed in iris recognition systems," *International Journal of Advances in Intelligent Paradigms*, 4(3/4):211–240, 2012.

[126]   F. Alonso-Fernandez, P. Tome-Gonzalez, V. Ruiz-Albacete, and J. Ortega-Garcia. "Iris recognition based on sift features," in *Proceedings of the International Conference on Biometrics, Identity and Security*, pp. 1–8. IEEE, Piscataway, NJ, 2009.

[127]   P. Wild, H. Hofbauer, J. Ferryman, and A. Uhl. "Segmentation-level fusion for iris recognition," in *International Conference of the Biometrics Special Interest Group (BIOSIG'15)*, pp. 61–72, 2015.

## Chapter 2

# An overview of periocular biometrics

*Fernando Alonso-Fernandez and Josef Bigun*

Despite being two well-studied areas, the performance of face or iris recognition algorithms degrades under certain perturbations inherent to uncontrolled or unconstrained environments. For example, partially covered faces due to hair, clothes or helmets severely affect performance of recognition systems based on full-facial analysis [1]. Regarding the iris modality, the iris texture may not have sufficient resolution in images captured at large stand-off distances, while closed eyes or off-angle gaze are factors also known to affect recognition performance [2]. In forensic analysis, faces of crime perpetrators may be covered, while in some religions, cultures or professions, face is intentionally covered. Therefore, in many of these cases, the region around the eyes might be the only visible biometrics trait from the facial region (Figure 2.1).

Periocular biometrics specifically refers to the externally visible skin region of the face that surrounds the eye socket [3], see Figure 2.2. Its utility is specially pronounced when the iris or the face cannot be properly acquired, being the ocular modality requiring the least constrained acquisition process [2]. Since 2009, the periocular modality has rapidly evolved [4], surpassing face in the case of occlusion or iris under low resolution. It appears over a wide range of distances, even under

*Figure 2.1   Example of images with facial occlusion and/or low resolution iris*

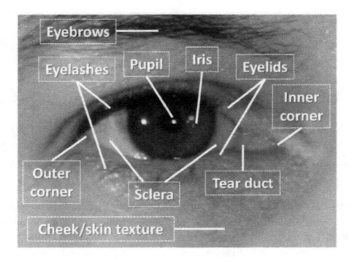

*Figure 2.2    Elements of the periocular region*

partial face occlusion (close distance) or low resolution iris (long distance), making it very suitable for unconstrained or uncooperative scenarios. It also avoids the need of iris segmentation, an issue in difficult images [5]. The periocular modality finds applicability, for example, in video surveillance, where recognition is still a challenge due to low quality, low resolution, changes in pose or expression, or partially masked faces. In such situation, identifying a suspect where only the periocular region is visible is one of the toughest real-world challenges in biometrics [6]. The richness of the periocular region in terms of identity is so high that the whole face can even be reconstructed only from images of the periocular region [6]. The technological shift to mobile devices has also resulted in many identity-sensitive applications becoming prevalent on these devices, but the inherent on the move conditions result in performance issues that are still unsolved [2].

## 2.1    Acquisition of periocular images

Acquisition setups of databases employed in periocular research include a variety of sensors in visible and near-infrared range capable of capturing still images or videos, such as: digital cameras, webcams, videocameras, smartphones, or close-up iris sensors. Table 2.1 summarizes the databases used in periocular research, with some sample images given in Figure 2.3. Although many databases have distance variability (e.g., FRGC, Compass, UBIRIS v2, UBIPr), acquisition is mostly done with the subject standing at several stand-off distances. A very few number of databases contain video data of subjects walking through an acquisition portal (MBGC, FOCS), or in hallways or atria (MBGC). There are databases captured to study particular problems too, such as ageing (MORPH, FG-NET), plastic surgery, gender transformation, expression changes (FaceExpressUBI), face occlusion (AR, Compass) or cross-spectral matching (CMU-H, IMP).

Table 2.1  Public databases used in periocular research

| Name | Subjects | Sessions | Data | Size | Illumination | Variability Cross-spectral | Distance | Expression | Lightning | Occlusion | Pose | Best accuracy EER | Rank-1 |
|---|---|---|---|---|---|---|---|---|---|---|---|---|---|
| **Facial databases** | | | | | | | | | | | | | |
| M2VTS [10] | 37 | 5 | 185 | 286 × 350 | V | N | N | N | Y | N | Y | 0.3% | n/a |
| AR [11] | 126 | 2 | 4,000 | 768 × 576 | V | N | N | Y | Y | Y | N | n/a | 76% |
| GTDB [12] | 50 | 2–3 | 750 | 640 × 480 | V | N | Y | Y | Y | N | Y | 0.25% | 89.2% |
| Caltech [13] | 27 | n/a | 450 | 896 × 592 | V | N | N | Y | Y | N | N | 0.12% | n/a |
| FERET [14] | 1,199 | 15 | 14,126 | 512 × 768 | V | N | N | Y | Y | Y | Y | 0.22% | 96.8% |
| CMU-H [15] | 54 | 1–5 | 764 | 640 × 480 | m | Y | Y | N | N | N | N | n/a | 97.2% |
| FRGC [16] | 741 | 1 | 36,818 | 1,200 × 1,400 | V | N | N | Y | Y | N | N | 0.09% | 98.3% |
| MORPH [17] | 515 | 2–5 | 1,690 | 400 × 500 | V | N | N | N | Y | Y | N | n/a | 33.2% |
| PUT [18] | 100 | n/a | 9,971 | 2,048 × 1,536 | V | N | N | N | Y | N | Y | 0.09% | 89.7% |
| MBGC v2 still [19] | 437 | n/a | 3,482 | variable | V | N | Y | Y | Y | Y | Y | 0.20% | 85% |
| MBGC v2 portal | 114 | n/a | 628 | 2,048 × 2,048 | N | Y | Y | N | Y | Y | N | 0.21% | 99.8% |
| | 91 | n/a | 571 | 1,440 × 1,080 | V | | | | | | | n/a | 98.5% |
| Plastic Surgery [20] | 900 | 2 | 1,800 | 200 × 200 | V | N | N | N | N | N | N | n/a | 63.9% |
| ND-twins [21] | 435 | n/a | 24,050 | 600 × 400 | V | N | N | Y | Y | Y | N | n/a | 98.3% |
| Compass [7] | 40 | n/a | 3,200 | 128 × 128 | V | N | Y | Y | N | Y | N | ~10% | n/a |
| FG-NET [22] | 82 | 12 | 1,002 | 400 × 500 | V | N | Y | Y | Y | Y | Y | 0.6% | 100% |
| CASIA Distance [23] | 142 | 1 | 2,567 | 2,352 × 1,728 | N | N | N | N | N | N | N | n/a | 67% |
| FaceExpressUBI [24] | 184 | 2 | 90,160 | 2,056 × 2,452 | V | N | N | Y | Y | N | N | 16% | n/a |

(*Continues*)

*Table 2.1 (Continued)*

| Name | Subjects | Sessions | Data | Size | Illumination | Cross-spectral | Distance | Expression | Lightning | Occlusion | Pose | EER | Rank-1 |
|---|---|---|---|---|---|---|---|---|---|---|---|---|---|
| | | | | | | **Variability** | | | | | | **Best accuracy** | |
| **Facial databases** | | | | | | | | | | | | | |
| BioSec [25] | 200 | 2 | 3,200 | 480 × 640 | N | N | N | N | N | N | N | 10.56% | 66% |
| CASIA Interval [23] | 249 | 2 | 2,655 | 280 × 320 | N | N | N | N | N | N | N | 8.45% | n/a |
| UBIRIS v2 [26] | 261 | 2 | 11,102 | 300 × 400 | V | N | Y | N | Y | N | Y | 9.5% | 87.62% |
| IIT Delhi v1.0 [27] | 224 | 1 | 2,240 | 240 × 320 | N | N | N | N | N | N | N | 1.88% | n/a |
| MobBIO [28] | 100 | 1 | 800 | 200 × 240 | V | N | N | N | Y | N | Y | 9.87% | 75% |
| **Periocular databases** | | | | | | | | | | | | | |
| UBIPr [29] | 261 | 1–2 | 10,950 | var. | V | N | Y | N | Y | Y | Y | 6.4% | 99.75% |
| FOCS [5] | 136 | var. | 9,581 | 750 × 600 | N | N | Y | N | Y | Y | Y | 18.8% | 97.75% |
| IMP [30] | 62 | n/a | 620 | 640 × 480 | N | Y | Y | N | N | N | N | 3.5% | n/a |
| | | | 310 | 600 × 300 | V | | | | | | | | |
| | | | 310 | 540 × 260 | n | | | | | | | | |
| CSIP [31] | 50 | n/a | 2,004 | var. | V | Y | Y | N | Y | Y | Y | 15.5% | n/a |
| MICHE-I [32] | 92 | n/a | 3,732 | var. | V | N | Y | N | Y | Y | Y | 0% | n/a |
| VSSIRIS [33] | 28 | 1 | 560 | var. | V | N | N | N | Y | Y | Y | 0.4% | n/a |

'Illumination': V = VW, N = NIR, n = night, m = multispectral.
All databases have images, except M2VTS, CMU-H and MBGC v2 portal, which have videos.
'Best accuracy' is the best performance reported in the literature [3].

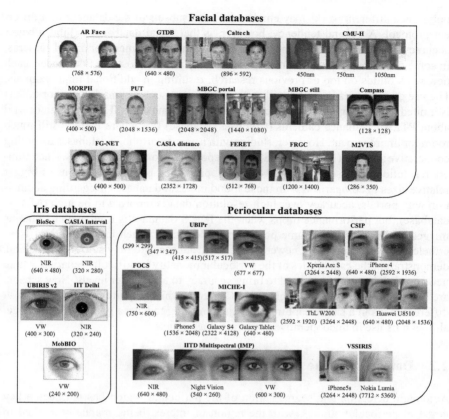

*Figure 2.3    Samples of databases used in periocular research*

Very few databases have been designed specifically for periocular research, with face and iris databases being predominantly used, since this region also appears in such databases. Nearly all of the face databases used are captured with video cameras or digital cameras in visible (VW) range, with some exceptions acquired with near-infrared (NIR) cameras, or even some multi-spectral databases captured at other wavelengths. Iris databases are either captured with dedicated close-up near-infrared iris sensors or digital cameras/webcams in visible range, but they have the limitation that eyebrows or other periocular parts may not appear in the image (see Figure 2.3, bottom left). Recently, some databases where the periocular region is specifically imaged and captured have been made public, such as FOCS (near-infrared portal), UBIPr (digital camera) or CSIP, MICHE-I and VSSIRIS (smartphones).

The most used databases to date in periocular research are face databases such as FERET, FRGC and MBGC. A number of studies also make use of UBIRIS v2 iris database, captured with a digital camera simulating a number of perturbations. The 'best accuracy' shown in Table 2.1 should be taken as an approximate indication only, since different works may employ different subsets of the database or a different protocol. A general tendency, however, is that facial databases exhibit a better accuracy. These are databases mostly acquired with video cameras or digital cameras, in some cases in high definition. Also, since they are the most used databases, each new work builds on top of previous research, resulting in additional improvements. The use of databases acquired with personal devices such as smartphones or tablets is limited, with recognition accuracy still behind in some cases. The same can be said about PTZ surveillance cameras (Compass database) [7], where there is still much room for improvement. However, studies with newer smartphone databases involving cooperative subjects (MICHE-I, VSSIRIS) report very competitive results, since camera resolution of these devices is very high, producing periocular regions of bigger relative sizes in comparison with those found in facial databases. Regarding acquisition with portals, accuracy over high-resolution data is reported to be high (MBGC database), but performance drops when the FOCS database is used, which contains images captured with the same portal principle, but of much less resolution. Nevertheless, the accuracy with newer periocular databases are only some steps behind, demonstrating the capabilities of the periocular modality in difficult scenarios, where new research leaps are expected to bring accuracy to even better levels. New sensors are being proposed, such as Light Field Cameras, which capture multiple images at different focus in a single capture [8,9], guaranteeing to have a good focused image, although its use in uncooperative scenarios is an issue.

## 2.2   Detection of the periocular region

Automatic detection and segmentation of the periocular region has not been a key target of periocular studies, with the region of interest being manually marked in most cases, or heuristically extracted after detecting the full face [7,42]. The latter strategy, however, is not reliable under face occlusion, which is likely to happen in some scenarios of interest where the periocular modality is intended to be used. Another strategy might be applying iris segmentation techniques to detect the eye region, but it may not be reliable under challenging conditions either [5].

In recent years, some works have dealt with the task of locating the eye position directly without relying on full-face or iris detectors, with a summary given in Table 2.2, and some graphic examples in Figure 2.4. The most popular approaches include:

- *Gabor features* proposed in the early work [43] for eye detection and face tracking purposes by performing saccades across the image
- *The Viola–Jones detector of face sub-parts* [45], employed more recently in [34,35]. The work [35] also experimented with the CMU hyperspectral database, which has images captured simultaneously at multiple wavelengths. Since the eye is centered in all bands, accuracy can be boosted by collective detection over all bands. Unfortunately, this type of illumination is only available in controlled scenarios. The Viola–Jones detector is known to perform well in frontal acquisition, but its performance degrades in non-frontal views of the face [46].

*Table 2.2*   *Eye/periocular detection and segmentation works. Acronyms are defined in the text or referenced papers*

| Approach | Features | Database | Accuracy |
|---|---|---|---|
| [43] | Gabor filters | M2VTS | 99.3% |
|  |  | XM2VTS | 99% |
| [7] | Active Shape Models (ASM) | Compass | n/a |
| [34] | Viola–Jones face parts detector | CASIA v4 distance | 96.4% |
|  |  | Yale-B (VW) | 99.2% |
| [41] | HSV + convex hull | UBIRIS v1 | n/a |
| [5] | Correlation filters | FOCS | 95% |
| [37] | LE-ASM + graph-cut | MBGC (VW still) | 99.4% |
| [44] | Correlation filters | HRT (VW images) | n/a |
| [39] | HSV | UBIRIS v1 | n/a |
| [40] | HSV+YCbCr | UBIRIS v2 / FRGC | n/a |
| [38] | Texture/shape | UBIRIS v2 | 97.5% |
| [36] | Symmetry filters | 4 NIR iris databases | 96% |
|  |  | 2 VW iris databases | 79% |
| [35] | Viola–Jones eyes detector + Hough | MBGC portal, UBIPr | n/a |
|  | Viola–Jones eyes detector + morphology | CMU-H | n/a |

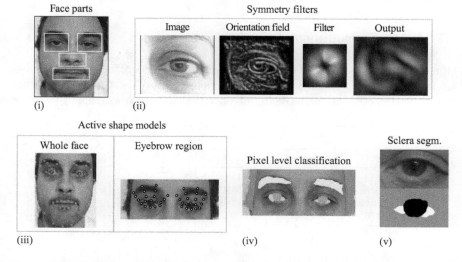

*Figure 2.4*   *Some strategies for detection of the periocular region or its parts: (i) Viola–Jones detector of face sub-parts [34,35]; (ii) symmetry filters [36]; (iii) Active Shape Models [7,37]; (iv) pixel-level classifiers [38]; (v) sclera segmentation [39–41]*

- *Symmetry filters* tuned to detect circular symmetries, as proposed in [36]. These have the advantage of not needing training, and detection is possible with a few 1D convolutions due to separability of the detection filters, built from derivatives of Gaussians. This method, however, still has to be fully tested in non-frontal images [46].
- *Correlation filters*, as used in [5] as preprocessing step prior to iris segmentation over the difficult FOCS database. Correlation filters were also used for eye detection in [44], but after applying a full-face detector. The filters used in these works, however, are trained with frontal images only.
- *Active Shape Models* (ASM), as used in [37] to detect the eyebrow region directly from a given face image, with eyebrow pixels segmented afterwards using graph-cut segmentation. ASMs were also used by Juefei-Xu and Savvides [7] to automatically extract the periocular region, albeit after the application of a full-face detector.

Recently, Proenca *et al.* [38] proposed a method to label seven components of the periocular region by using seven classifiers at the pixel level, with each classifier specialized in one component. The purpose of such region segmentation is to allow more accurate estimation of pose and gaze, and to enable the use of specific features that are more suitable for each region. It also allows to discard components not relevant for identity recognition, such as hair or glasses that might erroneously bias the process. Some works have also proposed the extraction of features from the sclera region only. For this purpose, [39], [40] and [41] have used the HSV/YCbCr color spaces for sclera segmentation. In these works, however, sclera detection is guided by a prior detection of the iris boundaries.

## 2.3   Feature encoding and comparison

Figure 2.5 gives a taxonomy of features employed in the literature for periocular recognition. They can be classified into global and local approaches. *Global approaches* extract properties of an entire region of interest (ROI), such as texture, shape or

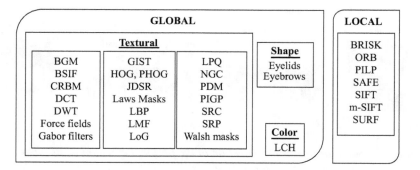

*Figure 2.5    Taxonomy of features used for periocular recognition. Acronyms are defined in the text or referenced papers*

color features. They are typically computed by dividing the image into a grid of local patches (Figure 2.6, top), and extracting features in each patch. A global descriptor representing the whole image is then obtained by concatenating features from each patch into a single vector. Matching with this type of features is very efficient via some distance measure, since feature vectors are of fixed length. Scale invariance is achieved by placing a grid whose size is proportional to the size of the eye (measured by the radius of the sclera circle, or the distance between inner and outer corners of the eye). Using the center of the pupil as reference point to center the grid is also the most widely used approach in global feature extraction methods. However, in unconstrained scenarios it has been proposed to employ the center of mass of the eye corners [29] or the cornea [38], since the pupil center is highly sensitive to off-angle gaze. In *local approaches*, on the other hand, a sparse set of characteristic points (called key points) is detected first, with features describing the neighborhood around key points only (Figure 2.6, bottom). The number of detected key points is not necessarily constant, therefore matching has to be done by comparing all possible pairs of key points between the two images. As a result, matching is more time-consuming. The output of the matching algorithm is usually the number of matched pairs, although some distance measurement between pairs may also be returned. Scale invariance is achieved by detecting key points at different scales. There are several key point detection algorithms in the literature, usually tied with each particular local feature extraction method. Other works, on the other hand, extract key point descriptors at selected sampling points in the center of image patches only [36,47,48], resembling the grid-like analysis of global approaches but using local features. This way, no key point detection is carried out, and the obtained 'local' feature vector is of fixed size.

The most widely used approaches in periocular recognition works include Local Binary Patterns (LBPs) and, to a lesser extent, Histogram of Oriented Gradients (HOG) and Scale-Invariant Feature Transform (SIFT) key points. Over the course of the years, many other descriptors have been proposed. The following is a brief review of existing works, highlighting the most important results or contributions. For the sake of page limitation, we omit the description of the feature extraction techniques, or references to the original works where they are presented. An extensive review can be found in [3,2]. Some preprocessing steps have been also used to cope with the difficulties found in unconstrained scenarios, such as pose correction by Active Appearance Models (AAM) [49], illumination normalization [50,51], correction of deformations due to expression change by Elastic Graph Matching (EGM) [52], Generic Elastic Models (GEM) [7], or color device-specific calibration [31]. The use of subspace representations is also becoming popular after feature extraction, either to improve performance or to reduce the feature set, as mentioned next during this section.

Periocular recognition started to gain popularity after studies [4,42]. Some pioneering works can be traced back to 2002 [43], although authors here did not call the local eye area 'periocular'. The good performance reported by [42] set the framework for the use of the periocular modality for recognition. Many works have followed this approach as inspiration, either improving it or introducing new perspectives, with LBPs and their variations being particularly extensive [47,53–57]. Some works have

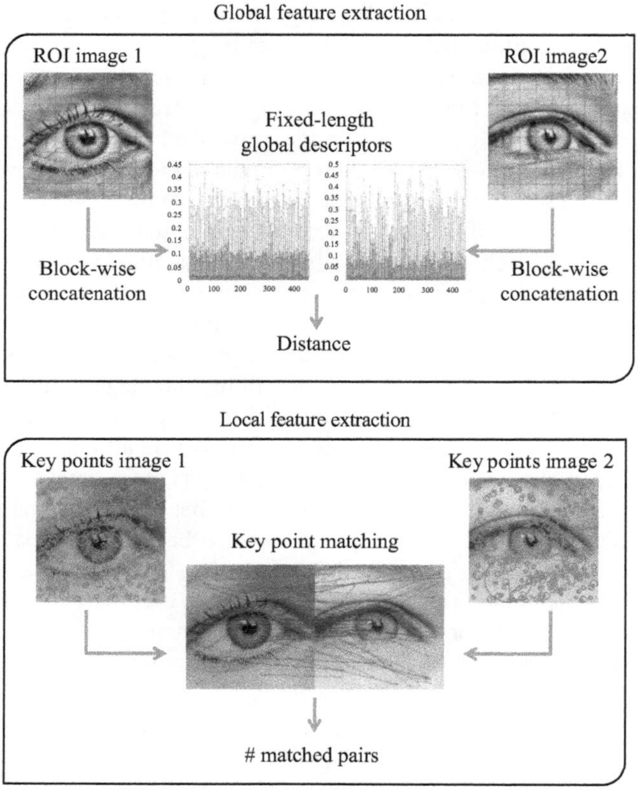

Global feature extraction

Local feature extraction

*Figure 2.6    Feature extraction approaches. Top: global feature extraction by local division of the image in patches. Bottom: local feature extraction by key point detection and matching. Actual size of green circles indicates the scale at which the key point has been detected*

also employed other features in addition to these [47,56]. The use of subspace representation methods after LBP extraction is also becoming a popular way either to improve performance or reducing the feature set [35,50,58–60].

Inspired by [4], the work [61] extended the experiments with additional global and local features to a significant larger set of the database employed (FRGC), including less ideal images (thus the lower accuracy observed with respect to previous studies with this database). They later addressed the problem of ageing degradation on periocular recognition [49], reported to be an issue even at small time lapses [42]. To obtain age-invariant features, they first performed preprocessing schemes, such as pose correction, illumination and periocular region normalization. LBP has also been used in other works analyzing, for example, the impact of plastic surgery [62] or gender transformation [44] on periocular recognition (see Section 2.6).

The framework set by [43] with Gabor filters served as inspiration for experiments with several databases in near-infrared and visible [36,63,64] spectra. Authors later proposed a matcher based on Symmetry Assessment by Feature Expansion (SAFE) descriptors [36,48], which describe neighborhoods around key points by estimating the presence of various symmetric curve families. Gabor filters were also used by Gangwar and Joshi [65] in their work presenting Local Phase Quantization (LPQ) as periocular descriptor. The work [66] also employed Gabor features over four different databases in the visible range, with features reduced by Direct Linear Discriminant Analysis (DLDA) and further classified by a Parzen Probabilistic Neural Network (PPNN).

The study [67] evaluated Circular Local Binary Patterns (CLBPs) and a global descriptor (GIST) consisting of five perceptual dimensions related with scene description (image naturalness, openness, roughness, expansion and ruggedness). They used the UBIRIS v2 database of uncontrolled iris images in visible range which includes a number of perturbations intentionally introduced. A number of subsequent works have also made use of UBIRIS v2 [38,40,68–71]. The work [70] proposed Phase Intensive Global Pattern (PIGP) features, and they later proposed the Phase Intensive Local Pattern (PILP) feature extractor [71], reporting the best Rank-1 periocular performance to date with UBIRIS v2. The study [40] extracted features from the eyelids region only, and the study [38] proposed a method to label seven components of the periocular region with the purpose of demonstrating that regions such as hair or glasses should be avoided since they are unreliable for recognition. Finally, the work [39] used the first version of UBIRIS in their study presenting directional projections or Structured Random Projections (SRP) as periocular features.

Other shape features have also been proposed, like eyebrow shape features [37,72], with surprisingly accurate results when the eyebrow is used as a stand-alone trait. Indeed, eyebrows have been used by forensic analysts for years to aid in facial recognition [37], and they are suggested to be the most salient and stable features in a face [73]. The study [72] also used the extracted eyebrow features for gender classification, and the work [37] proposed a eyebrow segmentation technique too.

The study [29] presented the first periocular database specifically acquired for periocular research (UBIPr). They also proposed to compute the ROI with respect to the midpoint of the eye corners (instead of the pupil center), which is less sensitive to gaze variations, leading to a significant improvement (EER from ~30% to ~20%). Previous studies have managed to improve performance over the UBIPr database using a variety of features [51,74]. The UBIPr database is also used by [35] in their extensive study evaluating data in visible (UBIPr, MBGC), near-infrared (MBGC) and multispectral (CMU-H) range, with the reported Rank-1 results being the best published performance to date for the four databases employed. A new database of challenging periocular images (CSIP) was presented recently by [31], the first one made public captured with smartphones, followed later by other studies with newer smartphone databases [33]. Another database captured specifically for cross-spectral periocular research (IMP) was also recently presented by [30], containing data in visible, near-infrared and night modalities. Cross-spectral recognition was also addressed by [75] using a proprietary database. Finally, [8] and [9] presented a database in the visible range acquired with a new type of camera, a Light Field Camera (LFC), which provides multiple images at different focus in a single capture. LFC overcomes one important disadvantage of sensors in the visible range, which is guaranteeing a good focused image. Unfortunately, the database has not been made available. Individuals were also acquired with a conventional digital camera, with a superior performance observed with the LFC camera. New features were also presented in the two studies.

## 2.4    Human performance evaluation

Understanding how humans recognize faces and which features find more useful for identity recognition can enable also researchers to develop better feature extraction techniques. The ability of (untrained) human observers to compare pairs of periocular images has been tested in several studies [76–78]. The first of these studies, which involved only near-infrared images, found that humans can correctly classify the pairs as belonging to the same person or different people with an accuracy of about 92% [76]. The study was extended later to understand the effect of different factors on human performance in periocular recognition [77]. In this case, queries were created by pairing subjects with the same gender, ethnicity and similar eye color, make-up, eye occlusion and eyelash length. In addition, viewing time of each query was limited to 3 s. As compared to the previous study where matching pairs were randomly selected and viewing time was not limited, accuracy dropped to about 79%. An exhaustive human study was carried out later involving images both in the near-infrared and visible spectra [78]. Human recognition accuracy was observed to be 88.4% with visible data and 78.8% with near-infrared data. They also tested the combination of three computer experts (LBP, HOG and SIFT), finding that the performance of humans and machines was similar. The study also identified which ocular elements humans find more useful for periocular recognition. With near-infrared images, eyelashes, tear ducts, eye shape and eyelids, were identified as the most useful, while skin was the less useful. But for visible data, blood vessels and skin were reported more helpful than eye shape and eyelashes. Similar studies have been done with automatic algorithms [74,79], with results in consonance with the study with humans, despite using several machine algorithms based on different features, and different databases. With near-infrared images, regions around the iris (including the inner tear duct and lower eyelash) were the most useful, while cheek and skin texture were the less important. With visible images, on the other hand, the skin texture surrounding the eye was found very important, with the eyebrow/brow region (when present) also favored in the visible range. This is in line with the assumption largely accepted in the literature that the iris texture is more suited to near-infrared illumination [80], whereas the periocular modality is best for visible illumination [78], suggesting that using visible spectrum periocular images instead of near-infrared images is a suitable direction for development of periocular biometrics research. This seems to be explained by the fact that near-infrared illumination reveals the details of the iris texture, while the skin reflects most of the light, appearing over-illuminated (see, e.g., 'BioSec' or other near-infrared iris examples in Figure 2.3); on the other hand, the skin texture is clearly

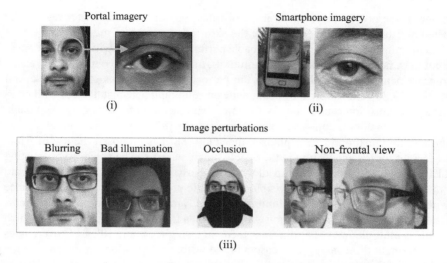

*Figure 2.7    Some scenarios of combination or comparison of periocular with other modalities: (i) fusion of iris and periocular using portal data [54,82,83]; (ii) fusion of iris and periocular using smartphone data [31,84]; (iii) effect of image perturbations [42,85]*

visible in visible range, but only irises with moderate levels of pigmentation image reasonably well in this range [81].

## 2.5   Comparison of periocular with other modalities

The periocular modality has evolved rapidly to competing with face or iris recognition. Since the periocular region appears in face or iris images, fusion with these modalities has been also proposed for improved accuracy or robustness. Some works have also made use of features from the sclera region [39], observing a significant improvement after the fusion with periocular features. This section provides a brief overview on these matters, with some visual examples provided in Figure 2.7. A more in-depth review of existing works can be found in [3].

Under difficult conditions, such as acquisition portals [54,82,83], distant acquisition [56], smartphones [31,84], webcams or digital cameras [36,86], the periocular modality is shown to be clearly superior to the iris modality, mostly due to the small size of the iris or the use of visible illumination. Small irises result in difficulty in segmenting the iris, and in a lack of sufficient information in the iris region, even if it is segmented successfully. Therefore, using the ocular region appears to be a better alternative. Regarding image lightning, visible illumination is predominant in relaxed or uncooperative setups due to the impossibility of using near-infrared illumination to date (e.g., distant acquisition or smartphones). Unfortunately, iris texture is more suited to the near-infrared spectrum, since this type of lightning better reveals the details of the iris texture [80], while the skin reflects most of the light, appearing over-illuminated. However, the reasonable performance shown by iris in visible spectrum in some cases, even at small resolutions, makes possible to obtain performance improvements when combined with the periocular modality, as shown by several studies [69,86]. Efforts are also being done on equipping personal devices with near-infrared acquisition capabilities, with major manufacturers following this approach, but no satisfactory solution has been provided yet [87]. As an alternative to near-infrared illumination in smartphones, the study [88] has proposed the use of white LED light for iris acquisition in the visible range, which can be achieved by existing LED flashes. The proposed setup allows to properly illuminate dark irises, which are not typically observable under visible light, while providing equivalent performance with respect to near-infrared illumination. The same authors also proposed a new type of camera in visible light for iris acquisition, a Light Field Camera (LFC), which provides multiple images at different focus in a single capture [8]. LFC overcomes one important disadvantage of sensors in the visible range, which is guaranteeing a good focused image.

Studies comparing periocular with face modality show similar trends that iris studies when it comes to robustness to perturbations in the acquisition. For example, face occlusions (even if partial) severely degrade performance of full-face matchers [42], pointing out the strength of periocular recognition when only partial face images are available (e.g., in criminal scenarios with surveillance cameras, where it is likely that the perpetrator masks parts of his face). At extreme values of blur or downsampling, it has been also found that periocular performs significantly better than face [85]. On the other hand, the same study found that both face and periocular matching using LBPs under uncontrolled lighting were very poor, indicating that this popular descriptor might not be well suited for this scenario. Periocular is shown to be more tolerant than face to expression variability, occlusion, and it has more capability of matching partial faces [50]. In addition, the periocular modality outperforms face matchers in case of undergoing plastic surgery [62] or gender transformation [44], as explored in the next section. It has been also shown that even the whole face can be reconstructed only from images of the periocular region, providing the possibility of using existing full-face commercial matchers, although at the expense of some drop in performance [6].

## 2.6    Other tasks using features from the periocular region

Features extracted from the periocular region have been also proposed for a number of other tasks, as shown in Table 2.3 and Figure 2.8, for example:

- *Soft-biometrics*, which refers to the classification of an individual in broad categories such as gender, ethnicity, age, height, weight, hair color, etc. While these are not sufficient to uniquely identify a subject, they can reduce the search space or provide additional information to improve recognition performance. Due to

*Table 2.3*  *Existing works on soft-biometrics, gender transformation and plastic surgery analysis using periocular features. Acronyms are fully defined in the text or referenced papers*

|  | Aim | Features | Database | Best accuracy |
|---|---|---|---|---|
| [92] | GC | raw pixels, LBP+ LDA-NN/PCA-NN/SVM | Prop. (936 VW images) | GC: 85% |
| [72] | GC | eyebrows shape+ MD/LDA/SVM | FRGC MBGC NIR portal | GC: 97% GC: 96% |
| [62] | PS | Periocular: SIFT, LBP Face: VeriLook (VL) Face: PittPatt (PP) | Plastic Surgery | Rank-1 periocular: 63.9%% Rank-1 face: 85.3% Rank-1 periocular + face: **87.4%** |
| [90] | GC | ICA + NN | FERET | GC: 90% |
| [89] | GC EC | LBP/HOG/DCT/LCH+ ANN/SVM | FRGC MBGC NIR portal | GC: 97.3%, EC: 94% GC: 90%, EC: 89% |
| [44] | GT | LBP, TPLBP, HOG applied to face parts and to the full face | HRT (>1.2 million VW images) | Perioc: EER=**35.21%**, Rank-1 = **57.79%** Nose: EER = 41.82%, Rank-1 = 44.57% Mouth: EER = 43.25%, Rank-1 = 39.24% Face: EER = 38.6%, Rank-1 = 46.69% |
|  |  | Face: PittPatt (PP) Face: FaceVACS (FV) |  | PP: EER = n/a, Rank-1 = 36.99% FV: EER = n/a, Rank-1 = 29.37% |

GC, Gender classification. EC, Ethnicity classification. PS, Plastic surgery. GT, Gender transformation.
GC, EC works report classification rates while PS, GT report recognition rates before/after transformation.

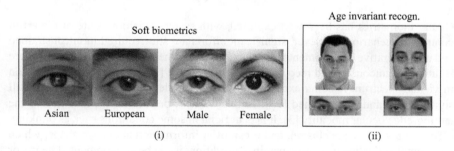

*Figure 2.8*  *Examples of other research involving periocular images: (i) soft-biometrics [89]; (ii) age-invariant recognition [49]. It can be observed that the periocular region is perceptually more stable for example across ages than the full face*

the popularity of facial recognition, face images have been frequently used to obtain both gender and ethnicity information, with high accuracy (>96%, for a summary see [89]). Recently, it has also been suggested that periocular features can be potentially used for soft-biometrics [89–92], with accuracies comparable to these obtained by using the entire face. The work [91] also showed that fusion of the soft-biometrics information with texture features from the periocular region can improve the recognition performance [91]. An interesting study by [72] with features extracted only from the eyebrow region showed very good results in gender classification. Following a different breadth, the work [49] addressed the problem of age-invariant face recognition using periocular images, with images from the FG-NET database. By using Walsh–Hadamard Local Binary Patterns (WLBPs) and Unsupervised Discriminant Projection, they obtain a 100% rank-1 identification rate and 98% verification rate at 0.1% FAR, proving the robustness of periocular biometrics for age-invariant biometric recognition.

• Effect of *plastic surgery* or *gender transformation* on the recognition performance. The work [44] studied the impact of gender transformation via Hormone Replacement Theory (HRT), which causes changes in the physical appearance of the face and body gradually over the course of the treatment. A database of >1.2 million face images from YouTube videos was built, observing that accuracy of the periocular region greatly outperformed other face components (nose, mouth) and the full face. Also, face matchers began to fail after only a few months of HRT treatment. The work [62] studied the matching of face images before and after undergoing plastic surgery, outperforming previous studies where only full-face matchers were used. Both [44] and [62] employed Commercial Off The Shelf (COTS) full face recognition systems in their experiments (VeriLook, PittPatt and FaceVACS).

## 2.7    Summary and future trends in periocular research

Ocular biometrics has been well explored, with systems yielding state of the art in controlled scenarios. However, significant efforts are still required in unconstrained and uncooperative environments [2]. The same phenomena is observed in face, where solutions to unconstrained recognition are still elusive [1]. Recently, iris recognition in visible spectrum is being explored due to mobile devices, but is a direction still significantly under-researched [93]. In this context, periocular has emerged as the cutting-edge ocular modality [3]. It is beneficial in some conditions such as blinking, off-angle iris or face occlusion, and it can offer information about eye shape, which also contains rich identity information. In addition, it can be complemented by iris or face features (if available), holding potential for improving unconstrained recognition. Fusion of multiple modalities using ocular data is a promising path forward that is receiving increasing attention due to unconstrained environments where switching between available modalities may be necessary [94].

The fast-growing uptake of face technologies in social networks and smartphones, as well as the widespread use of surveillance cameras and forensic applications, arguably increases the interest of the periocular modality. However, several problems arise when attempting to perform biometrics recognition in mobile or uncooperative environments. Images acquired in such heterogeneous scenarios have large intra-class variations, producing a significant drop in performance [95]. The wide variability of cameras produce image discrepancies in terms of color distortion, image resolution, illumination sources, field of view, etc.; and on-the-go acquisition also results in variabilities in pose, expression, illumination or scale and the acquisition environment may have inadequate lightning [31]. Unfortunately, matching of samples captured with different sensors will be common if, for example, people are allowed to use their own smartphone (cross-sensor matching). Exchange of information between law enforcement agencies also poses similar problems. Images from surveillance videos need to be matched against a gallery of still/mugshot images, which are usually the only images available during preliminary investigations. Therefore, heterogeneous ocular recognition is in great demand for a huge number of recognition applications [2]. A further difficulty that emerges is images are acquired with sensors working in different spectra (cross-spectral matching) [30,35]. For example, near-infrared is the choice of commercial iris systems, so cross-spectral matching with data from face images in visible light (from smartphones or surveillance cameras) holds significant practical value to ensure backward compatibility [75]. This *interoperability* problems, with developments capable of handling poor quality data from heterogeneous environments and a wide range of sensing devices working in same/different spectra, are key issues that current biometric technologies still need to solve [1].

Since the periocular modality is the ocular modality requiring the least constrained acquisition process, it is likely that the research community will move towards exploring ocular recognition at a distance and on the move in more detail as compared to previous studies [2]. Automatic detection and segmentation of the periocular region has been increasingly addressed, but existing works employ frontal semi-controlled data only. In this sense, newer machine learning algorithms such as deep learning promising outstanding performance in challenging detection [96] as well as other recognition tasks [1]. The periocular modality has also the potential to allow ocular recognition at large stand-off distances [97], with applications, for example, in surveillance. A limiting factor in recognizing people from a distance is low resolution, but reconstruction of images via super-resolution techniques is an under-explored trend in biometrics. There is a vast literature of super-resolution schemes aimed to enhance the visual appearance of general-scene images, but it does not necessarily correlate with better recognition performance [98]. Tailoring super-resolution approaches to the specificities of biometric images is still in its infancy, with most of the research concentrated in the reconstruction of full face images [99], a much more limited amount for iris [100], and completely lacking for periocular images. In addition, existing super-resolution works in biometrics have relied entirely upon the use of high quality images that are artificially down-sampled, therefore effective reconstruction approaches to cope with image variations found in unconstrained scenarios are still missing.

# Acknowledgments

F. A.-F. thank the Swedish Research Council for funding his research. Authors acknowledge the CAISR program of the Swedish Knowledge Foundation and the E U COST Action IC1106.

# References

[1]   A. K. Jain, K. Nandakumar, and A. Ross, "50 years of biometric research: Accomplishments, challenges, and opportunities," *Pattern Recognition Letters*, vol. 79, pp. 80–105, 2016.

[2]   I. Nigam, M. Vatsa, and R. Singh, "Ocular biometrics: A survey of modalities and fusion approaches," *Information Fusion*, vol. 26, pp. 1–35, 2015.

[3]   F. Alonso-Fernandez and J. Bigun, "A survey on periocular biometrics research," *Pattern Recognition Letters*, vol. 82, pp. 92–105, 2016.

[4]   U. Park, A. Ross, and A. K. Jain, "Periocular biometrics in the visible spectrum: A feasibility study," *IEEE International Conference on Biometrics: Theory, Applications, and Systems, BTAS*, Sep. 2009.

[5]   R. Jillela, A. A. Ross, V. N. Boddeti, *et al.*, "Iris segmentation for challenging periocular images," in *Handbook of Iris Recognition*, pp. 281–308. Berlin: Springer, 2013.

[6]   F. Juefei-Xu, D. K. Pal, and M. Savvides, "Hallucinating the full face from the periocular region via dimensionally weighted k-svd," *Proceedings of the IEEE Computer Vision and Pattern Recognition Biometrics Workshop, CVPRW*, June 2014.

[7]   F. Juefei-Xu and M. Savvides, "Unconstrained periocular biometric acquisition and recognition using cots PTZ camera for uncooperative and non-cooperative subjects," *Proceedings of the IEEE Workshop on Applications of Computer Vision, WACV*, Jan. 2012.

[8]   R. Raghavendra, K. B. Raja, B. Yang, and C. Busch, "Combining iris and periocular recognition using light field camera," *Proceedings of the Asian Conference Pattern Recognition, ACPR*, Nov. 2013.

[9]   K. B. Raja, R. Raghavendra, and C. Busch, "Binarized statistical features for improved iris and periocular recognition in visible spectrum," *Proceedings of the International Workshop Biometrics and Forensics, IWBF*, Mar. 2014.

[10]   S. Pigeon and L. Vandendorpe, "The M2VTS multimodal face database (release 1.00)," *Proceedings of the International Conference on Audio- and Video-Based Biometric Person Authentication, AVBPA*, Mar. 1997.

[11]   A. M. Martinez and R. Benavente, "The AR face database," *CVC Technical Report 24* – http://www2.ece.ohio-state.edu/~aleix/ARdatabase.html, 1998.

[12]   Georgia Tech face database (GTDB), *http://bit.ly/2fqMaYs*.

[13]   Caltech face database, *http://www.vision.caltech.edu/html-files/archive.html*.

[14]   P. J. Phillips, H. Moon, S. A. Rizvi, and P. J. Rauss, "The FERET evaluation methodology for face-recognition algorithms," *IEEE Transactions on Pattern Analysis and Machine Intelligence*, vol. 22, no. 10, pp. 1090–1104, 2000.

[15]   L. J. Denes, P. Metes, and Y. Liu, "Hyperspectral face database," Technical Report CMU-RI-TR-02-25, Robotics Institute, Pittsburgh, PA, Oct. 2002.

[16]   P. J. Phillips, P. J. Flynn, T. Scruggs, *et al.*, "Overview of the face recognition grand challenge," *Proceedings of the International Conference on Computer Vision and Pattern Recognition, CVPR*, Jun. 2005.

[17]  K. Ricanek and T. Tesafaye, "MORPH: A longitudinal image database of normal adult age-progression," *Proceedings of the International Conference on Automatic Face and Gesture Recognition, FG*, Apr. 2006.

[18]  A. Kasinski, A. Florek, and A. Schmidt, "The PUT face database," *Image Processing and Communication*, vol. 13, pp. 59–64, 2008.

[19]  P. J. Phillips, P. J. Flynn, J. R. Beveridge, *et al.* "Overview of the multiple biometrics grand challenge," *Proceedings of the International Conference on Biometrics, ICB*, Jun. 2009.

[20]  R. Singh, M. Vatsa, H. S. Bhatt, S. Bharadwaj, A. Noore, and S. S. Nooreyezdan, "Plastic surgery: A new dimension to face recognition," *IEEE Transactions on Information Forensics and Security*, vol. 5, no. 3, pp. 441–448, 2010.

[21]  P. J. Phillips, P. J. Flynn, K. W. Bowyer, *et al.*, "Distinguishing identical twins by face recognition," *Proceedings of the International Conference on Automatic Face and Gesture Recognition, FG*, Mar. 2011.

[22]  H. Han, C. Otto, X. Liu, and A. Jain, "Demographic estimation from face images: Human vs. machine performance," *IEEE Transactions on Pattern Analysis and Machine Intelligence*, vol. 37, no. 6, pp. 1148–1161, 2014.

[23]  CASIA databases, "http://biometrics.idealtest.org/".

[24]  E. Barroso, G. Santos, and H. Proenca, "Facial expressions: Discriminability of facial regions and relationship to biometrics recognition," *Proceedings of the IEEE Workshop on Computational Intelligence in Biometrics and Identity Management, CIBIM*, Apr. 2013.

[25]  J. Fierrez, J. Ortega-Garcia, D. Torre-Toledano, and J. Gonzalez-Rodriguez, "BioSec baseline corpus: A multimodal biometric database," *Pattern Recognition*, vol. 40, no. 4, pp. 1389–1392, 2007.

[26]  H. Proenca, S. Filipe, R. Santos, J. Oliveira, and L. A. Alexandre, "The ubiris.v2: A database of visible wavelength iris images captured on-the-move and at-a-distance," *IEEE Transactions on Pattern Analysis and Machine Intelligence*, vol. 32, no. 8, pp. 1529–1535, 2010.

[27]  A. Kumar and A. Passi, "Comparison and combination of iris matchers for reliable personal authentication," *Pattern Recognition*, vol. 43, no. 3, pp. 1016–1026, 2010.

[28]  A. F. Sequeira, J. C. Monteiro, A. Rebelo, and H. P. Oliveira, "MobBIO: A multimodal database captured with a portable handheld device," *Proceedings of the International Conference on Computer Vision Theory and Applications, VISAPP*, Jan. 2014.

[29]  C. N. Padole and H. Proenca, "Periocular recognition: Analysis of performance degradation factors," *Proceedings of the International Conference on Biometrics, ICB*, pp. 439–445, Mar. 2012.

[30]  A. Sharma, S. Verma, M. Vatsa, and R. Singh, "On cross spectral periocular recognition," *Proceedings of the International Conference on Image Processing, ICIP*, 2014.

[31]    G. Santos, E. Grancho, M. V. Bernardo, and P. T. Fiadeiro, "Fusing iris and periocular information for cross-sensor recognition," *Pattern Recognition Letters*, vol. 57, pp. 52–59, 2015.

[32]    M. Marsico, M. Nappi, D. Riccio, and H. Wechsler, "Mobile iris challenge evaluation (miche)-i, biometric iris dataset and protocols," *Pattern Recognition Letters*, vol. 57, pp. 17–23, 2015, Mobile Iris {CHallenge} Evaluation part I (MICHE I).

[33]    K. B. Raja, R. Raghavendra, V. K. Vemuri, and C. Busch, "Smartphone based visible iris recognition using deep sparse filtering," *Pattern Recognition Letters*, vol. 57, no. C, pp. 33–42, 2015.

[34]    A. Uhl and P. Wild, "Combining face with face-part detectors under gaussian assumption," *Proceedings of the International Conference on Image Analysis and Recognition, ICIAR*, Jun. 2012.

[35]    M. Uzair, A. Mahmood, A. S. Mian, and C. McDonald, "Periocular region-based person identification in the visible, infrared and hyperspectral imagery," *Neurocomputing*, vol. 149, pp. 854–867, 2015.

[36]    F. Alonso-Fernandez and J. Bigun, "Near-infrared and visible-light periocular recognition with Gabor features using frequency-adaptive automatic eye detection," *IET Biometrics*, vol. 4, no. 2, pp. 74–89, 2015.

[37]    T. H. N. Le, U. Prabhu, and M. Savvides, "A novel eyebrow segmentation and eyebrow shape-based identification," *Proceedings of the International Joint Conference on Biometrics, IJCB*, Sep. 2014.

[38]    H. Proenca, J. C. Neves, and G. Santos, "Segmenting the periocular region using a hierarchical graphical model fed by texture/shape information and geometrical constraints," *International Joint Conference on Biometrics, IJCB*, Sep. 2014.

[39]    K. Oh, B.-S. Oh, K.-A. Toh, W.-Y. Yau, and H.-L. Eng, "Combining sclera and periocular features for multi-modal identity verification," *Neurocomputing*, vol. 128, pp. 185–198, 2014.

[40]    H. Proenca, "Ocular biometrics by score-level fusion of disparate experts," *IEEE Transactions on Image Processing*, vol. 23, no. 12, pp. 5082–5093, 2014.

[41]    Z. Zhou, E. Y. Du, N. L. Thomas, and E. J. Delp, "A new human identification method: Sclera recognition," *IEEE Transactions on Systems, Man and Cybernetics, Part A*, vol. 42, no. 3, pp. 571–583, 2012.

[42]    U. Park, R. R. Jillela, A. Ross, and A. K. Jain, "Periocular biometrics in the visible spectrum," *IEEE Transactions on Information Forensics and Security*, vol. 6, no. 1, pp. 96–106, 2011.

[43]    F. Smeraldi and J. Bigün, "Retinal vision applied to facial features detection and face authentication," *Pattern Recognition Letters*, vol. 23, no. 4, pp. 463–475, 2002.

[44]    G. Mahalingam, K. Ricanek, and A. M. Albert, "Investigating the periocular-based face recognition across gender transformation," *IEEE Trans Information Forensics and Security*, vol. 9, no. 12, pp. 2180–2192, 2014.

[45] P. Viola and M. Jones, "Rapid object detection using a boosted cascade of simple features," *Proceedings of the Computer Vision and Pattern Recognition Conf, CVPR*, Dec. 2001.

[46] D. Teferi and J. Bigun, "Multi-view and multi-scale recognition of symmetric patterns," *Proceedings of the Scandinavian Conference on Image Analysis, SCIA*, Jun. 2009.

[47] S. Karahan, A. Karaoz, O. F. Ozdemir, A. G. Gu, and U. Uludag, "On identification from periocular region utilizing SIFT and SURF," *Proceedings of the European Signal Processing Conference, EUSIPCO*, Sep. 2014.

[48] A. Mikaelyan, F. Alonso-Fernandez, and J. Bigun, "Periocular recognition by detection of local symmetry patterns," *Proceedings of the Workshop Insight on Eye Biometrics, IEB, in conjunction with the International Conference on Signal Image Technology and Internet Based Systems, SITIS*, Nov. 2014.

[49] F. Juefei-Xu, K. Luu, M. Savvides, T. Bui, and C. Suen, "Investigating age invariant face recognition based on periocular biometrics," *Proceedings of the International Joint Conference on Biometrics, IJCB*, Oct. 2011.

[50] F. Juefei-Xu and M. Savvides, "Subspace-based discrete transform encoded local binary patterns representations for robust periocular matching on nist face recognition grand challenge," *IEEE Transactions on Image Processing*, vol. 23, no. 8, pp. 3490–3505, 2014.

[51] L. Nie, A. Kumar, and S. Zhan, "Periocular recognition using unsupervised convolutional RBM feature learning," *Proceedings of the International Conference on Pattern Recognition, ICPR*, Aug. 2014.

[52] H. Proenca and J. C. Briceno, "Periocular biometrics: Constraining the elastic graph matching algorithm to biologically plausible distortions," *Biometrics, IET*, vol. 3, no. 4, pp. 167–175, 2014.

[53] P. E. Miller, A. W. Rawls, S. J. Pundlik, and D. L. Woodard, "Personal identification using periocular skin texture," *Proceedings of the ACM Symposium on Applied Computing, SAC*, pp. 1496–1500, Mar. 2010.

[54] D. Woodard, S. Pundlik, P. Miller, R. Jillela, and A. Ross, "On the fusion of periocular and iris biometrics in non-ideal imagery," *Proceedings of the IAPR International Conference on Pattern Recognition, ICPR*, Aug. 2010.

[55] D. L. Woodard, S. J. Pundlik, J. R. Lyle, and P. E. Miller, "Periocular region appearance cues for biometric identification," *Proceedings of the IEEE Computer Vision and Pattern Recognition Biometrics Workshop, CVPRW*, Jun. 2010.

[56] C.-W. Tan and A. Kumar, "Human identification from at-a-distance images by simultaneously exploiting iris and periocular features," *Proceedings of the International Conference on Pattern Recognition, ICPR*, Nov. 2012.

[57] G. Mahalingam and K. Ricanek, "LBP-based periocular recognition on challenging face datasets," *EURASIP Journal Image and Video Processing*, vol. 36, pp. 1–13, 2013.

[58]    M. Uzair, A. Mahmood, A. Mian, and C. McDonald, "Periocular biometric recognition using image sets," *Proceedings of the Workshop Applications of Computer Vision, WACV*, Jan. 2013.

[59]    B.-S. Oh, K. Oh, and K.-A. Toh, "On projection-based methods for periocular identity verification," *Proceedings of the Conference on Industrial Electronics and Applications, ICIEA*, Jul. 2012.

[60]    J. Adams, D. L. Woodard, G. Dozier, P. Miller, K. Bryant, and G. Glenn, "Genetic-based type II feature extraction for periocular biometric recognition: Less is more," *Proceedings of the International Conference on Pattern Recognition, ICPR*, Aug. 2010.

[61]    F. Juefei-Xu, M. Cha, J. Heyman, S. Venugopalan, R. Abiantun, and M. Savvides, "Robust local binary pattern feature sets for periocular biometric identification," *Proceedings of the IEEE Conference on Biometrics: Theory, Applications and Systems, BTAS*, Sep. 2010.

[62]    R. Jillela and A. Ross, "Mitigating effects of plastic surgery: Fusing face and ocular biometrics," *Proceedings of the International Conference on Biometrics: Theory, Applications and Systems, BTAS*, Sep. 2012.

[63]    F. Alonso-Fernandez and J. Bigun, "Periocular recognition using retinotopic sampling and Gabor decomposition," *Proceedings of the International Workshop "What's in a Face?" WIAF, in conjunction with European Conference on Computer Vision, ECCV*, Oct. 2012.

[64]    F. Alonso-Fernandez and J. Bigun, "Eye detection by complex filtering for periocular recognition," *Proceedings of the International Workshop on Biometrics & Forensics, IWBF*, Mar. 2014.

[65]    A. Gangwar and A. Joshi, "Robust periocular biometrics based on local phase quantisation and Gabor transform," *Proceedings of the International Congress on Image and Signal Processing, CISP*, Oct. 2014.

[66]    A. Joshi, A. Gangwar, R. Sharma, A. Singh, and Z. Saquib, "Periocular recognition based on Gabor and Parzen PNN," *Proceedings of the International Conference on Image Processing, ICIP*, Oct. 2014.

[67]    S. Bharadwaj, H. S. Bhatt, M. Vatsa, and R. Singh, "Periocular biometrics: When iris recognition fails," *Proceedings of the IEEE Conference on Biometrics: Theory, Applications and Systems, BTAS*, Sep. 2010.

[68]    A. Joshi, A. K. Gangwar, and Z. Saquib, "Person recognition based on fusion of iris and periocular biometrics," *Proceedings of the International Conference on Hybrid Intelligent Systems, HIS*, Dec. 2012.

[69]    G. Santos and E. Hoyle, "A fusion approach to unconstrained iris recognition," *Pattern Recognition Letters*, vol. 33, no. 8, pp. 984–990, 2012.

[70]    S. Bakshi, P. K. Sa, and B. Majhi, "Phase intensive global pattern for periocular recognition," *Proceedings of the Annual IEEE India Conference, INDICON*, Dec. 2014.

[71]    S. Bakshi, P. K. Sa, and B. Majhi, "A novel phase-intensive local pattern for periocular recognition under visible spectrum," *Biocybernetics and Biomedical Engineering*, vol. 35, no. 1, pp. 30–44, 2015.

[72] Y. Dong and D. L. Woodard, "Eyebrow shape-based features for bio-metric recognition and gender classification: A feasibility study," *Proceedings of the International Joint Conference on Biometrics, IJCB*, Oct. 2011.

[73] J. Sadr, I. Jarudi, and P. Sinha, "The role of eyebrows in face recognition," *Perception*, vol. 32, no. 3, pp. 285–293, 2003.

[74] J. M. Smereka and B. V. K. V. Kumar, "What is a good periocular region for recognition?," *Proceedings of the IEEE Conference on Computer Vision and Pattern Recognition Workshops, CVPRW*, Jun. 2013.

[75] R. R. Jillela and A. Ross, "Matching face against iris images using periocular information," *Proceedings of the International Conference on Image Processing, ICIP*, Oct. 2014.

[76] K. Hollingsworth, K. W. Bowyer, and P. J. Flynn, "Identifying useful features for recognition in near-infrared periocular images," *Proceedings of the International Conference on Biometrics: Theory Applications and Systems, BTAS*, Sep. 2010.

[77] K. Hollingsworth, K. W. Bowyer, and P. J. Flynn, "Useful features for human verification in near-infrared periocular images," *Image and Vision Computing*, vol. 29, no. 11, pp. 707–715, 2011.

[78] K. Hollingsworth, S. S. Darnell, P. E. Miller, D. L. Woodard, K. W. Bowyer, and P. J. Flynn, "Human and machine performance on periocular biometrics under near-infrared light and visible light," *IEEE Transactions on Information Forensics and Security*, vol. 7, no. 2, pp. 588–601, 2012.

[79] F. Alonso-Fernandez and J. Bigun, "Best regions for periocular recognition with NIR and visible images," *Proceedings of the International Conference on Image Processing, ICIP*, Oct. 2014.

[80] J. Daugman, "How iris recognition works," *IEEE Transactions on Circuits and Systems for Video Technology*, vol. 14, pp. 21–30, 2004.

[81] K. W. Bowyer, K. Hollingsworth, and P. J. Flynn, "Image understanding for iris biometrics: A survey," *Computer Vision & Image Understanding*, vol. 110, pp. 281–307, 2007.

[82] V. N. Boddeti, J. M. Smereka, and B. V. K. V. Kumar, "A comparative evaluation of iris and ocular recognition methods on challenging ocular images," *Proceedings of the International Joint Conference on Biometrics, IJCB*, Oct. 2011.

[83] A. Ross, R. Jillela, J. M. Smereka, *et al.*, "Matching highly non-ideal ocular images: An information fusion approach," *Proceedings of the International Conference on Biometrics, ICB*, Mar. 2012.

[84] K. B. Raja, R. Raghavendra, and C. Busch, "Empirical evaluation of visible spectrum iris versus periocular recognition in unconstrained scenario on smartphones," in *Proceedings of the Asia-Pacific Signal and Information Processing Association Annual Summit and Conference, APSIPA*, Dec. 2014.

[85]    P. E. Miller, J. R. Lyle, S. J. Pundlik, and D. L. Woodard, "Performance evaluation of local appearance based periocular recognition," *Proceedings of the IEEE International Conference on Biometrics: Theory, Applications, and Systems, BTAS*, Oct. 2010.

[86]    F. Alonso-Fernandez, A. Mikaelyan, and J. Bigun, "Comparison and fusion of multiple iris and periocular matchers using near-infrared and visible images," *Proceedings of the International Workshop Biometrics and Forensics, IWBF*, Mar. 2015.

[87]    S. Thavalengal, I. Andorko, A. Drimbarean, P. Bigioi, and P. Corcoran, "Proof-of-concept and evaluation of a dual function visible/nir camera for iris authentication in smartphones," *IEEE Transactions on Consumer Electronics*, vol. 61, no. 2, pp. 137–143, 2015.

[88]    K. B. Raja, R. Raghavendra, and C. Busch, "Iris imaging in visible spectrum using white led," *Proceedings of the International Conference on Biometrics Theory, Applications and Systems, BTAS*, Sep. 2015.

[89]    J. R. Lyle, P. E. Miller, S. J. Pundlik, and D. L. Woodard, "Soft biometric classification using local appearance periocular region features," *Pattern Recognition*, vol. 45, no. 11, pp. 3877–3885, 2012.

[90]    S. Kumari, S. Bakshi, and B. Majhi, "Periocular gender classification using global ICA features for poor quality images," *International Conference on Modelling Optimization and Computing, ICMOC*, Apr. 2012.

[91]    J. R. Lyle, P. E. Miller, S. J. Pundlik, and D. L. Woodard, "Soft biometric classification using periocular region features," *Proceedings of the International Conference on Biometrics: Theory Applications and Systems, BTAS*, Sep. 2010.

[92]    J. Merkow, B. Jou, and M. Savvides, "An exploration of gender identification using only the periocular region," *Proceedings of the International Conference on Biometrics: Theory Applications and Systems, BTAS*, Sep. 2010.

[93]    R. R. Jillela and A. Ross, "Segmenting iris images in the visible spectrum with applications in mobile biometrics," *Pattern Recognition Letters*, vol. 57, pp. 4–16, 2015, Mobile Iris {CHallenge} Evaluation part I (MICHE I).

[94]    F. Alonso-Fernandez, J. Fierrez, D. Ramos, and J. Gonzalez-Rodriguez, "Quality-based conditional processing in multi-biometrics: Application to sensor interoperability," *IEEE Transactions on Systems, Man and Cybernetics-Part A: Systems and Humans*, vol. 40, no. 6, pp. 1168–1179, 2010.

[95]    F. Alonso-Fernandez, J. Fierrez, and J. Ortega-Garcia, "Quality measures in biometric systems," *IEEE Security and Privacy*, vol. 10, no. 6, pp. 52–62, 2012.

[96]    J. Long, E. Shelhamer, and T. Darrell, "Fully convolutional networks for semantic segmentation," *Proceedings of the International Conference on Computer Vision and Pattern Recognition, CVPR*, June 2015.

[97]    Z. X. Cao and N. A. Schmid, "Matching heterogeneous periocular regions: Short and long standoff distances," *Proceedings of the International Conference on Image Processing, ICIP*, Oct. 2014.

[98]   K. Nguyen, S. Sridharan, S. Denman, and C. Fookes, "Feature-domain super-resolution framework for Gabor-based face and iris recognition," *Proceedings of the IEEE Conference on Computer Vision and Pattern Recognition, CVPR*, Jun. 2012.

[99]   N. Wang, D. Tao, X. Gao, X. Li, and J. Li, "A comprehensive survey to face hallucination," *International Journal of Computer Vision*, vol. 106, no. 1, pp. 9–30, 2014.

[100]  F. Alonso-Fernandez, R. A. Farrugia, and J. Bigun, "Eigen-patch iris super-resolution for iris recognition improvement," *Proceedings of the European Signal Processing Conference, EUSIPCO*, Sep. 2015.

*Part II*

**Issues and Challenges**

*Chapter 3*

# Robust iris image segmentation

*Peter Wild, Heinz Hofbauer,*
*James Ferryman, and Andreas Uhl*

Despite numerous biometric challenges promoting iris and ocular recognition from less controlled images, iris localisation and segmentation are still challenging topics [1]. Examples include the Multiple Biometrics Grand Challenge (MBGC), Face and Ocular Challenge Series (FOCS), Noisy Iris Challenge Evaluation (NICE) and Iris Challenge Evaluation (ICE). For controlled environments and as long as iris segmentation algorithms can be tuned for the employed set, the segmentation problem can be managed. However, independent studies [2,3] have shown that algorithms may fail considerably if preconditions are not met. For datasets like FOCS segmentation performance ranging from 51% accuracy (using an integro-differential operator), to 90% (for active contours) is reported in [2].

Violations of preconditions or assumptions of controlled conditions such as particular hardware challenge the generalisation of successful segmentation techniques. Especially non-circularity of boundaries, non-frontal acquisition, or blurred images can impose difficulties. On the other hand, very flexible parameterless solutions allowing high degrees of freedom (e.g., off-axis acquisition) are sensitive to noise or reflections. They sometimes deliver inferior results to classical techniques for images where certain preconditions hold. Examples of challenging iris images include weak boundaries or reflections in visible range iris images, making it harder for iris localisation and segmentation algorithms to properly segment acquired images. Further undesirable conditions are on-the-move motion blur, out-of-focus images, narrowed eyes to a slit, or images with weak contrast. Figure 3.1 illustrates some examples.

Besides accuracy, also speed and usability play important factors for efficient iris segmentation [3]. For quick rejection of frames in video streams containing no iris image or unlikely leading to a successful verification, quality measures as predictors of recognition or segmentation accuracy can be used. This is useful to save precious processing time, as processing errors at early stages usually also result in classification errors.

This chapter reviews recent approaches of iris segmentation techniques and discusses approaches towards robust segmentation, highlighting the following topics:

- **Segmentation performance assessment**: Performance assessments are usually done with regards to recognition accuracy (thus evaluating an entire processing

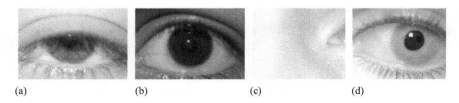

(a)                    (b)                    (c)                    (d)

*Figure 3.1    Challenging samples in the VIS and NIR iris database UTIRIS [4]:*
*(a) narrow eyes, (b) weak contrast, (c) false detection and (d) off-axis*

chain rather than individual algorithms). However, systematic segmentation errors can have a minor impact on performance if the same algorithm is used to segment gallery and probe images. This scenario is not realistic in cross-sensor applications. However, ground-truth only evaluations are restrictive, as they do not account for the tolerance of small segmentation inaccuracies by the employed feature extraction and comparison algorithm (e.g., via pooling operations). Different forms of segmentation accuracy evaluation are presented. There are separate segmentation-only challenges (e.g., NICE) and also ground-truth sets available (see [5]), to which this chapter contributes a manual segmentation of the multispectral iris database UTIRIS [4], see Section 3.3 for details.

- **Near infrared (NIR) vs. visible range (VIS) iris segmentation and the tuning problem**: Algorithms developed for specific image datasets or characteristics are not agnostic towards the type of imagery used. NIR images, for example, exhibit a pronounced inner pupillary boundary, while VIS images tend to have better outer limbic contrast. This is usually exploited by segmentation algorithms to achieve better accuracy on particular datasets. While more generic techniques offering the same processing chain for VIS and NIR images have been suggested [3], tuning can have a strong impact on performance. This chapter presents modifications to the open source USIT – University of Salzburg Iris Toolkit [6] for better performance, if prior knowledge on iris data characteristics (NIR vs. VIS, iris-to-pupil ratio, size of iris, etc.) is available. Source code of the modifications is released, see Section 3.3.2.
- **Iris fusion at segmentation-level**: Instead of developing completely new segmentation algorithms, the idea behind segmentation-level fusion is to make efficient use of different existing optimised approaches estimating their accuracy. The best individual segmentation technique can be used as the final segmentation result. Techniques can even be combined at a lower level leading to new segmentation results. For example, over- and under-segmentation methods can be fused to balance errors towards more robust iris segmentation. A learning-based approach predicting ground-truth segmentation performance [7] is compared to non-quality based combinations [8] in a multispectral setup assessing the impact of NIR and VIS spectras on performance.

New techniques are analysed with regards to a challenging multispectral NIR and VIS iris database involving the same subjects. This allows a direct comparison of

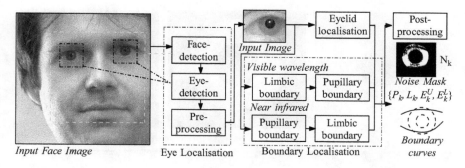

*Figure 3.2 . Segmentation processing chain*

robustness of segmentation results (localisation of pupillary and limbic boundaries) across the two spectra and illustrates the high versatility of the proposed methods.

## 3.1   Introduction

In order to recognise persons from iris patterns, a *feature vector* is computed from raw image data and used as compact representation of individual characteristics. It compresses identity-related information such that vectors calculated from the same person are close to each other with regards to some similarity metric while vectors originating from different individuals are further apart. The role of segmentation as a part of the process generating a feature vector is not entirely sharp and refers to specific models. Sometimes segmentation steps are executed as part of the feature extraction process. In iris recognition we can identify three preprocessing stages:

1. **(Face-) and eye localisation**, which for an input image $J$ detects a set of sub-windows containing a human (face) eye image (potentially nested);
2. **Iris boundary localisation**, which is at the heart of "segmentation" finding parameterisations of inner and outer iris boundaries $P$, $L$: $[0, 2\pi) \rightarrow [0, m] \times [0, n]$ enclosing the iris within a $m \times n$ subwindow. These are used within the subsequent "normalisation" step following the common iris model of Daugman's [9] Faberge coordinates: a rubbersheet transform $R(\theta, r) := (1 - r) \cdot P(\theta) + r \cdot L(\theta)$ is employed to obtain a representation with angle $\theta$ and pupil-to-limbic radial distance $r$ rectifying an iris pattern; and
3. **Iris noise detection**, which based on the detected boundaries generates a binary iris mask $N$ finalising "segmentation" (i.e., pixel-based iris classification). All pixels occluded by eyelids and reflections to be suppressed in the feature extraction stage are excluded. The final mask $N$ is eventually also mapped into Faberge coordinates.

Figure 3.2 illustrates these processing stages, which are commonly also referred to under the topics of "segmentation". The actual unrolling (iris mapping) and further

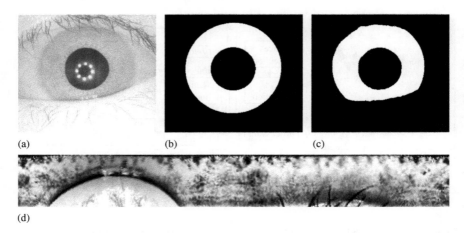

(a)　　　　　　　　　　　　　(b)　　　　　　　　　　　　　(c)

(d)

*Figure 3.3　Intermediate segmentation outputs for a CASIA [10] iris sample:*
*(a) input eye, (b) boundary mask, (c) iris mask and (d) normalised*
*and enhanced iris texture*

enhancement of the iris texture (contrast, anti-blur, etc.) as normalisation preprocessing are not the main topic of this chapter and the interested reader is referred
to [9]. The normalised representation is independent of pupillary dilation, the iris
texture is unified along the radial axis. Head rotations, which are as pronounced
as $\pm 10°$ degrees [11], can be accounted for by horizontal shifts in the rectified representation. ISO/IEC 19794-6:2011 defines a segmentation-only *cropped and masked*
exchange format, which generally allows for arbitrary segmentation approaches and
defines an interface between preprocessing and feature extraction in iris recognition.
Figure 3.3 illustrates (a) an input sample with corresponding (b) boundary mask indicating detected $P, L$. These are evolved into (c) an iris mask (via active contours)
masking out eyelids. Subfigure (d) shows the resulting iris texture after Rubbersheet
transform and local histogram enhancement.

## 3.1.1 Segmentation accuracy

There are two common ways to evaluate segmentation accuracy: (1) assessing pixel-
based classification or (2) investigating impact on recognition accuracy. The first
approach is based on an algorithm's $(m \times n)$ binary iris mask estimating the number
of true/false positive (in-iris) classifications $tp_i, fp_i$ and negatives (out-of-iris) classifications $tn_i, fn_i$ of the $i$th sample with regards to ground-truth yielding errors [12]:

$$E_1 := \frac{1}{k} \sum_{i=1}^{k} \frac{fp_i + fn_i}{mn}; \quad E_2 := \frac{1}{2} \left( \frac{1}{k} \sum_{i=1}^{k} \frac{fp_i}{fp_i + tn_i} \right) + \frac{1}{2} \left( \frac{1}{k} \sum_{i=1}^{k} \frac{fn_i}{fn_i + tp_i} \right) \quad (3.1)$$

Without proper ground-truth segmentation $E_1$ and $E_2$ cannot be computed. A further
drawback is that errors in boundary localisation are often more severe than missing

isolated occlusions. Especially if the centre for unrolling during normalisation is not accurate [13] the rubbersheet mapping may introduce distortions that cannot be accounted for in the rectified representation. Finally, a drawback of this type of assessment is the lacking sense for capturing the impact of outliers. Sometimes it is exactly the outliers that cause an irreversible impact leading to system errors, while minor accuracies can be accounted for in the natural pooling operations of classical feature extraction techniques.

Another set of measures are from the field of information retrieval, e.g., [14]: The precision ($\mathscr{P}$) gives the percentage of retrieved iris pixels which are correct. The recall ($\mathscr{R}$) gives the percentage of iris pixels in the ground-truth which were correctly retrieved. Since the target is to optimise both recall and precision these two scores are combined by the $\mathscr{F}$-measure, which is the harmonic mean of $\mathscr{P}$ and $\mathscr{R}$. Recall ($\mathscr{R}$), precision ($\mathscr{P}$) and the $\mathscr{F}$-measure are defined as follows:

$$\mathscr{P} := \frac{1}{k} \sum_{i=1}^{k} \frac{tp_i}{tp_i + fp_i}, \qquad \mathscr{R} := \frac{1}{k} \sum_{i=1}^{k} \frac{tp_i}{tp_i + fn_i}, \qquad \mathscr{F} := \frac{2\mathscr{R}\mathscr{P}}{\mathscr{R} + \mathscr{P}}. \quad (3.2)$$

The $\mathscr{F}$-measure, or more precisely $\mathscr{F}_1$-measure, is equivalent to the Dice and Sørensen measures in our case, which are commonly used for information retrieval.

Note, that these measures test the conformity of a segmentation to corresponding ground-truth. They are not a predictor for the overall performance of a biometric system, as shown in Hofbauer *et al.* [13]. Furthermore, the measures are capped, i.e., might not reach 100%, depending on the shapes used for segmentation, for example, elliptical ground-truth and circular masks used by an algorithm will generally not fully align.

The alternative to classification-based assessment is directly looking at an ROC-based evaluation, evaluating the entire preprocessing chain of an algorithm. By testing multiple processing chains with all elements fixed but segmentation, the impact of segmentation on biometric performance of a system can be estimated. As a drawback of this approach systematic errors might have a less pronounced effect than they would have if gallery and probe are segmented using the same algorithm. Therefore, ideally a ground-truth gallery image should be available for such assessments, yet there are no clear protocols available for this type of assessment. Further, assessments refer to specific feature extractors and comparison protocols. This makes an independent comparison of accuracies difficult. Also segmentation errors due to low image quality are not necessarily revealed by comparison-based assessment [15].

## 3.1.2 Iris segmentation quality

A good way to save processing time in iris biometric recognition is to avoid processing unsuccessfully segmented iris images. The key to early rejection of images which are difficult to segment, and also for unsuccessful segmentations (assessing their accuracy) without access to any ground-truth, is via quality measures. Alonso-Fernandez *et al.* [15] showed that segmentation and recognition performance are affected by different quality factors. Recently, Wild *et al.* [16] showed that integrated quality-based filtering can potentially shadow any temporal effects. ISO/IEC 29794-6 establishes a standard on iris quality. Common measures range from global measures, including histogram-related (spread, scalar), sharpness, motion blur, signal-to-noise ratio, to semantic measures like iris size (total, relative), pupil–iris ratio, boundary contrast (iris–sclera, iris–pupil), shape (pupil, iris), gaze-angle, and margin [17]. A predictor of segmentation accuracy can be easily formed using machine learning from these factors modelling regression (e.g., predicting ground-truth errors) or classification tasks (i.e., successful vs. unsuccessful segmentation). For example, Kalka *et al.* [18] employ a probabilistic model and Naive-Bayes tree predicting segmentation accuracy based on iris boundaries only and excess of iris and pupil centres. A recent method to estimate the correctness of iris segmentation results (iris masks) is presented by Mahadeo *et al.* [19] for fast rejection of difficult images. Their basic feature selection considered geometric features (pupillary and limbic radius, iris–pupil ratio, pupil and iris centre, occluded/clear iris and pupil area ratios) and texture features (mean pupil and iris intensity, variance, entropy in upper and lower half of clear region, quadrant intensities, intensities in eyelid regions and iris boundary contrast). Composite features are formed out of these basic features by linear combination and a Support Vector Machine (SVM) is used for building the classifier. The SVM with radial kernel is shown to outperform Naive-Bayes tree, decision tree and SVM with linear kernel approaches on the MBGC dataset with 98.7% prediction accuracy. They also stress the mapping impact of boundary segmentation (no eyelash removal) compared to the full segmentation task (with occlusion detection). Another recent study [20] considered standard deviation, local contrast, Gabor features and spectral magnitude Fourier transform as measures with best results attained by the Fourier measure.

Zuo and Schmid [21] are among the first to investigate automated evaluations of segmentation accuracy independently of the algorithm. This evaluation is used for feedback loops running more sophisticated segmentation routines on demand. They use minimum pupil size, the cumulative gradient magnitude of the iris and pupil boundaries, and an intensity test for iris, pupil and sclera for predicting segmentation accuracy. In their paper they analyse several image degradation effects on segmentation accuracy (comparing against ground-truth) on different challenging datasets. They showed that strong reflections, eyelid occlusions, different illumination conditions (especially poor or uneven illumination), very long eyelashes indicating wrong iris boundaries, low contrast, out-of-focus, off-angle, and motion blur have negative effects. More recently, Matveev *et al.* [22] suggest a Hough transform-based segmentation technique with subsequent refinements and additional attempts to locate boundaries, if they are not accurately found.

## 3.2   Advances in iris segmentation

Iris recognition literature has seen a vast amount of papers on iris segmentation and especially advances like on-the-move systems [23] and visible range iris recognition [24] have triggered innovative solutions. This section presents an overview of related works towards robust iris segmentation.

## 3.2.1 From circular models to active contours

As first groundbreaking achievement the original works by Daugman [9] introduce an integro-differential operator for iris segmentation. The operator behaves like a circular edge detector and looks for a circular boundary centred at $(x_0, y_0)$ with radius $r$ that maximises the blurred (Gaussian $G_\sigma$) pixel differences between inside and outside the circle in $I$. In partly modified form, for example, extended to off-axis images [25], or accelerated by partial evaluation in a ring [26] it influences segmentation techniques until today. As an alternative to the introduced operator, circular Hough transform (HT) iris segmentation as introduced by Wildes *et al.* [27] has seen broad application (e.g., in the iris segmentation method CAHT, part of the open source iris software USIT [6]).

More recent applications embed these core techniques at particular stages in the entire processing chain, often augmented by early coarse region-based clustering, or reflection removal and inpainting [28] to boost results. For example Tan *et al.*'s [26] NICE.I winning reference technique for colour iris segmentation employs clustering-based coarse iris localisation followed by a novel integro-differential operator and learned curvature model for eyelid detection. Recent examples from this group of clustering-based approaches, often aiming at real-time segmentation, are $k$-means clustering [29] or watershed and region merging with total variation flow [30].

Modern iris segmentation techniques often do no longer employ circular or elliptic models, but more general boundaries implementing active contours (AC). Active contour approaches can be classified as snake (or parametric) models using an image gradient to optimise the curve fitting, and level sets (geometric models) fitting regional image statistics to partition the image into different sets. The latter have advantages especially for images with weak edges. Good reference works in this area are introduced by Mumford–Shah [31,32] who extended the former model. These approaches inspired many researchers to add energies inside/outside a region for segmentation. Early iris segmentation paper in this category are [33] implementing an elliptical active shape model or [34] using a variant of Mumford–Shah energy function to fit the true boundary within a narrow band in a second stage after thresholding at first stage. Another popular approach is the algorithm by Viterbi [35], which is implemented in OSIRIS [36], a common open source iris segmentation algorithm. The critical prerequisite is to find a good initial detection of the coarse iris boundary.

Recently, Chai *et al.* [37] enhance a local Chan–Vese (CV) active contour model using B-splines, specifically targeted for visible range iris databases. The smoothness of the contour can be enforced through the B-spline representation making the regularisation term unnecessary. Experiments on NICE.I outperform global CV active contours and reduce the computational demands (separability of B-spline function). Abdullah *et al.* [38] also employ a CV active contour model in the second stage. Their first stage uses anisotropic diffusion for reflection removal, Otsu's thresholding for binarisation, morphologic operations for eyelash removal and centre-of-mass based centre detection. The second stage uses both edge-based as well as region-based elements in the energy function. Their overall segmentation method is shown to outperform geodesic AC, balloon AC, snake AC in terms of accuracy (2.37% EER

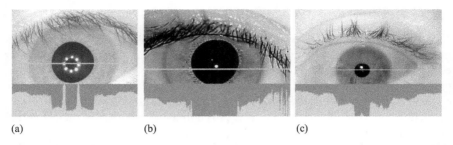

(a)                           (b)                           (c)

*Figure 3.4    Luminance along the cross-section of iris images from NIR databases:
(a) CASIA v4 – interval, (b) IIT Delhi and (c) Notredame*

(a)                           (b)                           (c)

*Figure 3.5    Luminance along the cross-section showing good pupillary contrast for
NIR, good limbic contrast for VIS brown and very low contrast overall
for VIS blue: (a) NIR, (b) VIS brown and (c) VIS blue*

vs. best 4.43%) and speed (581 ms vs. best 1,304 ms). Traditional circular HT and integro-differential operation were reported with far worse results.

## 3.2.2    Near infrared vs. visible range segmentation

Iris recognition for visible range has certain limitations especially for heavily pigmented irises [6] but nonetheless attracted several researchers in the past. An important difference between images in near infrared (NIR) and visible range (VIS) for segmentation is pupillary and limbic contrast. In NIR images, see Figure 3.4, there is a strong contrast between pupil and iris. This combined with the roughly circular shape of the pupil allows for robust detection of the pupil, and consequently the centre for the unrolling during normalisation. The detection of the iris to sclera (limbic) boundary then uses the pupillary boundary as a prior. When looking at Figure 3.4(a) the importance of the preprocessing step also becomes apparent when we look at the significant flanks due to reflection of the illumination in the pupil. When this is not controlled it might lead to a miss-detection of the circular light array as the pupil. A high quality acquisition overall can help in the detection of the pupillary boundary. Compare Figure 3.4(b), no blur and very sharp contrast at the pupillary boundary, with Figure 3.5(a), blurry acquisition and a smoother contrast. Good contrast and large depth of field is beneficial.

When looking at VIS images, compare Figure 3.5, the contrast at the pupillary boundary is strongly reduced. For dark iris colours the limbic boundary exhibits a higher contrast Figure 3.5(b). For brightly coloured irises however there is neither a strong contrast at the pupillary nor at the limbic boundary Figure 3.5(c). This makes iris segmentation in VIS images more challenging than for NIR images. The limbic boundary is frequently occluded by eyelids and further obscured by eyelashes protruding into the texture.

Efficient VIS segmentation algorithms take these particular characteristics for this type of imagery into account: Proenca *et al.* [24] suggest sclera detection as an important pre-step in iris localisation identifying certain types of configurations (dimension of sclera in each direction). Uhl and Wild's weighted adaptive Hough and ellipsopolar transforms [3] are agnostic with regards to the first boundary found and look for the second boundary on both sides, segmenting VIS and NIR images. Zhao and Kumar [39] present a novel iris segmentation framework for VIS and NIR image processing based on a total variation regulariser, suppressing noise and effective self-correcting post-processing, using dynamic thresholding. Leo *et al.* [40] use randomised circle detection to find the outer iris boundary in VIS images first, after which the pupillary boundary is detected within a pre-defined radius window. Randomised circle detection analyses the curvature of isophones, which are series of connected pixels of similar intensity.

In recent years, colour iris segmentation has attracted more interest. Hu *et al.* [41] presents a good survey of colour iris segmentation, especially focusing on iris segmentation for smart phones. They also present a multi-stage approach relying on simple linear iterative clustering super-pixel location of the iris and 1-norm regression for fine boundary fitting. Segmentation algorithms operate on the normalised, thresholded and contrast-enhanced red channel and correlation histograms are used as super-pixel features. Thavalenga *et al.* [42] present an iris segmentation technique aimed for high-frame-rate capture at smart phones, especially for their introduced VIS-NIR imaging system. They avoid active contours and instead of shape fitting, one-dimensional (1-D) processing is employed. Eye positions are tracked and the eye region is automatically cropped and smoothed along one direction. By thickening edge responses the iris is found among the largest connected areas, after which the pupil is found inside the iris. Results on a proprietary database reveals good results (3.49% errors vs 10.66% for OSIRIS 4.1). Trokielewicz [43] presents a new VIS iris database collected via smart phones (the first one with sufficient quality as of ISO/IEC 19794-6) and evaluated using commercial and open source software (IriCode ranked before OSIRIS, MIRLIN and VeriEye). Images were collected using iPhone 5s and achieved genuine match rates (GMR) in excess of 99.5% at zero false-match rate (FMR) for the best algorithm. Failures to enroll were as low as 1.27%.

### 3.2.3 *Learning-based techniques*

Very recently, iris segmentation has also embraced new trends to learn features employing latest machine learning approaches for this task. Happold [44] presents Structured Random Forests in anisotropic diffusion input and intensity variants as learning-based edge detectors for the problem of iris boundary detection. They out-performed classical Canny edge detectors on CASIA and Notredame datasets and are remarkably resistant towards wrongly detecting eyelashes as iris boundaries. Liu *et al.* [45] use hierarchical convolutional neural networks (HCNN) as well as multi-scale fully convolutional networks (MFCN). The HCNN use three layers which are then combined in a fully convolutional layer. The main drawback with respect to the MFCN is that the three separate chains have a large overlap of information. The MFCN on the other hand are basically a single chain going from fine to coarse representation, but at each pooling layer the results are extracted and further refined with three con-volutional layers. All extracted scales are then combined in a final fusion and softmax layer. The performance of the MFCN surpasses the HCNN result. Gangwar *et al.* [46] use a coarse to fine iterative segmentation process using information from Cartesian and polar space. Basically they preprocess the image and generate a bank of pupil candidate segmentations. They select the best candidate by region-based features and refine the boundary of the pupillary region. Based on the best pupillary region they estimate a coarse limbic boundary approximation in the polar space. They estimate eyelids and further refine the outer iris boundary. Since the whole segmentation hinges on the accurate finding of the pupil the approach is best fit for NIR images. Gangwar *et al.* mention in the conclusion that they limited their experiments to NIR images and will focus on adapting the method to visible spectrum images in the future.

### 3.2.4   *Segmentation fusion*

Instead of refining individual segmentation algorithms, an option to cope with diffi-cult iris images is to employ fusion methods ideally combining strengths of individual techniques. There is only little research on the combination of segmentation algo-rithms. Uhl and Wild [47] suggested a theoretical approach for such combination of independent segmentation algorithms, evaluated as proof of concept using differ-ent ground-truth segmentations. The approach was refined in [8] where a fusion framework for the automated combination of segmentation algorithms was sug-gested. However, all results of individual algorithms were combined without rejection of unsuccessful segmentations of individual algorithms. In [7], we extended this approach by introducing quality. Multiple noise masks are combined into a single mask, more specifically the algorithm traces and combines pupillary $P$ and limbic $L$ boundaries as key elements in the segmentation phase. Different approaches with the same goal are combining multiple normalised iris textures at image-level obtained by different segmentation algorithms [48], or combining multiple iris normalisations (requiring multiple execution of this step) [49]. In the following, the basics of [8] with extension in [7] are outlined. The approach consists of the phases *tracing, model fusion, prediction* and *selection*.

In *Tracing* the boundaries $P_i, L_i$ for each ($i$th) candidate are derived through scan-ning masks and pruning outliers. This step is useful as typical iris software lacks the ability to output boundaries and there is no interchange format due to varying (e.g., elliptical vs. circular or spline-based) boundary models. Note, that the model ide-ally has noise masks without masked eyelids available to re-trace the true possible

Table 3.1   *Employed parameters in [7] to indicate segmentation accuracy*

| Parameters | Description |
| --- | --- |
| $p_x, p_y$ | Pupil centre coordinates |
| $p_r$ | Pupil radius from pupil centre to pupillary boundary |
| $c_P$ | Contrast along the pupillary boundary |
| $l_x, l_y$ | Iris centre coordinates |
| $l_r$ | Iris radius from the iris centre to limbic boundary |
| $c_L$ | Contrast along the limbic boundary |
| $a_I$ | Iris area contained between pupillary and limbic boundary |
| $\mu, \sigma$ | Mean and standard deviation of the iris intensity |

occluded boundaries. The typical approach in [8] intersects a mesh grid with the binary noise mask $N$ locating binary transitions to re-trace boundaries. Topological inconsistencies are avoided by heuristical processing (morphologic operators, assuming 2 boundaries, etc.). Outliers are removed in a pruning phase (deviation radius with $z$-score $\geq 2.5$ from the centre of gravity $C_r$). The result of tracing is a set of limbic $L_i$ and pupillary boundary points $P_i$.

In *Model fusion* candidate boundaries, that is, $k$ sets of points $B_1, B_2, \ldots, B_k$ are merged into a joint set resulting in a continuous curve $B : [0, 2\pi) \rightarrow [0, m] \times [0, n]$. Strategies for this task are [47]:

$$\text{Aug Rule: } B(\theta) := ModelFit\left(\bigcup_{i=1}^{k} B_i\right)(\theta). \tag{3.3}$$

$$\text{Sum Rule: } B(\theta) := \frac{1}{k}\sum_{i=1}^{k} B_i(\theta) \tag{3.4}$$

In the first method the final curve is obtained via fitting a model (e.g., least-squares circular or ellipse fitting – e.g., via Fitzgibbon's method [50]) on the union of boundary points. For the latter (sum rule) the final curve is simply the interpolated boundary points. Note, that methods can combine any number of available candidates.

While in [8,47] all results are combined, rejection of bad segmentations can significantly strengthen the approach. In a recent addition to this concept [7] therefore an additional *prediction* phase is employed. Given an input image and corresponding traced segmentation (as specified via its trace $P, L$ derived from noise mask $N$), a set of quality features is computed, see Table 3.1. The set of $n = 11$ quality-based parameters $x_0, \ldots, x_n \in [0, 1]$ is fed into a neural network, which is trained to predict $E_2$ segmentation errors as quality score $q(P, L) : = h_W(x(P, L)) = f(W_1^T f(W_2^T x))$ for a segmentation result $(P, L)$, that is, a two-layer network is employed. As activation function the sigmoid $f(z) = 1/1 + \exp(-z)$ is used and $W_1, W_2$ are trained via the limited memory Broyden–Fletcher–Goldfarb–Shanno algorithm [51] on a training set containing comparisons of algorithm segmentations with ground-truth masks.

Finally, *Fusion Selection* acts as a multiplexer returning the (combined or individual) segmentation candidate $s$ out of $m$ individual candidates and combinations with lowest predicted segmentation error $q(P_i, L_i)$:

$$s := \arg \min_{1 \leq i \leq m} (q(P_i, L_i)). \tag{3.5}$$

The corresponding boundaries ($P_s$ and $L_s$) are used as the segmentation result and for the rubbersheet transform. Note, that a noise mask can be further generated by drawing boundary curves and/or merging results from the corresponding occlusion noise masks.

## 3.3   Experiments

In order to illustrate the impact of NIR vs. VIS iris segmentation, we employ the UTIRIS database by Hosseini *et al.* [4], consisting of 1,540 challenging images in near infrared and visible range spectra. The database contains 158 eyes from 79 individuals (5 images per eye) recorded in 2 different sessions. Since images originate from the same individuals it is particularly interesting to compare the two spectra. Prior to processing, visible range images (2,048 × 1,360 pixels) were resized by factor 2 to match the resolution of near infrared images (1,000 × 776 pixels). Ground-truth segmentations (made available online with this chapter at http://www.cvg.reading.ac.uk/utirisgt) were obtained using USIT's [6] manual iris segmentation software with an elliptical model supported by at least five labelled boundary locations for inner and outer iris curves through which an ellipse (using Fitzgibbion's fitting algorithm) was approximated. For the mapping process, a target texture size of 512 × 64 pixels was chosen and contrast-limited adaptive histogram equalisation (CLAHE) was applied for enhancement. The following algorithms are employed in the test:

- **CAHT** [6]: Contrast-adjusted Hough Transform is USIT's classical Wildes-based [27] circular model, using contrast-enhancement and strict pupil-before-limbic near infrared processing order. We extended the implementation providing a possibility to tune the algorithm towards specific databases: boundary detection order is reversible for the visible range, target pupil and iris size for scaling support and contrast thresholds can be set tolerating different gain.
- **IFPP** [28]: Iterative Fourier-based Pulling and Pushing (IFPP) is a k-means clustering-based technique with elliptical model in USIT building on an iterative Fourier boundary approximation and pulling-and-pushing methods.
- **OSIRIS** [36]: We employed the well-known Open Source Iris Recognition Toolkit (OSIRIS) as a representative of an active contours approach (Viterbi) after circular Hough Transform.
- **REF**: As baseline reference, a commercial IREX-ranked reference algorithm based on active shape models is employed. The software claims to be robust against illumination variability, off-gaze acquisitions, reflections, occlusions and noise with support for non-circular boundaries. Vendor anonymity is kept as the

aim of the comparison is not a commercial assessment, but an industrial reference is chosen to set open source algorithms' performance into context.

- **WAHET** [3]: Weighted Adaptive Hough and Ellipsopolar Transforms (WAHET) is a two-stage segmentation with elliptical models. It features a multi-scale HT and ellipsopolar transform for assisted second boundary detection claiming better robustness for cross-spectral application. Like for CAHT we extended the algorithm enhancing usability by allowing to indicate an expected target pupil and iris size (scaling support) considered in the weighting phase.

For the recognition-based assessment we employ the feature extractors Quality Assessment and Selection of Spatial Wavelets (QSW) by Ma *et al.* [52] and 1-D Log-Gabor (LG) by Masek [53] from the USIT 2.0 toolkit [6]. Both algorithms yield 10,240 bit codes and are compared using hamming distances normalised to [0, 1]. QSW is based on dyadic wavelet transform with min–max coding, LG is a 1-D implementation of Daugman's reference Gabor phase quantisation.

### 3.3.1 Individual NIR vs. VIS performance

Segmentation using the VIS and NIR sets in UTIRIS is quite challenging. Figure 3.6 illustrates good and bad segmentation results for different algorithms. From manual inspection of NIR masks we can see that there are some total failures for CAHT where no circular boundary is found. With challenging images, IFPP sometimes generates irregular non-circular-shaped boundaries, while OSIRIS has some under-segmentation problems overshooting the outer limbic boundary. WAHET easily runs into oversegmentation getting the inner boundary right, but returning collarette or radial furrows within the iris as outer boundary. The REF algorithm also delivers some false detections where a small iris is detected, distracted by strong edges of eyelids.

For VIS images the algorithm CAHT has clearly problems with detecting the very weak pupillary boundary and preferred smaller circles in its configuration while often getting the outer boundary right. IFPP and OSIRIS have pronounced problems – the first with many failures and irregular boundaries and the second with many false detections. Compared to the other algorithms REF and WAHET deliver more consistent masks, however WAHET shows some preference for masks with a high pupil-to-iris ratio, while REF sometimes squeezes masks to elliptical shape distracted by the upper eyelid. Figure 3.7 plots the $\mathscr{F}$-measure as ground-truth performance indicator supplementing this analysis of segmentation failures, illustrating the segmentation accuracy across the employed set of samples.

In a previous work [8] most of the algorithms were tested on the well-known NIR reference database CASIA Version 4 Interval test set [10]: EERs were 1.04% for OSIRIS, followed by 1.22% for CAHT, 1.89% for WAHET and 8.1% for IFPP using LG feature extraction, and 0.73% for OSIRIS, 0.99% for CAHT, 1.72% for WAHET and 8.78% for IFPP with QSW feature extraction. This clearly illustrates the low impact of the feature extractor and that accuracies are mainly determined by the segmentation algorithm, which is supported in our experiments. For the multispectral UTIRIS database, Table 3.2 lists recognition-based performance of individual

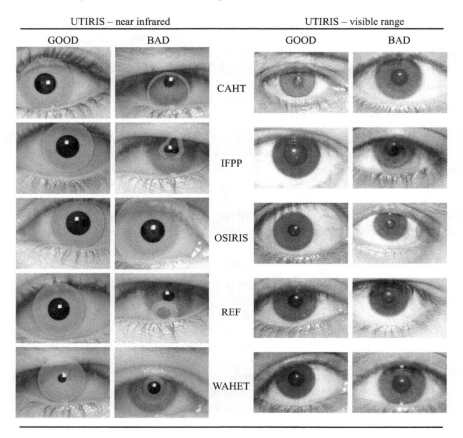

*Figure 3.6    Examples of good and bad segmentations for all algorithms over the UTIRIS [4] database*

algorithms along the diagonal. With most NIR images in UTIRIS being frontal the circularity assumption largely holds, yet weaker boundaries impose constraints. Reference EER performance for ground-truth segmentations is 2.66% (LG) and 2.13% (QSW) for NIR and 6.91% (LG) and 6.36% (QSW) for VIS.

For the NIR set (like for a different NIR database in [8]) there is only a minor change in the relative EER performance (ranking) of algorithms: 9.34% CAHT, 12.51% WAHET, 13.73% OSIRIS, 27.89% REF and 28.95% IFPP. A very different result is obtained for VIS with the commercial REF achieving best (lowest) 15.82% EER, WAHET 20.36%, CAHT 23.39%, IFPP 35.29% and OSIRIS 39.23%. The low accuracies can be explained, all open source algorithms were developed for NIR and no quality restrictions are imposed.

Table 3.3 lists results for individual ground-truth-based assessments along the diagonal, delivering a similar result keeping the order of algorithms with $E_1$ segmentation errors from 3.7% to 9.9% for NIR and 3.4% to 13.6% for VIS and $E_2$

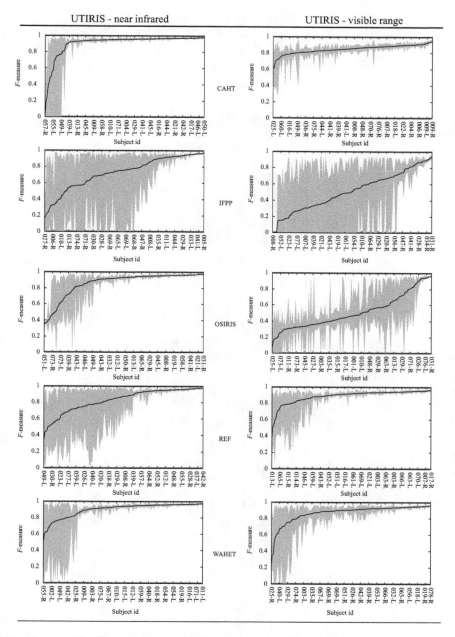

*Figure 3.7   The $\mathscr{F}$-measure grouped by user (values are minimum, maximum and average $\mathscr{F}$-measure) with entries sorted by average $\mathscr{F}$-measure. The measure is the given algorithm in comparison to the ground-truth on the UTIRIS [4] database.*

Table 3.2   *Recognition evaluation for segmentation methods and pairwise basic combined [8] segmentation fusion*

**(a) UTIRIS – near infrared**

Equal-error rate (%) for LG

| | CAHT | IFPP | OSIRIS | REF | WAHET |
|---|---|---|---|---|---|
| CAHT | 9.34 | 12.67 | **8.10** | 16.92 | **6.83** |
| IFPP | | 28.95 | 16.03 | **22.14** | 13.91 |
| OSIRIS | | | 13.73 | 18.82 | **8.50** |
| REF | | | | 27.89 | 17.52 |
| WAHET | | | | | 12.51 |

**(b) UTIRIS – visible range**

Equal-error rate (%) for LG

| | CAHT | IFPP | OSIRIS | REF | WAHET |
|---|---|---|---|---|---|
| CAHT | 23.39 | 33.13 | 35.45 | **14.97** | **19.33** |
| IFPP | | 35.29 | 36.52 | 25.42 | 25.84 |
| OSIRIS | | | 39.23 | 30.69 | 31.01 |
| REF | | | | 15.82 | **13.85** |
| WAHET | | | | | 20.36 |

Table 3.3  Ground-truth classification evaluation of individual and basic combined [8] segmentation methods

**(a) UTIRIS – near infrared**

Average $E_1$ (%)

|        | CAHT | IFPP | OSIRIS | REF | WAHET |
|--------|------|------|--------|-----|-------|
| CAHT   | 3.7  | 4.3  | **3.7** | 4.1 | **2.7** |
| IFPP   |      | 9.9  | **5.7** | 6.3 | 4.7   |
| OSIRIS |      |      | 6.1    | **5.3** | 4.0 |
| REF    |      |      |        | 7.0 | 4.4   |
| WAHET  |      |      |        |     | 4.1   |

Average $E_2$ (%)

|        | CAHT | IFPP | OSIRIS | REF | WAHET |
|--------|------|------|--------|-----|-------|
| CAHT   | 5.1  | 5.8  | **4.7** | 5.3 | **4.0** |
| IFPP   |      | 15.7 | **8.2** | 9.0 | 7.3   |
| OSIRIS |      |      | 9.4    | 7.4 | 6.1   |
| REF    |      |      |        | 10.5 | **6.7** |
| WAHET  |      |      |        |     | 7.2   |

Average $\mathscr{F}$ (%)

|        | CAHT | IFPP | OSIRIS | REF | WAHET |
|--------|------|------|--------|-----|-------|
| CAHT   | 90.5 | **90.9** | **92.5** | **91.5** | **94.3** |
| IFPP   |      | 72.8 | **87.9** | **86.1** | **89.7** |
| OSIRIS |      |      | 86.1   | **89.0** | **91.7** |
| REF    |      |      |        | 82.9 | **90.5** |
| WAHET  |      |      |        |      | 89.8 |

**(b) UTIRIS – visible range**

Average $E_1$ (%)

|        | CAHT | IFPP | OSIRIS | REF | WAHET |
|--------|------|------|--------|-----|-------|
| CAHT   | 5.4  | 9.3  | 9.2    | **3.1** | **3.5** |
| IFPP   |      | 13.1 | 14.5   | 8.4 | 8.8   |
| OSIRIS |      |      | 13.6   | 8.6 | 8.7   |
| REF    |      |      |        | **3.4** | **3.4** |
| WAHET  |      |      |        |     | 4.0   |

Average $E_2$ (%)

|        | CAHT | IFPP | OSIRIS | REF | WAHET |
|--------|------|------|--------|-----|-------|
| CAHT   | 5.5  | 15.6 | 15.5   | **5.0** | **5.0** |
| IFPP   |      | 27.5 | **26.6** | 15.0 | 15.1 |
| OSIRIS |      |      | 28.6   | 15.9 | 15.7  |
| REF    |      |      |        | 6.7 | **5.9** |
| WAHET  |      |      |        |     | 6.7   |

Average $\mathscr{F}$ (%)

|         | CAHT | IFPP | OSIRIS | REF | WAHET |
|---------|------|------|--------|-----|-------|
| CAHTVIS | 84.9 | 73.2 | 72.3   | **90.7** | **89.3** |
| IFPP    |      | 51.5 | **65.2** | 76.0 | 75.0 |
| OSIRIS  |      |      | 51.6   | 72.9 | 72.6  |
| REF     |      |      |        | 88.6 | **89.3** |
| WAHET   |      |      |        |      | 87.0 |

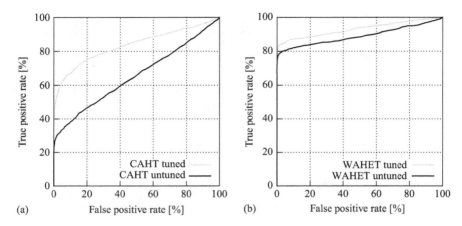

*Figure 3.8   Receiver-operator characteristic (ROC) plots for optimised and non-optimised versions of an algorithm. (a) An untuned CAHT used in the visible range and a version of the algorithm which is tuned to VR images on UTIRIS set VIS and (b) comparison of the untuned wahet and tuned WAHET on the UTIRIS set NIR.*

segmentation errors slightly higher from 5.1% to 15.7% for NIR and 5.5% to 28.6% for VIS. Summarising, WAHET provides the best overall cross-spectral performance, while CAHT is the best NIR and REF the best VIS algorithm. Systematic errors like CAHT's difficulties to detect pupils in VIS are hard to read directly from rates.

## 3.3.2   Impact of tuning

The impact of tuned vs. untuned versions for the variants of WAHET and CAHT turned out to be very pronounced, see Figure 3.8 for a direct comparison. Modifications were undertaken in a way that the original implementation is preserved, but allowing for a modification of parameter settings. Especially, the expected iris and pupil diameter means and variance can now be set to scale input images and derive weights for the model fitting process (extensions to USIT are made available online with this chapter at http://www.wavelab.at/sources). For WAHET an option to suppress eyelids masking out upper and lower thirds of the image, which was activated for NIR and VIS, is introduced. Besides, a border weight as multiplicative factor for the inner vs. outer boundary weighting phase is made available for VIS images to account for stronger limbic boundaries. For CAHT, an external parameter to modify minimum and maximum contrast is added and an option to change the order of boundary detection (starting with outer limbic boundary before pupillary boundary detection) is introduced.

### 3.3.3 Combinations of segmentation performance

We further looked into basic combinations as in [8] (where the concept is introduced) without considering segmentation quality using augmented model fusion. Tables 3.2 and 3.3 show obtained accuracies for pairwise segmentations in terms of EER and ground-truth errors $E_1$, $E_2$ and $\mathscr{F}$-measure. Combining CAHT and WAHET as in [8] returns best results for NIR (6.83% EER, 2.7% $E_1$, 4.0% $E_2$) confirming the assumption that improvements are primarily happening when over- and undersegmentation are combined. For VIS a different result is obtained, now with REF and WAHET (resp. CAHT for ground-truth) being the best combinations (13.85% EER, 3.4% $E_1$, 5.9% $E_2$ and 14.97%, 3.1% $E_1$, 5% $E_2$, respectively). It also becomes clear that improvements in the visible range are more sparse and less pronounced than fusion in the NIR range. This lack of clear improvement suggests that the lower quality of segmentation results has also a large impact on the fusion. An approach which takes estimation of segmentation accuracy of the combined algorithms into account is motivated.

Next, we evaluated the quality-based multi-segmentation fusion method in [7] (see also Section 3.2.4). As this method needs training, we subdivided UTIRIS sets NIR and VIS in two subsets, classes 1–40 for training and 41–79 for testing. Table 3.4 lists the recognition results for this combination and Table 3.5 the ground-truth-based errors. With regards to EER performance all combinations in NIR and all but two combinations in VIS, namely CAHT+REF and CAHT+WAHET, deliver more accurate results. The failed fusion with CAHT can be explained by good outer limbic boundary detections by CAHT favouring this result despite a weak pupil boundary therefore affecting recognition accuracy. Note the subtle changes in EER for individual algorithms as now the test set is a subset of the database. EERs as good as 4.25% for NIR and 9.77% for VIS (both for WAHET + REF) are obtained for the LG feature. The combination is less dependent on the type of segmentation errors, but yields a better improvement for different approaches (failing for different types of images). Rates are overall much better for this type of fusion even across different features (LG and QSW). Ground-truth evaluation in terms of $E_1$ errors are in the range of 2.82% to 5.15% (NIR) and 2.44% to 8.93% (VIS) and $E_2$ errors in the range of 5.41% to 11.27% (NIR) and 7.35% to 32.54% (VIS).

## 3.4 Conclusion and future work

In this chapter we presented current trends in iris segmentation, summarising the advances in state-of-the-art individual segmentation, shedding light on the pitfalls of NIR vs. VIS iris segmentation and illustrating means to tune existing iris segmentation towards specific datasets. We highlighted different forms of segmentation accuracy assessment and evaluated different approaches to combine segmentation algorithms. To get more stable results we integrated segmentation quality prediction in the fusion process. Individual algorithms can fail considerably for specific iris images (9.64%–29.18% EER for LG in NIR and 13.81%–46.22% EER for LG in VIS). Pairwise combinations of algorithms, taking quality into account, reduced

Table 3.4  Test set EER performance for individual algorithms (diagonal) vs. pair-wise quality-based [7] segmentation fusion

**(a) UTIRIS – near infrared**

Equal-error rate (%) for LG

|        | CAHT | IFPP  | OSIRIS | REF   | WAHET |
|--------|------|-------|--------|-------|-------|
| CAHT   | 9.63 | 5.36  | 7.18   | 4.56  | 4.70  |
| IFPP   |      | 29.18 | 12.22  | 12.50 | 5.88  |
| OSIRIS |      |       | 20.23  | 9.98  | 5.12  |
| REF    |      |       |        | 23.35 | 4.25  |
| WAHET  |      |       |        |       | 8.69  |

Equal-error rate (%) for QSW

|        | CAHT | IFPP  | OSIRIS | REF   | WAHET |
|--------|------|-------|--------|-------|-------|
| CAHT   | 9.84 | 5.45  | 7.34   | 4.64  | 4.15  |
| IFPP   |      | 30.06 | 12.78  | 13.85 | 5.68  |
| OSIRIS |      |       | 22.53  | 11.17 | 5.06  |
| REF    |      |       |        | 23.94 | 3.85  |
| WAHET  |      |       |        |       | 8.35  |

**(b) UTIRIS – visible range**

Equal-error rate (%) for LG

|        | CAHT  | IFPP  | OSIRIS | REF   | WAHET |
|--------|-------|-------|--------|-------|-------|
| CAHT   | 31.05 | 29.06 | 29.61  | 21.83 | 21.47 |
| IFPP   |       | 40.63 | 35.51  | 13.08 | 13.85 |
| OSIRIS |       |       | 46.22  | 13.35 | 13.65 |
| REF    |       |       |        | 13.81 | 9.77  |
| WAHET  |       |       |        |       | 13.98 |

Equal-error rate (%) for QSW

|        | CAHT  | IFPP  | OSIRIS | REF   | WAHET |
|--------|-------|-------|--------|-------|-------|
| CAHT   | 36.47 | 32.60 | 33.44  | 23.48 | 24.67 |
| IFPP   |       | 40.84 | 35.68  | 14.59 | 14.76 |
| OSIRIS |       |       | 46.68  | 15.69 | 14.73 |
| REF    |       |       |        | 15.93 | 11.06 |
| WAHET  |       |       |        |       | 14.87 |

Table 3.5   Test set segmentation errors (comparison with ground-truth) for individual algorithms (diagonal) vs. pair-wise quality-based [7] segmentation fusion

**(a) UTIRIS – near infrared**

Average $E_1$ (%)

|        | CAHT  | IFPP  | OSIRIS | REF  | WAHET |
|--------|-------|-------|--------|------|-------|
| CAHT   | 4.15  | 3.42  | 3.33   | 2.66 | 2.82  |
| IFPP   |       | 11.19 | 5.05   | 5.15 | 3.54  |
| OSIRIS |       |       | 7.54   | 3.50 | 3.06  |
| REF    |       |       |        | 8.10 | 3.03  |
| WAHET  |       |       |        |      | 4.06  |

Average $E_2$ (%)

|        | CAHT  | IFPP  | OSIRIS | REF   | WAHET |
|--------|-------|-------|--------|-------|-------|
| CAHT   | 11.25 | 7.52  | 6.61   | 5.41  | 5.88  |
| IFPP   |       | 29.89 | 10.69  | 11.27 | 7.76  |
| OSIRIS |       |       | 15.39  | 7.19  | 6.43  |
| REF    |       |       |        | 21.22 | 6.27  |
| WAHET  |       |       |        |       | 9.30  |

**(b) UTIRIS – visible range**

Average $E_1$ (%)

|        | CAHT | IFPP  | OSIRIS | REF  | WAHET |
|--------|------|-------|--------|------|-------|
| CAHT   | 5.91 | 5.27  | 5.47   | 3.42 | 3.77  |
| IFPP   |      | 11.99 | 8.93   | 2.71 | 3.40  |
| OSIRIS |      |       | 14.12  | 2.80 | 3.48  |
| REF    |      |       |        | 2.90 | 2.44  |
| WAHET  |      |       |        |      | 3.78  |

Average $E_2$ (%)

|        | CAHT  | IFPP  | OSIRIS | REF  | WAHET |
|--------|-------|-------|--------|------|-------|
| CAHT   | 15.06 | 13.83 | 14.30  | 9.52 | 9.95  |
| IFPP   |       | 46.86 | 32.54  | 8.39 | 9.88  |
| OSIRIS |       |       | 48.12  | 8.56 | 10.13 |
| REF    |       |       |        | 9.21 | 7.35  |
| WAHET  |       |       |        |      | 11.35 |

EERs to as low as 4.25% for NIR and 9.77% for VIS (combining REF and WAHET). Tests on ground-truth confirm this positive effect across the board. For the goal of developing more robust iris segmentation techniques, multi-segmentation fusion and quality prediction can be a versatile tool to obtain more stable and more accurate results. However, performing fusion and running multiple segmentation algorithms impacts the processing time. This was not explicitly covered and further research on frame throughput might be necessary for video-based iris segmentation.

# References

[1]    Jillela, R., Ross, A., Flynn, P. "Information fusion in low-resolution iris videos using principal components transform," in: *IEEE Workshop on Applications of Computer Vision (WACV)*, pp. 262–269. IEEE, Piscataway, NJ (2011). DOI 10.1109/WACV.2011.5711512

[2]    Jillela, R., Ross, A., N. Boddeti, B.V.K., Hu, X., Plemmons, R., Pauca, P. "Iris segmentation for challenging periocular images," in: *Handbook of Iris Recognition*, pp. 281–308. Springer (2013).

[3]    Uhl, A., Wild, P. "Weighted adaptive hough and ellipsopolar transforms for real-time iris segmentation," in: *Proceedings of the International Conference on Biometrics (ICB)*, pp. 283–290. IEEE, Piscataway, NJ (2012). DOI 10.1109/ICB.2012.6199821

[4]    Hosseini, M., Araabi, B., Soltanian-Zadeh, H. "Pigment melanin: pattern for iris recognition," *IEEE Transactions on Instrumentation and Measurement* **59**(4), 792–804 (2010). DOI 10.1109/TIM.2009.2037996

[5]    Hofbauer, H., Alonso-Fernandez, F., Wild, P., Bigun, J., Uhl, A. "A ground truth for iris segmentation," in: *Proceedings of the International Conference on Pattern Recognition (ICPR)*, pp. 527–532. IEEE, Piscataway, NJ (2014). DOI 10.1109/ICPR.2014.101

[6]    Rathgeb, C., Uhl, A., Wild, P. "Iris recognition: from segmentation to template security," *Advances in Information Security*, vol. 59. Springer, New York, NY, USA (2012).

[7]    Wild, P., Hofbauer, H., Ferryman, J., Uhl, A. "Quality-based iris segmentation-level fusion," *EURASIP Journal on Information Security* (25), 12 (2016). DOI 10.1186/s13635-016-0048-x

[8]    Wild, P., Hofbauer, H., Ferryman, J., Uhl, A. "Segmentation-level fusion for iris recognition," in: *Proceedings of the International Conference of the Biometrics Special Interest Group (BIOSIG)*, pp. 61–72. IEEE, Piscataway, NJ (2015). DOI 10.1109/BIOSIG.2015.7314620

[9]    Daugman, J. "How iris recognition works," *IEEE Transactions on Circuits Systems and Video Technology* **14**(1), 21–30 (2004). DOI 10.1109/TCSVT.2003.818350

[10]   CASIA-IrisV4 Interval Database. http://biometrics.idealtest.org/

[11]   Rathgeb, C., Hofbauer, H., Uhl, A., Busch, C. "TripleA: accelerated accuracy-preserving alignment for iris-codes," in: *Proceedings of the International Conference on Biometrics (ICB)*, p. 8. IEEE, Piscataway, NJ (2016).

[12]    Proença, H., Alexandre, L. "Toward covert iris biometric recognition: experi-
        mental results from the NICE contests," *IEEE Transactions on Information on
        Forensics & Security* **7**(2), 798–808 (2012). DOI 10.1109/TIFS.2011.2177659
[13]    Hofbauer, H., Alonso-Fernandez, F., Bigun, J., Uhl, A. "Experimental analy-
        sis regarding the influence of iris segmentation on the recognition rate," *IET
        Biometrics*, **5**(3), 200–211 (2016). DOI 10.1049/iet-bmt.2015.0069
[14]    van Rijsbergen, C.J. *Information Retrieval*, second edn. Butterworth-
        Heinemann, Newton, MA, USA (1979).
[15]    Alonso-Fernandez, F., Bigun, J. "Quality factors affecting iris segmentation
        and matching," in: *Proceedings of the International Conference on Biometrics
        (ICB)*, pp. 1–6. IEEE, Piscataway, NJ (2013). DOI 10.1109/ICB.2013.6613016
[16]    Wild, P., Ferryman, J., Uhl, A. "Impact of (segmentation) quality on long vs.
        short-timespan assessments in iris recognition performance," *IET Biometrics*
        **4**(4), 227–235 (2015). DOI 10.1049/iet-bmt.2014.0073
[17]    Tabassi, E. "Large scale iris image quality evaluation," in: *Proceedings of
        the International Conference on Biometrics Special Interest Group (BIOSIG)*,
        pp. 173–184. Lecture Notes in Informatics (2011).
[18]    Kalka, N., Bartlow, N., Cukic, B. "An automated method for predicting iris
        segmentation failures," in: *Proceedings of the International Conference on Bio-
        metrics: Theory, Applications and Systems (BTAS)*, pp. 1–8. IEEE, Piscataway,
        NJ (2009). DOI 10.1109/BTAS.2009.5339062
[19]    Mahadeo, N., Haffari, G., Paplinski, A. "Predicting segmentation errors in an
        iris recognition system," in: *Proceedings of the International Conference on
        Biometrics (ICB)*, pp. 23–30. IEEE (2015). DOI 10.1109/ICB.2015.7139071
[20]    Makinana, S., Merwe, J.J.V.D., Malumedzha, T. "A Fourier transform quality
        measure for iris images," in: *Proceedings of the International Symposium on
        Biometrics and Security Technologies (ISBAST)*, pp. 51–56. IEEE, Piscataway,
        NJ (2014). DOI 10.1109/ISBAST.2014.7013093
[21]    Zuo, J., Schmid, N.A. "An automatic algorithm for evaluating the precision
        of iris segmentation," in: *Proceedings of the International Conference on Bio-
        metrics: Theory, Applications and Systems (BTAS)*, pp. 1–6. IEEE, Piscataway,
        NJ (2008). DOI 10.1109/BTAS.2008.4699358
[22]    Matveev, I., Gankin, K. "Iris segmentation system based on approximate
        feature detection with subsequent refinements," in: *Proceedings of the Inter-
        national Conference on Pattern Recognition (ICPR)*, pp. 1704–1709. IEEE,
        Piscataway, NJ (2014). DOI 10.1109/ICPR.2014.300
[23]    Matey, J., Naroditsky, O., Hanna, K., *et al.* "Iris on the move: acquisition of
        images for iris recognition in less constrained environments," *Proceedings of
        the IEEE* **94**(11), 1936–1947 (2006). DOI 10.1109/JPROC.2006.884091
[24]    Proença, H. "Iris recognition: on the segmentation of degraded images acquired
        in the visible wavelength," *IEEE Transactions on Pattern Analysis and Machine
        Intelligence* **32**(8), 1502–1516 (2010). DOI 10.1109/TPAMI.2009.140
[25]    Schuckers, S., Schmid, N., Abhyankar, A., Dorairaj, V., Boyce, C., Hornak,
        L. "On techniques for angle compensation in nonideal iris recognition," *IEEE
        Transactions on Systems, Man, and Cybenetics, Part B* **37**(5), 1176–1190
        (2007). DOI 10.1109/TSMCB.2007.904831

[26]    Tan, T., He, Z., Sun, Z. "Efficient and robust segmentation of noisy iris images for non-cooperative iris recognition," *Image and Vision Computing* **28**(2), 223–230 (2010). DOI 10.1016/j.imavis.2009.05.008

[27]    Wildes, R.P. "Iris recognition: an emerging biometric technology," in: *Proceedings of the IEEE*, vol. 85, pp. 1348–1363. IEEE (1997). DOI 10.1109/5.628669

[28]    Uhl, A., Wild, P. "Multi-stage visible wavelength and near infrared iris segmentation framework," in: A. Campilho, M. Kamel (eds.) *Proceedings of the International Conference on Image Analysis and Recognition. (ICIAR)*, Lecture Notes in Computer Science, pp. 1–10. Springer (2012). DOI 10.1007/978-3-642-31298-4_1

[29]    Du, Y., Arslanturk, E., Zhou, Z., Belcher, C. "Video-based noncooperative iris image segmentation," *IEEE Transactions on Systems, Man, and Cybernetics, Part B* **41**(1), 64–74 (2011). DOI 10.1109/TSMCB.2010.2045371

[30]    Yan, F., Tian, Y., Wu, H., Zhou, Y., Cao, L., Zhou, C. "Iris segmentation using watershed and region merging," in: *Proceedings of the International Conference on Industrial Electronics and Applications (ICIEA)*, pp. 835–840. IEEE, Piscataway, NJ (2014). DOI 10.1109/ICIEA.2014.6931278

[31]    Chan, T., Vese, L. "Active contours without edges," *IEEE Transactions on Image Processing* **10**(2), 266–277 (2001). DOI 10.1109/83.902291

[32]    Mumford, D. "Optimal approximation by piecewise smooth functions and associated variational problems," *Communications in Pure and Applied Mathematics* **42**(5), 577–685 (1989).

[33]    Abhyankar, A., Schuckers, S. "Active shape models for effective iris segmentation," in: *Proceedings of SPIE*, p. 62020H. SPIE (2006). DOI 10.1117/12.666435

[34]    Vatsa, M., Singh, R., Noore, A. "Improving iris recognition performance using segmentation, quality enhancement, match score fusion, and indexing," *IEEE Transactions on Systems, Man, and Cybernetics, Part B* **38**(4), 1021–1035 (2008). DOI 10.1109/TSMCB.2008.922059

[35]    Sutra, G., Garcia-Salicetti, S., Dorizzi, B. "The Viterbi algorithm at different resolutions for enhanced iris segmentation," in: *Proceedings of the International Conference on Biometrics (ICB)*, pp. 310–316. IEEE (2012). DOI 10.1109/ICB.2012.6199825

[36]    Petrovska, D., Mayoue, A. "Description and documentation of the biosecure software library," Technical Report, Project No IST-2002-507634 – BioSecure (2007).

[37]    Chai, T., Goi, B., Tay, Y., Chin, W., Lai, Y. "Local Chan–Vese segmentation for non-ideal visible wavelength iris images," in: *Proceedings of the Conference on Technologies and Applications of Artificial Intelligence (TAAI)*, pp. 506–511. IEEE, Piscataway, NJ (2015). DOI 10.1109/TAAI.2015.7407059

[38]    Abdullah, M., Dlay, S., Woo, W., Chambers, J. "Robust iris segmentation method based on a new active contour force with a noncircular normalization," *IEEE Transactions on Systems, Man, and Cybernetics*, pp. 1–14 (2016). DOI 10.1109/TSMC.2016.2562500

[39]   Zhao, Z., Kumar, A. "An accurate iris segmentation framework under relaxed imaging constraints using total variation model," in: *Proceedings of the International Conference on Computer Vision (ICCV)*, pp. 3828–3836. IEEE, Piscataway, NJ (2015). DOI 10.1109/ICCV.2015.436

[40]   Leo, M., Marco, T.D., Distante, C. "Highly usable and accurate iris segmentation," in: *Proceedings of the International Conference on Pattern Recognition (ICPR)*, pp. 2489–2494. IEEE, Piscataway, NJ (2014). DOI 10.1109/ICPR.2014.430

[41]   Hu, Y., Sirlantzis, K., Howells, G. "A robust algorithm for colour iris segmentation based on 1-norm regression," in: *Proceedings of the International Joint Conference on Biometrics (IJCB)*, pp. 1–8. IEEE, Piscataway, NJ (2014). DOI 10.1109/BTAS.2014.6996233

[42]   Thavalengal, S., Bigioi, P., Corcoran, P. "Efficient segmentation for multi-frame iris acquisition on smartphones," in: *Proceedings of the International Conference on Consumer Electr. (ICCE)*, pp. 198–199. IEEE (2016). DOI 10.1109/ICCE.2016.7430578

[43]   Trokielewicz, M. "Iris recognition with a database of iris images obtained in visible light using smartphone camera," in: *Proceedings of the International Conference on Identity, Security and Behavior Analysis (ISBA)*, pp. 1–6. IEEE, Piscataway, NJ (2016). DOI 10.1109/ISBA.2016.7477233

[44]   Happold, M. "Structured forest edge detectors for improved eyelid and iris segmentation," in: *Proceedings of the Conference on Biometrics Special Interest Group (BIOSIG)*, pp. 1–6. IEEE, Piscataway, NJ (2015). DOI 10.1109/BIOSIG.2015.7314622

[45]   Liu, N., Li, H., Zhang, M., Liu, J., Sun, Z., Tan, T. "Accurate iris segmentation in non-cooperative environments using fully convolutional networks," in: *Proceedings of the International Conference on Biometrics (ICB)*, p. 6. IEEE, Piscataway, NJ (2016).

[46]   Gangwar, A., Joshi, A., Singh, A., Alonso-Fernandez, F., Bigun, J. "IrisSeg: a fast and robust iris segmentation framework for non-ideal iris images," in: *Proceedings of the International Conference on Biometrics (ICB)*, p. 6. IEEE, Piscataway, NJ (2016).

[47]   Uhl, A., Wild, P. "Fusion of iris segmentation results," in: J. Ruiz-Shulcloper, G.S. di Baja (eds.) *Proceedings of the 18th Iberamerican Congress on Pattern Recognition, (CIARP)*, Lecture Notes in Computer Science, vol. 8259, pp. 310–317. Springer, Heidelberg (2013). DOI 10.1007/978-3-642-41827-3_39

[48]   Llano, E., Vargas, J., García-Vázquez, M., Fuentes, L., Ramírez-Acosta, A. "Cross-sensor iris verification applying robust fused segmentation algorithms," in: *Proceedings of the International Conference on Biometrics (ICB)*, pp. 17–22 (2015). DOI 10.1109/ICB.2015.7139042

[49]   Sanchez-Gonzalez, Y., Cabrera, Y., Llano, E. "A comparison of fused segmentation algorithms for iris verification," in: E. Bayro-Corrochano, E. Hancock (eds.) *Proceedings of the Iberoamerican Congress on Pattern Recognition, (CIARP)*, Lecture Notes in Computer Science, pp. 112–119. Springer, Heidelberg (2014). DOI 10.1007/978-3-319-12568-8_14

[50]    Fitzgibbon, A., Pilu, M., Fisher, R.B. "Direct least square fitting of ellipses," *IEEE Transactions on Pattern Analysis and Machine Intelligence* **21**(5), 476–480 (1999).

[51]    Zhu, C., Byrd, R., Lu, P., Nocedal, J. "Algorithm 778: L-bfgs-b: Fortran subroutines for large-scale bound-constrained optimization," *ACM Transactions on Mathematical Software* **23**(4), 550–560 (1997). DOI 10.1145/279232.279236

[52]    Ma, L., Tan, T., Wang, Y., Zhang, D. "Efficient iris recognition by characterizing key local variations," *IEEE Transactions on Image Processing* **13**(6) (2004). DOI 10.1109/TIP.2004.827237

[53]    Masek, L. "Recognition of human iris patterns for biometric identification," MSc thesis, University of Western Australia, 2003.

*Chapter 4*

# Iris image quality metrics with veto power and nonlinear importance tailoring

*John Daugman and Cathryn Downing*

Linear combinations of metrics for assessing biometric sample quality are weak, because they lack veto power. For example, a good score for a sharp focus of an ocular image would 'compensate' in an additive combination for the fact that the eyelids are fully closed; or fully open eyelids would compensate for the image being many diopters out-of-focus. Normalised multiplicative quality factors are better because they are punitive, and thereby confer veto powers. This chapter explains the basis for the product of power functions which underlie the ISO/IEC 29794-6 Iris Image Sample Quality Standard, in particular how the exponents of the power functions allow importance tailoring of each element.

## 4.1 Introduction

US National Institute of Standards and Technology (NIST) papers by Grother and Tabassi [1], and also by Phillips and Beveridge [2], defined *quality measures* as measurable covariates that are both predictive of biometric recognition performance, and actionable. Quality measures include *subject covariates* that are attributes of a person and which may be transient, such as expression, eyelid occlusion or the wearing of eyeglasses. Quality measures also include *image covariates* that depend on the sensor and acquisition conditions, such as focus, resolution, and illumination effects. Some subject covariates may not be capable of change or improvement, such as permanent injury or physical deformation; other subject covariates may be improvable but not under voluntary control, such as the degree of pupil dilation. In this chapter we do not explore those distinctions further, but rather our concern is how best to combine multivariate quality measures into one actionable quantity: e.g., to decide whether to enroll a given biometric sample into a database, or to reject it and acquire a new one. In the context of having a plurality of measures, this amounts to computing a single actionable scalar – a *Quality Score QS* – from a vector consisting of several elements. We thus explore a new definition resembling a norm, or rather a semi-norm, of a biometric quality vector. Because of the resemblance, we call it a *pseudonorm*. Its attributes include the ability to treat the different elements of the vector differently and to control their compensatory trade-offs. We construct and demonstrate an algorithmic procedure for determining the parameters of this pseudonorm on quality vectors, using a public test database of iris images.

### 4.1.1   Related work

The most comprehensive treatment of *quality-based fusion* – incorporating quality measures together with comparison scores for multi-biometric classifier combination to achieve optimal fusion and decisions within a general Bayesian framework – was by Poh and Kittler [3]. A related proposal for comparator fusion based on quality in multi-biometric systems was by Nandakumar *et al.* [4]. They showed that overall performance of multi-biometric systems could be improved by dynamically assigning weights to the outputs of individual comparators, using the sample quality, in a likelihood-ratio fusion scheme for combining their comparison scores.

Schmid and her co-authors [5] proposed an adaptive iris authentication approach based on quality measures that were selected for their utility in separating genuine from impostor score distributions, in labelled data. A feedforward neural network (two hidden layers, having 16 neurons and 2 neurons) was trained to learn a nonlinear mapping of the measures onto an overall QS, with good results. But it is unknown what actual nonlinear combinations developed in the hidden layers. In other work by Schmid *et al.* [6], the Dempster–Shafer theory of evidence was deployed as a basis for combining quality factors.

Another adaptive use of quality metrics was demonstrated by Li *et al.* [7] for the dynamic selection of the actual recognition strategy. They fused several quality factors into an overall score by products of likelihood ratios, using conditional probabilities of score values given good, versus given poor, quality. The samples may then be processed differently, based on their quality. Belcher and Du [8] used quality measures to calculate a confidence level associated with a comparison score, an idea also present in [5]. Galbally *et al.* [9] proposed that quality-related features could be used for liveness detection in an anti-spoofing strategy.

Hofbauer *et al.* [10] even proposed that quality metrics themselves could be used directly for comparing iris images and estimating their similarities. Phillips and Beveridge [2] argued that biometric comparison and quality assessment are actually equivalent, in a formal sense which they call 'biometric completeness'. Their argument is that a perfect method for quality assessment would predict whenever a one-to-one verification algorithm would give the wrong result. The wrong decision could then be reversed, producing error-free verification.

## 4.2   Normalisation of individual quality measures

We assume that individual quality measures are non-negative numbers which might have no inherent upper bound, with larger values implying higher quality. We are agnostic here about what they actually measure, but we will illustrate using actual examples. Some measures may already be inherently normalised, such as those that

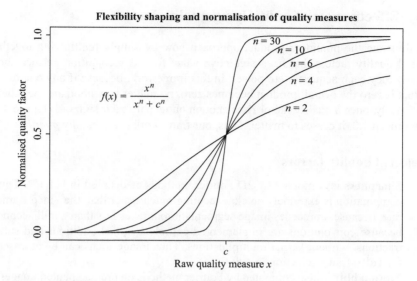

**Figure 4.1** *Family of functions normalising quality measures into quality factors, and shaping their sensitivity to differences. Larger n means greater rigidity, and the value x = c is the quality factor halfway point*

are defined as a percentage. But the goal of our *unity-based normalisation* is not merely to re-scale, but also to embed a semantics of acceptability and flexibility. We map any measure $x$ onto the [0, 1] unit interval using a sigmoidal function $f(x)$ having two parameters: $n$ controlling the measure's rigidity or flexibility, and $c$ specifying a 'set point' at which the measure is deemed to correspond to 50% quality:

$$f(x) = \frac{x^n}{x^n + c^n} \tag{4.1}$$

Parameter $c$ has the same units as $x$, whether those be dimensionless (such as the percentage of the iris that is visible between the eyelids), or dimensional (such as square-millimetres of visible iris area). The choice of parameter $n$ should reflect how 'non-negotiable' are the demands on $x$, with larger $n$ meaning greater sensitivity to small differences in the value of $x$ around $c$. For the measures actually used, we find that a soft $n = 2$ is often a better choice than larger values which begin to effect a rigid 'brick wall' threshold. It is easily shown that the slope of $f(x)$, its sensitivity to differences, at the 50% point ($x = c$) is $n/4c$. A family of curves representing such functions for $n \in \{2, 4, 6, 10, 30\}$ are shown in Figure 4.1. Once a quality measure $x$ is normalised onto the [0, 1] unit interval and shaped in this way for appropriate sensitivity to differences, we call it a *quality factor* because it will combine into the overall QS product as a multiplicative factor. (Later we shall tailor the relative importance of quality factors, using empirically tuned power functions of them.)

## 4.3   Effectiveness of multiplying quality factors

We illustrate the nonlinear product approach now by simply multiplying together several quality factors that have predictive value for iris recognition performance. Clearly they each acquire 'veto power' in this framework, because if any one of the factors is zero the overall product becomes zero, no matter how good are the others. Conversely, once a quality factor rises enough on its particular sigmoid (Figure 4.1) to approach 1.0, it ceases to matter. Thus, our framework is essentially *punitive*.

### Selected quality factors

1. **Sharpness** is computed by 2D Fourier methods as detailed in [11,12]. This computation is extremely quick, executing much faster than the video frame rate, because it precedes image segmentation. It takes less than a millisecond because convolutions are implemented entirely by pixel additions and subtractions, without kernel multiplications. Thus image focus can be assessed for video frame selection in real-time.

2. **Motion blur** is also computed by Fourier methods on pre-segmented images, within a millisecond. It examines local image structure across the temporal interval between the two fields of a (non-progressive scan) video frame, detecting any interlace shear between even and odd lines.

3. **Texture energy** is a measure of iris contrast, signal-to-noise ratio, and salience, based on the distribution of 2D Gabor wavelet coefficients [11] computed when encoding the iris pattern into an IrisCode. The lowest quartile (in absolute value) of this distribution of coefficients are ignored as being unreliable, when the IrisCodes are compared.

4. **Gaze angle** is estimated as per [13] using Fourier-based trigonometry. Gaze on-axis with the camera's optical axis is preferred, as the iris is not strictly planar so simple affine transformation to compensate for off-axis gaze is imperfect.

5. **Shaped pupil contrast** is a post-segmentation measure of image focus. Once the actual shape of the pupil has been estimated by a Fourier series expansion [13] of its boundary, a nonlinear Weber contrast is computed using a series of contour integrals that follow various dilations of this shape, ranging from inside to outside the pupil.

6. **Limbus contrast** also integrates various dilations of the outer boundary of the iris once its shape has been estimated by a Fourier series expansion [13]. But unlike the pupil contrast estimate, which is both edge-based and region-based (because the inside of the pupil should be dark), this outer boundary factor uses a linear Michelson contrast measure.

7. **IrisCode entropy** is an estimate of how much useful information the image contains. It takes into account the percentage of iris area that is not occluded by the eyelids, the total number of bits that are unmasked (deemed to be uncorrupted, e.g., by eyelashes or corneal specular reflections), and it penalises excessive pupil dilation.

Image quality as a predictor of Hamming distance

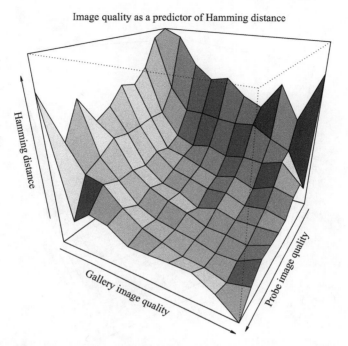

*Figure 4.2    Effect of combined image quality factors on same-eye comparison
scores. Hamming distance is the fraction of bits that disagree when two
IrisCodes are compared [11]. Poor image quality elevates Hamming
distances [14]*

For the publically available images in the NIST *Iris Challenge Evaluation* (ICE)
database, we computed quality factors using five of these measures and multiplied
them together to obtain a provisional quality score for every image. We then studied
how these scores predicted Hamming distance for all possible 'probe' and 'gallery'
pairings of images that came from a given eye and thus ought to match. Distances are
plotted as a surface function of the quality pairings (after quantisation into ten bins
for each) in Figure 4.2, showing a clear bivariate effect. Figure 4.3 shows the effect
of these scores on False non-Match Rates, plotted both as a surface and as a contour
map, when using a 0.32 Hamming distance match threshold.

## 4.4    Importance tailoring

We have seen in Figures 4.2 and 4.3 that multiplying quality factors together can
produce a strong predictor of match performance. This punitive approach gives veto
power to every factor, but intuitively, not every veto should necessarily be equally
insistent. Therefore we wish to tailor the relative importance of the quality factors

Image quality as a predictor of False non–Match Rate

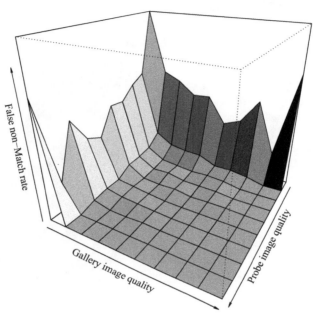

Image quality as a predictor of False non–Match Rate

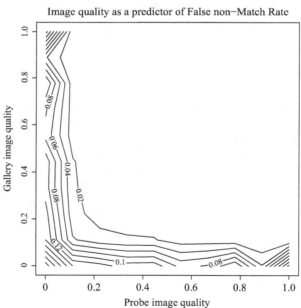

*Figure 4.3    Effect of combined image quality factors on False non-Match Rates,*
*plotted as surface and contour maps, at a 0.32 Hamming distance*
*threshold [14]*

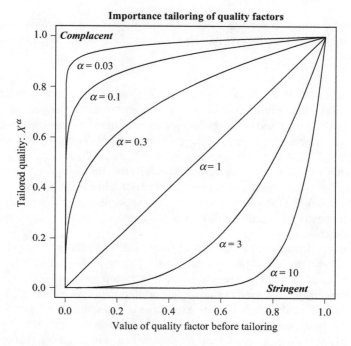

**Importance tailoring of quality factors**

*Figure 4.4*  *Quality factors can be given complacent or stringent exponents, which determine how quickly or how reluctantly they rise to 1.0 and cease to be punitive*

by raising them to different powers, using exponents determined by an empirical multivariate regression procedure.

Using various quality factors $X, Y, Z, \ldots$, we define an overall Quality Score (QS) as the product of *power functions* of them, with positive exponents $\alpha, \beta, \gamma, \ldots$:

$$QS = X^{\alpha} Y^{\beta} Z^{\gamma} \cdots \quad \text{for } \{\alpha, \beta, \gamma, \ldots\} > 0 \tag{4.2}$$

Component quality factors $X, Y, Z, \ldots$ are the normalised functions defined in (4.1). Figure 4.4 shows that exponents larger than $\alpha = 1$ make a quality factor stringent, whereas small exponents make it complacent because it rises to 1.0 very quickly.

## Pseudonorm

There exist many alternative definitions for the norm of a vector, using various power functions of its elements, of which the most familiar is the Euclidean norm: $\|(x, y, z, \ldots)\| = \sqrt{x^2 + y^2 + z^2 + \cdots}$. A subset of norms are semi-norms, which have the property of equaling 0 if any element is 0, thereby providing veto power. Our QS has properties similar to these but it does not actually satisfy the formal

properties of norms or semi-norms, so we call it instead a *pseudonorm*, defined
on elements in the unit interval [0, 1] and mapping also to the unit interval [0, 1],
with parameters $\alpha, \beta, \gamma, \ldots$.

In order to select which quality factors to include in the QS and determine the
exponents that optimally tailor their relative importance, we submitted Hamming
distances (HDs) for a fixed probe-gallery partition of the same-eye images in the ICE
2,953 image database to a series of regression analyses. The goal was to discover
which factors $X, Y, Z, \ldots$ and corresponding exponents $\alpha, \beta, \gamma, \ldots$ in QS make it the
best predictor of HD and thus False non-Match Rates. By taking logarithms of both
sides of (4.2), the optimisation acquires the form of a linear multivariate regression
problem in which the calculated slopes are the exponents sought. In the analyses we
used this framework to predict HD, in order to construct empirically an optimal QS
product of factors.

Figure 4.5 illustrates the method we used. We first examined the correlations
between HD and the log of the smaller of each pair of probe and gallery values for
various quality factors (QFs). These are shown in the top panel of Figure 4.5, with bar
height representing the absolute value of the correlation between the log(QF) and HD.
Many of these candidate factors were experimental variables developed for a NIST
trial [15] that launched the ISO/IEC iris quality standardisation project [16]. We have
labelled only those that were selected for further analysis and those that correspond
either to required (Clause 6.2) or to recommended (Clause 6.3) factors in the ISO/IEC
29794-6 Standard that eventually emerged [16].

The top panel of Figure 4.5 shows that several quality factors were reasonably
well correlated with HD. We chose IrisCode entropy, which accounted for 26% of
the variability in HD ($R^2 = .26$), as the best single factor and used it in a simple
linear regression to predict HD. This quality factor is named and indicated in black
in the top panel. Subsequently selected factors for the product sequence are similarly
highlighted in the next panels.

The second panel shows the correlations between the log(QFs) and the residuals
from Regression 1 (the difference between observed and predicted HDs). Note that
because many of the quality factors were correlated with IrisCode entropy, they have
much smaller correlations with the residuals of Regression 1, shown in the second
panel, than with HD, shown in the first. Texture energy, however, stands out from
the rest in having a substantial correlation that is independent from IrisCode entropy.
Therefore, to construct the best two-factor QS, we used IrisCode entropy and Texture
energy in a multiple regression to predict HD. This, Regression 2, accounted for 41%
of the variance in HD. We repeated this process of selecting quality factors to add
to the regression predicting HD, as shown in the subsequent panels of Figure 4.5,
until the increase in $R^2$ (the proportion of variance accounted for) by adding another
variable to the regression equation became nugatory.

Using the five selected quality factors in the multiple regression analysis we can
account for almost 51% of the variance in HD, whereas, as shown in Figure 4.5, the
best formula with only one or two variables could account for only 26% or 41% of

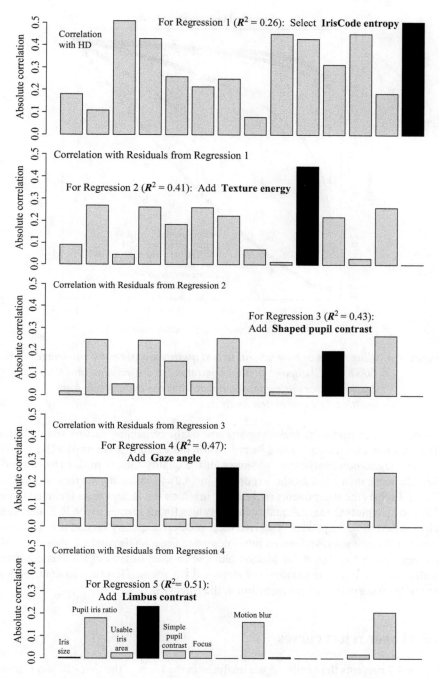

*Figure 4.5    Illustration of the correlational method used to select quality factors and their exponents for a multifactorial QS. Bar heights represent absolute correlation. The factors incorporated sequentially into QS are highlighted in black*

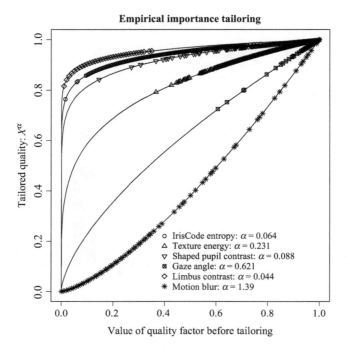

*Figure 4.6   Illustration of how several actual quality measures for the images in the NIST Iris Challenge Evaluation database are mapped into tailored quality factors using the exponents derived empirically from multivariate regression analysis*

the variance, respectively. Including any one of the remaining factors shown in the fifth panel in a six-variable multiple regression increased $R^2$ to at most 52%.

The regression coefficients obtained for the quality factors in this five-variable formula were then used as the exponents in (4.2) to tailor the factors in the QS. Figure 4.6 lists these exponents in the order in which the factors were included in the QS, and also plots the actual quality factor values for all images in the ICE database, when raised to these powers. The five most predictive quality factors are: IrisCode entropy; Texture energy; Shaped pupil contrast; Gaze angle; and Limbus contrast. The regression coefficient for Motion blur (when used instead of Limbus contrast) is also given to illustrate the range of empirical fits found. However, its contribution was not sufficient to warrant inclusion in the final QS.

## 4.5   Error reject curves

Figure 4.7 presents the results of our analyses in the form of the Error Reject Curves (ERC) developed by Grother and Tabassi [1]. These rank-order HDs according to the worst of the two image quality measures corresponding to each HD and define

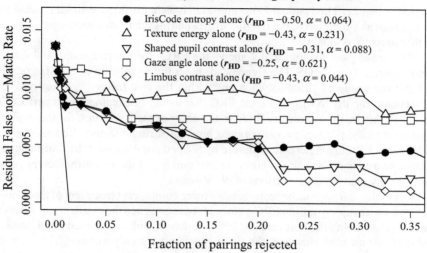

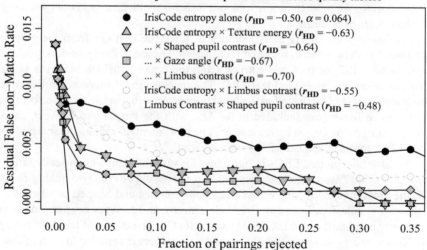

*Figure 4.7* *Error Reject Curves for single quality factors, upper panel, and multi-factorial QS, lower panel, showing the decline in False non-Match Rate as the worst image pairings are successively dismissed. Ideal performance is indicated by the solid line. Combinations of factors are greatly superior to single factors alone*

successively smaller subsets of HDs by rejecting those where quality falls below a requisite quantile. The residual False non-Match Rate for these shrinking subsets is then plotted as a function of the fraction of HD pairings that have been rejected based on that measure of quality.

The upper panel of Figure 4.7 shows the ERCs for the five individual quality factors selected for our QS. The figure also shows the correlation of each factor with HD and the exponent fitted for each one within the five-factor QS. The solid line (descending rapidly to 0) is the ideal ERC that would be obtained for a perfect QS ($r = -1$). Again, the factors are listed in the order of their selection for the multiple regression analyses. Over the entire range, among the single factors, Limbus contrast had the best ERC characteristic, followed by Shaped pupil contrast, IrisCode entropy, Texture energy and Gaze angle. However, the ranking of these quality factors varied considerably over different sections of the abscissa.

Generally, our approach recommends a more complacent treatment of those quality factors with rapidly falling ERCs, since we have assigned the smallest exponents to those factors having the best-ranking ERCs: Limbus contrast, Shaped pupil contrast and IrisCode entropy. However, the fitted exponents are only meaningful within the context of a multiple factor QS. The ERCs for tailored and untailored quality factors are identical (except for degenerate cases) since the ordering of the factor is preserved in the tailoring. We must therefore look at ERCs for multi-factor QSs to gain insight into the effects of tailoring.

The lower panel of Figure 4.7 plots ERCs for the best QS from the previous regressions with two, three, four and five factors. These are shown in dark gray. The single factor ERC for IrisCode entropy from the upper panel (in black) is plotted again for reference, as is the ideal ERC. The increase in $R^2$ reported earlier in the regression analyses is reflected in the notable improvements in the ERC performance as successive factors are included in the QS, with the largest gap between curves corresponding to the largest increase in $R^2$. The movement of each successive multiple factor ERC, toward the ideal ERC curve, is clear.

We have also plotted ERCs in light gray for two alternative empirically tailored two-factor QSs. These illustrate the point that the product of the two quality factors with the most impressive single ERCs (Limbus contrast and Shaped pupil contrast) or the two smallest values of $\alpha$ (IrisCode entropy and Limbus contrast) does not necessarily produce the best ERC for a two-factor QS. The choice of IrisCode entropy and Texture energy suggested by the regression analyses is superior to both of these, especially in the initial range of the ERCs where fewer pairings are rejected.

## 4.6  Discipline in punishment

The rationale for our multiplicative approach to scoring quality was that it confers punitive veto power, which is absent in the additive approach of piece-meal accretion of reward for quality. But discipline must be exercised in the punitive approach, lest the quality factors be either too stringent or too complacent. Empirical tailoring of the exponents based on performance is a way to determine the optimal balance.

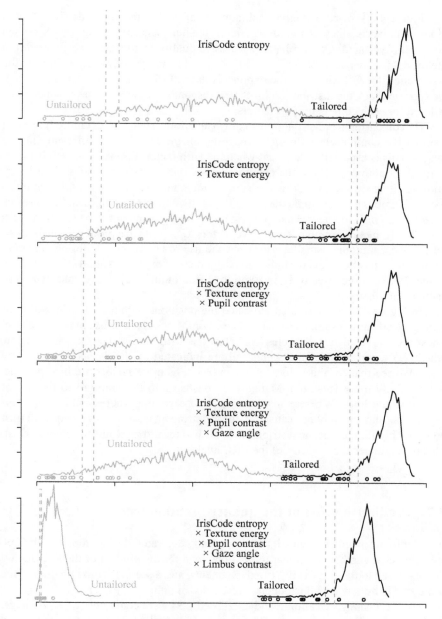

*Figure 4.8    Comparison between tailored (black) and untailored (gray) approaches*
*to quality scoring using product sequences with increasing numbers of*
*factors. The small circles indicate the QS value associated with each*
*False non-Match, revealing why such failures are better predicted by*
*the multi-factorial tailored QS*

Figure 4.5 showed that empirical tailoring of quality factors made five of them more complacent: they acquired exponents smaller than 1.0. In Figure 4.8 we see how both five-factor QSs evolve as successive quality factors are incorporated into the product sequence: $X^\alpha$; $X^\alpha Y^\beta$; $X^\alpha Y^\beta Z^\gamma$ …. In each panel, we show in gray the distribution for QS where all the exponents are 1.0 (the untailored approach), and we show in black the distribution for the corresponding QS where the exponents are tailored for each combination of factors.

By comparing the distributions in gray (untailored) and black (tailored), we can see that the complacent tailoring is mapping all the QSs into significantly higher ranges. This is especially important when the fifth factor, Limbus contrast, is incorporated (bottom panel). For the untailored case, this pushes all images toward very poor quality scores, making this measure far too punitive. But the complacent treatment of Limbus contrast in the tailored QS greatly elevates its values and restores its usefulness. This illustrates the importance of knowing when not to be too punitive.

Clearly, incorporation of successive factors $X_i \in [0, 1)$ in a product sequence can only make the product smaller. However, the use of complacent tailoring generally counteracts this effect, and the judicious assignment of exponents allows the full range of the QS to be used effectively, in addition to fine-tuning the relative contribution of each quality factor.

In Figure 4.8, we have also indicated the positions within each distribution of each of the datapoints linked to the 18 False non-Matches that arose in this database. These are marked by small circles just below the distributions. The QS values below which 2.5% and 5.0% of the scores fall are indicated by the dashed vertical gray lines. We see that for the tailored case, in black, as more factors are incorporated, progressively more False non-Matches are pushed into the long tail to the left of these quantile lines. Whereas for the untailored case, the progression is much less remarkable and the problems caused by the too stringent treatment of Limbus contrast are again highlighted, as the distribution is pushed toward zero and selectivity is lost. This recapitulates the findings of the ERC analyses.

## 4.7    Predictive value of the quality pseudonorm

In order to determine how much this empirically derived tailoring improves the QS, we conducted another two linear regressions: one for the product of the five quality factors raised to the empirically fitted exponents, and a second for the product of the same five factors when all exponents were fixed at unity.

For these analyses, tailored and untailored QSs were computed for each image. The smallest *tailored* five-factor QS for each probe-gallery image pair was entered into a simple linear regression to predict HD. Then the smallest *untailored* five-factor QS for each pair was used in a second simple linear regression on HD.

In Figure 4.9, we plot predicted HD vs observed HD for both of these regressions. The analyses show that by tailoring the QS exponents, we were able to account for 22% more of the variance: $R^2$ increased from 0.27 for the untailored QS to 0.49

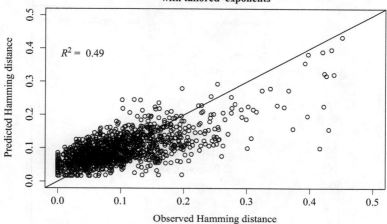

Hamming distance prediction using five–factor quality score with tailored exponents

$R^2 = 0.49$

Predicted Hamming distance

Observed Hamming distance

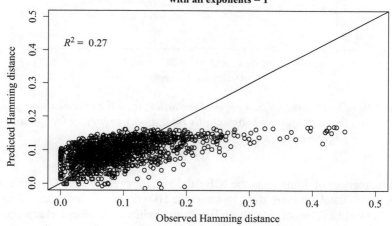

Hamming distance prediction using five–factor quality score with all exponents = 1

$R^2 = 0.27$

Predicted Hamming distance

Observed Hamming distance

*Figure 4.9   Demonstration that importance tailoring of exponents on quality factors improves substantially their ability to predict Hamming distances*

for the empirically tailored QS. The scatterplots reveal a substantial improvement in predictions for the larger HDs, which matter the most for False non-Match Rates.

Because the empirically tailored QS is better able to predict HD than untailored QS, it can be used to flag images that are more likely to produce False non-Matches when same-eye images are compared. On this basis, operational decisions may be taken to repeat image acquisition, or else to proceed with an image. Figure 4.10 shows

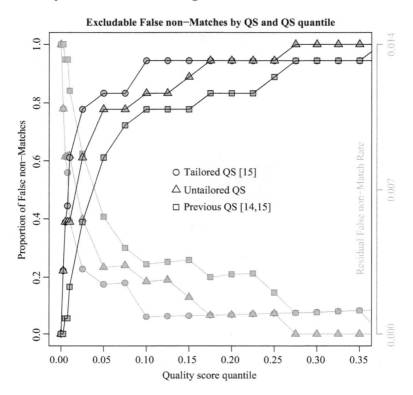

*Figure 4.10*   *Demonstration that importance tailoring of the exponents on quality factors improves substantially their ability to predict recognition performance*

the consequences of doing so in the ICE database, for each of three QSs considered above. In this database there were 18 same-eye HDs that exceeded 0.32, a threshold that would yield a False non-Match, and the figure indicates (in black) what proportion of these would be excluded as a function of the proportion of images flagged on the basis of poor QS. This is similar to the ERC 'exclusion' methodology for evaluating quality assessment algorithms that was originally proposed by Grother and Tabassi [1] and used by NIST in [15], but it is perhaps more straightforward because monotonicity is ensured. ERC curves according to their methodology are also shown in this figure (in gray) for comparison.

The tailored QS would allow the exclusion of a significantly larger number of False non-Matches than the untailored QS, over the entire quantile range. A *t*-test confirms significance over the full range but especially up to 12.5% (pairwise *t*-test (df = 8) 3.78, $p = 0.005$), and particularly when considering the poorest 1%–2% of images. It is also noteworthy that both of the new QSs, tailored and untailored, are substantially better at predicting and excluding False non-Matches than the QS described in [14] and shown in Figures 4.2 and 4.3.

## 4.8 Testing and adoption into an ISO/IEC quality standard

NIST invited submissions of iris image quality assessment algorithms both from academic and industry providers, and conducted tests of them during 2009–2011. The methods described in this chapter were largely developed and refined in response to the NIST initiative, which was called *IQCE: Image Quality Calibration and Evaluation* [15]. The final report on these methods concluded: 'Implementations from Cambridge University are the most effective in predicting recognition performance of their own native iris recognition algorithm, as well as others' [15, p. 6]. It also reported that these quality assessment algorithms 'are the fastest implementations, with an average 30 milliseconds per image across all four datasets' [15, p. 7].

Consequently an international standardisation project for assessing iris image quality, led by NIST's Elham Tabassi, incorporated these methods into the ISO/IEC 29794-6 International Standard on iris sample quality [16] during the workprogram of ISO/IEC Joint Technical Committee 1 (SC 37). The pseudonorm QS described here (4.2) became the framework of nonlinear combination of vector elements for generating an actionable scalar, specified now in Clause 6.5.2 of the Standard.

## 4.9 A possible criticism of the method

It should be noted that the power function form (4.2) defining the QS is monotonic in each individual quality element. Thus the better any one of them is, the better is the overall QS. Our approach may be criticised for this property, because one can imagine interactions of factors for which this 'collective monotonicity' may not be optimal. For example, suppose that one quality element is a focus score and another is signal-to-noise ratio, which is degraded by high frequency noise of external origin (not arising in the sensor array itself but present pre-optically, say in a grainy photograph being digitised). If the recognition algorithm is sensitive to such noise, then its performance may actually be enhanced by a degree of defocus, removing the noise. This violates the assumption of collective monotonicity among the elements.

## Acknowledgement

Statistical analyses and generation of figures were performed using the 'R' package: https://cran.r-project.org/.

## References

[1]  Grother, P., Tabassi, E., "Performance of biometric quality measures," *IEEE Transactions on Pattern Analysis and Machine Intelligence* **29**, (4), pp. 531–543, Apr. 2007.

[2]  Phillips, P.J., Beveridge, J.R., "An introduction to biometric-completeness: the equivalence of matching and quality," *IEEE Third International Conference*

on Biometrics: Theory, Applications, and Systems, Washington, pp. 1–5, Sep. 2009.

[3]    Poh, N., Kittler, J., "A unified framework for biometric expert fusion incorporating quality measures," *IEEE Transactions on Pattern Analysis and Machine Intelligence* **34**, (1), pp. 3–18, Jan. 2012.

[4]    Nandakumar, K., Chen, Y., Jain, A.K., "Quality-based score level fusion in multibiometric systems", *18th International Conference on Pattern Recognition (ICPR)*, **4**, pp. 473–476, Hong Kong, 2006.

[5]    Schmid, N.A., Zuo, J., Nicolo, F., Wechsler, H., "Iris quality metrics for adaptive authentication," Chapter 4 in *Handbook of Iris Recognition* (Burge and Bowyer, eds.), Springer-Verlag, London, 2013, pp. 67–84.

[6]    Kalka, N.D., Zuo, J., Schmid, N.A., Cukic, B., "Estimating and fusing quality factors for iris biometric images," *IEEE Transactions on Systems, Man, and Cybernetics – Part A*, 2010, **40**, (3), pp. 509–524.

[7]    Li, X., Sun, Z., Tan, T., "Comprehensive assessment of iris image quality," *18th IEEE International Conference on Image Processing*, pp. 3117–3120, 2011.

[8]    Belcher, C., Du, Y., "A selective feature information approach for iris image-quality measure," *IEEE Transactions on Information Forensics and Security*, **3**, (3), pp. 572–577, Sep. 2008.

[9]    Galbally, J., Marcel, S., Fierrez, J., "Image quality assessment for fake biometric detection: application to iris, fingerprint, and face recognition," *IEEE Transactions on Image Processing*, **23**, (2) pp. 710–724, Feb. 2014.

[10]   Hofbauer, H., Rathgeb, C., Uhl, A., Wild, P., "Iris recognition in image domain: quality-metric based comparators," *International Symposium on Visual Computing*, Springer, 2012.

[11]   Daugman, J., "How iris recognition works," *IEEE Transactions on Circuits and Systems for Video Technology*, **14**, (1), pp. 21–30, Jan. 2004.

[12]   Daugman, J., *US Patent 6,753,919: Fast Focus Assessment System and Method for Imaging*. US Patent Office, 22 Jun. 2004.

[13]   Daugman, J., "New methods in iris recognition," *IEEE Transactions on Systems, Man, and Cybernetics – Part B*, **37**, (5), pp. 1167–1175, Oct. 2007.

[14]   Daugman, J., Downing, C., "Effect of severe image compression on iris recognition performance," *IEEE Transactions on Information Forensics and Security*, **3**, (1), pp. 52–61, Mar. 2008.

[15]   Tabassi, E., Grother, P., Salamon, W., *Iris Quality Calibration and Evaluation: Performance of Iris Image Quality Assessment Algorithms*. NIST Interagency Report **7820**, Bethesda, 30 Sept. 2011.

[16]   International Standard ISO/IEC 29794-6: *Biometric Sample Quality – Part 6: Iris Image*. ISO Copyright Office, Geneva, 2015. See also the overall quality framework document: International Standard ISO/IEC 29794-1: *Biometric Sample Quality – Part 1: Framework*. ISO Copyright Office, Geneva, 2016.

*Chapter 5*

# Iris biometric indexing

*Hugo Proença and João C. Neves*

Indexing/retrieving sets of iris biometric signatures has been a topic of increasing popularity, mostly due to the deployment of iris recognition systems in nationwide scale scenarios. In these conditions, for each identification attempt, there might exist hundreds of millions of enrolled identities and is unrealistic to match the probe against all gallery elements in a reasonable amount of time. Hence, the idea of indexing/retrieval is – upon receiving one sample – to find in a quick way a subset of elements in the database that most probably contains the identity of interest, i.e., the one corresponding to the probe. Most of the state-of-the-art strategies to index iris biometric signatures were devised to decision environments with a clear separation between genuine and impostor matching scores. However, if iris recognition systems work in low quality data, the resulting decision environments are poorly separable, with a significant overlap between the distributions of both matching scores. This chapter summarizes the state-of-the-art in terms of iris biometric indexing/retrieval and focuses in an indexing/retrieval method for such low quality data and operates at the *code* level, i.e., after the signature encoding process. Gallery codes are decomposed at multiple scales, and using the most reliable components of each scale, their position in a *n*-ary tree is determined. During retrieval, the probe is decomposed similarly, and the distances to multi-scale centroids are used to penalize paths in the tree. At the end, only a subset of branches is traversed up to the last level.

## 5.1 Introduction

Iris biometrics is now used in various scenarios with satisfactory results (e.g., refugee control, security assessments and forensics) and nationwide deployment of iris recognition systems has already begun. In the last information update about the UAE system [1], the number of enrolled identities was over 800,000, and more than $2e^{12}$ iris cross-comparisons have been performed. The Unique Identification Authority of

India [2] is developing the largest-scale recognition system in the world (over 1,200 million persons).

Though matching *IrisCodes* primarily involves the accumulation of bitwise XOR operations on binary sequences, an increase in turnaround time occurs in national or continental contexts, which motivated growing interest in iris indexing strategies able to reduce the turnaround time without substantially affecting precision. As noted by Hao *et al.* [3], the indexing problem is a specific case of the more general nearest neighbor search problem, and motivated several proposals in the last few years (Section 5.2). However, most of these methods were devised to decision environments of good quality, with a clear separation between the matching scores of genuine and impostors comparisons.

In this chapter, not only we summarize the state-of-the-art in terms of iris biometric indexing/retrieval, but we focus particularly on the problem of indexing in decisions environments of poor quality, with a significant overlap between the matching scores of genuine and impostor comparisons. This kind of environment is likely when iris recognition systems operate in non-controlled data acquisition protocols (e.g., automated surveillance scenarios, using COTS hardware). We analyze a method [4] that operates at the code level, i.e., after the *IrisCode* encoding process. We decompose the codes at multiple levels and find their most reliable components, determining their position in an *n*-ary tree. During retrieval, the probe is decomposed similarly, and distances to the multi-scale centroids are obtained, penalizing paths of the tree and traversing only a subset up to the leaves.

The remainder of this chapter is organized as follows: Section 5.2 summarizes the state-of-the-art in terms of the published indexing/retrieval strategies. Section 5.3 provides a description of the method we focus in the chapter. Section 5.4 presents and discusses the results with respect to state-of-the-art techniques. Finally, the conclusions are given in Section 5.5.

## 5.2    State of the art

Table 5.1 summarizes the iris indexing methods that were reported recently in the literature, which can be coarsely classified using two criteria: (1) the light spectrum used for the data acquisition (either at near-infrared or visible wavelengths) and (2) the methods' input, which is either the raw iris texture or the biometric signature (*IrisCode*).

Yu *et al.* [5] represented the normalized iris data in the polar domain, dividing it radially into sixteen regions, and obtaining the fractal dimension for each one. Using first-order statistics, a set of semantic rules indexes the data into one of four classes. During retrieval, each probe is matched exclusively against gallery data in the same class. Fu *et al.*'s [6] use color information and suggest that artificial

Table 5.1 *Overview of the most relevant recently published iris indexing methods. NIR stands for near-infrared and VW for visible wavelength data*

| Method | Type | Spectrum | Preprocessing | Summary |
|---|---|---|---|---|
| Fu et al. [6] | Color | Own (9 images) | Segmentation | Artificial color filters, pre-tuned to a range of colors. C-means to define classes. |
| Gadde et al. [7] | Texture, *IrisCode* | CASIA-V3 (NIR) | Segmentation, normalization | Pixel-by-pixel Euclidean distance to clusters used in indexing |
| Hao et al. [3] | *IrisCode* | 632 500 UAE *IrisCodes* | Segmentation, normalization, feature extraction | estimation of intensity distribution, binarization, counting binary patterns with less coefficient of variation, division into radial bands, density estimation selection of most reliable bytes, bits decorrelation (interleaving and rotations), partition of identities into beacons, detection of multiple collisions |
| Jayaraman and Prakash [8] | Color, Texture | UBIRIS.v1, UPOL (VW) | Segmentation | Color analysis in YCbCr space. SURF keypoint description, Kd-tree indexing |
| Mehrotra et al. [9] | Texture | CASIA.1, ICE, WVU (NIR) | Segmentation | Keypoints localization, geometric analysis, hash table construction |
| Mehrotra et al. [10] | Texture | CASIA, Bath, IITK (NIR) | Segmentation, normalization | Multi-resolution decomposition (DCT). Energy of sub-bands extraced in Morton order, B-tree indexing |
| Rathgeb et al. [11] | *IrisCode* | IITD (NIR) | Segmentation, normalization | 1) sub-blocks division, Bloom filters, tree partition and traversal |
| Mukherjee and Ross [12] | Texture, *IrisCode* | CASIA-V3 (NIR) | Segmentation, normalization | 1) sub-blocks division, top-n similarity between blocks, tree partition; 2) sublocks partition, k-means clustering |
| Puhan and Sudha [13] | Color | UBIRIS.v1, UPOL (VW) | Segmentation | Conversion to YCbCr, semantic decision tree |
| Qiu et al. [14] | Texture | CASIA.1, ICE, WVU (NIR) | Segmentation, normalization | Extraction of texton histograms, Chi-square dissimilarity, K-means clustering |
| Vatsa et al. [15] | Texture | CASIA.1, ICE, WVU (NIR), UBIRIS.v1 (VW) | Segmentation, normalization | 8-bit planes of the masked polar image, extraction of topological information (Euler numbers), nearest neighbor classification |
| Yu et al. [5] | Texture | CASIA.1, ICE, WVU (NIR) | Segmentation, normalization | Definition of radial ROIs, extraction of local fractal dimensions, semantic decision tree |
| Zhao [16] | Color | UBIRIS.v2 (VW) | Segmentation, noise detection | Estimation of luminance, color compensation, average color, projection and quantization into three 1D feature spaces, union of identities enrolled from corresponding bins |

color filters provide an orthogonal discriminator of the spatial iris patterns. Each color filter is represented by a discriminator that operates at the pixel level. Gadde *et al.* [7] analyzed the distribution of intensities and selected patterns with low coefficients of variation (CVs) as indexing pivots. For each probe represented in the polar domain, a radial division of n-bands was performed and indexed using the highest density of CV patterns. Hao *et al.* [3] exclusively relied on the *IrisCode*. Using the spatial spread of the most reliable bits, they were based on the notion of multi-collisions. In the retrieval process, a minimum of $k$ collisions between the probe and gallery samples is required to identify a potential match. Jayaraman and Prakash [8] fused texture and color information: they estimated the iris color in the YCbCr space and determined an index to reduce the search space. Texture was encoded using SURF [17] keypoint detection and description. Mehrotra *et al.* [9] used SIFT [18] descriptors and their spatial distribution. To overcome the effect of non-uniform illumination and partial occlusions caused by eyelids, keypoints were extracted from angularly constrained regions of the iris. During retrieval, the geometric hashed location determined from the probe data accesses the appropriate bin of a hash table, and for every entry found, a vote is cast. The identities that receive more than a certain number of votes are considered possible candidates. Mehrotra *et al.* [10] divided the polar iris data into sub-bands using a multi-resolution Discrete Cosine transformation. Energy-based histograms were extracted for each band, divided into fixed-size bins, and iris images with similar energy values were grouped. A B-tree in which instances with the same key appear in the same leaf node was built. For a query, the corresponding key was generated, and the tree was traversed until a leaf node was reached. The templates stored at the leaf node were retrieved and compared with the query template to find the best match. Mukherjee and Ross [12] approached the problem from two different perspectives: by analyzing the iris texture and the *IrisCode*. The best results in the latter case were attained when each code was split into fixed-size blocks. First-order statistics for each block were used as the primary indexing value. A $k$-means strategy was used to divide the feature space into different classes. Qiu *et al.* [14] created a small finite dictionary of visual words (clusters in the feature space), called *textons*, to represent visual primitives of iris images. Then, texton histograms were used to represent the global features, and the $k$-means algorithm was used to classify the irises into five categories. Vatsa *et al.* [15] represented pixels of sub-regions of the unwrapped iris data in an 8-D binary feature space. The four most significant bits were used to build four corresponding maps from which the Euler numbers were extracted. Retrieving was performed using the nearest neighbor technique for the topological data. Zhao [16] determined the average RGB values inside the iris, weighted them by the luminance component to form a 3D feature space, and subsequently projected them into independent 1D spaces. Probes were matched only against gallery samples that corresponded to the union of the identities in the bins of each space. A similar approach was due to Puhan and Sudha [13]: they obtained the color index (in the YCbCr color space) and used a semantic decision tree to index the database.

## 5.3 Indexing/retrieving poorly separated data

### 5.3.1 Indexing

#### 5.3.1.1 Codes decomposition/reconstruction

Let $s_i$ denote a signature (*IrisCode*) $s$ from subject $i$. As illustrated in Figure 5.1, the key insight of the method we describe here is to obtain coarse-to-fine representations of $s_i$ as a function of the level $l$ in the tree ($s_i^{(l)}$). These representations are grouped according to their similarity in the $L_2$ metric space, and stored in tree nodes. A node is considered a leaf when a cluster centroid is at a sufficiently small distance from $s_j^{(l)}$, $\forall j$.

Let $\phi(x) = \sum_{k \in \mathbb{Z}} h(k)\sqrt{2}\phi(2x - k)$ and $\psi(x) = \sum_{k \in \mathbb{Z}} g(k)\sqrt{2}\phi(2x - k)$ be two filters, where $h(\cdot)$ and $g(\cdot)$ are low-pass and high-pass filters. According to Mallat's multi-resolution analysis [19], the operator representation of these filters is defined by

$$(H_a)_k = \sum_n h(n - 2k)a_n$$

$$(G_a)_k = \sum_n g(n - 2k)a_n,$$

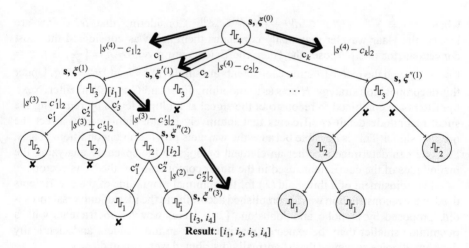

**Result:** $[i_1, i_2, i_3, i_4]$

*Figure 5.1* *Cohesive perspective of the indexing structure and of the retrieval strategy focused in this chapter. For a query $s$ with residual $\xi^{(0)}$, the distance between the decomposition of $s$ at top level ($s^{(4)}$) to the centroids is used to generate the new generation of residuals ($\xi^{(1)}$). For any branch with negative values, the search is stopped, meaning that subsequent levels in the tree are not traversed (illustrated by gray nodes). When traversing the tree, every identity found at any level where $\xi^{(\cdot)} > 0$ is included in the retrieved set*

where $H$ and $G$ correspond to one-step wavelet decomposition. Let $s^{(n)}$ denote the original signal of length $N = 2^n$ (in our experiments, $n = 11$). $s^{(n)}$ is represented by a linear combination of $\phi$ filters:

$$s^{(n)} = \sum_n a_k^{(n)} \phi_{nk}.$$

At each iteration, a coarser approximation $s^{(j-1)} = H\, s^{(j)}$ for $j \in \{1, \ldots, n\}$, is obtained; $d^{(j-1)} = G\, s^{(j)}$ are the residuals of the transformation $s^{(j)} \to s^{(j-1)}$. The discrete wavelet transformation of $s^{(j)}$ is

$$s^{(n)} \equiv [d^{(n-1)}, d^{(n-2)}, \ldots, d^{(0)}, s^{(0)}],$$

where $\left( \sum_{i=0}^{n-1} len(d^{(i)}) \right) + len(s^{(0)}) = len(s^{(n)}) = 2^n$; $len(.)$ is the number of signal coefficients. $s^{(n)}$ are approximated at different levels $l$ using $H^*$ and $G^*$ reconstruction filters:

$$(H_a^*)_l = \sum_k h(l - 2k) a_k$$

$$(G_a^*)_l = \sum_k g(l - 2k) a_k,$$

where $s^{(n)} := \sum_{j=0}^{n-1} (H^*)^{(j)} G^* d^{(j)} + (H^*)^{(n)} G^* s^{(0)}$. Considering that *IrisCodes* are binary, the Haar wavelet maximally correlates them and was considered the most convenient for this purpose: the filter coefficients are given by $h = [\frac{1}{\sqrt{2}}, \frac{1}{\sqrt{2}}]$, $g = [\frac{1}{\sqrt{2}}, -\frac{1}{\sqrt{2}}]$ and the reconstruction coefficients are similar $h^* = h$ and $g^* = -g$. Under this decomposition strategy, $H$ acts as a smoothing filter and $G$ as a detail filter. Next, the filters are combined to reconstruct the signal at multiple levels by removing the small-magnitude detail coefficients that intuitively do not significantly affect the original signal. This is possible because the wavelets provide an unconditional basis, i.e., one can determine whether an element belongs to the space by analyzing the magnitudes of the coefficients used in the linear combination of the basis vectors.

The adjustment of a threshold ($\lambda$) for the minimal magnitude of the coefficients used in the reconstruction was accomplished according to the idea of universal threshold, proposed by Donoho and Johnstone [20]. Here, wavelet coefficients with a magnitude smaller than the expected maximum for an independent and identically distributed noise sequence that is normally distributed were ignored:

$$\lambda = \sqrt{2\log{(n)}}\hat{\sigma}, \tag{5.1}$$

where $2^n$ is the length of the original signal and $\sigma$ is estimated using

$$\sigma = \sqrt{\frac{1}{N/2 - 1} \sum_{i=1}^{N/2} (d_{i,j} - \bar{d})^2}, \tag{5.2}$$

where $d_{i,j}$ denotes the $i$th wavelet coefficient at level $j$ and $\bar{d}$ is the mean of coefficients. Figure 5.2 illustrates an *IrisCode* $s_i$ and its representations at different levels ($n = \{1, 2, 10\}$). The coarsest representation $s^{(10)}$ retains the lowest frequency components

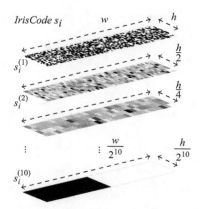

*Figure 5.2    Representation of an IrisCode (upper image) at different levels,
retaining coarse (bottom image) to fine information from an input code.
The $s_i^{(10)}$ representation is used in the root of the indexing tree and the
remaining representations at the deeper levels of the tree. Intensities
and sizes are stretched for visualization purposes*

of the input code (intensities are stretched for visualization purposes) and is used in
the root of the indexing tree, whereas the finest representation $s^{(1)}$ is used at the leaves.

As illustrated in Figure 5.3, $s^{(i)}$ corresponds to increasingly smoothed versions
of $s$. They were used at each level of the $n$-ary tree, starting from the coarsest reconstruction (root of the tree) and iteratively adding detail coefficients at the deeper
levels. The top plot shows the average residuals between the original signal and the
reconstruction with respect to the levels used (horizontal axis); being evident that –
on average – residuals decrease directly with respect to the decomposition level. The
plots at the bottom row show histograms of the residuals for the coarsest (center)
and finest scales (right); enabling to perceive that the coarsest-scale reconstruction is
essentially a mean of the original signal.

### 5.3.1.2    Determining the number of branches per node

Having a set of reconstructed signals $\{s_i^{(l)}\}$, a clustering algorithm was used to find
centroid that corresponds to a node in the tree and a partition of $\{s_i^{(l)}\}$, according to
the distances of elements to that centroid. Also, if the distance between $\{s_i^{(l)}\}$ and the
cluster centroid is less than a residual ($\upsilon \approx n \cdot 0.1, \forall i$), the indexing process stops at
that level for that branch, and the node is considered a leaf.

The number of clusters determines the number of branches in each node of the
tree. In order to determine the *optimal* value, a comparison between the proportion of
variance in the data with respect to the number of clusters was carried out. Intuitively,
if the number of clusters is too low, new partitions reduce the variance significantly,
but – at the other extreme – if the number of clusters is too high, adding a new one
almost does not reduce variance. Hence, the ideal number of clusters was considered

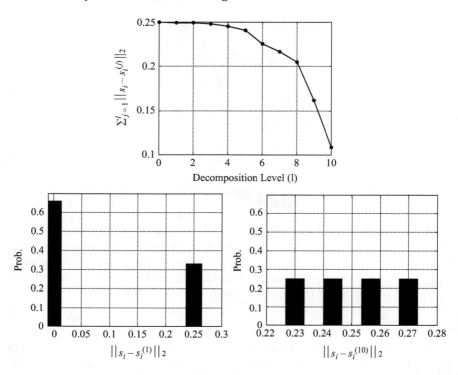

*Figure 5.3*   *Average sum of residuals between an IrisCode $s_i$ and its representations at different levels ($s_i^{(l)}$, top image). The images in the bottom row give the histograms of the residuals observed for decompositions/reconstructions at the coarsest (left) and finest (right) levels*

to be reached when this marginal gain decreases significantly, Let $k$ be the number of clusters. The proportion of the variance explained is characterized by a $F$-test:

$$F(k) = \frac{(n-k)\sum_{i=1}^{k} n_i(\overline{Y_i} - \overline{Y})^2}{(k-1)\sum_{i=1}^{k}\sum_{j=1}^{k}(\overline{Y_{ij}} - \overline{Y_i})^2}, \tag{5.3}$$

where $\overline{Y_i}$ is the sample mean in the cluster, $n_i$ is the number of codes and $\overline{Y}$ the overall mean. Considering $(k_i, F(k_i))$ as points on a curve, we seek the value with minimal curvature, which corresponds to the number of clusters at which the marginal gain drops more. Parameterizing the curve $(x(t), y(t)) = (k_i', F(k_i'))$ using quadratic polynomials yields a polygon with segments defined by

$$\begin{cases} x(t) = a_3 t^2 + a_2 t + a_1 \\ y(t) = b_3 t^2 + b_2 t + b_1 \end{cases} \tag{5.4}$$

The $x(t)$ and $y(t)$ polynomials were fitted via the least squares strategy using the previous and next points at each point to find the $a_i$ and $b_i$ coefficients:

$$\Upsilon_a = \sum_{i=1}^{3} \left[ y_i - a_1 + a_2 x_i + a_3 x_i^2 \right]^2 . \tag{5.5}$$

Setting $\partial \Upsilon / \partial a_i = 0$ yields:

$$\begin{bmatrix} 3 & \sum x_i & \sum x_i^2 \\ \sum x_i & \sum x_i^2 & \sum x_i^3 \\ \sum x_i^2 & \sum x_i^3 & \sum x_i^4 \end{bmatrix} \begin{bmatrix} a_1 \\ a_2 \\ a_3 \end{bmatrix} = \begin{bmatrix} \sum y_i \\ \sum x_i y_i \\ \sum x_i^2 y_i \end{bmatrix} \tag{5.6}$$

By solving the system of linear equations for $a_i$, the coefficients of the interpolating polynomials are obtained. The $b_i$ values are obtained similarly. The curvature $\kappa$ at each point $k_i$ is given by:

$$\kappa(k_i) = \frac{x(t)'y(t)'' - y(t)'x(t)''}{\sqrt{(x(t)'^2 + y(t)'^2)^3}}, \tag{5.7}$$

where primes denote derivatives with respect to $t$. In our case, $x'(t) = 2ta_3 + a_2$, $x''(t) = 2a_3$, $y'(t) = 2tb_3 + b_2$ and $y''(t) = 2b_3$. Hence, (5.7) was rewritten as:

$$\kappa(k_i) = \frac{(2ta_3 + a_2)2b_3 - 2a_3(2b_3 + b_2)}{((2ta_3 + a_2)^2 + (2tb_3 + b_2)^2)^{\frac{3}{2}}}. \tag{5.8}$$

Because we are primarily interested in the curvature at each point, $t$ was replaced by 0, yielding:

$$\kappa(k_i) = \frac{2(a_2 b_3 - a_3 b_2)}{(a_2^2 + b_2^2)^{\frac{3}{2}}}. \tag{5.9}$$

Finally, the position with minimal curvature was deemed to be the optimal number of clusters for that node:

$$\hat{k} = \arg\min_i \kappa(k_i) \tag{5.10}$$

Figure 5.4 shows an example of the described strategy to find the number of clusters per node. Here, the $F(k_i)$ values were tested for $k_i \in \{2, \ldots, 11\}$ (continuous lines). The dashed line corresponds to the $\kappa(k_i)$ values. The minimum curvature of the interpolating polynomials was observed at $\hat{k} = 8$.

## 5.3.2 Retrieval

The retrieval process receives a query signature $s$ and a residual value $\xi > 0$. The idea is to traverse only a subset of the paths in the tree, by iteratively decreasing $\xi$ and stopping when $\xi < 0$. At each node, the $L_2$ distance between the reconstructed version of the query at level $(l)$ and a cluster centroid is subtracted from $\xi$, considering the maximum distance between that centroid and the identities stored in the branch. Formally, let $q(s, \xi^{(0)})$ be the query parameters at the tree root (level $l$). $s^{(l)}$ is the

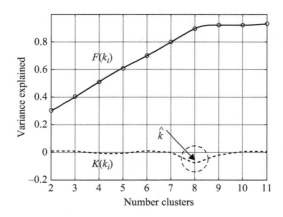

*Figure 5.4*    *Illustration of the strategy used to determine the number of clusters at each node of the n-ary tree. For ($k_i \in \{2, \ldots, 11\}$), the amount of variance explained $F(k_i)$, is denoted by circular data points. Quadratic polynomials were fitted to interpolate this data (continuous lines), from where the curvature at each point was found (dashed line). The number of clusters $\hat{k} = 8$ corresponds to the point where the gain in the explained variance drops, i.e., where the curvature value attains a minimum*

reconstruction of $s$ at the highest scale. The next generation of residual values $\xi^{(l-1)}$ at the child nodes is given by

$$\xi^{(l-1)} = \xi^{(l)} - \max\left\{0, \left\|s^{(l)} - c_i^{(l)}\right\|_2 - \max\left\{\left\|s_j^{(l)} - c_i^{(l)}\right\|_2, j \in \{1, \ldots, t_i\}\right\}\right\},$$

(5.11)

being $c_i$ the $i$th cluster and $s_j^{(l)}$ the reconstruction at scale $l$ of the signatures in that branch of the tree. The set of identities retrieved is obtained by

$$q(s, \xi^{(l)}) = \begin{cases} [\{i.\}, q(s, \xi^{(l-1)})], & \text{if } \xi^{(l)} > 0 \wedge l > 1 \\ \{i.\}, & \text{if } \xi^{(l)} > 0 \wedge l = 1 \\ \emptyset, & \text{if } \xi^{(l)} \leq 0 \end{cases}$$

(5.12)

where $[,]$ denotes vector concatenation and $\{i.\}$ is the set of identities in each node. Because of the intrinsic properties of wavelet decomposition/reconstruction, the distance values at the higher scales should be weighted, as they represent more signal components. This was done by the erf function:

$$w(l) = \frac{1 + \text{erf}(\alpha(l - n))}{2},$$

(5.13)

$\alpha$ being a parameter that controls the shape of the sigmoid. Figure 5.5 shows examples of histograms of the cuts in residuals $\xi$ with respect to the level in the tree. On the horizontal axis, note the decreasing magnitudes with respect to level. The dashed

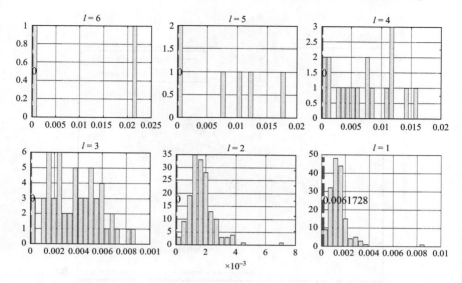

*Figure 5.5* *Histograms of the cuts in residuals $\xi^{(l)}$ per level during retrieval. The vertical dashed lines give the cumulative distribution values of the cuts, in the paths that correspond to the matching identity of the query. Gray bars express the frequencies of the cuts occurred in the remaining paths of the tree*

vertical lines indicate the residual cuts in the tree path that contained the true identity. Note that, with exception to the leaf level ($l = 1$), no cuts in the residual were performed for the paths that correspond to the true identity. This is in opposition to the remaining paths on the tree, where cuts in residual occurred at all levels.

## 5.3.3 Time complexity

Here we are primarily interested in analyzing the time complexity of the retrieving algorithm, and how the turnaround time depends on the number of identities enrolled. Let $T_s$, $T_c$ and $T_m$ denote the average elapsed time in the segmentation, coding and matching stages. Without indexing, the average turnaround time for an exhaustive search $T_e$ is given by:

$$T_e = T_s + T_c + N\,0.5\,T_m, \tag{5.14}$$

where $N$ is the number of identities enrolled by the system. When indexing at the *IrisCodes* phase, the average turnaround time $T_i$ corresponds to:

$$T_i = T_s + T_c + N\,T_r + (h\,p + (1 - h))\,0.5\,N\,T_m, \tag{5.15}$$

$T_r$ being the average turnaround time for retrieval and $h$ and $p$ the hit and penetration rates. Figure 5.6 compares the expected values for the $T_i$ and $T_e$ turnaround times with

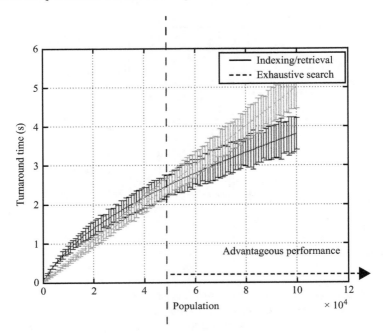

*Figure 5.6*    *Comparison between the turnaround times of an exhaustive search
(grey line) and when using the indexing/retrieval strategy analyzed in
this chapter (black line), with respect to the number of identities
enrolled in the system*

respect to the number of identities enrolled. $T_s$ and $T_c$ were disregarded because they do not affect the comparison. The values were obtained by repeatedly assessing the turnaround times of the analyzed method and of exhaustive searches. The horizontal bars near each point give the range of values observed, enabling to conclude that the indexing/retrieving starts to be advantageous when more than 54,000 identities are enrolled (vertical dashed line). Note that this value depends of the hit/penetration rates considered, which are functions of data quality. Even though, it serves as an approximation of the minimum number of identities that turn the indexing process advantageous in terms of turnaround time.

## 5.4  Performance comparison

Performance comparison was carried out at three different levels: (1) a set of synthetic signatures was generated to perceive performance with respect to slight changes in classes separability, which will be extremely hard to obtain using real data; (2) a data set of relatively well separated near infra-red data (CASIA.v4 Thousand) was used, in order to predict performance on scenarios that correspond to the currently deployed iris recognition systems; and (3) a data set of visible wavelength data with poor classes separability was used (UBIRIS.v2), which fits closely the purposes of the method described in this chapter. Four methods were selected for comparison, based on their property of operating at the *IrisCode* level: Gadde *et al.* [7], Hao *et al.* [3] and Mukherjee and Ross [12]. All the results correspond to our implementations of these techniques. In an appendix, detailed instructions to access the source code implementations are given.

To summarize performance by a single value, the proposal of Gadde *et al.* [7] was used, combining the hit and penetration rates. Similarly, a new measure $\tau$ corresponding to the Euclidean distance between an operating point $(h, p)$ and the optimal performance(hit = 1, penetration $\approx$ 0), was defined:

$$\gamma(h, p) = \sqrt{h(1 - p)} \tag{5.16}$$

$$\tau(h, p) = \sqrt{(h - 1)^2 + p^2}, \tag{5.17}$$

where $(h, p)$ express the hit and penetration rates.

### 5.4.1 Synthetic IrisCodes

A set of synthetic binary signatures was generated as described in http://www.di.ubi .pt/~hugomcp/doc/TR_VWII.pdf. This method is based on data correlation and simulates signatures extracted from data with broad levels of quality, ranging from extremely degraded to optimal. This is illustrated in Figure 5.7, showing various decision environments from optimal (Env. A) to extremely poor separated (Env. C).

When applied to good-quality data, the effectiveness of the Hao *et al.* [3] method is remarkable (see the top left plot of Figure 5.8). In this case, this method outperforms by more than one order of magnitude. However, its effectiveness decreases significantly in the case of degraded codes (bottom plot), which might be due to the concept of multiple collisions that becomes less effective as the probability of a given collision (of a minimum of $n$ bits) approaches for genuine and impostor comparisons. The approach of Gadde *et al.* [7] had the poorest performance for all the environments, whereas the method of Mukherjee and Ross [12] ranked third for the range of the performance space in good-quality environments. However, this was the unique technique that did not attain hit values above 0.9, either for good-quality or degraded data.

The analyzed method ranked second on good-quality data and showed the least decrease in performance for degraded data. The higher robustness was particularly evident for very high hit rates, which is the most important range for biometrics scenarios. Table 5.2 summarizes the performance indicators and the corresponding 95% confidence intervals for the three types of environments. Each cell contains two values: the top value regards the full operating range, and the bottom values are for the hit $\geq$ 95% range. Again, the values confirm the above observations about the relative performance of the techniques analyzed.

Figure 5.9 shows a statistic of the penetration rates (vertical axes) observed for queries that returned the true identity, for five kinds of environments, ranging from poorly separated (Env. C) to good quality data (Env. A). This plot emphasizes the good

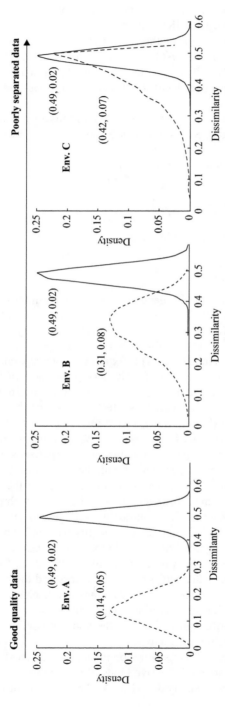

*Figure 5.7   Illustration of the separation between genuine (dashed lines) and impostor (continuous lines) comparisons, for different levels of quality. At the far left, histograms corresponding to data acquired in heavily controlled scenarios are shown (A). Classes separability decreases in the right direction*

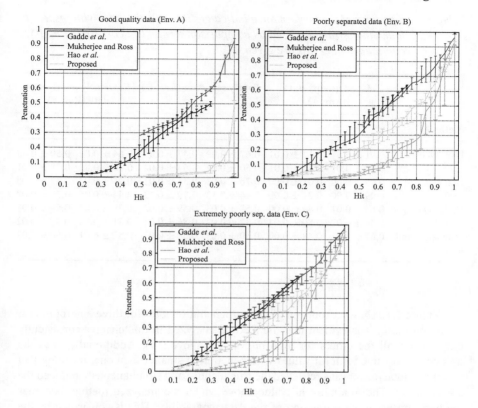

*Figure 5.8    Comparison between the hit/penetration rates observed for the strategy focused in this chapter and three state-of-the-art methods used as baselines. Results are expressed for three levels of data quality*

performance of the Gadde *et al.* method for good-quality data, obtaining penetration values close to 0. For poor-quality data, though the median value of the analyzed method's data penetration is higher than that of Hao *et al.* ($\approx 0.52$ versus $0.13$), it should be stressed that this statistic only accounts for cases in which the true identity was returned, which is more frequent for our proposal than for any other. Additionally, the interquartile range of Proenca's method penetration values was narrower than that of Hao *et al.*'s method, which is a signal of the stability of its performance with respect to different queries. For all methods tested, the penetration values decrease substantially for good separable data, though this is less evident for Mukherjee and Ross's proposal. This decrease is explained by the intrinsic properties of the clustering process involved here: clusters tend to have similar number of elements, and for every query, all identities inside a given cluster are returned. This prevents only a small set of identities from being returned even for highly separable data.

*Table 5.2    Summary of the performance indicators in synthetic data, with respect to four other strategies used as comparison terms. The corresponding 95% confidence intervals are given*

| Method | Good quality data (Env. A) | | Poorly sep. data (Env. D) | | Extremely poorly sep. data (Env. E) | |
|---|---|---|---|---|---|---|
| | $\gamma$ | $\tau$ | $\gamma$ | $\tau$ | $\gamma$ | $\tau$ |
| Proença [4] | $0.91 \pm 0.01$ | $0.12 \pm 0.01$ | $0.67 \pm 0.02$ | $0.47 \pm 0.01$ | $0.64 \pm 0.02$ | $0.50 \pm 0.03$ |
| | $0.90 \pm 0.01$ | $0.15 \pm 0.01$ | $0.50 \pm 0.02$ | $0.54 \pm 0.02$ | $0.46 \pm 0.03$ | $0.78 \pm 0.01$ |
| Hao *et al.* [3] | $0.99 \pm 0.00$ | $0.01 \pm 0.00$ | $0.76 \pm 0.03$ | $0.33 \pm 0.01$ | $0.74 \pm 0.03$ | $0.37 \pm 0.05$ |
| | $0.99 \pm 0.00$ | $0.01 \pm 0.00$ | $0.44 \pm 0.13$ | $0.79 \pm 0.13$ | $0.44 \pm 0.05$ | $0.79 \pm 0.02$ |
| Gadde *et al.* [7] | $0.65 \pm 0.01$ | $0.49 \pm 0.00$ | $0.58 \pm 0.03$ | $0.59 \pm 0.02$ | $0.58 \pm 0.02$ | $0.59 \pm 0.01$ |
| | $0.44 \pm 0.07$ | $0.80 \pm 0.04$ | $0.37 \pm 0.07$ | $0.86 \pm 0.01$ | $0.31 \pm 0.05$ | $0.90 \pm 0.02$ |
| Mukherjee and Ross [12] | $0.67 \pm 0.01$ | $0.48 \pm 0.00$ | $0.59 \pm 0.03$ | $0.58 \pm 0.03$ | $0.57 \pm 0.01$ | $0.60 \pm 0.01$ |
| | – | – | – | – | – | – |

Figure 5.10 shows a zoom-in of the hit/penetration rates for three environments. Based on these, it is evident that the method analyzed in this chapter consistently outperforms all the others for high hit values (above 0.9). Additionally, it is the unique that obtained full hit values with penetration smaller than one, meaning that was the unique that always retrieved the true identity and simultaneously reduced the search space. The minimum hit value above which the analyzed method becomes the best appears to be a function of the data separability. This is confirmed by the rightmost plot, which relates the quality of data and the minimum hit value. For the worst kind of data (Env. C, quality $= 0.0$), the analyzed method outperforms any other for hit values above 0.88. As data separability increases, the minimum hit value varies roughly linearly, and for environments with a quality index higher than 0.2, the method of Hao *et al.* starts to be the best and should be used instead of ours.

## 5.4.2    Well separated near infra-red data

The CASIA-Iris-Thousand[1] was used in performance evaluation, to represent reasonably well separated data. This data set contains 20,000 images from both eyes of 1,000 subjects, yielding the evaluation with 2,000 different classes (eyes).

The noise-free regions of the irises were segmented according to the method of He *et al.* [21] and an elliptical parameterization was chosen for both iris boundaries, according to the random elliptic Hough Transform algorithm. Next, the reasonability of the segmentation was manually adjusted, 110 images were discarded due to bad quality and the remaining data translated into a pseudo-polar coordinate system using the Daugman's *rubber sheet* model. Next, three different configurations for Gabor

---

[1]CASIA Iris Image Database: http://biometrics.idealtest.org/

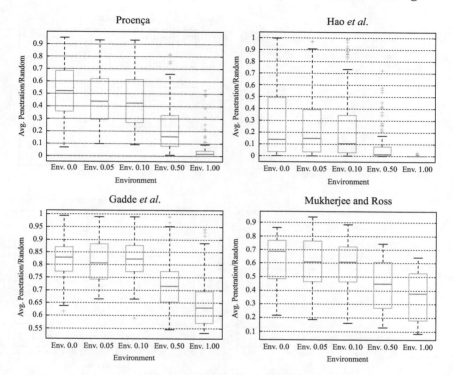

*Figure 5.9    Boxplots of the penetration rates observed in cases where the true
identity was retrieved. Values are shown for different levels of data
separability, starting from data of poorest quality (Env. 0.0) to good
quality data (Env. 1.0)*

kernels were used in signature encoding (wavelength $\omega$ and orientation $\theta$ were varied,
phase $\phi$ and ratio $r$ were not considered). The optimal parameters for the Gabor kernels
$g$ were obtained by maximizing the decidability index $d' = |\mu_I - \mu_G|/\sqrt{\frac{1}{2}(\sigma_G^2 + \sigma_I^2)}$,
$\mu_G$ and $\mu_I$ being the means of the genuine and impostors distributions and $\sigma_G$ and $\sigma_I$
their standard deviations.

$$g(x,y,\omega,\theta,\sigma_x,\sigma_y) = \frac{1}{2\pi\sigma_x\sigma_y}e^{-\frac{1}{2}\left(\Phi_1^2/\sigma_x^2 + \Phi_2^2/\sigma_y^2\right)}e^{i2\pi\Phi_1/\omega}, \tag{5.18}$$

where $\Phi_1 = x\cos(\theta) - y\sin(\theta)$, $\Phi_2 = -x\cos(\theta) + y\sin(\theta)$, $\omega$ the wavelength, $\theta$ the
orientation and $\sigma_x = \sigma_y = \omega/2$. The parameters found were obtained by exhaustive
evaluation in a training data set of 200 images randomly sampled from the data set:
$(\omega,\theta) = \{(0.33, \pi/4), (0.28, 3\pi/4), (0.51, \pi/2)\}$. Figure 5.11 gives some examples of
the noise-free iris masks and the iris boundaries for the CASIA.v4 Thousand images.

Results are given in Figure 5.12. The top left plot gives the decision environment,
according to the recognition scheme used. At the center, a comparison between the

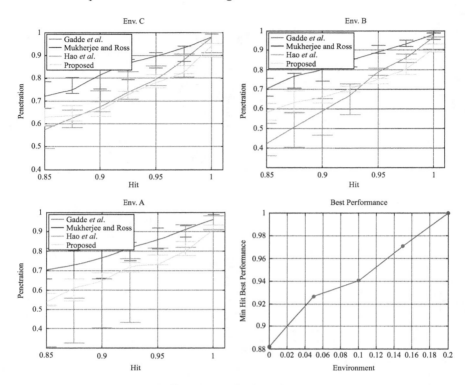

*Figure 5.10   Comparison between the hit/penetration plots in the performance range that was considered most important for biometric recognition purposes (hit values above 0.85). In poorly separable data the analyzed method outperforms all the remaining ones in this performance interval, and the minimal hit value above which our method becomes the best varies roughly linearly with the data separability (bottom right plot)*

hit/penetration values for the four techniques is shown, whereas the plot given at the top right corner summarizes the penetration rates in cases where the true identity was retrieved. Results confirm the previously obtained for synthetic data (for environments of average to good quality) and the approach of Hao *et al.* largely outperformed. The analyzed method got a consistent second rank, followed by that of Mukherjee and Ross and Gadde *et al.*'s. The boxplots confirm these observations, being also notorious the smaller variance of the analyzed method and Hao *et al.*'s in the number of retrieved identities, when compared to Gadde *et al.*'s and Mukherjee and Ross.

## 5.4.3   Poorly separated visible wavelength data

The UBIRIS.v2 [22] data set constitutes the largest amount of iris data acquired from large distances (4–8 m) at visible wavelengths, containing images of degraded

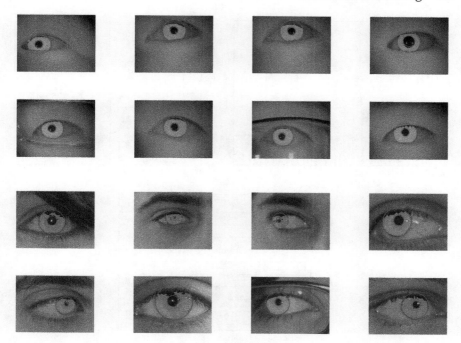

Figure 5.11    Examples of the real iris images used in performance evaluation,
segmented according to the method of He et al. [21]. The upper rows
regard the CASIA.v4 Thousand data set, and the bottom rows give
images of the UBIRIS.v2 data set

quality that lead to poor separability between the matching scores of *genuine* and
*impostors* comparisons. It has 11,102 images from 522 classes, from which 285
were not considered due to their extreme low quality level (e.g., out of iris or almost
completely occluded data). Similarly to the process described for the CASIA.v4
Thousand set, images were segmented according to the method of He *et al.* [21]
and followed the same processing chain, using the Gabor filters $G$ with parameters
$(\omega, \theta) = \{(0.18, \pi/6), (0.35, 4\pi/6), (0.20, 7\pi/8)\}$. The bottom rows of Figure 5.11
illustrate some examples of the images used.

These results were regarded in a particularly positive way, as they correspond to
the environments where this method was devised for. As illustrated by the decision
environment of Figure 5.13, classes have extremely poor separability, that can only be
used in cases where no human effort is put in the recognition process (e.g., automated
surveillance). For this type of data, the analyzed method outperformed in the most
interesting performance range, i.e, for hit values above 90% (plot at the center). The
rightmost plot gives a complementary perspective of results, by comparing the penetration rates in queries where the true identity was retrieved. In this case, the Proença's
method got clearly higher penetration rates than Hao *et al.*'s, and the value for the

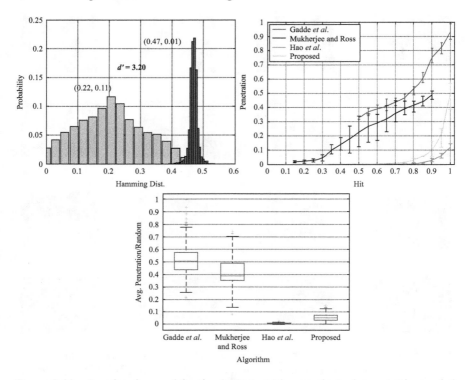

*Figure 5.12    Results observed for the CASIA.v4 Thousand iris data set. The top left plot gives an illustration of the decision environment yielded by the used recognition techniques is given. Plot at the top right corner compares the hit/penetration rates and the bottom plot summarizes the penetration rates observed in cases where the true identity was retrieved*

upper whisker is particularly important: for all queries the former method was able to reduce the set of identities retrieved, which did not happened in any of the other methods. Confirming the previous results, Hao *et al.*'s was the best for low hit values and got a solid second rate in the remaining performance range. Also, the smaller interquartile range of our method when compared to Hao *et al.*'s was also positively regarded as an indicator of the smaller variability with respect to different queries. Mukherjee and Ross' slightly better results than Gadde *et al.*'s, but in the former no hit values above 0.9 were obtained.

Table 5.3 summarizes the results observed in the CASIA.v4 Thousand and UBIRIS.v2 data sets. The upper value in each cell regards the full operating range and the bottom value regards the meaningful range for biometrics scenarios (hit values above 95%). The values in bold confirm the suitability of the Proença's method to work on poorly separable data (UBIRIS.v2, $\Delta \gamma = +0.11$ of ours with respect to

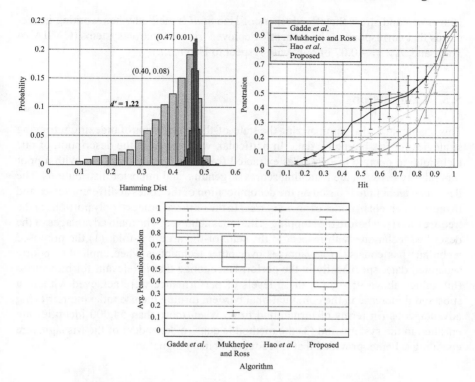

*Figure 5.13    Results observed for the UBIRIS.v2 iris data set. The top left plot gives
an illustration of the decision environment yielded by the used
recognition techniques is given. Plot at the top right corner compares
the hit/penetration rates and the right plot summarizes the penetration
rates observed in cases where the true identity was retrieved*

*Table 5.3    Summary of the performance indicators (5.17) observed in the CASIA.v4
Thousand and UBIRIS.v2 data sets, with respect to four strategies used as
comparison terms. The corresponding 95% confidence intervals are given*

| Method | CASIA.v4 Thousand (NIR) | | UBIRIS.v2 (VW) | |
|---|---|---|---|---|
| | $\gamma$ | $\tau$ | $\gamma$ | $\tau$ |
| Proença [4] | $0.91 \pm 0.02$ | $0.12 \pm 0.01$ | $0.71 \pm 0.02$ | $0.36 \pm 0.02$ |
| | $\mathbf{0.88 \pm 0.02}$ | $\mathbf{0.14 \pm 0.02}$ | $\mathbf{0.53 \pm 0.03}$ | $\mathbf{0.78 \pm 0.02}$ |
| Hao *et al.* [3] | $0.96 \pm 0.01$ | $0.04 \pm 0.01$ | $0.75 \pm 0.03$ | $0.34 \pm 0.02$ |
| | $\mathbf{0.95 \pm 0.01}$ | $\mathbf{0.05 \pm 0.01}$ | $\mathbf{0.42 \pm 0.06}$ | $\mathbf{0.82 \pm 0.04}$ |
| Gadde *et al.* [7] | $0.62 \pm 0.01$ | $0.51 \pm 0.02$ | $0.60 \pm 0.02$ | $0.47 \pm 0.02$ |
| | $0.40 \pm 0.07$ | $0.82 \pm 0.02$ | $0.37 \pm 0.04$ | $0.88 \pm 0.03$ |
| Mukherjee and Ross [12] | $0.76 \pm 0.02$ | $0.43 \pm 0.02$ | $0.61 \pm 0.02$ | $0.46 \pm 0.02$ |
| | – | – | – | – |

Hao *et al.*) and stress the effectiveness of Hao *et al.*'s method when working in scenarios that correspond to the currently deployed iris recognition systems (CASIA.v4 Thousand, $\Delta\gamma = -0.07$ of ours with respect to Hao *et al.*).

## 5.5   Conclusions

This chapter aimed at summarizing the state-of-the-art in terms of indexing/retrieving strategies for iris biometric data. In particular, we focused in the description of one technique to operate in *IrisCodes* extracted from low quality data, i.e., with a poor separability between the matching scores of genuine and impostor distributions. The described technique is based on the decomposition of the codes at different scales and in their placement in nodes of an *n*-ary tree. In the retrieval process, only portions of the tree are traversed before the stopping criterion is achieved. The main advantages of the described technique with respect to the state-of-the-art are 3-fold: (1) the proposed technique has consistent advantages over other techniques when applied to poorly separated data, specifically in the performance range that is relevant for biometrics (hit values above 95%); (2) these levels of performance were achieved without a substantial increase in the computational burden, turning the use indexing/retrieving advantageous (in terms of turnaround time) when more than 54,000 identities are enrolled in the system; and (3) the method is quasi-independent of the iris signature encoding scheme, provided that it produces a binary signature.

## Acknowledgment

This work was supported by PEst-OE/EEI/LA0008/2013 research program.

## References

[1]   J. Daugman and I. Mallas. "Iris recognition border-crossing system in the UAE," *International Airport Review*, vol. 8, no. 2, pp. 49–53, 2004.

[2]   Unique Identification Authority of India. [online], http://uidai.gov.in/about-uidai.html, accessed on June 2012.

[3]   F. Hao, J. Daugman and P. Zielinski. "A fast search algorithm for a large fuzzy database," *IEEE Transactions on Information Forensics and Security*, vol. 3, no. 2, pp. 203–211, 2008.

[4]   H. Proença. "Indexing and retrieving heavily degraded data," *IEEE Transactions on Information Forensics and Security*, vol. 8, no. 12, pp. 1975–1985, 2013.

[5]   L. Yu, K. Wang and D. Zhang. "A novel method for coarse iris classification," *Proceedings of the International Conference on Biometrics, Lecture Notes on Computer Science*, vol. 3832, pp. 404–410, 2006.

[6]   J. Fu, H. Caulfield, S. Yoo and V. Atluri. "Use of artificial color filtering to improve iris recognition and searching," *Pattern Recognition Letters*, vol. 26, pp. 2244–2251, 2005.

[7]   R. Gadde, D. Adjeroh and A. Ross. "Indexing iris images using the Burrows–Wheeler transform," *Proceedings of the IEEE International Workshop on Information Forensics and Security (WIFS)*, pp. 1–6, 2010.

[8]   U. Jayaraman and S. Prakash. "An iris retrieval technique based on color and texture," *Proceedings of the Seventh Indian Conference on Computer Vision, Graphics and Image Processing*, pp. 93–100, 2010.

[9]   H. Mehrotra, B. Majhi and P. Gupta. "Robust iris indexing scheme using geometric hashing of SIFT keypoints," *Journal of Network and Computer Applications*, vol. 33, pp. 300–313, 2010.

[10]  H. Mehrotra, B. Srinivas, B. Majhi and P. Gupta. "Indexing iris biometric database using energy histogram of DCT subbands," *Journal of Communications in Computer and Information Science*, vol. 40, pp. 194–204, 2009.

[11]  C. Rathgeb, F. Breitinger, H. Baier and C. Busch. "Towards bloom filter-based indexing of iris biometric data," *Proceedings of the Eighth IAPR International Conference on Biometrics (ICB 2015)*, pp. 422–429, 2015.

[12]  R. Mukherjee and A. Ross. "Indexing iris images," *Proceedings of the 19th International Conference on Pattern Recognition (ICPR 2008)*, pp. 1–4, 2008.

[13]  N. Puhan and N. Sudha. "Coarse indexing of iris database based on iris color," *International Journal on Biometrics*, vol. 3, no. 4, pp. 353–375, 2011.

[14]  X. Qiu, Z. Sun and T. Tan. "Coarse iris classification by learned visual dictionary," *Proceedings of the International Conference on Biometrics, Lecture Notes on Computer Science*, vol. 4642, pp. 770–779, 2007.

[15]  M. Vatsa, R. Singh and A. Noore. "Improving iris recognition performance using segmentation, quality enhancement, match score fusion, and indexing," *IEEE Transactions on Systems, Man and Cybernetics – Part B: Cybernetics*, vol. 38, no. 4, pp. 1021–1035, 2008.

[16]  Q. Zhao. "A new approach for noisy iris database indexing based on color information," *Proceedings of the Sixth International Conference on Computer Science and Education (ICCSE 2011)*, pp. 28–31, 2011.

[17]  H. Bay, A. Ess, T. Tuytelaars and L. Van Gool. "Speeded-Up Robust Features (SURF)," *Computer Vision and Image Understanding*, vol. 110, no. 3, pp. 346–359, 2008.

[18]  D. Lowe. "Object recognition from local scale-invariant features," *Proceedings of the International Conference on Computer Vision*, pp. 1150–1157, 1999.

[19]  S. Mallat. *A Wavelet Tour of Signal Processing*. Salt Lake City, UT: Academic Press, ISBN: 0-12-466606-X, 1999.

[20]  D. Donoho and I. Johnstone. "Ideal spatial adaptation by wavelet shrinkage," *Biometrika*, vol. 81, pp. 425–455, 1994.

[21]    Z. He, T. Tan, Z. Sun and X. Qiu "Towards accurate and fast iris segmentation for iris biometrics," *IEEE Transactions on Pattern Analysis and Machine Intelligence*, vol. 31, no. 9, pp. 1617–1632, 2009.

[22]    H. Proença, S. Filipe, R. Santos, J. Oliveira and L. A. Alexandre. "The UBIRIS.v2: a database of visible wavelength iris images captured on-the-move and at-a-distance. *IEEE Transactions on Pattern Analysis and Machine Intelligence*, vol. 32, no. 8, pp. 1502–1516, 2010.

*Chapter 6*

# Identifying the best periocular region for biometric recognition

*Jonathon M. Smereka and B.V.K. Vijaya Kumar*

Periocular images contain iris as well as other near-by regions such as eyelids and eyebrows. It is useful to know which regions of ocular images are important for achieving good recognition performance. This chapter will present observations from experiments with regard to the effect of changing the size or selection of the periocular region on biometric recognition performance. One can use a single definition of the periocular region to identify areas that contain significant discriminative textural information in an attempt to identify such regions. In this chapter, we investigate sub-region effects over varying cropping sizes to determine an appropriate periocular image representation.

## 6.1 Introduction

Face and iris recognition are both well-established biometric modalities with systems being able to discriminate among a large number of subjects. Accordingly, significant research has been conducted in the area of representation for both face [1] and iris [2,3,4] modalities. In particular, the region of interest for face recognition can vary, e.g., include or not-include hair, chin, ears, etc., however the consensus is that separating the face/head from the other parts of the image is beneficial to system performance. Some work [5] even goes as far as detailing necessary features that should be extracted individually for 'local' components (e.g., eyes, nose, mouth, etc.) in addition to the 'global' facial image.

In contrast to the definition of an appropriate facial region, the iris region to be used for recognition is inherently well defined. The process of performing iris recognition essentially boils down to first extracting the iris region from the image and then mapping the segmented iris region into a polar coordinate system to be compared against a database of training images [6]. However, previous work has shown that system accuracy can be greatly affected by image compression [2] (i.e., image format), quality [3] (e.g., harsh illumination or off-angle gaze) and the system used for image collection [4]. Thus, in order to achieve the high levels of accuracy desired, a high-quality, non-occluded iris image is required.

*Figure 6.1    Examples of where ocular recognition can be beneficial (harsh
           illumination, occlusion, etc.) when attempting iris and/or face
           recognition*

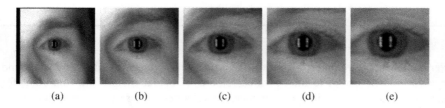

(a)              (b)              (c)              (d)              (e)

*Figure 6.2    Examples of potential periocular image representations using different
           cropped regions*

One potential solution is to use area containing the eye and iris, known as the
ocular region of the face (referred to as periocular biometrics when comparing a single eye region and bi-ocular when using both eye regions). Defined as the area of
the face that includes the eyelids, eyelashes, eyebrow and the skin surrounding the
eye, the periocular region is a discriminative area that can be obtained from images
traditionally used for face and iris recognition. Ocular modality benefits include
avoiding iris segmentation [7] and an increased resistance to the effects of ageing [8].
In addition, the periocular region has been shown to offer improved recognition performance when fused with iris images [9,10,11,12] and when fused with face images
[13,14]. Some examples of where the use of the ocular region can be beneficial are
shown in Figure 6.1. In addition, it has been shown that periocular recognition can
outperform iris and face recognition in challenging environments [7,15,16,17].

Unfortunately, the definition of a periocular image is unclear compared to the
face or iris. That is, there is no commonly accepted consensus of what exactly constitutes 'the best periocular region', i.e., the size or selection of the ocular region
which offers improved biometric recognition performance. However, several efforts
have been devoted to providing insights and evidence towards identifying discriminative features of the periocular region. For visible light (VL) images, it is generally
accepted that the eyebrow [15,18,19,20,21] and skin texture [18,20,21,22,23] are particularly important. While near-infrared (NIR) images have received considerably
less attention, previous work [20,21,23] has found eye shape, eyelashes and tear duct
information to be important.

Towards the goal of shedding further light on the effect of the ocular region on
recognition performance, this chapter discusses the question of what is the best periocular region for recognition. As shown in Figure 6.2, the definition of a periocular

region can correspond to several interpretations of how much information is available around (note that 'peri' is a Greek root for 'around' or 'surrounding') the eye. Specifically, we identify a distinct region of interest that appears to provide the most support in comparing NIR and VL ocular images.

## 6.2  Experimental setup

Several previous efforts have evaluated general system design decisions (e.g., masking the eye/iris is detrimental to system performance [12,18,22,24]) and generally provide some evidence of how to obtain improved performance (e.g., in VL images, the eyebrow is beneficial for subject recognition while the cheek regions may not be [17,21,25]). However, until recently [20,23] there had been an insufficient evaluation of differing periocular image representations in NIR and VL images at a sub-region level. Through experimentation, we investigate the image representation for providing the best discrimination between users by examining match and non-match scores in depth over sub-regions of the ocular image.

We test two fundamentally different algorithms on challenging periocular datasets of contrasting type. By testing two different comparison methods, we ensure that our observations about the sub-regions are not unique to a particular method:

- **Modified SIFT (m-SIFT)** [26]: A comparison score is obtained using a modified version of the standard Scale-Invariant Feature Transform (SIFT) [27] features. The keypoints to compare are spatially constrained to lie in the same region for the gallery and the probe image. To determine a comparison score, the number of similar keypoints are counted, with the expectation that authentic match pairs will have more similar keypoints than impostors.
- **Correlation filters (CFs):** A CF is a spatial-frequency array that is specifically synthesized from a single training image or a set of training images to produce a sharp correlation output peak at the location that the trained template is most similar to the test image for an authentic image comparison and no such peak for an impostor image comparison [28]. After feature extraction, each image is divided into patches where a CF template is built for each gallery patch to be compared with the corresponding probe image patch. Each patch score is the correlation output peak height and the final comparison score is then the sum of the patch scores.

The two databases investigated are the Face and Ocular Challenge Series (FOCS) [29] database captured in NIR light from moving subjects in an unconstrained environment, and the University of Beira Interior Periocular (UBIPr) [25] database which is composed of VL images of varying scale, pose and illumination. In each experiment, each image of each dataset is compared with every other image within the respective dataset (i.e., one-to-one). Thus, over the FOCS and UBIPr datasets, each test on each periocular region involves ~23 million comparisons and ~26 million comparisons, respectively. Refer to Table 6.1 for more information on each dataset.

*Table 6.1　Periocular dataset information*

| Dataset | Number of images (right/left) | Number of subjects | Number of samples per subject | Resolution (pixels) | Number of comparisons per experiment |
|---|---|---|---|---|---|
| FOCS (NIR) | 4,789/4,792 | 136 | 2–236 | $600 \times 750$ | ~23 million |
| UBIPr (VL) | 5,126/5,126 | 344 | 27–30 | $501 \times 401$ to $1,001 \times 801$ | ~26 million |

Note: 'VL' and 'NIR' represent the spectrum in which the images are acquired, VL, visible light and NIR, near infrared

The images from each dataset have had their eye corners labelled; we use the pre-packaged labellings included with the UBIPr set and hand label the FOCS set as no labels are provided, to perform experiments over varying periocular regions shown in Figure 6.2. The periocular region ranges from the case where the eye corners are at the edge of each image with very little cheek or brow information (Figure 6.2(e)), to images showing the complete area around the eye available in both datasets (Figure 6.2(a)). Experiments over these regions allows for the results to be different from each other with respect to the amount of periocular information available while also considering the level of granularity in determining an appropriate periocular image representation for recognition.

We use equal error rate (EER) and verification rate (VR) at 0.001 False Acceptance Rate (FAR) as the quantitative metrics of overall system performance. Receiver operating characteristic (ROC) curves are then used to display the system performance at varying score thresholds with FAR on the $x$-axis and VR on the $y$-axis. Sub-region effects are derived from the average score of each CF patch and recorded locations of any keypoints remaining after applying the m-SIFT spatial constraints. We do not employ cross-validation (instead generating one large score matrix for each periocular region) since no training is involved and we are interested in performance over a particular image representation (rather than algorithm).

Finally, for computational purposes all periocular images are re-sized to $128 \times 128$ pixels. To preserve the image quality and address the slightly varying size of the eye within each image (which is measured from corner to corner), we compute each normalized periocular image representation when down-sampling each image. This is done such that the desired image representation size ($F_h$ and $F_w$) is present when cropped (from the centre of the image) at $128 \times 128$ pixels (see Figure 6.3). For example, if the final height and width are $F_h$ and $F_w$, respectively ($128 \times 128$ pixels), and the original height and width are $O_h$ and $O_w$, respectively ($600 \times 750$ pixels), then the image is processed as described in Algorithm 6.1 with the cropping ratio, $S_c$, starting at 0 and increasing in increments of 0.25 with each change in image representation. Lastly, in addition to equalizing each image histogram for a basic normalization against harsh illumination, we use the eye corners to align each image individually in order to prevent possible effects from scale and rotation from biasing the results.

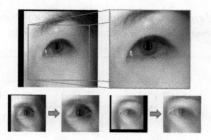

*Figure 6.3   Examples of periocular region extraction*

---

**Algorithm 6.1** Periocular Image Cropping

---

**Require:** *Im* ($O_h \times O_w$ pixels), scale adjustment of $S_c$, and desired region size $F_h \times F_w$

1: $Max_h = \dfrac{F_h}{\text{Distance between eye corners}}, Min_h = \dfrac{F_h}{O_h}$

2: $Max_w = \dfrac{F_w}{\text{Distance between eye corners}}, Min_w = \dfrac{F_w}{O_w}$

3: $Sz_h = O_h * (Min_h + (Max_h - Min_h) * S_c)$

4: $Sz_w = O_w * (Min_w + (Max_w - Min_w) * S_c)$

5: $Im_{new}$ = resize($Im,[Sz_h, Sz_w]$)

6: Crop $F_h \times F_w$ region from the centre of $Im_{new}$

---

## 6.3   Results

The VRs from each test are shown in Table 6.2 with corresponding ROCs provided in Figure 6.4 (EERs are displayed in the legend of each ROC). The performance of each algorithm over the varied periocular image representations are summarized for each dataset/spectrum.

### FOCS (NIR wavelength)

- *m-SIFT*: Representations generated from $S_c = 0.25$ and 0.50 provide the best performance (similar EERs and VRs) on both left and right ocular images.
- *CFs*: Best performance on the left ocular images is obtained using representations generated from $S_c = 0.00$ and 0.25, whereas both produce comparable EERs with $S_c = 0.25$ showing a ~15% VR improvement. For the right ocular images, best results are found at representations $S_c = 0.25$ (EER) and $S_c = 0.50/0.75$ (VR).

*Table 6.2 Left (L) and right (R) periocular recognition VRs at 0.001 FAR over each periocular image representation for the FOCS and UBIPr datasets*

|  |  | FOCS | | UBIPr | |
| --- | --- | --- | --- | --- | --- |
|  |  | m-SIFT (%) | CF (%) | m-SIFT (%) | CF (%) |
| $S_c = 0.00$ | R | 9.02 | 9.50% | 35.73 | 81.12 |
|  | L | 9.30 | 30.60 | 31.14% | 76.31 |
| $S_c = 0.25$ | R | 34.56 | 23.84 | 46.63 | 84.33 |
|  | L | 32.33 | 46.44 | 38.92 | 77.95 |
| $S_c = 0.50$ | R | 33.16 | 38.55 | 40.03 | 74.72 |
|  | L | 29.33 | 43.40 | 31.94 | 66.55 |
| $S_c = 0.75$ | R | 27.73 | 37.73 | 32.47 | 61.51 |
|  | L | 23.94 | 37.31 | 25.53 | 52.26 |
| $S_c = 1.00$ | R | 22.77 | 32.62 | 25.52 | 50.44 |
|  | L | 19.86 | 30.71 | 20.61 | 41.29 |

We first briefly discuss the cause of the difference in VR over the right ocular images from the CF results ($S_c < 0.75$ shows a significant drop in VR compared to results on the left ocular images) by examining comparisons misclassified by the CFs at $S_c = 0.00$ and $S_c = 0.25$, but are correctly classified at $S_c = 0.50$ and $S_c = 0.75$. We collected the corresponding gallery and probe images from impostor comparisons, aligned and cropped each, and then recorded any areas which do not possess any image information, i.e., the RGB values are all zeros (also referred to as 'incomplete') from shifting or aligning the periocular image. Figure 6.5 shows the frequency (areas of more occurrence are on the left side of the left column figures, while areas of lower occurrence are on the right side) of where each misclassified gallery image is incomplete (probe images have similar results) as well as the full score distributions with labelled EER and VR thresholds. The score distributions clearly indicate that some high-scoring impostors are degrading the VR for representations generated from $S_c = 0.00$ and $S_c = 0.25$. In contrast, examining the frequency of where each misclassified gallery image is incomplete (left column images) indicates that the cause is related to the amount of available data to process during a comparison.

Since this discrepancy is not occurring on the left ocular images to the same degree as seen on the right ocular images, we include the frequency of missing data from the left ocular images. Figure 6.6 shows graphically how often misclassified gallery images are incomplete (probe images have similar results) for the left and right ocular regions. However, unlike in Figure 6.5, here we *only* cropped the collected images and did *not* perform any alignment. From Figures 6.5 and 6.6 it is clear that the right ocular images from the FOCS dataset inherently contain more incomplete areas than the left ocular images, and further, the process of aligning the images may result in additional information being lost.

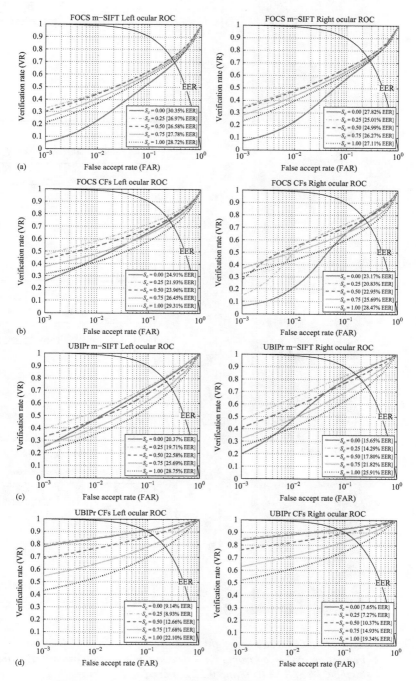

*Figure 6.4* CF and m-SIFT ROC curves over varying periocular image representations from the FOCS and UBIPr datasets. Corresponding EERs are displayed in the legend for each image representation. (a) FOCS m-SIFT ROCs, (b) FOCS CF ROCs, (c) UBIPr m-SIFT ROCs and (d) UBIPr CF ROCs

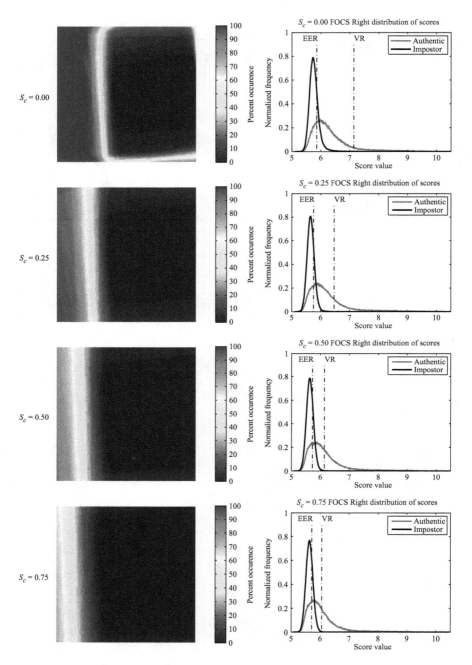

*Figure 6.5    Examining CF results on right ocular images from the FOCS dataset.
Images are collected from impostor comparisons that are misclassified
by the CFs at $S_c = 0.00$ and 0.25 but are correctly classified at
$S_c = 0.50$ and 0.75. Left column: The frequency (areas of more
occurrence are on the left side of the left column figures, while areas of
lower occurrence are on the right side) of where each corresponding
gallery image is incomplete due to alignment (probe images have
similar results). Right column: Score distributions from the full set of
comparisons with labelled EER and VR thresholds*

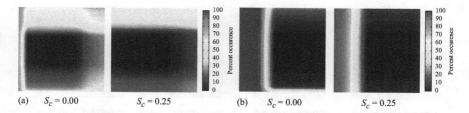

(a)    $S_c = 0.00$         $S_c = 0.25$        (b)    $S_c = 0.00$         $S_c = 0.25$

*Figure 6.6   Frequency of where each misclassified gallery image (cropped at*
*$S_c = 0.00$ and $S_c = 0.25$, but not aligned) is incomplete for the left and*
*right ocular regions from the FOCS dataset (probe images have similar*
*results and are thus omitted). (a) Left ocular and (b) right ocular*

The phenomenon highlights one of the difficulties of comparing images obtained from poor acquisition environments. *Challenges such as scale and rotational changes occurring in often incomplete images can cause a large amount of image data to be lost during alignment.* Since the design of the m-SIFT algorithm only examines areas which are complete (in addition to providing robustness to transformations like scale and rotation), it can compensate for missing data (provided much of the discriminative texture is still available in both images). However, too much data was missing in the FOCS right periocular images such that the correlation in the remaining patches was not sufficient, and thus, resulting in a considerable decrease in the VR.

Lastly, we discuss the average patch scores from the CFs as well as the recorded locations of any keypoints remaining after applying m-SIFT spatial constraints from all of the authentic and impostor comparisons over each cropping region. Shown within Figures 6.7–6.11, each set of sub-region scores is scaled to allow for an easier visual comparison between neighbouring regions with each sub-region score set also overlaid onto the respective average periocular region.

The left and right ocular image sub-regions from CFs exhibit the highest scores for both authentic and impostor comparisons across the eye itself with consistently smaller scores on the cheek and brow areas when present. The only exceptions are the impostor scores from $S_c = 0.00$ in Figure 6.7, where scores are significantly larger along the left edge of the image (further substantiating the previously discussed effects from missing image data). The m-SIFT keypoint graphs reflect a similar result with best locations often being found on and around the eye itself than in the surrounding areas. However graphs from $S_c = 0.00$ and 0.25 have several keypoints outside of the eye (skin texture), a phenomenon that is not as frequently observed in the other image representations. Specifically in both CF and m-SIFT sub-region results, the area near the inner tear duct of the eye, the pupil/iris and overall eye shape appear to contribute the most to the final comparison scores for both authentic and impostor comparisons.

Based on high sub-region score locations along with the corresponding VRs and EERs for the image representations tested, the best cropping for NIR images requires

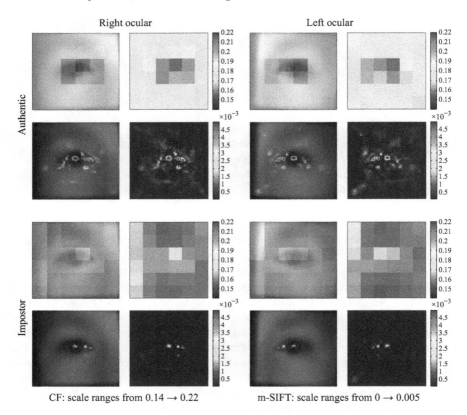

*Figure 6.7   Average authentic (top two rows) and impostor (bottom two rows) patch
scores from CFs (rows one and three) and remaining keypoint locations
after applying m-SIFT constraints (rows two and four) for $S_c = 0.00$ on
the FOCS dataset*

a close cut around the eye itself with a small amount area just outside of eye corners
($S_c \geq 0.25$). Including brow information has also shown to be useful ($S_c < 0.75$),
though cheek and skin texture may not be completely necessary (due to increased
skin reflectance at smaller wavelengths).

## UBIPr (VL wavelength)

- *m-SIFT*: Best results are obtained using image representations generated from
  $S_c = 0.00$ and 0.25. However, the VRs between $S_c = 0.00$ and 0.25 contain a
  ~8%–10% difference favouring $S_c = 0.25$.
- *CFs*: Best results are also obtained using image representations from $S_c = 0.00$
  and 0.25.

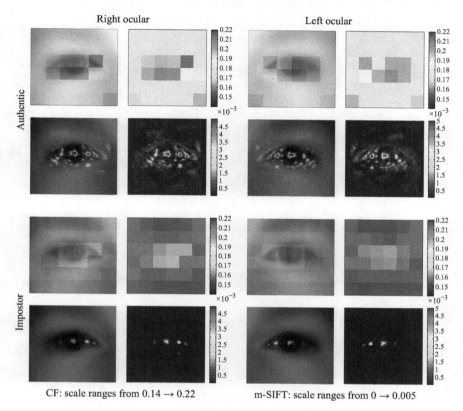

*Figure 6.8* *Average authentic (top two rows) and impostor (bottom two rows) patch scores from CFs (rows one and three) and remaining keypoint locations after applying m-SIFT constraints (rows two and four) for $S_c = 0.25$ on the FOCS dataset*

We briefly discuss the cause of the difference in VR from m-SIFT by examining comparisons misclassified by m-SIFT at $S_c = 0.00$ but are correctly classified at $S_c = 0.25$. From the score sets we found that ~75%–80% of the false negatives (FNs) are shared between $S_c = 0.00$ and $S_c = 0.25$, but only ~5%–7% of the false positives (FPs) are shared. Accordingly the VR discrepancy is largely caused by an increased robustness to impostor comparisons.

Recall that the m-SIFT algorithm applies spatial constraints to reduce the initial set of keypoints used for comparison based on the assumption that authentic matches will have more similar keypoints than impostors. Examining the sets further we found that ~25% of the FPs at $S_c = 0.00$ were correctly classified at $S_c = 0.25$ due to having *no* remaining keypoints after applying the spatial constraints (thus assigning the lowest score possible) with the other FPs showing at least some reduction in the number

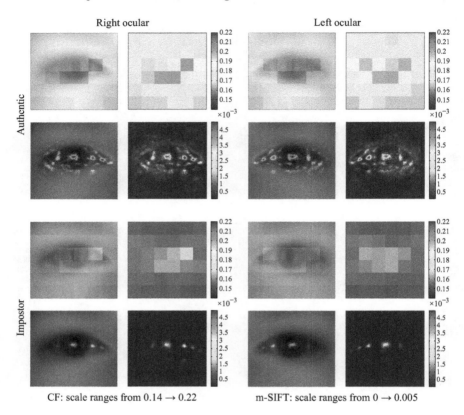

*Figure 6.9    Average authentic (top two rows) and impostor (bottom two rows) patch scores from CFs (rows one and three) and remaining keypoint locations after applying m-SIFT constraints (rows two and four) for $S_c = 0.50$ on the FOCS dataset*

of keypoints. When including the number of keypoints over all cropping sizes, a trend emerges, in which increasing the cropping region also increases the standard deviation of the number of remaining keypoints (while the mean is roughly constant). The dispersion indicates that as the cropping increases, the stability of the algorithm decreases. That is, in comparisons at a larger cropping region, the distribution of the number of remaining keypoints is becoming increasingly bimodal. In this case, the effect benefited the algorithm between regions cropped at $S_c = 0.00$ and $0.25$.

Next we discuss the sub-region scores shown in Figures 6.12–6.16 (the periocular images labelled 'left' in the UBIPr dataset are the equivalent of the FOCS 'right' ocular images and vice-versa). Unlike the NIR data from the FOCS set, the UBIPr sub-region scores for both methods are able to help explain the value of using the surrounding skin texture when performing periocular recognition with VL images.

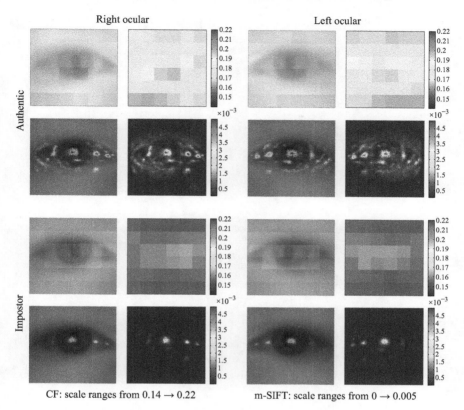

*Figure 6.10*   *Average authentic (top two rows) and impostor (bottom two rows)*
*patch scores from CFs (rows one and three) and remaining keypoint*
*locations after applying m-SIFT constraints (rows two and four) for*
$S_c = 0.75$ *on the FOCS dataset*

The authentic CF sub-regions distinctly favour the eyebrow and cheek regions (where
they exist) while the impostors are focused solely on the eye region for all cropped
representations. In addition, while m-SIFT keypoint locations primarily congregate
across the eye, the graphs show that there is also a substantial presence of keypoints
outside of the eye. Though similar to the NIR results, the UBIPr impostor keypoints
focus solely on the eye itself.

The sub-region scores for both m-SIFT and CFs largely indicate that as more of
the periocular region is cropped, the less distinguishable the individual sub-regions
become. With CFs, the variance of the patch score values decrease, and with m-SIFT
the density of the keypoint locations becomes increasingly less focused. Based on high
sub-region score locations along with the corresponding VRs and EERs for the image
representations tested, the best cropping for VL images requires a large cut around

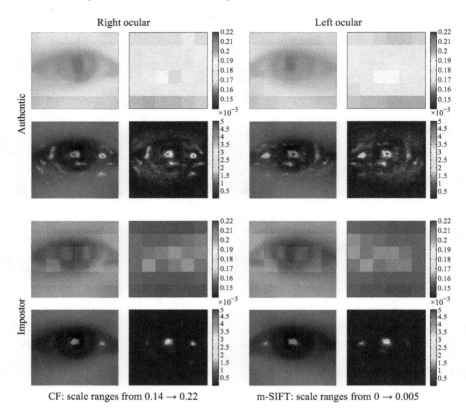

*Figure 6.11    Average authentic (top two rows) and impostor (bottom two rows)
patch scores from CFs (rows one and three) and remaining keypoint
locations after applying m-SIFT constraints (rows two and four) for
$S_c = 1.00$ on the FOCS dataset*

the eye that includes the eyebrow and surrounding skin texture (cheek) for necessary discriminative information ($S_c \leq 0.25$). This is distinctly different from what was seen for NIR images which favour a smaller region; and suggests that the shape of the eye is more important, specifically the inner tear duct with less contribution from the brow.

Our results are consistent with similar work performed by Alonso-Fernandez and Bigun [23], which also evaluated sub-regions of NIR and VL periocular images, but from the BioSec [30] and MobBIO [31] datasets, respectively. The employed method first divides each image into non-overlapping rectangular patches, then samples the local power spectrum from the centre points of each patch (after applying a bank of Gabor filters) and computes the $\chi^2$ distance between corresponding points of the grid where the final comparison score is the average of each 'sub-distance'. Using a

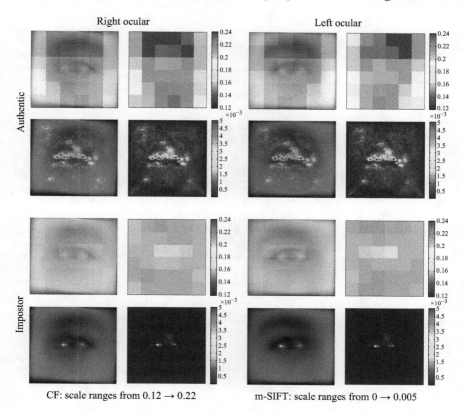

Right ocular          Left ocular

Authentic

Impostor

CF: scale ranges from 0.12 → 0.22          m-SIFT: scale ranges from 0 → 0.005

*Figure 6.12   Average authentic (top two rows) and impostor (bottom two rows)
patch scores from CFs (rows one and three) and remaining keypoint
locations after applying m-SIFT constraints (rows two and four) for
$S_c = 0.00$ on the UBIPr dataset*

verification scenario, the authors evaluated varying combinations of grid points and
selected the best by use of Sequential Forward Floating Selection (SFFS) [32].

While Alonso-Fernandez and Bigun [23] do not examine the effects of varying the
periocular image representation, they identify sub-regions that consistently perform
best in their experiments. However, since the images from both datasets were collected with the sole intent of evaluating iris recognition (thus many of the periocular
images do not contain eyebrows or sufficient surrounding skin texture), the subsequent analysis conforms most directly with periocular representations that enclose
only the eye (e.g., $S_c \geq 0.75$). From the tested NIR images in the BioSec dataset,
selected points from SFFS include the iris, sclera and eyelashes, with surrounding
skin being discarded. Meanwhile, from the tested VL images in the MobBIO dataset,
selected points include surrounding skin, sclera and eyelashes. Alonso-Fernandez and

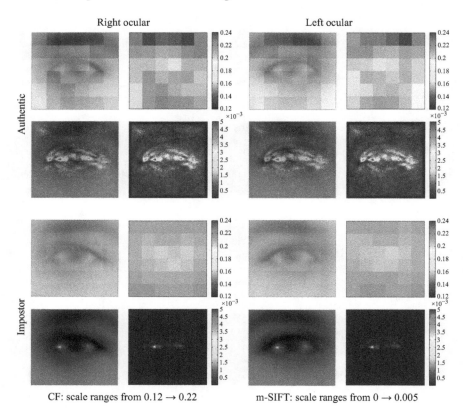

*Figure 6.13    Average authentic (top two rows) and impostor (bottom two rows)
patch scores from CFs (rows one and three) and remaining keypoint
locations after applying m-SIFT constraints (rows two and four) for
$S_c = 0.25$ on the UBIPr dataset*

Bigun [23] note that the appropriate selection of grid points can consequently reduce the EER by more than 25%.

## 6.4    Summary

Comparatively little work has been done to examine the periocular modality as a biometric, but some effort has been devoted to determining some of the fundamental recommendations required for successful matching. For instance, Park *et al.* [19] demonstrated that masking the iris and/or eye, such as was done in early works [12,22], will degrade system performance. We evaluate the conclusions of previous work in conjunction with what is presented in this chapter to answer several questions.

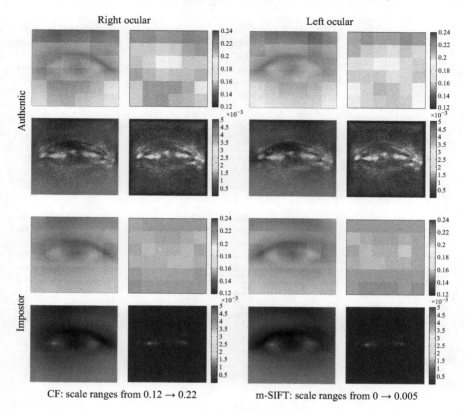

Right ocular                         Left ocular

CF: scale ranges from 0.12 → 0.22       m-SIFT: scale ranges from 0 → 0.005

*Figure 6.14*   *Average authentic (top two rows) and impostor (bottom two rows) patch scores from CFs (rows one and three) and remaining keypoint locations after applying m-SIFT constraints (rows two and four) for $S_c = 0.50$ on the UBIPr dataset*

How can we define what constitutes the best periocular region for biometric recognition?

In summary: *It depends on the sensor used to acquire the image. VL images require a larger region which needs to encompass the eyebrow and a significant amount of skin texture, while NIR images require a smaller region also encompassing the eyebrow, but with little to no skin texture.*

– Previous efforts [18,19,22,24,25] examined a single definition of the periocular region in VL images to identify areas that contain significant discriminative textural information in an attempt to obtain an answer to this question. Using sub-region effects over varying cropping sizes, we found that

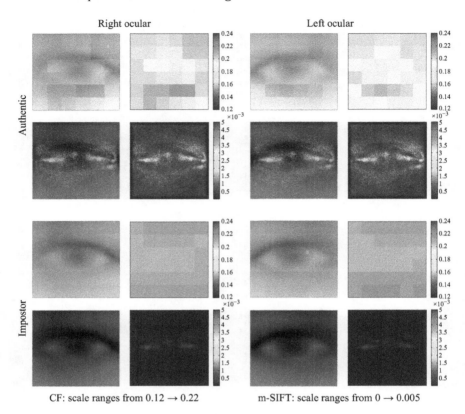

*Figure 6.15    Average authentic (top two rows) and impostor (bottom two rows)
patch scores from CFs (rows one and three) and remaining keypoint
locations after applying m-SIFT constraints (rows two and four) for
$S_c = 0.75$ on the UBIPr dataset*

in general, the eyebrow and skin texture play an important role in recognition performance and thus requires an image representation that encompasses significant portions of both.

—   Comparatively much less work [21,23,24] has been done to examine NIR images (again, only evaluating a single image representation). Our results over varying definitions of a periocular region also found that cheek and skin texture information is the least helpful in recognition, while the most important features are eye shape and inner tear duct. Thus an image representation that is more tightly fitted around the eye itself is best. However, since the eyebrow contribution was shown to be less important, but not negligible, an appropriate image representation will also include the brow.

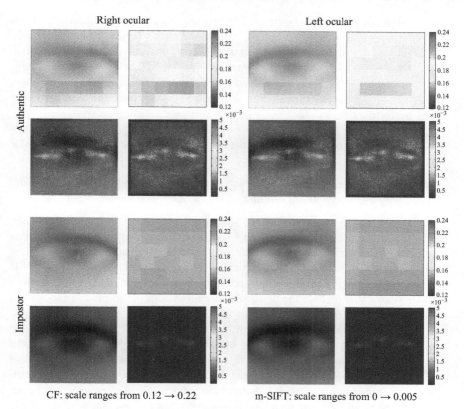

CF: scale ranges from 0.12 → 0.22          m-SIFT: scale ranges from 0 → 0.005

*Figure 6.16    Average authentic (top two rows) and impostor (bottom two rows)
patch scores from CFs (rows one and three) and remaining keypoint
locations after applying m-SIFT constraints (rows two and four) for
$S_c = 1.00$ on the UBIPr dataset*

Face recognition traditionally uses visible light (VL) images and iris recognition uses near-infrared (NIR) images, but which is better for periocular?

In summary: *Visible light images are better due to the availability of more discriminative texture.*

–   Hollingsworth *et al.* [21] made the initial conclusion using human studies that VL reveals more discriminatory information for improved recognition performance and that moving from VL to NIR in controlled acquisition environments will decrease recognition rates. This is due to the saturation in the eye and skin regions from high gain settings of NIR cameras (to account for the skin and sclera being less reflective at longer wavelengths [33]) hiding potential discriminative information as both human and computer performance increases with VL illumination.

- From our experiments over challenging images of unconstrained users from NIR and VL sensors, we have also found that VL images are preferable for periocular recognition. In particular, the increased skin reflectance at smaller wavelengths (i.e., the wavelengths that NIR sensors operate at) significantly reduces the amount of discriminatory information in the image, leaving the cheeks/surrounding skin texture to contribute the least to system performance.

How should we best perform periocular recognition when we must use NIR images?

In summary: *Reduce the amount of cheek/skin texture in the image while leveraging the discriminative nature of the eye itself (shape, tear duct, etc.).*

- Unfortunately many real-world scenarios employ NIR sensors, e.g., when extracting the periocular region from surveillance images of faces, such as found using a CCTV (Closed Circuit Television), or using images originally captured for iris recognition (which is traditionally best performed using NIR images [34,35]). In addition, there is significant amount of literature showing systems that include the periocular region in addition to the full face or iris can improve overall performance [9,10,11,12,13,14].
- In these situations, Hollingsworth *et al.* [21] found that humans use the eyelashes, eyelid, tear duct and eye shape as discriminating features from cooperative subjects. We found that computer performance in uncooperative subjects also indicated areas near the inner tear duct of the eye and overall eye shape appear to contribute the most recognition performance. However, we failed to find evidence identifying eyelashes and eyelids as substantially effective, instead identifying the pupil/iris and brow information as useful. In all cases, cheek and skin texture were found to be not completely necessary. Thus, as shown in [23,36], by leveraging these meaningful regions it is possible to increase system performance.

Does the periocular region need to be aligned to effectively compare a pair of periocular images?

In summary: *In unconstrained acquisition environments alignment alone may hurt performance in some cases, however, it is necessary in order to properly measure distortions.*

- While there has been much work on periocular recognition, most assume that periocular patterns do not undergo local in-plane deformations. We showed that periocular regions indeed undergo local in-plane deformations, especially under challenging imaging conditions, and accounting for these local distortions in the image comparison process can result in significant improvements in performance.
- We also showed that aligning the periocular region can sometimes hurt performance when comparing to the nearest neighbour despite the

recommendations of Padole and Proenca [25], because aligning incomplete periocular images can introduce additional incomplete regions within the image. However, the benefits of aligning an image during distortion tolerant matching outweigh any possible losses from alignment itself.

# References

[1]    E. Sariyanidi, H. Gunes, and A. Cavallaro. "Automatic analysis of facial affect: A survey of registration, representation, and recognition," *IEEE Transactions on Pattern Analysis and Machine Intelligence*, 37(6):1113–1133, Jun. 2015.

[2]    P. Grother, E. Tabassi, G. W. Quinn, and W. Salamon. "IREX I: Performance of iris recognition algorithms on standard images," Technical report, NIST Interagency Report 7629, Oct. 2009.

[3]    E. Tabassi, P. Grother, and W. Salamon. "IREX II – Iris Quality Calibration and Evaluation: Performance of iris image quality assessment algorithms," Technical report, NIST Interagency Report 7820, Sept. 2011.

[4]    G. W. Quinn, J. Matey, E. Tabassi, and P. Grother. "IREX V: Guidance for iris image collection," Technical report, NIST Interagency Report 8013, July 2014.

[5]    K. Bonnen, B. F. Klare, and A. K. Jain. "Component-based representation in automated face recognition," *IEEE Transactions on Information Forensics and Security*, 8:239–253, Jan. 2013.

[6]    J. G. Daugman. "How iris recognition works," *IEEE Transactions on Circuits and Systems for Video Technology*, 14(1):21–30, Jan. 2004.

[7]    V. N. Boddeti, J. M. Smereka, and B. V. K. Vijaya Kumar. "A comparative evaluation of iris and ocular recognition methods on challenging ocular images," in *International Joint Conference on Biometrics*. IEEE, Piscataway, NJ, Oct. 2011.

[8]    F. Juefei-Xu, K. Luu, M. Savvides, T. D. Bui, and C. Y. Suen. "Investigating age invariant face recognition based on periocular biometrics," in *International Joint Conference on Biometrics*, pp. 1–7. IEEE, Piscataway, NJ, Oct. 2011.

[9]    C.-W. Tan and A. Kumar. "Human identification from at-a-distance images by simultaneously exploiting iris and periocular features," in *International Conference on Pattern Recognition,"* pp. 553–556. IEEE, Piscataway, NJ, Nov. 2012.

[10]    A. Joshi, A. K. Gangwar, and Z. Saquib. "Person recognition based on fusion of iris and periocular biometrics," in *International Conference on Hybrid Intelligent Systems*, pp. 57–62. IEEE, Piscataway, NJ, Dec. 2012.

[11]    V. Gottemukkula, S. K. Saripalle, S. P. Tankasala, R. Derakhshani, R. Pasula, and A. Ross. "Fusing iris and conjunctival vasculature: Ocular biometrics in the visible spectrum," in *IEEE Conference on Technologies for Homeland Security*, pp. 150–155, Nov. 2012.

[12]    D. L. Woodard, S. Pundlik, P. Miller, R. Jillela, and A. Ross. "On the fusion of periocular and iris biometrics in non-ideal imagery," in *International*

*Conference on Pattern Recognition*, pp. 201–204. IEEE, Piscataway, NJ, Aug. 2010.

[13] H. S. Bhatt, S. Bharadwaj, R. Singh, and M. Vatsa. "Recognizing surgically altered face images using multiobjective evolutionary algorithm," *IEEE Transactions on Information Forensics and Security*, 8(1):89–100, Jan. 2013.

[14] R. Jillela and A. Ross. "Mitigating effects of plastic surgery: Fusing face and ocular biometrics," in *IEEE International Conference on Biometrics: Theory, Applications and Systems*, pp. 402–411, Sep. 2012.

[15] S. Bharadwaj, H. S. Bhatt, M. Vatsa, and R. Singh. "Periocular biometrics: When iris recognition fails," in *IEEE International Conference on Biometrics: Theory Applications and Systems*, pp. 1–6, Sep. 2010.

[16] F. Juefei-Xu and M. Savvides. "Unconstrained periocular biometric acquisition and recognition using COTS PTZ camera for uncooperative and non-cooperative subjects," in *IEEE Workshop on Applications of Computer Vision*, pp. 201–208, Jan. 2012.

[17] P. E. Miller, J. R. Lyle, S. J. Pundlik, and D. L. Woodard. "Performance evaluation of local appearance based periocular recognition," in *IEEE International Conference on Biometrics: Theory Applications and Systems*, pp. 1–6, Sep. 2010.

[18] J. Merkow, B. Jou, and M. Savvides. "An exploration of gender identification using only the periocular region," in *IEEE International Conference on Biometrics: Theory Applications and Systems*, pp. 1–5, Sep. 2010.

[19] U. Park, R. R. Jillela, A. Ross, and A. K. Jain. "Periocular biometrics in the visible spectrum," *IEEE Transactions on Information Forensics and Security*, 6(1):96–106, Mar. 2011.

[20] J. M. Smereka and B. V. K. Vijaya Kumar. "What is a 'good' periocular region for recognition?" in *IEEE Conference on Computer Vision and Pattern Recognition Workshops*, pp. 117–124, June 2013.

[21] K. P. Hollingsworth, S. S. Darnell, P. E. Miller, D. L. Woodard, K. W. Bowyer, and P. J. Flynn. "Human and machine performance on periocular biometrics under near-infrared light and visible light," *IEEE Transactions on Information Forensics and Security*, 7(2):588–601, Apr. 2012.

[22] D. L. Woodard, S. J. Pundlik, J. R. Lyle, and P. E. Miller. "Periocular region appearance cues for biometric identification," in *IEEE Conference on Computer Vision and Pattern Recognition Workshops*, pp. 162–169, Jun. 2010.

[23] F. Alonso-Fernandez and J. Bigun. "Best regions for periocular recognition with NIR and visible images," in *IEEE International Conference on Image Processing*, pp. 4987–4991, Oct. 2014.

[24] B.-S. Oh, K. Oh, and K.-A. Toh. "On projection-based methods for periocular identity verification," in *IEEE Conference on Industrial Electronics and Applications*, pp. 871–876, Jul. 2012.

[25] C. N. Padole and H. Proenca. "Periocular recognition: Analysis of performance degradation factors," in *IAPR International Conference on Biometrics*, pp. 439–445. IEEE, Piscataway, NJ, Mar. 2012.

[26] A. Ross, R. Jillela, J. M. Smereka *et al.* "Matching highly non-ideal ocular images: An information fusion approach," in *IAPR International Conference on Biometrics.* IEEE, Piscataway, NJ, March 2012.

[27] D. G. Lowe. "Distinctive image features from scale-invariant keypoints," *International Journal of Computer Vision*, 60(2):91–110, Nov. 2004.

[28] B. V. K. Vijaya Kumar, A. Mahalanobis, and R. Juday. *Correlation Pattern Recognition.* New York: Cambridge University Press, 2005.

[29] NIST. Face and Ocular Challenge Series (FOCS). Online. http://www.nist.gov/itl/iad/ig/focs.cfm.

[30] J. Fierrez, J. Ortega-Garcia, D. T. Toledano, and J. Gonzalez-Rodriguez. "BioSec baseline corpus: A multimodal biometric database," *Pattern Recognition*, 40(4):1389–1392, Apr. 2007.

[31] A. F. Sequeira, J. C. Monteiro, A. Rebelo, and H. P. Oliveira. "MobBIO: A multimodal database captured with a portable handheld device," *International Conference on Computer Vision Theory and Applications*, 3:133–139, Jan. 2014.

[32] P. Pudil, J. Novovičová, and J. Kittler. "Floating search methods in feature selection," *Pattern Recognition Letters*, 15(11):1119–1125, Nov. 1994.

[33] J. G. Daugman. Absorption spectrum of melanin. Online. http://www.cl.cam.ac.uk/ jgd1000/melanin.html, July 2009.

[34] G. Molenberghs and Y. E. Du. "Review of iris recognition: Cameras, systems, and their applications," *Sensor Review*, 26(1):66–69, Jan. 2006.

[35] C. K. Boyce. *Multispectral Iris Recognition Analysis: Techniques and Evaluation.* PhD thesis, West Virginia University, Dec. 2006.

[36] J. M. Smereka, B. V. K. Vijaya Kumar, and A. Rodriguez. "Selecting discriminative regions for periocular verification," in *IEEE International Conference on Identity, Security and Behavior Analysis*, pp. 1–8, Feb. 2016.

*Chapter 7*

# Light field cameras for presentation attack resistant robust biometric ocular system

*Raghavendra Ramachandra, Kiran B. Raja,*
*and Christoph Busch*

Over the last decade one can observe an increasing interest in visible spectrum iris recognition. Current iris recognition systems that operate in visible spectrum suffer from many factors such as out-of-focus images, motion blurred images and images with low texture visibility. Specifically the out-of-focus images are a consequence of the limited depth-of-field property of a conventional 2D camera. In this chapter, we present a new paradigm for solving the problem of out-of-focus imaging in visible spectrum iris recognition by employing a Light Field Camera (LFC). The ability of LFC, to refocus the parts of interest in a scene after the image capture, is explored to solve the out-of-focus problem in iris imaging in visible spectrum. This chapter also extends the use of LFC for periocular recognition and thereby presents an ocular biometric system using LFC in visible spectrum. Further, the chapter also presents the weighted score level fusion of iris and periocular characteristics to improve the overall performance.

A major threat to the visible spectrum ocular recognition systems arises from the presentation attacks against the capture device. The availability of high quality images in social media can be exploited to present ocular samples to a visible spectrum ocular biometric system with the intent to gain unauthorized access. Thus, the inherent property of light-field imaging to provide multiple depth images can be used to detect such presentation attacks. In this chapter, we also present an intuitive and simple way of employing the depth or focus images to detect presentation attacks by estimating the energy values.

In this chapter, we first demonstrate the applicability of LFC for ocular recognition in visible spectrum over using a conventional 2D camera. In a set of experiments, we present the improved verification results on a dataset of 42 subjects collected from both conventional 2D camera and LFC camera. In a second set of experiments, we demonstrate the applicability of LFC cameras to detect presentation attacks (electronic and printed attacks) by exploring the depth images from LFC. In the end, we establish the usability of LFC for ocular biometrics with good verification purpose and as a method, which is also robust against the printed and electronic attacks.

## 7.1 Introduction

Iris has been a preferred biometric characteristics owing to the high degree of genuine accepts with very low degree of false rejects. Widely deployed commercial systems use Near-Infra-Red (NIR) illumination to reveal the textural characteristics of the iris even for the subjects with high pigment (melanin) density [1]. Motivated by the success of iris recognition in NIR spectrum, recent works have investigated the feasibility of iris recognition in visible spectrum [2,3]. The biggest hurdle in achieving reliable biometric performance for iris recognition in visible spectrum is fully based on the ability to resolve the iris texture which often is not effective for subjects with highly pigmented iris patterns [3]. Intuitively, this problem can, for visible spectrum images, be handled by employing the periocular region, which is inherently captured during the acquisition of iris images [4]. While the periocular region can contribute complementary information, a higher biometric performance of a system can be obtained by fusing the two characteristics [5,6].

Employing both the iris and periocular characteristics in visible spectrum helps a biometric system to operate a walk through mode (a.k.a "on-the-fly" or "on-the-move") and thereby can reduce the interaction time of data subjects with the biometric system. The benefit of using the on-the-move paradigm comes with an additional demerit of out-of-focus imaging. Due to non-standard capture distance under at-a-distance capture process, the images obtained may result in significantly degraded image quality consisting of out-of-focus images, motion blurred images. The conventional imaging systems can capture the scene only in a fixed focus distance due to the operational principles of 2D imaging and the limited field-of-view of the camera. Thus, the images cannot be focused after the acquisition and the only way to capture sharply focused periocular-iris images is by adjusting the focus during the acquisition, which is not feasible for an on-the-move capture process.

Numerous approaches have been proposed to address the problem of out-of-focus images. In [7,8], extended depth of field is achieved using a wavefront coding in a fixed nature. Here, the wavefront coding technique has been applied on the captured image to improve the overall sharpness of the image. In [9], the spectral reflection generated by the IR-LED illuminator is used to focus/zoom the lens to achieve the best focus. In [10], auto focus property is explored to automatically align the focus of the camera to capture the best focus image. In [1], the 2D Fourier spectrum of the captured image sequence is analysed to select the best focused frame in the video by maximizing the power in both middle and upper frequencies. Recently, a video-based hyper-focal imaging scheme is proposed to capture the Iris in visible spectrum [11]. Here, video frames are captured by varying the focus such that each video frame is focused differently. Finally, these multiple focus frames are fused to form a single improved focus image.

Thus, from the existing schemes it can be observed that most of them are proposed for the near infrared sensor [1,7,8,9,10]. Further these schemes are more concentrated in extending the depth of focus, which in-turn results in images of low dynamic range and low Signal-to-Noise-Ratio (SNR). Finally, the use of video frames to generate multiple focus images requires accurate variation of the focus during acquisition and also demands very high computational power.

In this chapter, we present an alternative approach to address the problem of focus in obtaining iris and periocular images by employing the light-field (or plenoptic) camera [12,13]. The Light Field Camera (LFC) captures the image by sampling the 4-Dimensional (4D) light field on its sensor in a single photographic exposure by inserting a micro-lens array [12] or a pin-hole array [14] or masks [15] between the camera sensor and the main lens. Thus, the presence of the micro lens (or array of pin-holes or masks) allows one to measure the total amount of light intensity deposited on the sensor along with the direction of each ray from incoming light. Finally, by re-sorting the measured rays of light to where they would have terminated one can obtain a number of sharp images focused at different depth [12]. Thus, the light-field camera exhibits interesting features as compared to a conventional camera which can be listed as

- generating images at different focus (or depth) in one shot,
- it is not required to adjust the lens to set the focus on the object in a scene,
- portable and hand-held,
- real-time exposure.

Thus by employing the LFC, one can obtain a set of images that can be used to perform: (1) refocus on specific parts of a scene, (2) all-in-focus image with all objects in scene to be in-focus, (3) depth estimation and (4) generate synthetic aperture.

The above-mentioned properties of the light-field camera can be used to obtain a set of depth (multi-focus) images from the iris and periocular characteristics to resolve the problem of out-of-focus imaging. The main objective of this chapter is to demonstrate the strengths of LFC imaging to address the problem of focus, especially for the iris capture process in visible spectrum. In addition, we also explore the feasibility of periocular biometrics [16] using LFC. Thus, in this chapter, we first present the comprehensive results of verification performance on the iris images obtained using a conventional 2D imaging sensor (SONY DSC S750 digital camera) and compare the results of verification performance obtained using iris images from a light-field camera developed by Lytro [17]. The database employed for the experiments in this chapter consist of 84 ocular images (iris and periocular) captured from 42 subjects collected using both SONY DSC S750 digital camera and Lytro light-field camera.

Along with the benefits such as on-the-move capture with no additional illumination, the visible spectrum ocular recognition also bears a new risk. As face images of a large proportion of the population are available over social media, such publicly available images can be used to print the ocular characteristics and present the printed artefact to the biometric capture device. The artefacts can also be presented using an electronic screen (e.g., a tablet, mobile phones or computer) to deceive the system. These kinds of attacks carried out at the sensor level are called presentation attacks or spoofing attacks. If such attacks are not detected, the security of the biometric system is compromised. Thus, it is essential to have a presentation attack detection (PAD) mechanism to detect such attacks. The feasibility of these attacks is acknowledged by the number of available research works [18–24] that strongly dictates the importance of detecting these attacks. Further, various iris PAD detection competitions [25,26] organized recently also show the importance of this problem.

Presentation Attack Detection (PAD) (or spoof detection or anti-spoofing) algorithms at the sensor level can be broadly classified into two groups [20]: (1) hardware based and (2) software based. The idea of the hardware-based approach is to include an additional hardware component as a supporting device with the actual iris sensor that can perform the PAD. In [23], the PAD is carried out by employing a multiple light source to measure the variation of reflectance between iris and sclera region. The attack detection is based on Oculomotor Plant Characteristics (OPC) to capture the saccadic eye movements using eye tracker [27]. A new hardware setup to capture a stereo image of visible spectrum iris to detect the attack especially by presenting a cosmetic contact lens is introduced in [24].

In the case of software-based approach, the presentation attacks are detected after the data was captured by the sensor. The majority of state-of-the-art schemes are especially devoted to near-infrared iris recognition. Most of these software-based PAD schemes are analysing the statistical characteristics of the observed iris pattern. These observed iris patterns are further quantified to reflect the quality of the captured image with the intent to possibly reveal the presence of a presentation attack. The potential of image quality assessment methods to reliably detect a presentation attack is presented in [20], in which 25 well-established image quality measures are successfully employed to detect the photo print attack. The use of the frequency spectrum [18], time–frequency analysis using wavelet packets [22], combination of different quality factors like contrast (both local and global), analysis of Purkinje reflection [28], analysis of pupil dilation [29] and statistical texture features obtained from the Grey-Level-Co-occurrence-Matrix [21] are also well explored.

In this chapter, we also present an alternative approach of using the inherent characteristics of the LFC to address the presentation attacks, which eliminates the need for special software-based attack detection mechanisms or the need for explicit additional hardware to detect the presentation attacks. We present an intuitive way of exploiting the multiple depth (focus) images rendered by the LFC to extract the nature of the presentation as bona-fide (live/normal) or artefact (or spoof) by measuring the variation of focus between multiple depth images rendered by the LFC. We employ a large-scale database comprised of both bona-fide and artefact samples of visible spectrum ocular images collected from 52 subjects constituting 104 unique ocular patterns. The artefacts are generated using both high-resolution image captures of the eye using Canon EOS 550D DSLR camera and also the all-in-focus image obtained using LFC camera. Through a series of experiments, we present the significance of using LFC for detecting presentation attacks and achieving at the same time a good biometric performance.

The key contributions of this chapter can be summarized as:

- Presenting the applicability of light-field camera to solve the out-of-focus imaging in visible spectrum ocular imaging.
- Validating the usefulness of employing a light-field camera as compared to a traditional 2D camera.
- Presenting a unique perspective to detect presentation attacks on ocular systems with no additional hardware.
- Presenting a simplistic method of exploring the variation of focus between multiple depth images rendered by the LFC to detect presentation attacks on visible spectrum ocular (iris and periocular) biometric system.
- Extensive experiments are presented to validate the PAD approach for iris and periocular characteristics employing LFC.

The rest of the chapter is organized as follows: Section 7.2 presents the brief introduction of the light-field imaging. Section 7.3 presents the approach of using the light-field camera for iris and periocular verification. We also present the set of experiments for demonstrating the superiority of LFC over conventional 2D cameras for visible spectrum ocular biometrics. In Section 7.4, we demonstrate the use of LFC for presentation attack detection with a simple approach. The section also lists the extensive set of experiments to validate the approach of detecting presentation attacks using LFC. Finally, Section 7.5 presents the concluding remarks of this chapter.

## 7.2 Light-field imaging

Each sample acquired using LFC results in a file that consists of set of images that are focused at different depths in the scene. These multi-focus images will have only one particular region or image area in focus. In addition to these multi-focus images, the raw file also hosts metadata that provide the information about the regions that are in focus in each of these multi-focus images. We then use this metadata to select the best focus region from each of these multi-focus images to obtain the all-in-focus image. Figure 7.1(a)–(d) shows the multi-focus images obtained in one shot using Lytro LFC and Figure 7.1(e) shows the all-in-focus image. The obtained all-in-focus image shows the rich features of iris textures when compared to multi-focus images. Figure 7.2 shows the sample images acquired using both conventional and Lytro LFC for the same subject. Here also one can acknowledge for the improved visible quality of the LFC imaging.

## 7.3 Scheme for iris and periocular verification

Figure 7.3 shows the block diagram for iris and periocular verification in visible spectrum. The core idea of the presented scheme is to explore both iris and periocular biometric samples captured using Lytro LFC. The framework can be considered in

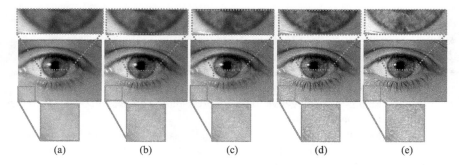

*Figure 7.1   Multi-focus image obtained using LFC. (a) Focus image 1, (b) focus image 2, (c) focus image 3, (d) focus image 4 and (e) all-in-focus image*

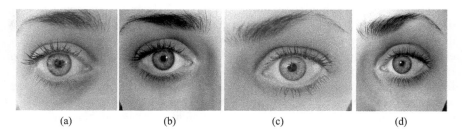

*Figure 7.2   Example of captured Iris samples: (a) and (c) conventional camera, (b) and (d) all-in-focus light-field image*

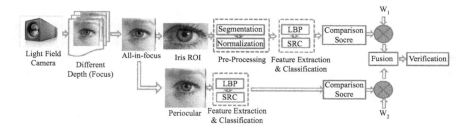

*Figure 7.3   The proposed scheme for biometric verification using iris and periocular characteristics with light-field camera*

three functional blocks, namely: (1) iris verification unit, (2) periocular verification unit and (3) comparison score level fusion unit.

## 7.3.1   Iris verification

The steps involved in the verification scheme based on the iris characteristics are explained in the following section.

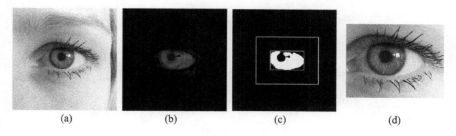

*Figure 7.4    Illustration of ROI extraction*

### 7.3.1.1   ROI extraction

The first step in iris recognition involves extracting the Region of Interest (ROI) consisting of only the eye region to reduce errors in the subsequent iris segmentation step. In this chapter, we employ a simple yet accurate eye region extraction scheme that is illustrated in Figure 7.4. The first step involves converting the *RGB* image into $YC_bC_r$ and then we obtain the difference between $C_b$ and $C_r$ so that information about the sclera region can be accurately determined (see Figure 7.4(b)). We then carry out the morphological operations to weed out small spurious noise and subsequently we determine the centroid to construct the bounding box by locating the outer edge of the obtained sclera region (see Figure 7.4(c) (inner rectangle)). Since this procedure will allow us to determine only the sclera part, we further extend the bounding box in both horizontal and vertical direction by 100 pixels (see Figure 7.4(c) (outer rectangle)) to complete an ROI that corresponds to the eye region (see Figure 7.4(d)).

### 7.3.1.2   Segmentation and normalization

In the experiments related to this chapter, we carry out the iris segmentation and normalization using OSIRIS v4.1 [30]. Figure 7.5 illustrates the qualitative results of the employed segmentation and normalization schemes of iris images obtained using light-field camera and conventional camera. These qualitative results further justify the effectiveness of the light-field imaging for accurate iris verification by overcoming the problem of focus.

### 7.3.1.3   Feature extraction and classification

We employ a feature extraction and classification scheme based on Local Binary Patterns (LBPs) [31] and Sparse Representation Classification (SRC) [32]. Since the LBP algorithm is well known to capture accurate texture information it appears as an elegant choice to accurately represent the iris texture features. The LBP operates by thresholding the differences of the centre value and its neighbourhood in a $3 \times 3$ grid window for one pixel. The output value for a pixel is regarded as an 8-bit binary number representing the pixel. The descriptor of the given image is constructed by the histogram of these binary numbers in the whole image [31]. We have employed the LBP operator with a radius of 2. This value is selected based on experimental trials.

In order to accurately classify the texture features, we employ the SRC by considering its exciting results in classifying biometric features [32]. Specifically, we carry out $L_1$ – minimization via $SPGL_1$ solver based on spectral gradient projection [32].

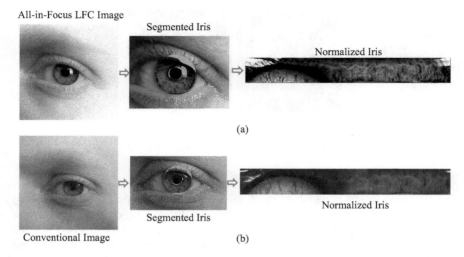

*Figure 7.5    Segmentation and normalization on: (a) LFC and (b) conventional image*

*Figure 7.6    Extracted ROI on: (a) Light-field camera and (b) conventional camera*

We then obtain the comparison scores that directly correspond to the residual errors obtained using SRC.

## 7.3.2   Periocular verification

Details of all steps involved in the verification scheme based on the periocular characteristics are presented in the following section.

Given the captured image, the first step in the periocular verification pipeline is to extract the ROI. We employ Haar cascade-based periocular detector to locate the exact region corresponding to the eye such that the extracted periocular ROI contains a small margin region of skin texture around the eye, which also includes eyebrows. Figure 7.6 shows the periocular ROI extracted for both light-field and conventional camera. In the next step, we carry out the feature extraction and classification by employing LBP-SRC combination as discussed in Section 7.3.1.3.

## 7.3.3 Combining iris and periocular characteristics

As both biometric characteristics iris and periocular can be used independently, we fuse the comparison scores of these two individual modalities using the weighted sum rule [33] to improve the verification accuracy. Let $CS_{Ir}$ and $CS_{Pe}$ corresponds to the comparison scores of iris and periocular modality, then the weighted sum fusion can be expressed as follows [33]:

$$Fusion = W_1 \times CS_{Ir} + W_2 \times CS_{Pe} \qquad (7.1)$$

where $W_1$ and $W_2$ represent the computed weights corresponding to iris and periocular modality, respectively. The weights in this chapter are computed according to the accuracy of the individual modalities as mentioned in [33], i.e., the characteristic providing higher verification accuracy gains more weight from the available weights.

## 7.3.4 Database, experiments and results

In this section, we present and discuss the database consisting of ocular images captured using light-field camera and the conventional camera. Further, the details of the experimental protocols are discussed along with the obtained results.

### 7.3.4.1 Light-field iris and periocular database

The whole database employed for experiments in this chapter is captured using both light-field (Lytro) and conventional (Sony DSC S750) camera. For each subject, there exists five samples corresponding to left and right eyes. Thus, each subject has 10 samples in total. Each sample is captured at a varying distance of 10–15 inches from the subject face in an uncontrolled lighting environment. The whole database consists of 42 subjects with both left iris and right iris resulting in a total of 84 unique iris patterns. It has to be noted that the database has a large number of light eyes (about 90%) and small number of dark eyes (amber eyes (about 7%) and brown eyes (about 3%)).

The Lytro light-field camera employed to capture the database has a resolution of 11 Mega rays with a working spatial resolution of $1,080 \times 1,080$ pixels, while the conventional camera is of 7 Mega pixels with a working resolution of $2,304 \times 3,072$. The conventional camera was set in an auto focus mode during the capture process. The database thus has 420 samples for each camera from 84 unique Iris patterns corresponding to 42 subjects.

### 7.3.4.2 Experiments and results

Since each iris pattern was captured with five different samples, we propose an evaluation protocol by partitioning four samples as the reference and remaining one sample as a probe. This reference and probe partition is repeated $m$ times ($m = 10$) using leave-one-out cross-validation. Thus, the experiments are carried out on all 10 different partitions and results are presented by taking the mean of all 10 different trials. Finally, results presented in this chapter are reported in terms of Equal Error Rate (ERR), which is defined as a point where the False Match Rate (FMR) [34] is equal to the False Non-Match Rate (FNMR) [34]. Thus the lower the values of the EER, the better is the biometric performance.

*Table 7.1    Quantitative evaluation of the proposed scheme*

| Modality | Methods | EER (%) | |
|---|---|---|---|
| | | **Light-field camera** | **Conventional camera** |
| Iris | Rathgeb *et al.* [35] | 22.21 | 27.38 |
| | Uhl *et al.* [36] | 12.39 | 18.24 |
| | Ko *et al.* [37] | 6.51 | 12.55 |
| | Ma *et al.* [38] | 6.27 | 12.07 |
| | Masek *et al.* [39] | 4.92 | 12.44 |
| | Daugman [1] | 3.61 | 8.53 |
| | LBP-SRC [6] | **1.20** | **8.24** |
| Periocular | LBP-SRC [6] | **12.04** | **16.21** |
| Fusion | LBP-SRC [6] | **0.81** | **7.45** |

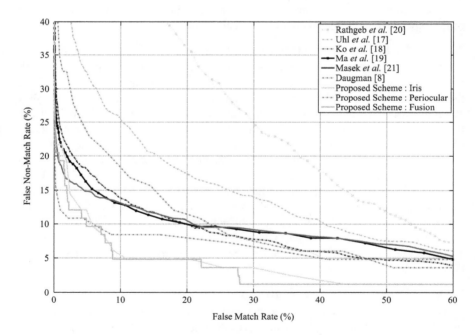

*Figure 7.7    Verification performance for conventional camera*

Table 7.1 shows the quantitative performance of the proposed light field based Iris and Periocular verification. Here, we present the quantitative performance of the light-field camera as compared with the conventional camera and also present the comprehensive comparison of the proposed LBP-SRC-based feature extraction

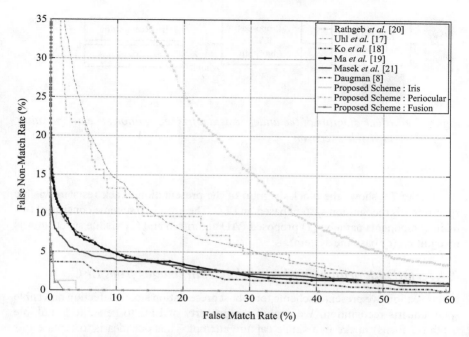

*Figure 7.8   Verification performance for light-field camera*

and classification scheme with well-known state-of-the-art iris verification schemes mentioned in [1,35–39]. Thus, from Table 7.1, the following facts can be observed:

- The improved performance of light-field camera recognition by an amount of 5% as compared to that of conventional camera.
- The superior performance of the proposed LBP-SRC based feature extraction and classification scheme with an improved accuracy of $EER = 2\%$ with light-field and about $EER = 3\%$ with conventional camera.
- Further, the fusion of iris and periocular using weighted sum rule shows the outstanding accuracy with $EER = 0.8\%$ with light-field camera and indicating the effectiveness of adopting the same for iris and periocular recognition.

Thus, from the above experiments, we found that the adoption of the light-field camera can improve the performance of the iris/periocular based biometric system in real-life surveillance scenarios.

## 7.4   Presentation attack resistant ocular biometrics in visible spectrum

Convinced by the benefit of light-field cameras for improving biometric systems, in this section, we present the applicability of light-field cameras for detecting presentation attacks.

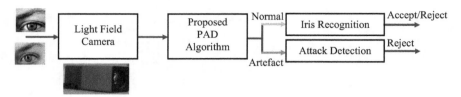

*Figure 7.9   Block diagram of the attack resistant iris recognition scheme operating in visible spectrum*

Figure 7.9 shows the block diagram of the presentation attack resistant visible spectrum iris recognition system based on LFC. The proposed scheme consists of two main components namely: (1) proposed PAD algorithm and (2) visible spectrum iris recognition (or baseline system).

## 7.4.1   PAD algorithm employing the depth images from LFC

In this section, we present a scheme for robust presentation attack detection on visible spectrum iris recognition. We exploit the property of LFC to generate a multiple depth (or focus) image in a single capture attempt. Thus our idea is to capture this variation of focus or depth exhibited between multiple depth images that can provide possible information about the presence of a presentation attack. This chapter not only focuses on cost-effective attacks, but also on the well adopted and accepted presentation attacks carried out using either a photo (printed using a laser printer) or an electronic screen (image displayed using a tablet). Hence, the artefacts that have been used in the experiments appear to represent an in-plane object when compared to the normal (or bona-fide) eye images. Therefore, the use of LFC as a biometric sensor to capture these visible spectrum eye artefacts will reflect less variation in depth when compared to that of bona-fide eye images. Thus, by accurately capturing these variations in terms of focus along multiple depth images rendered by the LFC reveals an attack presentation.

Figures 7.10 and 7.11 illustrate the multiple depth images rendered by the LFC on both normal (or real) and artefact (photo print) eye image, respectively. The artefact image shown in Figure 7.11 is generated from the normal (or real) image shown in Figure 7.10 by printing it on a good quality paper using a laser printer which in turn is captured using LFC. Here, one can observe two interesting facts: (1) a larger number of depth images can be obtained from the normal (or live captured) biometric characteristic capture process when compared with the artefact capture process. (2) Variation of the focus (or depth) information between multiple depth images is noted more with a normal (or real) image when compared to that of an artefact image. These observations further justify our idea of capturing the variation of focus between multiple depth images to reliably carry out the PAD.

Figure 7.12 shows the block diagram of the PAD algorithm based on capturing the variation of focus between multiple depth images. The presented scheme can be

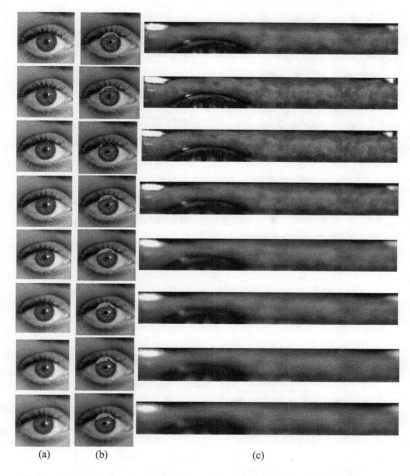

*Figure 7.10*   *Multiple depth (or focus) image rendered by the LFC on bona-fide (or normal) image capture. (a) Bona-fide eye image, (b) iris segmentation and (c) iris normalization*

structured in three important units, namely: (1) pre-processing and focus estimation, (2) estimation of variation in focus and (3) classification.

### 7.4.1.1   Pre-processing and focus estimation

Given the eye image $I$ captured using LFC a series of pre-processing steps are carried out to obtain the normalized iris image on which the focus is determined. Similar to experiments in previous sections, OSIRIS V 4.1 [30] is employed for iris segmentation and normalization. Figure 7.10(b) and (c) shows the qualitative results of iris segmentation and normalization obtained using OSIRIS V 4.1 on bona-fide eye

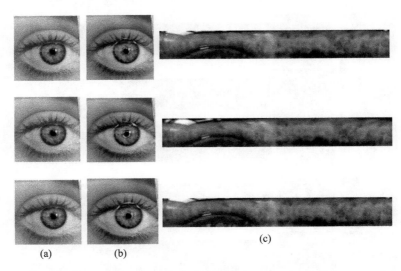

(a)        (b)                                           (c)

*Figure 7.11    Multiple depth (or focus) image rendered by the LFC on the artefact
image capture. (a) Artefact eye image, (b) iris segmentation and
(c) iris normalization*

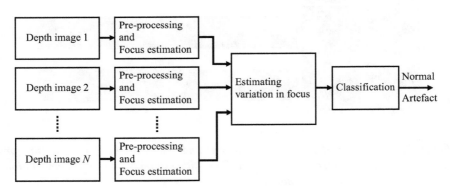

*Figure 7.12    Block diagram of the proposed PAD algorithm*

images. Similar results can also be visualized for the artefact eye image in Figure 7.11. The use of LFC to capture an eye image $I$ will result in multiple depth images such that $I = I_1, I_2, \ldots, I_k$, where $k$ corresponds to the number of depth images. We then perform the pre-processing (segmentation and normalization) on each of these depth images rendered by the LFC to obtain the corresponding normalized iris image as $I_{Nor} = \{I_{Nor1}, I_{Nor2}, \ldots, I_{Nork}\}$.

In the next step, we measure the focus by computing the energy using the Discrete Wavelet Transform (DWT) with a Haar mother wavelet [40]. We choose this approach by considering its many advantages that include: (1) monotonic quantitative value with respect to the focus. Thus a higher degree of focus in the image will result in a greater value of the energy. (2) This measure is robust to noise. (3) Low computational costs [41]. Given the normalized iris depth image $I_{Nor1}$, we first carry out the DWT to obtain the four different sub-images namely: approximate ($Ia_{Nor1}$), horizontal ($Ih_{Nor1}$), vertical ($Iv_{Nor1}$) and diagonal ($Id_{Nor1}$). We then compute the energy corresponding to three different DWT sub-images $Ih_{Nor1}$, $Iv_{Nor1}$ and $Id_{Nor1}$ as:

$$Eh_{Nor1} = \sum_{x=1}^{R} \sum_{y=1}^{C} (Ih_{Nor1}(x,y))^2 \tag{7.2}$$

$$Ev_{Nor1} = \sum_{x=1}^{R} \sum_{y=1}^{C} (Iv_{Nor1}(x,y))^2 \tag{7.3}$$

$$Ed_{Nor1} = \sum_{x=1}^{R} \sum_{y=1}^{C} (Id_{Nor1}(x,y))^2 \tag{7.4}$$

Finally, we compute the total energy that can measure the focus of the normalized image $I_{Nor1}$ as: $FE_{Nor1} = Eh_{Nor1} + Ev_{Nor1} + Ed_{Nor1}$. We repeat this procedure on the remaining normalized depth images to obtain their corresponding energy values as: $FE = \{FE_{Nor1}, FE_{Nor2}, \ldots, FE_{Nork}\}$.

### 7.4.1.2 Estimation of variation in focus

After computing the energy value corresponding to multiple depth images rendered by the LFC, we proceed further to compute the quantitative value that can reflect the variation in focus between these depth images. We first normalize the energy value corresponding to each depth image in $FE$ using the sigmoid normalization [42] to obtain the corresponding normalized energy as: $NFE = \{NFE_{Nor1}, NFE_{Nor2}, \ldots, NFE_{Nork}\}$. In the next step, we sort the normalized energy $NFE$ in the descending order as $S_N = \{S_{n1}, S_{n2}, \ldots, S_{nk}\}$. We then compute the variation in focus by obtaining the consecutive difference of these normalized and sorted energy value corresponding to the multiple depth images as follows:

$$VF = S_{n1} - S_{n2} - \cdots - S_{nk} \tag{7.5}$$

where $VF$ represents the qualitative value that can describe the variation in focus of the normalized iris image $I_{Nor}$.

### 7.4.1.3 Classification

We employ a simple yet accurate classifier based on an empirically determined threshold to classify the bona-fide presentations against artefact presentations. The value of $VF$ is compared against the threshold value to classify $I_{Nor}$ as either an artefact (or spoof) or a bona-fide (or normal) image. The value of threshold is obtained on the development database as explained in Section 7.4.4.

## 7.4.2   Visible spectrum iris recognition system

The baseline visible spectrum iris recognition system employed in this chapter is based on the Local Binary Patterns (LBP$_{3\times3}^{u2}$) and Sparse Representation Classifier (SRC). We employed this algorithm by considering its improved performance over existing state-of-the-art algorithms on the visible spectrum light field iris recognition [6]. Since, the LFC will render multiple depth images and employing all these images for recognition is not very useful from a computation perspective. In recent literature two ways in which the information from these depth images can be combined have been proposed namely: (1) All-in-focus image construction [6]. (2) Selection of the best focus image based on the highest energy [43]. Since the performance accuracy of these two schemes are quite similar, in this chapter, we adopt the second scheme which is based on selecting the best focus image corresponding to the highest energy. As we already computed the energy on each of these focus images rendered by the LFC (see Section 7.4.1) we can select the best focus image without any additional computational cost. Thus our final baseline visible spectrum iris recognition system is based on the best focus image on which the combination of LBP$_{3\times3}^{u2}$ and SRC is carried out to obtain the comparison score.

## 7.4.3   Presentation attack database

The database employed in the set of experiments in this section consists of both bona-fide (or normal) and artefact samples collected from 52 subjects resulting in 104 unique eye samples. In the following section, we present the data collection protocols adopted to collect both bona-fide and artefact iris samples. Further we also present the evaluation protocols that can be used to benchmark the effectiveness of the presented PAD algorithm on the visible spectrum iris recognition.

### 7.4.3.1   Bona-fide iris database

The bona-fide visible spectrum iris data in the employed database is acquired in natural (sun) and artificial (room) lighting. The data was collected over a period of 6 months in a single session. Figure 7.13(a) shows the bona-fide image acquisition setup. Each subject was asked to stand in front of the capture device at a distance of 9–15 inches from the camera. The data was then captured using two different cameras, namely: (1) Lytro LFC with 1.2 Megapixel. (2) Canon EOS 550D DSLR camera with 18.1 Megapixel. For each subject, we collected five different samples using both Lytro LFC and Canon DSLR camera independently. The key motivation behind using Canon DSLR camera is only to generate the high quality artefacts to carry out presentation attacks. The whole database consists of 520 (=104 × 5) high resolution DSLR eye images and 4,327 normal (or real) light field eye images (including multiple depth images).

### 7.4.3.2   Artefact iris database

The artefact iris database is captured using only Lytro LFC by reflecting a similar acquisition conditions that was followed while capturing the bona-fide iris image data set. Figure 7.13(b) illustrates the collection of artefact images captured by the

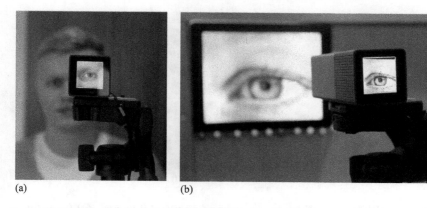

(a)                (b)

*Figure 7.13*    *Illustration of data collection set up. (a) Bona-fide (or normal) image acquisition setup and (b) artefact image acquisition setup*

electronic screen attack. The artefacts were generated using both high resolution DSLR camera samples as well as the images captured using light-field camera.

For high-resolution samples captured using the DSLR camera, the artefacts are generated using three different methods, namely: (1) *Photo print artefact:* Here, every high quality sample captured using DSLR camera is first printed on a good quality A4 paper using RICOH ATICIO MP C4502 laser printer and then recaptured using Lytro LFC. This sub-corpus of attack presentations has 1,122 light-field samples (each consisting of multiple depth images) corresponding to 520 high resolution DSLR samples. (2) *Screen artefact using iPad:* Here the artefacts are recorded by displaying each DSLR captured image using iPad (fourth generation) with retina display to the Lytro LFC as shown in Figure 7.13(b). This artefact database consists of 1,444 light-field samples (each consisting of multiple depth images) corresponding to 520 high resolution DSLR camera samples. (3) *Screen artefact using Samsung Galaxy Note 10.1:* In order to more effectively analyze the effect of electronic screen attacks, the artefact samples are collected using Samsung Galaxy note 10.1. The high quality samples are captured using DSLR in the Samsung tablet that in turn are presented to the LFC. This database is comprised of 1,208 light field samples including multiple depth images. Figure 7.14 shows the examples of the artefact images captured using iPad (Figure 7.15(b)), Samsung tablet (Figure 7.14(c)) and photo print (Figure 7.14(d)). While Figure 7.14(a) shows the corresponding bona-fide image.

In addition to the above-mentioned artefact generation using high quality print-outs of the enrolled iris samples, we also consider the artefacts using normal (or bona-fide) samples collected using LFC. As the baseline system in this chapter is based on only the best focus image selected from the set of multiple focus image rendered by the LFC, the artefact dataset is generated using the set of best focus images. The artefact samples are generated by using best focus sample displayed on both iPad and Samsung tablet, one at a time to the Lytro LFC. The artefact image

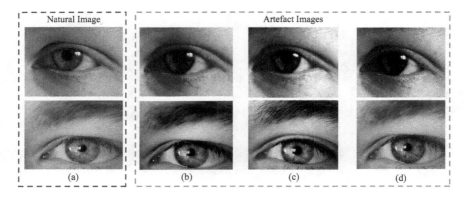

*Figure 7.14   Illustration of artefacts generated from DSLR images. (a) Bona-fide images, (b) iPad, (c) Samsung tablet and (d) photo print*

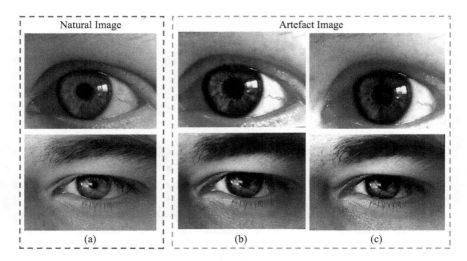

*Figure 7.15   Illustration of artefacts generated from best focus images obtained from the Lytro LFC. (a) Bona-fide (or normal) image, (b) iPad and (c) Samsung tablet*

database collected using an iPad has a total of 1,860 light-field samples while the use of Samsung tablet resulted in a total of 1,973 light-field samples. Figure 7.15 shows the examples of the artefacts captured from LFC using iPad (Figure 7.15(b)) and Samsung tablet (Figure 7.15(c)).

## 7.4.4   Performance evaluation protocol

In this section, we describe the performance evaluation protocol to effectively analyse the prominence of our generated visible spectrum iris artefacts and also to benchmark the performance of our proposed PAD algorithm. The whole database of 104 unique iris is divided into three independent non-overlapping groups, namely: training, development and testing. The training set consists of 4 unique eyes, development set consist of 20 unique eyes and testing set consists of 80 unique eyes. In the set of experiments, we have employed the development set to tune the PAD algorithm and also to fix the value of the threshold, which is further used for the PAD classification. The value of the threshold adopted in this chapter corresponds to the Equal Error Rate (EER) calculated on the development dataset. The test dataset is solely used to report the performance of the proposed PAD algorithm. Further, in order to evaluate the vulnerability of the visible spectrum iris recognition system for the artefacts in the employed dataset, we further divide both development and testing dataset into reference and probe samples. We have used five samples for each subject and thus, we choose the first four samples as reference and the last sample as the probe. With this setting of reference and probe, we report the baseline performance with normal (or bonafide) visible spectrum iris biometric samples. While, to study the vulnerability of the system for the presentation attacks we employ the artefact samples (generated as mentioned in Section 7.4.3.2) as the probe samples.

### 7.4.5 Experiments and results

Experimental results presented in this chapter are carried out according to the protocol presented in Section 7.4.4. The performance of the presented PAD algorithm is measured using two kinds of errors [44] namely: (1) Attack Presentation Classification Error Rate (APCER), which reports the proportion of attack presentations (with a fake or artefact) that are incorrectly classified as bona-fide (normal) presentation. (2) Bona-fide Presentation Classification Error Rate (BPCER), which reports the proportion of normal presentations incorrectly classified as attack. Finally the performance of the overall PAD algorithm is presented in terms of Average Classification Error Rate (ACER), which is an indicative metric such that,

$$ACER = \frac{APCER + NPCER}{2} \tag{7.6}$$

The lower the values of the ACER, the better is the performance. The values of APCER, BPCER and ACER are computed with the classifier threshold. The classifier threshold is calculated on the development dataset such that the threshold value corresponds to the point where APCER and BPCER are equal (nothing but an EER as mentioned in Section 7.4.4) on the development dataset.

Figure 7.16 shows the distribution of the comparison scores obtained using the baseline visible spectrum iris system on both bona-fide (or normal) and artefact (photo print attack) samples. Here, one can observe the strong overlapping of the artefact comparison scores with the genuine comparison scores obtained on the bona-fide samples. With the situation illustrated in Figure 7.16 the baseline system has falsely accepted 96.78% of artefact samples as the genuine samples. This illustrates not only the quality of the artefact employed in this chapter but also the strong need for an efficient and robust PAD algorithm.

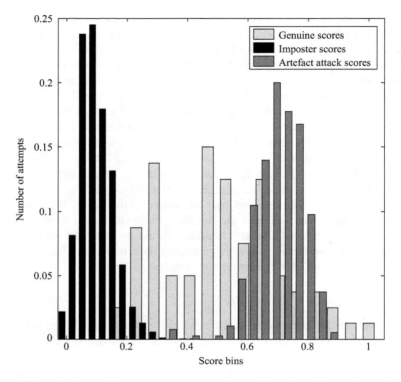

*Figure 7.16    Overlapped comparison scores obtained on bona-fide and artefact (photo print)*

Figure 7.17 shows the verification performance of the baseline visible spectrum iris system under attack (photo print). It can be observed here that despite the good performance of the baseline system (with EER = 2.26%) on normal (or bona-fide) samples, when exposed to the attack the performance is heavily deceived. This fact can be attributed to the strong overlapping of genuine and artefact comparison scores as shown in Figure 7.16. Further, it can also be observed from Figure 7.17 that the implication of the proposed PAD algorithm, which can mitigate the presentation attacks, will bring back the baseline performance to the normal level (indicated by the overlapping of the performance curves of the baseline and the baseline with proposed PAD in Figure 7.17). Similar results can also be observed with the iPad attack that uses Lytro samples as shown in Figure 7.18. We illustrated these two cases for simplicity, however a similar observation can also be made with all five kinds of attacks that are addressed in this chapter.

Table 7.2 presents the quantitative performance of the PAD scheme based on light-field imaging on all five kinds of attacks discussed in this chapter. Quantitative results shown in Table 7.2 are obtained by performing 10-fold cross validation to partition the whole database into training, development and testing set. Here, it can

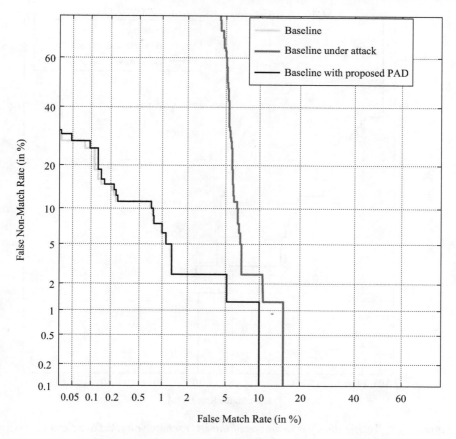

*Figure 7.17  Verification performance of the PAD scheme on photo print artefact corresponding to DSLR samples*

be observed that the presented PAD algorithm has shown the outstanding performance especially on the print attack with ACER of 0.5%. In addition, a similar performance can also be acknowledged on the remaining four different kinds of attacks. Thus, based on the above experiments the presented PAD algorithm that explores the variation of focus between multiple depth images rendered by the LFC provides a new dimension for biometric applications.

In addition the presented PAD algorithm offers the following advantages: (1) Since we are exploring an inherent characteristics of the LFC, the PAD scheme will work as an integral component rather than a stand-alone unit that are normally used in the available state-of-the-art schemes. (2) Overcomes the need of additional feature extractions scheme based on LBP, quality defining parameters and also complex classifiers. Further there is no need for additional hardware components and the proposed scheme is computationally efficient.

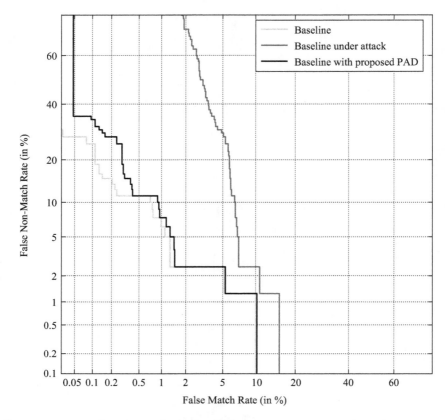

*Figure 7.18   Verification performance of the PAD scheme on iPad artefacts corresponding to Lytro samples*

*Table 7.2    Performance of the PAD algorithm*

| Presentation attacks | | PAD performance measure (%) | | |
|---|---|---|---|---|
| | | **APCER** | **BPCER** | **ACER** |
| Artefacts from DSLR samples | Photo print | 0.25 | 0.75 | 0.5 |
| | iPad | 1.50 | 0.50 | 1.00 |
| | Samsung pad | 0.75 | 0.50 | 0.62 |
| Artefacts from Lytro samples | iPad | 1.20 | 3.70 | 2.45 |
| | Samsung pad | 1.40 | 0.90 | 1.15 |

## 7.5 Conclusion

In this chapter, we have systematically demonstrated the applicability of the light-field camera for biometric applications, specifically for ocular biometrics. The key advantage of using such a camera is the ability to refocus the scene after the acquisition is completed. This property has been exploited to handle the problem of out-of-focus imaging in iris recognition where iris image can be refocused after the capture. Thus, the problem of out-of-focus can be easily handled with the help of multiple depth (focus) images. Further, we have demonstrated the use of LFC for periocular recognition and thereby established the use of LFC for complete ocular system. Additionally, we have exploited the inherent property of multiple depth images from single capture to determine a presentation attack. The varying number of depth images in the case of bona-fide presentations and artefact presentations are used to reject presentation attacks on ocular biometric system. The presented methodology of determining the artefact presentation is a simple approach, which is based on computing energy from a number of depth images. The approach was validated on both printed attacks and electronic screen attacks. As a concluding remark, this chapter has demonstrated the use of LFC for visible spectrum ocular recognition system which is resistant to presentation attacks.

## References

[1] John Daugman. "How iris recognition works," *IEEE Transactions on Circuits and Systems for Video Technology*, 14(1):21–30, 2004.

[2] Chun-Wei Tan and Ajay Kumar. "Automated segmentation of iris images using visible wavelength face images," in *2011 IEEE Computer Society Conference on Computer Vision and Pattern Recognition Workshops (CVPRW)*, pp. 9–14, 2011.

[3] Hugo Proenca, Silvio Filipe, Ricardo Santos, Joao Oliveira, and Luis A. Alexandre. "The ubiris. v2: A database of visible wavelength iris images captured on-the-move and at-a-distance," *IEEE Transactions on Pattern Analysis and Machine Intelligence*, 32(8):1529–1535, 2010.

[4] Unsang Park, Raghavender Reddy Jillela, Arun Ross, and Anil K. Jain. "Periocular biometrics in the visible spectrum," *IEEE Transactions on Information Forensics and Security*, 6(1):96–106, 2011.

[5] Damon L Woodard, Shrinivas Pundlik, Philip Miller, Raghavender Jillela, and Arun Ross. "On the fusion of periocular and iris biometrics in non-ideal imagery," in *2010 20th International Conference on Pattern Recognition (ICPR)*, pp. 201–204. IEEE, Piscataway, NJ, 2010.

[6] Ramachandra Raghavendra, Kiran B. Raja, Bian Yang, and Christoph Busch. "Combining iris and periocular recognition using light field camera," in *2013 Second IAPR Asian Conference on Pattern Recognition*, pages 155–159. IEEE, Piscataway, NJ, 2013.

[7]    Ramkumar Narayanswamy, Paulo E.X. Silveira, Harsha Setty, V. Paul Pauca, and Joseph van der Gracht. "Extended depth-of-field iris recognition system for a workstation environment," in *Defense and Security*, pp. 41–50, 2005.

[8]    N. Boddeti and B.V.K.V. Kumar. "Extended depth of field iris recognition with correlation filters," in *Second IEEE International Conference on Biometrics: Theory, Applications and Systems, 2008. BTAS 2008*, pp. 1–8, 2008.

[9]    Kang Ryoung Park and Jaihie Kim. "A real-time focusing algorithm for iris recognition camera," *IEEE Transactions on Systems, Man, and Cybernetics, Part C: Applications and Reviews* , 35(3):441–444, 2005.

[10]   Yuqing He, Jiali Cui, Tieniu Tan, and Yangsheng Wang. "Key techniques and methods for imaging iris in focus," in *18th International Conference on Pattern Recognition, 2006. ICPR 2006*, vol. 4, pp. 557–561, 2006.

[11]   Sriram Pavan Tankasala, *et al.* "A video-based hyper-focal imaging method for iris recognition in the visible spectrum," in *2012 IEEE Conference on Technologies for Homeland Security (HST)*, pp. 214–219, 2012.

[12]   Ren Ng, *et al.*, "Light field photography with a hand-held plenoptic camera," in *Stanford University Computer Science Technical Report CSTR 2005-02*, pp. 1–11, 2005.

[13]   Mark Harris. "Light-field photography revolutionizes imaging," in *IEEE Spectrum*, pp. 46–50, 2012.

[14]   R. Horstmeyer, G. Euliss, R. Athale, and M. Levoy. "Flexible multimodal camera using a light field architecture," in *International Conference on Computational Photography (ICCP)*, pp. 1–8, 2009.

[15]   Ashok Veeraraghavan, Ramesh Raskar, Amit Agrawal, Ankit Mohan, and Jack Tumblin. "Dappled photography: Mask enhanced cameras for heterodyned light fields and coded aperture refocusing," *ACM Transactions on Graphs*, 26(3):1–12, Jul. 2007.

[16]   Unsang Park, Arun Ross, and Anil K. Jain. "Periocular biometrics in the visible spectrum: A feasibility study," in *IEEE Third International Conference on Biometrics: Theory, Applications, and Systems, 2009. BTAS'09*, pp. 1–6, 2009.

[17]   Homepage of lytro. "www.lytro.com".

[18]   J.Daugman. "Iris recognition and anti-spoofing countermeasures," in *Seventh International Biometrics Conference*, 2004.

[19]   Xiaofu He, Yue Lu, and Pengfei Shi. "A fake iris detection method based on FFT and quality assessment," in *Chinese Conference on Pattern Recognition*, pp. 1–4, Oct. 2008.

[20]   J. Galbally, S. Marcel, and J. Fierrez. "Image quality assessment for fake biometric detection: Application to iris, fingerprint, and face recognition," *IEEE Transactions on Image Processing*, 23(2):710–724, Feb. 2014.

[21]   Ana F. Sequeira, Juliano Murari, and Jaime S. Cardoso. "Iris liveness detection methods in mobile applications," in *Ninth International Conference on Computer Vision Theory and Applications*, pp. 1–5, 2013.

[22]   Xiaofu He, Yue Lu, and Pengfei Shi. "A new fake iris detection method," in Massimo Tistarelli and MarkS. Nixon, editors, *Advances in Biometrics*, vol. 5558, pp. 1132–1139. Springer, Berlin, 2009.

[23] Sung Joo Lee, Kang Ryoung Park, and Jaihie Kim. "Robust fake iris detection based on variation of the reflectance ratio between the iris and the sclera," in *Biometric Consortium Conference*, pp. 1–6, Sep. 2006.

[24] Ken Hughes and Kevin W. Bowyer. "Detection of contact-lens-based iris biometric spoofs using stereo imaging," in *46th Hawaii International Conference on System Sciences*, pp. 1763–1772, Jan. 2013.

[25] Livdet-iris competition. "http://people.clarkson.edu/projects/biosal/iris/index.php".

[26] Mobile iris liveliness detection competition (mobilive 2014). "http://mobilive 2014.inescporto.pt".

[27] O.V. Komogortsev and A. Karpov. "Liveness detection via oculomotor plant characteristics: attack of mechanical replicas," in *International Conference on Biometrics (ICB)*, pp. 1–8, June 2013.

[28] A. Pacut and A. Czajka. "Aliveness detection for iris biometrics," in *Carnahan Conferences Security Technology, Proceedings of the 2006 40th Annual IEEE International*, pp. 122–129, Oct. 2006.

[29] Adam Czajka, Przemek Strzelczyk, and Andrzej Pacut. "Making iris recognition more reliable and spoof resistant," in *SPIE*, 2007.

[30] Guillaume Sutra, Bernadette Dorizzi, Sonia Garcia-Salicetti, and Nadia Othman. A biometric reference system for iris, OSIRIS version 4.1. 2012.

[31] T. Ojala, M. Pietikainen, and T. Maenpaa. "Multiresolution gray-scale and rotation invariant texture classification with local binary patterns," *IEEE Transactions on Pattern Analysis and Machine Intelligence*, 24(7):971–987, 2002.

[32] J. Wright, A.Y. Yang, A. Ganesh, S.S. Sastry, and Y. Ma. "Robust face recognition via sparse representation," *IEEE Transactions on Pattern Analysis and Machine Intelligence*, 31(2):210–227, 2009.

[33] R Raghavendra, Hemantha Kumar, and Ashok Rao. "Qualitative weight assignment for multimodal biometric fusion," in *2009 Seventh International Conference on Advances in Pattern Recognition*, pp. 193–196, 2009.

[34] ISO/IEC TC JTC1 SC37 Biometrics. *ISO/IEC 19795-1:2006. Information Technology – Biometric Performance Testing and Reporting – Part 1: Principles and Framework*. International Organization for Standardization and International Electrotechnical Committee, Mar. 2006.

[35] C. Rathgeb and A. Uhl. "Context-based biometric key generation for iris," *IET Computer Vision*, 5(6):389–397, 2011.

[36] Christian Rathgeb and Andreas Uhl. "Secure iris recognition based on local intensity variations," in *Image Analysis and Recognition*, pp. 266–275, 2010.

[37] Jong-Gook Ko, Youn-Hee Gil, Jang-Hee Yoo, and Kyo-IL Chung. "A novel and efficient feature extraction method for iris recognition," *ETRI Journal*, 29(3):399–401, 2007.

[38] Li Ma, Tieniu Tan, Yunhong Wang, and Dexin Zhang. "Personal identification based on iris texture analysis," *IEEE Transactions on Pattern Analysis and Machine Intelligence*, 25(12):1519–1533, 2003.

[39] Libor Masek and Peter Kovesi. "MATLAB source code for a biometric identification system based on iris patterns," *The School of Computer Science and Software Engineering, The University of Western Australia*, 2(4), 2003.

[40] Jaroslav Kautsky, Jan Flusser, Barbara Zitová, and Stanislava Šimberová. "A new wavelet-based measure of image focus," *Pattern Recognition Letters*, 23(14):1785–1794, 2002.

[41] R. Raghavendra, Bian Yang, Kiran B. Raja, and Christoph Busch. "A new perspective: Face recognition with light-field camera," in *2013 International Conference on Biometrics (ICB)*, pp. 1–8, 2013.

[42] Anil Jain, Karthik Nandakumar, and Arun Ross. "Score normalization in multimodal biometric systems," *Pattern Recognition*, 38(12):2270–2285, 2005.

[43] Kiran B. Raja, R. Raghavendra, Faouzi Alaya Cheikh, Bian Yang, and Christoph Busch. "Robust iris recognition using light field camera," in *The Colour and Visual Computing Symposium 2013*, 2013.

[44] ISO/IEC JTC1 SC37 Biometrics. *ISO/IEC DIS 30107-3 Information Technology – Biometric presentation attack detection – Part 3: Testing and reporting*. International Organization for Standardization, 2016.

*Part III*

# Soft Biometric Classification

*Chapter 8*

# Gender classification from near infrared iris images

*Juan Tapia*

## 8.1 Introduction

At the beginning of this century, one of the most important problems in biometric research was to identify people using iris images. Today we can get more information from the iris such as emotions, age, gender and maybe others. This new scope of research area was called 'Soft biometrics'.

One active area of Soft biometrics research involves classifying the gender of the person from the biometric sample [1]. Most work done on gender classification has involved the analysis of face images [2–4]. Various types of feature extraction, selection and classifiers have been used in gender classification.

In terms of comparing iris-codes for identity recognition, iris-codes of different individuals, and even of the left and right eyes of the same individual, have been shown to be independent. At the same time, several authors have reported that, using an analysis of iris texture different from that used for identity recognition, it is possible to classify the gender of the person with an accuracy much higher than chance.

Nowadays, essentially all commercial systems to identify people from iris are based on the iris-code proposed by Daugman [5]. Therefore, the iris-code is already being computed in iris recognition systems and could be used for other purposes such as gender prediction, either to help speed up the matching process, and/or to know something about people who are not recognized. Commercial iris recognition systems typically do not also acquire face images or fingerprint images, and so gender-from-iris is the only option for gender information in an iris recognition system. If the gender is computed before a search for a match to an enrolled iris-code, then the average search time can potentially be cut in half. In instances where the person is not recognized, it may be useful to know the gender and other information about people trying to gain entry.

Several aftermarket accessories and products have been developed to facilitate reliable iris recognition on smartphones. An example of such a product is Aoptix Stratus,[1] a wraparound sleeve that facilitates NIR iris recognition on the iPhone. Also

---

[1] http://www.ngtel-group.com/files/stratusmxds.pdf

Samsung Electronics requested patent for the system that unlocks the cell-phone with the eyes.[2] In general, the problem of iris recognition is also understudied in the context of smartphones and certainly we can apply gender classification as a second stage.

There are several reasons why gender-from-iris is an interesting and potentially useful problem [6]. One possible use arises in searching an enrolled database for a match. If the gender of the sample can be determined, then it can be used to order the search and reduce the average search time. Another possible use arises in social settings where it may be useful to screen entry to some area based on gender, but without recording identity. Gender classification is also important for demographic information collection, marketing research and real time electronic marketing [7–9]. For example, displays at retail stores and web pages could offer products or benefits according to the person's gender. Another possible use is that in high-security scenarios, there may be value to knowing the gender of the people who attempt entry but are not recognized as any of the enrolled persons. And, at a basic science level, it is of value to more fully understand what information about a person can be extracted from analysis of their iris texture.

Soft biometrics using iris information is a challenging research to estimate demographic information, such as, gender, ethnicity, age and emotions [1]; currently, we can find two main branches of research. The first one, uses iris images captured from Near Infrared cameras (NIR) and the second one, uses images captured from Visible Range Cameras (VRC). This chapter tells us about the first approach.

## 8.2    Anatomy structure of the eye

"The iris is so unique that no two irises are alike, even among identical twins or even between the left and the right eyes of the same person, in the entire human population" [10]. The human iris is an annular part between the pupil and the white sclera, has an extraordinary structure and provides many interlacing minute characteristics such as freckles, coronas, stripes, furrows, crypts and so on. These visible characteristics, generally called the texture of the iris, are unique to each subject [5,10,11,12], see Figure 8.1.

Individual differences that exist in the development of anatomical structures in the body result in such uniqueness. Some research work has also stated that the iris is essentially stable throughout a persons' life. The biometric literature indicates that the iris exhibits substantial diversity in its texture across the population. 'The uniqueness of each iris is assumed to be a consequence of the random morphogenesis of its textural relief during prenatal growth. Even the irises of monozygotic twins exhibit differences in their texture, thereby suggesting that these patterns are determined epigenetically by random events during development that impact the morphogenesis of the tissue' [10]. Nevertheless, to our understanding there is not scientific evidence about the relationship among these patterns and the gender of the people.

---

[2]US Patent Application 20150269419.

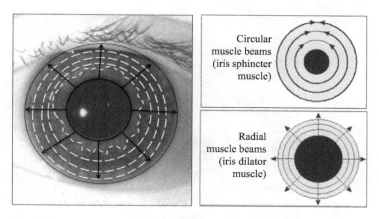

*Figure 8.1    Iris representation with the main structures*

## 8.3    Feature extraction

Since the iris is an internal organ as well as externally visible, iris-based personal identification systems can be non-invasive to their users [5,11], which is greatly important
for practical applications. All these desirable properties (i.e., uniqueness, stability
and non-invasiveness) make the gender classification suitable and complementary
for highly reliable personal identification.

The iris feature extraction process involves four steps: Acquisition, Segmentation,
Normalization and Encoding.

In the acquisition, a camera captures a near infrared image of the eyes at the
same time or separately according to the sensor used. All commercial iris recognition
systems use near infrared illumination, to be able to image iris texture of both 'dark'
and 'light' eyes.

In the segmentation stage, the iris and the pupil region are located within the
image and in the normalization stage, the annular region of the iris is transformed
from raw image coordinates to normalized polar coordinates, See Figure 8.2. This
results in what is sometimes called an 'unwrapped' or 'rectangular' iris image.

The radial resolution ($r$) and angular resolution ($\theta$) used during the normalization
or 'unwrapping' stage determine the size of the rectangular iris image, and can significantly influence the iris recognition rate and the gender classification [13]. This
unwrapping is referred to as using Daugman's rubber sheet model [5].

In the encoding stage a texture filter is applied at a grid of locations on this
unwrapped iris image, and the filter responses are quantized to yield a binary
iris-code [14]. For the encoding stage, the outputs of each filter should be independent, so that there are no correlations in the encoded template, otherwise the
filters would be redundant. For maximum independence, the bandwidths of each

Cartesian Coordinates                                        Polar Coordinates

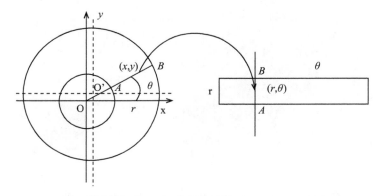

*Figure 8.2  Transformation of Cartesian coordinates (x, y) to Polar coordinate (r, θ) in order to generate the normalized image*

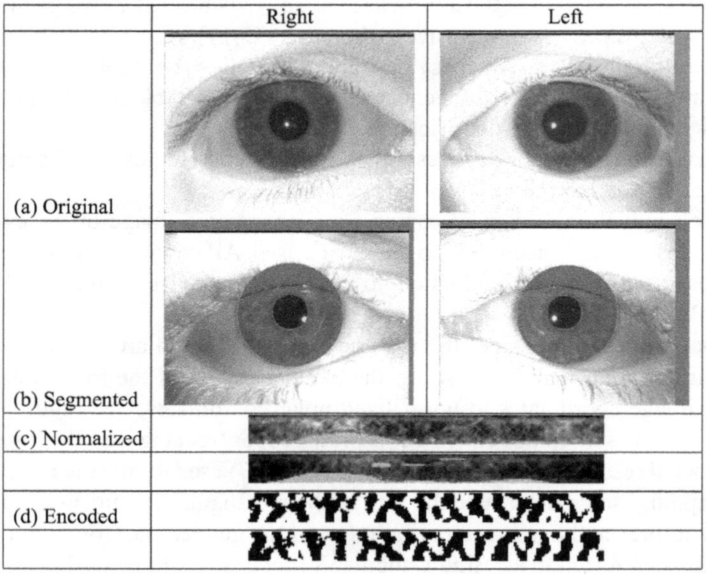

|                  | Right | Left |
|------------------|-------|------|
| (a) Original     |       |      |
| (b) Segmented    |       |      |
| (c) Normalized   |       |      |
|                  |       |      |
| (d) Encoded      |       |      |

*Figure 8.3  Two original images from right and left eye (a). Segmented images subject with eyelids and eyelashes detection using IrisBee implementation. The images (b) and (c) represent segmented and normalized images with the masks, (d) encoded images*

filter must not overlap in the frequency domain, as well as center frequencies must be spread out.

The output of the Gabor filters is transformed into the binary iris-code by quantizing the phase information into four levels, for each possible quadrant in the complex plane. Taking only the phase information allows encoding of the most discriminating information from the iris, while discarding redundant or noisy information, which is represented by the amplitude component [5].

Iris recognition systems operating on these principles are widely used in a variety of applications around the world [15–17]. Currently, we can find several softwares that implement all the stages to get the normalized image or iris-code such as: Masek Implementation [18], IrisBee Implementation [19], Osiris Implementation [20] and others. Figure 8.3 shows examples of the original eye image with the corresponding segmentation, normalized and encoded image.

The Masek implementation generates an iris-code that is $20 \times 240 \times 2 = 9,600$ bits and uses a single scale. While IrisBee is an improvement of Masek implementation and also achieves an iris-code of 9,600 bits. Osiris is an open source iris recognition system based on Daugman's works. The feature extraction stage is based on Gabor phase demodulation. Each iris-code is saved as a binary image of size $W \times (n \times H)$, where $W \times H$ is the size of the normalized image, and $n$ is the number of Gabor filters. In fact, there are $n/2$ Gabor filters, with real and imaginary parts, but Osiris considers the real and imaginary parts as two independent filters thus we get six images as results. Therefore, the implementations deliver different number of encoded images according to the number of Gabor filters applied. This should influence the gender classification rate. Most of the works use only the real part of the image in order to classify gender according to [21].

All implementations also create a segmentation mask of the same size as the rectangular image. The segmentation mask indicates the portions of the normalized iris image that are not valid due to occlusion by eyelids, eyelashes or specular reflections. For each image we have a unique mask. This is an important point because for one iris image a particular pixel maybe masked while in other images not.

To classify gender-from-iris we can use the normalized image or we can use the binary iris-code. See Figure 8.4. Therefore, we can explore different kinds of texture analysis techniques (intensity, shape or texture information) and parameters $((r), (\theta))$. The goal is to explore the texture of iris from images of male and female for the left and right eyes separately and try to find the best method for describing a given texture, linking the morphology patterns present in the iris using intensity, shape or texture information and separate the two classes and identify particular patterns.

## 8.4 State of the art

Gender classification using NIR iris information is a rather new topic, with only a few papers published [2,6,22–25].

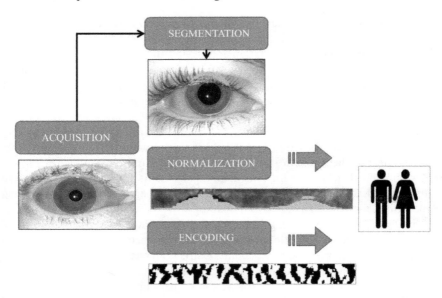

*Figure 8.4    Block diagram with the four stages of the feature extraction process of iris recognition: acquisition, segmentation, normalization, and encoding. The arrows show the stages from where the information was taken for gender classification*

Most gender classification methods reported in the literature use all iris texture features for classification. As a result, gender-irrelevant information might be fed into the classifier which may result in poor generalization, especially when the training set is small. It has been shown both theoretically and empirically that reducing the number of irrelevant or redundant features increases the learning efficiency of the classifier [26].

Thomas *et al.* [24] were the first to explore gender-from-iris, using images acquired with an LG 2200 sensor. They segmented the iris region and employed machine-learning techniques to develop models that predict gender based on the iris texture features. They segmented the iris region, created a normalized iris image, and then a log-Gabor filtered version of the normalized image. In addition to the log-Gabor texture features, they used seven geometric features of the pupil and iris, and were able to reach an accuracy close to 80%.

Lagree *et al.* [23] experimented with iris images acquired using an LG 4000 sensor. They computed texture features separately for eight five-pixel horizontal bands, running from the pupil–iris boundary out to the iris sclera boundary, and ten twenty-four-pixel vertical bands from a $40 \times 240$ image. The rectangular image is not processed by the log-Gabor filters that are used by IrisBee software [27] to create the iris-code for biometrics purpose and do not use any geometrics features to develop models that predict gender and ethnicity based on the iris texture features.

These are the differences from features computed by Thomas *et al.* in [24]. This approach reached an accuracy close to 62% for gender and close to 80% for ethnicity.

Bansal *et al.* [22] experimented with iris images acquired with a CrossMatch SCAN-2 dual-iris camera. A statistical feature extraction technique based on correlation between adjacent pixels was combined with a 2D wavelet tree based on feature extraction techniques to extract significant features from the iris image. This approach reached an accuracy of 83.06% for gender classification. Nevertheless, the database used in this experiment was very small (300 images) compared to other studies published in literature.

Tapia *et al.* [2] experimented with iris images acquired using an LG 4000 sensor to classify gender of a person based on analysis of features of the iris texture. They explored using different implementations of Local Binary Patterns [28] from the iris image using the masked information. Uniform LBP with concatenated histograms significantly improves accuracy of gender prediction relative to using the whole iris image. They were able to achieve over 91% correct gender prediction but using a non person-disjoint dataset and only the left texture of the iris.

Costa-Abreu *et al.* [25] explored the gender prediction task with respect to three approaches using only geometric features, only texture features and both geometric and texture features extracted from iris images. This work used a BioSecure Multimodal DataBase (BMDB) and these images were taken using a LG Iris Access EOU-3000. They were able to achieve over 89.74% correct gender prediction using the texture of the iris with only 200 subjects. Nevertheless this dataset is not available.

Bobeldik *et al.* [29] explored the gender prediction accuracy associated with four different regions from NIR iris images: the extended ocular region, the iris-excluded ocular region, the iris-only region and the normalized iris-only region. They used a Binarized Statistical Image Feature (BSIF) texture operator to extract features from the regions previously defined. The ocular region reached the best performance with 85.7% while the normalized images exhibits the worst performance, with almost a 20% difference in performance over the ocular region. Thereby we can understand that the normalization process may be filtering out useful information. This database is not available. See the summary of previous work in Table 8.1.

All previous work on gender-from-iris tried to find the best feature extraction techniques to represent the information of the iris texture for gender classification. However, none of the previous works looking for the quality and quantity of the features selected for the problem of gender-from-iris.

Most gender classification methods reported in the literature use all of the features extracted for classification purposes. In image understanding, raw input data often has very high dimensionality and a limited number of samples. In this area, feature selection plays an important role in improving accuracy, efficiency and scalability of the object identification process [30]. According to Bengio *et al.* [31], 'The success of machine learning algorithms generally depends on data representation, and we hypothesize that this is because different representations can entangle and hide more or less the different explanatory factors of variation behind the data'.

Recently, Tapia *et al.* [6] predict gender directly from the same binary iris-code that could be used for recognition. They found that information for gender prediction

Table 8.1    *Gender classification summary of previously published paper*

| Paper | Sensor | Size image | Number of images | Number of subjects | N | E | Obs |
|-------|--------|-----------|------------------|--------------------|---|---|-----|
| Thomas *et al.* [24] | LG 2200 | 20 × 240 | 16,469 | N/A | X | – | Only left eye |
| Lagree *et al.* [23] | LG 4000 | 40 × 240 | 600 | 300 | X | – | |
| Bansal *et al.* [22] | Cross Match SCAN-2 | 10 × 500 | 400 | 200 | X | – | |
| Tapia *et al.* [2] | LG 4000 | 20 × 240 | 1,500 | 1,500 | X | – | Only left eye |
| Costa-Abreu *et al.* [25] | LG EOU-3000 | 20 × 240 | 1,600 | 200 | X | – | |
| Bobeldyk *et al.* [29] | NIR Sensor | 20 × 240 | 3,314 | 1,083 | X | – | Periocular images |
| Tapia *et al.* [6] | LG 4000 | 20 × 240 | 3,000 | 1,500 | – | X | |

N represents: Normalized Image, E represents: Encoded Image

is distributed across the iris, rather than localized in particular concentric bands. They also found that using selected features representing a subset of the iris region achieves better accuracy than using features representing the whole iris region achieving 89% correct gender prediction using the fusion of the best features of iris-code from the left and the right eyes.

## 8.5    Databases

One of the most challenging problem in gender classification from iris research is related to databases, because they were created focusing on an iris identification problem, thus we can find many sessions by the same subjects and many of them do not have the gender information available. Two databases used to classify gender are described as follows:

### 8.5.1    *BioSecure multimodal database*

The BioSecure Multimodal Database (BMDB) was used in [25]. This database was collected as part of an extensive (and commercially available) multi-modal database by 11 European institutions participating in the BioSecure Network Excellence [32]. The eye images were acquired in a standard office environment managed by a supervisor and using the LG Iris Access EOU 3000 setup. During the acquisition, spectacles were not allowed to be worn by subject, although contact lenses were allowed. Four eye images (two left and two right) were acquired in two different sessions with a resolution of 640 × 480 pixels in total. See Figure 8.5.

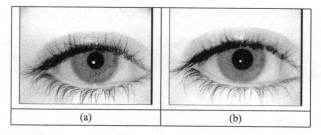

Figure 8.5    BMDB database. Two images taken from sensor LG Iris access
EOU 3000

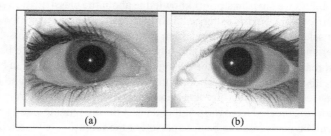

Figure 8.6    ND-GFI database. Images taken from sensor LG 4000

According to [25] 10 subjects were found to be incorrectly labeled in this database and were discarded. Therefore, the available number of images is 1,600 from 200 subjects. This database is not publicly available anymore, more information on http://biosecure.it-sudparis.eu/AB/getdatabase.php.

## 8.5.2   Gender from iris dataset (ND-GFI)

The Dataset (ND-GFI) was used in [6]. This database contains 3,000 images: 750 left-iris images from men, 750 right-iris images from men, 750 left-iris images from female and 750 right-iris images from female. Of the 1,500 distinct persons in the ND-GFI dataset, visual inspection of the images indicates that about 1/4 are wearing clear contact lenses. These images were taken with an LG 4000 sensor.

The ND-GFI database also contain a validation set called UND_Validation. It contains 1,944 images: three left eye images and three right eye images for each of 175 males and 149 females. It is known that some subjects are wearing clear contact lenses, and evidence of this is visible in some images. Also, a few subjects are wearing cosmetic contact lenses in some images. See Figure 8.6.

Both databases are person-disjoint set, that is, no person has an image in both the training images and the test images. Results computed for person-disjoint train and test are generally expected to be lower than for train and test that is not person-disjoint.

However, person-disjoint train and test should be more realistic and should generalize better to new data.

Both the original database and this additional validation database are available at the URL http://www3.nd.edu/~cvrl/CVRL/Data_Sets.html – 'The Gender from Iris Dataset (ND-GFI)'.

## 8.6 Feature selection

Most gender classification methods reported in the literature use all of the features extracted for classification purposes using the texture of the iris. Therefore it is usually difficult to decide how many and which features are needed to improve the classification rate; unless we have an a priori knowledge about the classification problem. In gender classification from iris using near infrared images, this is an unsolved problem.

Feature selection is a process in which features from a dataset are selected with the characteristics such as face, iris and others. Biometric systems are based on the premise that many of the physical or behavioral attributes of people can be uniquely associated with an individual thus improving classification accuracy and decreasing computational complexity. It is closely related to feature extraction, a process in which feature vectors are created from the original dataset through manipulations of the data space, and can be considered to be a 'superset' of the feature selection techniques.

Feature selection can be classified into three main groups: Filters, Wrappers and Embedded [26]. Filter methods use independent techniques to select features from the feature space. These techniques are based on a number of different statistical tests. Wrapper methods embed the feature selection into a classifier, where the classification performance is used to measure the quality of the currently selected feature set. Embedded methods are similar to wrappers, but instead of using the classifier as a black box, they integrate the classifier into the feature selection algorithm.

Feature selection methods can also be categorized on the basis of the search strategies used [26]. The following search strategies are commonly used: Forward selection that starts with an empty set and adds features greedily one at a time; Backward elimination that starts with a feature set containing all the features and removes features greedily one at a time.

In order to show one example of gender classification from normalized iris images, a simple and fast method to analyze the most relevant features from the iris is used. This allows to reduce the computational time and, at a basic science level, it is of value to more fully understand what information about a person can be extracted from analysis of their iris texture. For this example we used the 'Gender from Iris Dataset' (ND-GFI).

We can apply the statistical $t$-test approach on each feature and compare the $p$-value for each feature as a measure of how effective it is at separating features from female and male iris images. To analyze this approach, we can use nine ensemble classifiers such as: AdaboostM1, LogitBoost, GentleBoost, RobustBoost, LPBoost, TotalBoost, RUSBoost with 'Tree learners' classifiers and learning rate of 0.1. Subspace and Random Forest classifier (RF) with 500 'Tree' and also a SVM classifier

Table 8.2   Gender classification rates for the
left and right normalized iris images
using the whole image

| Method | Left (%) | Right (%) |
|---|---|---|
| SVM | 62.66 | 61.33 |
| AdaboostM1 | 63.67 | 64.33 |
| LogitBoost | 63.33 | 61.00 |
| GentleBoost | 60.00 | 62.33 |
| RobustBoost | 57.00 | 62.33 |
| LPBoost | 60.33 | 56.00 |
| TotalBoost | 57.33 | 62.67 |
| RUSBoost | 54.67 | 56.00 |
| Subspace | 63.00 | 66.00 |
| RF | 63.33 | 67.33 |

with Gaussian kernel with LIBSVM implementation [33]. The results are shown in Table 8.2.

Filters are usually used in a pre-processing stage as they are simple and fast, thus reducing the dimension of the data by finding a small set of important features which can give good classification performance. A widely used filter method for data is to apply an univariate criterion separately on each feature, assuming that there is no interaction between features, we apply the statistical $t$-test approach on each feature and compare the $p$-value ($p < 0.05$) for each feature as a measure of how effective it is at separating features from female and male iris images. In order to get a general idea of how well-separated the two groups (female and male) are by each feature. We can calculate and plot the empirical Cumulative Distribution Function (CDF) of the $p$-values for left normalized iris images, Figure 8.7 and right normalized iris images according to the values of $p$, Figure 8.8.

Figure 8.7, shows that there are about 10% of features having $p$-values close to zero and close to 22% of features having $p$-values smaller than 0.05 meaning there are more than 1,056 features from normalized images among the original 4,800 features that have strong discriminant power. We can sort these features according to their $p$-values and select some features from the sorted list. However, if we used the encoded images, the number of relevant features that have strong discriminant power increase 5% compared to normalized images, particularly for this example. Now we have 1,200 features.

Figure 8.8 shows that there are about 20% of features having $p$-values close to zero and close to 40% of features having $p$-values smaller than 0.05 meaning there are more than 1,920 features from normalized images among the original 4,800 features that have strong discriminant power. We can sort these features according to their $p$-values and select some features from the sorted list. When we used encoded images, the number of features with $p$-values lower to 0.05 decrease close to 32% particularly for

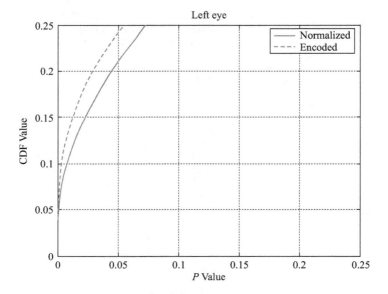

*Figure 8.7    Normalized vs encoded images. Empirical cumulative distribution function of the p-values for the left iris*

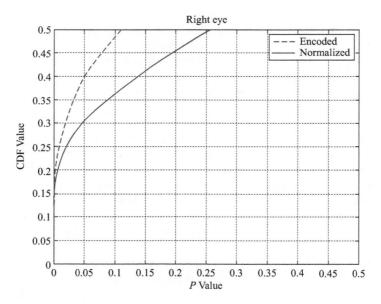

*Figure 8.8    Normalized vs encoded images. Empirical cumulative distribution function of the p-values for the right iris*

this example. Now we have 1,536 features. Therefore, it is very important to analyze the data statistically previously to classify gender in relation with the information available on the database and the quality of the images (occlusions, size of the mask and segmentation).

However, it is usually difficult to decide how many features are needed unless one has domain knowledge or the maximum number of features that can be considered has been dictated in advance based on outside constraints. Nevertheless the number of relevant features are not the same when we used Normalized images (pixels values) and Encoded images (binary images), for this example. More relevant information from the features can be captured from iris-code rather than normalized image for the left eye.

One quick way to decide the number of needed features is to estimate the plot of the Miss-Classification-Error (MCE) on the test set as a function of the number of features rather than Resubstitution Error (RE), since RE is not a good error estimate of the test error.

We computed MCE for different number of features in steps of 10 features up to 250 features. The Resubstitution MCE values are overly optimistic here. Most are smaller than the cross-validation MCE values and the Resubstitution MCE values goes to the lower value when using over 100 features.

The previous example of selection algorithm does not consider the interaction between features, which means features selected from a list based on their individual ranking may also contain redundant information, therefore not all the features are needed. One example of this is the linear correlation between the features. This kind of simple feature selection approach is usually used as a pre-processing step since it is fast. More advanced feature selection algorithm using complementarity or synergistic properties might improve the performance [30]. Sequential feature selection selects a subset of features by sequentially adding (forward search) or removing (backward selection) until the stop condition criterion is reached.

For this example, we used a forward sequential feature selection in a wrapper approach to find the most important features. More specifically, since the typical goal of classification is to minimize the MCE, the feature selection procedure performs a sequential search using a MCE of the learning algorithm Quadratic Discriminant Analysis (QDA) on each candidate feature subset as the performance indicator for that subset. QDA fits multivariate normal densities with covariance estimates stratified by group, in this case female and male. The training set is used to select the features and to fit the QDA model, and the test set is used to evaluate the performance of the finally selected features.

During the feature selection procedure, we evaluated and compared the performance of each candidate feature subset. We applied 10-fold-cross-validation to the training set. Again, the Resubstitution MCE values were overly optimistic here. Most were smaller than the cross-validation MCE values and the Resubstitution MCE values went to the lower values when using over 50 features. Finally we used the 100 most relevant features to re-estimated the accuracy of the gender classification, in order to compare the result with Table 8.2, using the nine ensemble classifiers and SVM again. We present the results in Table 8.3.

*Table 8.3   Gender classification rates for the left and right normalized iris images using the most relevant features based on p-values*

| Method | Left (%) | Right (%) |
|---|---|---|
| SVM | 72.66 | 70.66 |
| AdaboostM1 | 69.67 | 68.33 |
| LogitBoost | 69.33 | 68.00 |
| GentleBoost | 68.00 | 66.33 |
| RobustBoost | 67.00 | 69.33 |
| LPBoost | 70.33 | 66.00 |
| TotalBoost | 67.33 | 69.67 |
| RUSBoost | 64.67 | 66.00 |
| Subspace | 63.00 | 69.00 |
| RF | 79.33 | 78.33 |

The results of Table 8.3 using the statistical information from the feature belonging to the normalized images improves the accuracy to the gender classification problem. On average the accuracy increases 10% points more than when we use the whole information from the iris. Therefore, we can conclude that we have irrelevant and redundant information and that high dimensional data may confuse the classifiers. Thus feature selection is a powerful tool to detect the most relevant features. A video that show examples of feature selection methods over normalized and iris-code image will be found online http://jtapiafarias.wordpress.com.

## 8.7   Research trends and challenges

In this section we describe some of the trends to continue this research and the main challenge to improve the gender classification from the iris images. The goal is to get general results independent of the sensor, illumination or database to understand the relationship between the results and the biological features extracted from the iris. The application of Deep Learning techniques using supervised and unsupervised neural networks may help with these task.

### 8.7.1   Segmentation

According to [34] regarding the feature extraction stages, the pupil dilation becomes an important and influential property. When age advances, the pupil size decreases and the recognition rate increases. Most of the algorithms are based on a linear normalization, which converts from a polar to a Cartesian scale. However, the results obtained by several authors point out that this normalization is not enough and therefore, a further study of how the iris muscle deforms should be carried out to model it and propose a better normalization algorithm. Researchers have demonstrated that the performance of iris biometrics decreases with time, and these initial results point out the necessity of continuing to study this phenomenon in order to develop long-time gender classification stage.

## 8.7.2 Accuracy

The estimated accuracy of an algorithm plays a central and essential role in the gender classification problem. The accuracy goal of gender classification is simple – the higher, the better. However, how confident is an estimated accuracy for a given dataset? Moreover, how generalizable is the proposed method for a wider variety of conditions or sensors? When we attempt to answer such questions, we typically focus on the elements of the dataset. What is the number of images used in the dataset? How many images by session were used? Is the database person-disjoint? And what is the number of partition in a cross-validation process? Therefore, results must be clearly reported indicating the parameters used on the train and the test set.

Results computed for a person-disjoint train and test are generally expected to be lower than for the train and test that is not person-disjoint. However, person-disjoint train and test should yield a more realistic accuracy estimate for generalization to new data.

## 8.7.3 Fragile bits

Hollingsworth *et al.* and Tan *et al.* [35,36] experimented with the concept of 'fragile bit' in a binary iris-code. Their researches were the first to present an experiment documenting that some bits are more consistent than others. Different regions in the iris-code were compared to evaluate their relative consistency in the iris-code. They concluded that not all bits in the iris-code contributed similarly [35,36]. While their work was done in the context of person recognition rather than gender classification, the idea that not all bits of the iris-code are equally useful may apply to both problems. The fragile bits results suggest that using a selected subset of the iris, rather than the whole iris region, may improve accuracy.

Therefore, it is important to evaluate the influence of these parameters in the accuracy of the gender classification task. For example, in [6] the authors only used a fragile bit theory on IrisBee software with default value of 25% fragile bits masked.

## 8.7.4 Sensors

'The issue of interoperability among iris sensors is an important topic in large-scale and long-term applications of iris biometric systems' [37–39]. Several factors which may impact single-sensor and cross-sensor performance might be analyzed, including changes in the acquisition environment and differences in dilation ratio between iris images [38]. To our understanding texture patterns gender must prevail independent of the camera/sensor used [37,39]. See Figure 8.9. It is also important to perform different tests and estimate accuracy separately for the left iris and right iris because some systems work with a single iris. Moreover, even systems that can work with two irises need to be able to work with only one if the other is covered for some reason.

IrisGuard Image 04233d3014        LG 4000 Image 04233d3004        TD100 Image 04233d3010

*Figure 8.9    Three images belong to the same subject, captured with different sensors on the same session. The left image was taken with Iris Guard AD100, Center with LG 4000 and the right the image was taken with the sensor TD 100. Each image has a size of 640 × 480 [38]*

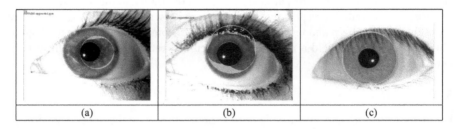

| (a) | (b) | (c) |

*Figure 8.10    Segmentation results: (a) an image with mascara and a wrong segmentation without influence of makeup, (b) an image with mascara and a wrong segmentation and (c) an image with straight eyelashes and with the occlusion mask*

### 8.7.5    Makeup

In order to analyze the influence of makeup in the gender classification task, we need to analyze the quality of the segmentation stage, because most of the software failed to segmented the pupil and the iris well. These errors are produced by the straight eyelashes, different illumination, occlusion of the texture, blur images or texture contact lenses. The iris texture exhibits a great degree of irregularity and its textural content varies substantially across eyes. Indeed, the iris could be viewed as a stochastic texture containing numerous 'edge' like features that are randomly distributed on its previous surface. The iris is an annular-like structure bounded by the pupil in its internal perimeter, and the sclera and eyelids in its external perimeter. Incorrectly estimating these boundaries (contours) can result in the over-segmentation or under-segmentation of the iris entity. In some eye images, these boundaries (especially the limbus boundary) may not be very sharp, thereby affecting the accuracy of the boundary estimation process. Furthermore, the boundary defined by the eyelids is irregularly shaped. The iris texture may be partially occluded by eyelashes. The eyelashes protruding into the iris image can generate spurious edges and affect the segmentation process. Generally, the female segmentation presents more iris miss-detection than the male images. This difference is because the female eyes present more occlusion from eyelid and eyelashes than the male. It appears that eyelash mascara and makeup in some images of female irises cause inaccuracy in the segmentation [6]. It should also be noted that the Asian people present predominantly straight eyelashes and occluded the texture therefore the gender with ethnicity should be a good research area. See Figure 8.10.

## 8.8   Concluding remarks

Gender classification is an important topic in a wide variety of applications ranging from surveillance to selective marketing. Several recent studies have shown the predominance of local matching approaches in gender classifications results.

Previous works in predicting gender-from-iris have relied on computing a separate set of textures representation. The state of the art shows that gender can be successfully predicted from the iris. There are clear computational advantages to predicting gender from the binary iris-code rather than computing another different texture representation.

The initial experiments suggest that information relevant to gender-from-iris is distributed throughout the whole iris region. However, literature also shows that selecting features that represent the areas of the iris that allow the most stable texture computations improve performance relative to using features from all areas of the iris. This is likely due to a combination of the high feature dimensionality when using the whole area of the iris, and the fact that segmentation inaccuracies can make some areas of the iris less stable for texture computation.

This topic brings new insights about the information present in the iris (and iris-code) to determine demographic information. The previous work adds evidence answering the fundamental question that the iris contains specific information about us, such as gender. The results, which show that gender classification from iris code is possible, will spur research to determine if other demographic factors (e.g., ethnicity, age, emotions) can also be predicted. This is an area of research that is overall in the early stages.

Possible avenues to pursue for improved accuracy in gender-from-iris include more accurate alignment of iris images, more accurate segmentation of iris images, training on larger datasets using Deep Learning methods, with different kinds of sensors, establish the reproducibility of these experiments and study the generalization or confidence with a new data sets.

## Acknowledgments

Thanks to Vince Thomas, Mike Batanian, Steve Lagree, Yingjie, Bansal and Costa-Abreu for the work they have previously done in this research topic and also to Professor Kevin Bowyer at University of Notre Dame for providing the databases and Professor Claudio A. Perez for their support. This work has been supported by Universidad Andres Bello, Faculty of Engineering, Department of Engineering Sciences (DCI).

# References

[1]   A. Dantcheva, P. Elia, and A. Ross, "What else does your biometric data reveal? A survey on soft biometrics," *IEEE Transactions on Information Forensics and Security, 2015*, ISSN: 1556-6013, 04 2015.

[2]   J. E. Tapia, C. A. Perez, and K. W. Bowyer, "Gender classification from iris images using fusion of uniform local binary patterns," *European Conference on Computer Vision-ECCV, Soft Biometrics Workshop*, 2014.

[3]   L. A. Alexandre, "Gender recognition: A multiscale decision fusion approach," *Pattern Recognition Letters*, vol. 31, no. 11, pp. 1422–1427, 2010.

[4]   E. Makinen and R. Raisamo, "Evaluation of gender classification methods with automatically detected and aligned faces," *IEEE Transactions on Pattern Analysis and Machine Intelligence*, vol. 30, no. 3, pp. 541–547, 2008.

[5]   J. Daugman, "How iris recognition works," *IEEE Transactions on Circuits and Systems for Video Technology*, vol. 14, no. 1, pp. 21–30, 2004.

[6]   J. Tapia, C. Perez, and K. Bowyer, "Gender classification from the same iris code used for recognition," *IEEE Transactions on Information Forensics and Security*, vol. 11, no. 8, pp. 1760–1770, 2016.

[7]   H. Han, C. Otto, X. Liu, and A. Jain, "Demographic estimation from face images: Human vs. machine performance," *IEEE Transactions on Pattern Analysis and Machine Intelligence*, vol. 37, no. 6, pp. 1148–1161, 2014.

[8]   C. Perez, J. Tapia, P. Estevez, and C. Held, "Gender classification from face images using mutual information and feature fusion," *International Journal of Optomechatronics*, vol. 6, no. 1, pp. 92–119, 2012.

[9]   J. Tapia and C. Perez, "Gender classification based on fusion of different spatial scale features selected by mutual information from histogram of LBP, intensity, and shape," *IEEE Transactions on Information Forensics and Security*, vol. 8, no. 3, pp. 488–499, 2013.

[10]  F. Adler, *Physiology of the Eye*. Clinical Application 4th Edition, London: C.V. Mosby Company, 1965.

[11]  J. Daugman, "Statistical richness of visual phase information: Update on recognizing persons by iris patterns," *International Journal of Computer Vision*, vol. 45, pp. 25–38, Oct. 2001.

[12]  M. Larsson and N. L. Pedersen, "Genetic correlations among texture characteristics in the human iris," *Molecular Vision*, vol. 10, pp. 821–831, 2004.

[13]  A. Kong, D. Zhang, and M. Kamel, "An analysis of iriscode," *IEEE Transactions on Image Processing*, vol. 19, no. 2, pp. 522–532, Feb. 2010.

[14]  K. W. Bowyer, K. Hollingsworth, and P. J. Flynn, "Image understanding for iris biometrics: A survey," *Computer Vision and Image Understanding*, vol. 110, no. 2, pp. 281–307, 2008.

[15]  UIDAI, "Unique identification authority of India," 2014.

[16]  CANPASS, "Canadian border services agency, CANPASS," 1996.

[17]  J. Daugman, "Iris recognition at airports and border-crossings," in *Encyclopedia of Biometrics*, 1st Edition, pp. 819–825, 2009.

[18]  P. K. Libor Masek, "MATLAB source code for a biometric identification system based on iris patterns," *The School of Computer Science and Software Engineering, The University of Western Australia*, 2003.

[19]  X. Liu, K. Bowyer, and P. Flynn, "Experiments with an improved iris segmentation algorithm," in *Fourth IEEE Workshop on Automatic Identification Advanced Technologies*, pp. 118–123, Oct. 2005.

[20]  S. G.-S. N. O. G. Sutra and B. Dorizzi, "A biometric reference system for iris, osiris version 4.1," 2012.

[21]  J. Daugman, "Information theory and the iriscode," *IEEE Transactions on Information Forensics and Security*, vol. 11, pp. 400–409, Feb. 2016.

[22]  A. Bansal, R. Agarwal, and R. K. Sharma, "SVM based gender classification using iris images," in *Fourth International Conference on Computational Intelligence and Communication Networks (CICN)*, pp. 425–429, Nov. 2012.

[23]  S. Lagree and K. Bowyer, "Predicting ethnicity and gender from iris texture," in *IEEE International Conference on Technologies for Homeland Security (HST)*, pp. 440–445, Nov. 2011.

[24]  V. Thomas, N. Chawla, K. Bowyer, and P. Flynn, "Learning to predict gender from iris images," in *First IEEE International Conference on Biometrics: Theory, Applications, and Systems, BTAS 2007*, pp. 1–5, Sep. 2007.

[25]  M. D. Costa-Abreu, M. Fairhurst, and M. Erbilek, "Exploring gender prediction from iris biometrics," in *International Conference of the Biometrics Special Interest Group (BIOSIG), 2015*, pp. 1–11, Sep. 2015.

[26]  I. Guyon, S. Gunn, M. Nikravesh, and L. A. Zadeh, *Feature Extraction, Foundations and Applications, Studies in Fuzziness and Soft Computing*. Secaucus, NJ, USA: Springer-Verlag New York, Inc., 2006.

[27]  P. J. Phillips, W. T. Scruggs, A. J. O'Toole, *et al.*, "FRVT 2006 and ICE 2006 large-scale results," *IEEE Transactions on Pattern Analysis and Machine Intelligence*, vol. 32, no. 5, pp. 831–846, 2010.

[28]  L. Nanni, A. Lumini, and S. Brahnam, "Survey on LBP based texture descriptors for image classification," *Expert Systems with Applications*, vol. 39, pp. 3634–3641, Feb. 2012.

[29]  D. Bobeldyk and A. Ross, "Iris or periocular? Exploring sex prediction from near infrared ocular images," in *Lectures Notes in Informatics (LNI), Gesellschaft fur Informatik, Bonn 2016*, 2016.

[30]  J. Vergara and P. Estevez, "A review of feature selection methods based on mutual information," *Neural Computing and Applications*, vol. 24, no. 1, pp. 175–186, 2014.

[31]  Y. Bengio, A. Courville, and P. Vincent, "Representation learning: A review and new perspectives," *IEEE Transactions on Pattern Analysis and Machine Intelligence*, vol. 35, no. 8, pp. 1798–1828, 2013.

[32]  J. Ortega-Garcia, J. Fierrez, F. Alonso-Fernandez, *et al.*, "The multiscenario multienvironment biosecure multimodal database (BMDB)," *IEEE*

*Transactions on Pattern Analysis & Machine Intelligence*, vol. 32, no. 6, pp. 1097–1111, 2009.

[33]   C.-C. Chang and C.-J. Lin, "LIBSVM: A library for support vector machines," *ACM Transactions on Intelligent Systems and Technology*, vol. 2, no. 3, pp. 1–27, 2011.

[34]   J. Liu-Jimenez and R. Sanchez-Reillo, *Age Factor in Biometric Processing – Chapter 7, Ageing in iris Biometric.* The Institution of Engineering and Technology, London, United Kingdom, 2013.

[35]   K. W. Bowyer and K. Hollingsworth, "The best bits in an iris code," *IEEE Transactions on Pattern Analysis and Machine Intelligence*, vol. 31, no. 6, pp. 1–1, 2009.

[36]   T. Tan, W. Dong, and Z. Sun, "Iris matching based on personalized weight map," *IEEE Transactions on Pattern Analysis & Machine Intelligence*, vol. 33, no. 9, pp. 1744–1757, 2010.

[37]   C. Boyce, A. Ross, M. Monaco, L. Hornak, and X. Li, "Multispectral iris analysis: A preliminary study51," in *Conference on Computer Vision and Pattern Recognition Workshop, 2006. CVPRW'06*, pp. 51–51, Jun. 2006.

[38]   R. Connaughton, A. Sgroi, K. Bowyer, and P. Flynn, "A multialgorithm analysis of three iris biometric sensors," *IEEE Transactions on Information Forensics and Security*, vol. 7, no. 3, pp. 919–931, Jun. 2012.

[39]   J. Pillai, M. Puertas, and R. Chellappa, "Cross-sensor iris recognition through kernel learning," *IEEE Transactions on Pattern Analysis and Machine Intelligence*, vol. 36, no. 1, pp. 73–85, Jan. 2014.

*Chapter 9*

# Periocular-based soft biometric classification

*Damon L. Woodard, Kalaivani Sundararajan, Nicole Tobias, and Jamie Lyle*

Soft biometric traits refer to characteristics that provide some information about an individual but do not possess the distinctiveness and permanence necessary to sufficiently differentiate between any two individuals. Examples of soft biometric traits include gender, ethnicity, age, weight, and height. As early as 1997, researchers suggested that soft biometrics could be used to improve biometric recognition performance [1]. Researchers later demonstrated the use of gender, ethnicity, and height to improve the performance of a fingerprint recognition system [2]. This chapter discusses the use of local appearance features extracted from the periocular region for gender and ethnicity classification.

## 9.1 Introduction

In the past decade, major advancements have been made in the area of biometric recognition. As a result, the number of potential applications which could benefit from the use of biometric technology have steadily increased. Despite technology advancements, the performance of biometric recognition in more challenging scenarios has not yet matched that achieved under research conditions. For example, two challenging scenarios which may be encountered during the operation of a biometric recognition system would be large-scale identification and identification using partial data. One possible approach to these challenges is the use of soft biometric classification.

The human face can reveal much about an individual such as their identity, age, gender, ethnicity, and emotional state. Various research efforts have involved the use of the face for gender and ethnicity classification. The commonly used features include Local Binary Patterns (LBPs), Principal Component Analysis (PCA) eigenvectors, Haar-like features, image intensities, Gabor wavelets, Active Shape Models (ASM), Difference of Gaussian (DoG), and Laplacian of Gaussian (LoG) filters. These features are used in conjunction with discriminative classifiers like Support Vector Machines (SVM), Neural Networks (NN), Linear Discriminant Analysis (LDA), AdaBoost, and Decision Trees (DT). A summary of some of these efforts is listed in Table 9.1.

Table 9.1 A summary of gender and ethnic classification approaches using face images

| Approach | Features | Classifier | Dataset | Ethnicity | Gender |
|---|---|---|---|---|---|
| Gutta et al. [3] | grayscale pixel intensities | Neural Networks, SVM and DTs | FERET | 92% (Caucasian, South Asian, East Asian, African) | 96% |
| Moghaddam and Yang [4] | low-res grayscale images | SVM | FERET (1755 total images) | – | 97% |
| Balci and Atalay [5] | PCA eigenvectors | Multi-layer perceptron | FERET (500 training, 260 testing) | – | 92% |
| Wu et al. [6] | grayscale pixel intensities | LUT weak classifier based Adaboost | FERET, WWW pictures for training, 2600 WWW pictures for testing | – | 88% |
| Hosoi et al. [7] | Gabor wavelet transform with retina sampling | SVM | HOIP dataset | Asian - 96%, European – 93%, African – 94% | – |
| BenAbdelkader and Griffin [8] | local and global features (eigenfaces) | SVM, LDA | FERET, PIE and Univ. of Essex (12964 total images) | – | 85% |
| Lapedriza et al. [9] | DoG, LoG filters on facial fragments | Adaboost, Jointboost | FRGC (controlled and uncontrolled lighting) | – | 92% |
| Lu et al. [10] | Range and pixel intensity | SVM | 376 subjects, 1240 scans | 98% (Asian, Non-Asian) | 91% |
| Yang et al. [11] | Normalized face images | SVM, LDA, Adaboost | 11500 Chinese snapshot images | – | 97% |
| Yang and Ai [12] | LBP, Haar-like features | Adaboost | FERET, PIE | 97% (Asian, Non-Asian) | 93% |
| Makinen and Raisamo [13] | Grayscale pixel intensity, Haar-like features, LBP | SVM, NN, Adaboost | IMM face dataset, FERET (304 training, 107 testing) | – | 84% |
| Xu et al. [14] | Haar (appearance) | SVM | FERET, AR, misc. web images (1000 test subjects) | – | 92% |
| Gao and Ail [15] | ASM based landmarks for normalization (grayscale intensities) | SVM, Adaboost, Probabilistic Boosting Trees (PBT) | 15300 Chinese snapshot, 10100 consumer images | Ethnicity specific gender classification | 97% |

Facial recognition research has suggested that the periocular region may be the most discriminating part of the face [16]. The periocular region can be defined as the facial area that extends from the top of the eyebrow to the cheekbone which includes the area from the midline of the nose to just inside of the ear. Features may include upper/lower eye folds, upper/lower eyelids, eye corners, eyebrows, various skin lesions (freckles, moles), as well as skin texture.

A motivating factor for the use of the periocular region as a soft biometric feature source is its possible presence in images determined non-ideal for use by other biometric recognition methods. For example, in order for iris recognition to be successful, a pristine image of the iris is required. It is widely accepted that facial images occluded by facial hair cause facial recognition performance degradation. In such cases, the periocular region could be visible and sufficient for soft biometric classification tasks. This additional ancillary information about the individual can then be used to improve biometric recognition performance. In the following sections, soft biometric classification, specifically gender and ethnicity are investigated using local appearance features extracted only from the periocular region.

## 9.2   Approach

### 9.2.1   Data

To determine the accuracy of soft biometric classification, a collection of booking images from the Pinellas County Sheriff's Office is used. This database is ideal for testing classification accuracy because it contains images of individuals from a more diverse set of ethnicity compared to many of the facial databases utilized in the research literature.

The Pinellas database is a collection of mug shots collected from the Pinellas County Sheriff's Office[1] in Florida. This set of data contains over 1.4 million 2D color still images from approximately 400 000 subjects acquired sporadically over 20 years. The image resolution varies, but is approximately 480 × 600 pixels. Figure 9.1 shows example images from this database. The image quality of the samples is an accurate representation of the image quality available in many biometric application scenarios. A subset of images were selected for use in experiments. This subset of images included images from 10 demographic groups which include Black Males, Black Females, Caucasian Males, Caucasian Females, Hispanic Males, Hispanic Females, Asian Males, Asian Females, Native American Males, and Native American Females. For each demographic group besides Native American Males and Native American Females, 1,000 samples are chosen to construct a balanced set of data. The lack of representation of Native Americans within the Pinellas database results

---

[1]The mug shot data used in these experiments is acquired in the public domain through Florida's "Sunshine" laws. Subjects shown in this manuscript may or may not have been convicted of a criminal charge, and thus should be presumed innocent of any wrongdoing.

*Figure 9.1   Pinellas database image samples*

in only 605 Native American Male and 168 Native American Female samples being available.

## 9.2.2   Preprocessing

Prior to their use in soft biometric classification the facial image requires processing. The processing steps of geometric normalization, histogram equalization, and periocular region extraction are described in the following sections.

### 9.2.2.1   Geometric normalization

An important first step in processing is image alignment. Similar to the yaw, pitch, and roll of an airplane, the face can be rotated along the axis from the face to the observer (in-plane rotation), or rotated along the vertical or horizontal axis (out-of-plane rotation), or both. In facial recognition, the generally accepted points of alignment to correct in-plane rotation are the eye centers. To perform a successful correction, the location of the eye centers must be known. A commercial system was used for detecting the eye centers in the Pinellas database. In the event that the commercial system fails to find the eye center coordinates, the eye centers are manually labeled.

Using the eye centers, a rotation matrix $R$ was generated as follows and applied to an image to correct in-place rotation.

$$R = \begin{bmatrix} \cos(\theta) & \sin(\theta) \\ -\sin(\theta) & \cos(\theta) \end{bmatrix}$$

where

$$\theta = \tan^{-1}\left( \sqrt{\frac{(r_y - l_y)^2}{(l_x - r_x)^2}} \right)$$

where $l$ is the pixel location of the left eye and $r$ is the pixel location of the right eye.

Left eye periocular region

Geometric
Normalization

Cropping

Scaling

Histogram
Equalization

Right eye periocular region

*Figure 9.2   Image preprocessing flow*

## 9.2.2.2   Histogram equalization

The next preprocessing step is to perform histogram equalization, in which the contrast of the image is enhanced. The purpose of this step is to normalize the relative levels of illumination between images by altering pixel intensity values so that the size of each bin of the histogram of pixel intensities is approximately equal. This ensures that images which are overall darker or brighter than a neutral image are adjusted to have approximately the same illumination intensity.

Histogram equalization is a three-step process consisting of calculating the histogram of the image, calculating the cumulative distribution function of the histogram, and updating the intensity values of each pixel. The histogram of an image is the count of the occurrences of each intensity value and is calculated by

$$H(I)[k] = \sum_{i=0}^{w} \sum_{j=0}^{h} \begin{Bmatrix} 1 & \text{if } I(i,j) = k \\ 0 & \text{otherwise.} \end{Bmatrix}$$

where $I$ is an image of $w \times h$ pixels and $k$ is an intensity value.

The cumulative distribution function (CDF) of the histogram is the accumulated sum of the histogram values up to a certain intensity level. The CDF is calculated and a transpose function is applied over it to get the equalized intensity values. All the pixel intensity values are then replaced by the calculated intensity values based on their corresponding values.

## 9.2.2.3   Periocular region extraction

After an image has been corrected for in-plane rotation and the histogram has been equalized, the periocular regions are extracted by placing a square bounding box around each eye center location with sides of length given by the distance between two eye centers. This distance is also known as the interocular distance. Figure 9.2 depicts the process of extracting out the periocular regions from a facial image.

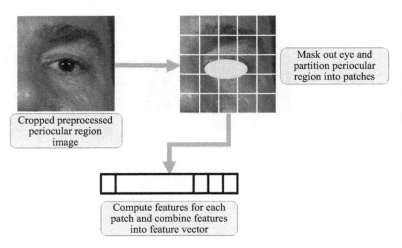

*Figure 9.3   Local appearance feature approach*

## 9.2.3   Feature representations

The periocular region of an individual is represented by first masking out the iris region of the periocular image to eliminate its effect on classification performance. The periocular region is then partitioned into patches and local feature representation are computed for each patch. The feature vectors for each patch are then concatenated into a feature vector for use in soft biometric classification, as depicted in Figure 9.3.

There are a number of advantages for using the local appearance based method for feature representation which include preservation of some spatial information, facilitation of feature level fusion, simple comparison of feature vectors, and robustness to small pose changes. Six local appearance feature representations which include Local Binary Patterns (LBP), Local Ternary Patterns (LTP), Local Salient Patterns (LSP), Local Phase Quantization (LPQ), Local Color Histograms (LCH), and Histograms of Gabor Ordinal Measures (HOGOM) are used for feature representation and described in the following sections.

### 9.2.3.1   Local binary patterns

Local Binary Patterns (LBP) [17] is a texture descriptor widely used in various applications due to its computational simplicity and discriminative power. Its robustness to monotonic grayscale changes caused by illumination variations makes it useful in real-world applications. LBP takes a local neighborhood around each pixel, thresholds the neighboring pixels with respect to the center pixel and labels each pixel as an integer using binary coding of the thresholded values. In the original approach, $3 \times 3$ neighborhoods were used for producing 8-bit codes using eight neighboring pixels around every pixel. The LBP operator can be represented as

$$\text{LBP}(x_c, y_c) = \sum_{n=0}^{7} 2^n \text{sgn}(g_n - g_c) \tag{9.1}$$

where $(x_c, y_c)$ denotes the center pixel, $g_n$ and $g_c$ represent the image intensities of the neighboring pixel and center pixel, respectively, and sgn$(u)$ represents the sign function in which sgn$(u) = 1$ if $u \geq 0$. A sample LBP code is shown in Figure 9.4(a). After the pixels are labeled by corresponding LBP code, the image is divided into non-overlapping regions and a histogram of LBP codes in each region is computed. The final LBP feature descriptor is obtained by concatenating the histogram of LBP codes from all the regions.

Two extensions were proposed to LBP computation in [18]. The first extension was proposed to handle textures at different scales by defining the local neighborhood as a set of $P$ evenly spaced points on a circle of radius $R$ centered at every pixel. Bilinear interpolation was used when the neighboring pixels lied between actual image pixels. The second extension defined *uniform patterns* which consist of at most two transitions from 0 to 1 and vice versa in the LBP code. While computing LBP histograms, each uniform pattern represented a bin while all non-uniform patterns were grouped into one bin. Ahonen *et al.* [19] inferred that 85.2% of texture patterns were uniform for FERET face images.

### 9.2.3.2 Local ternary patterns

Since LBP codes threshold exactly using the center pixel intensity, LBP can be sensitive to noise especially in near-uniform image regions. To provide a texture representation that is more robust to such noises, Local Ternary Patterns (LTP) [20] was proposed. LTP uses three-valued ternary codes as against the binary codes used in LBP. Given a center pixel intensity $g_c$, neighboring pixel intensities in the zone $g_c \pm t$ are set to 0 while those outside this zone are set to $\pm 1$. LTP thresholding can be represented by

$$s(g_n, g_c, t) = \begin{cases} -1 & g_n \leq g_c - t \\ 0 & |g_n - g_c| < t \\ 1 & g_n \geq g_c + t \end{cases} \tag{9.2}$$

where $g_n$ represents intensity of neighboring pixel.

To allow coding of ternary LTP codes, they are divided into positive and negative parts treating those as separate descriptors. A sample LTP code is shown in Figure 9.4(b). The image is divided into non-overlapping regions and a histogram of positive and negative LTP codes in each region is computed. The final LTP feature descriptor is obtained by concatenating the histogram of positive and negative LTP codes from all the regions.

### 9.2.3.3 Local salient patterns

Local Salient Patterns (LSPs) [21] was proposed to handle noise sensitivity by encoding only salient comparisons of neighborhood differential information. Two main steps for salient comparisons include:

- Computing local comparisons for a certain order – For a $3 \times 3$ neighborhood, let $g_1, \ldots, g_8$ represent the image intensities of neighboring pixels of a center pixel with intensity $g_c$. The 0-order local comparisons are given by $V_{0,i} = g_i - g_c$

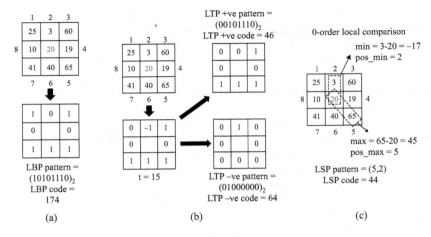

*Figure 9.4    Coding for (a) LBP, (b) LTP, and (c) LSP*

where $i$ denotes the neighboring pixel index. Similarly, 1-order local comparisons can be represented as $V_{1,i} = V_{0,i} - V_{0,i+1}$, 2-order local comparisons as $V_{2,i} = V_{1,i} - V_{1,i+1}$, and so on. LBP uses only 0-level local comparisons.

- Finding extremum values of comparisons for each order – After computing $k$-order local comparisons, the tuple $(pos_{max}, pos_{min})$ representing the index of maximum and minimum values of the local comparisons is computed. For 8 neighboring pixels, this represents 57 $(A(8,2) + 1)$ patterns where A represents the number of possible permutations such that $pos_{max} \neq pos_{min}$ and 1 pattern where all comparisons are the same. The pattern corresponding to this tuple containing the positions of extremum values of local comparisons constitutes the LSP code for a certain order.

For a given image, each pixel is labeled using the LSP code for different orders. A sample LSP code is shown in Figure 9.4(c). The image is divided into non-overlapping regions and histogram of LSP codes is computed for each region. The final LSP feature descriptor is obtained by concatenating the histogram of LSP codes from all regions for different orders.

### 9.2.3.4    Local phase quantization

Local Phase Quantization (LPQ) [22] takes advantage of the blur invariant property of Fourier phase spectrum to provide a feature representation robust to image blurring and illumination changes. It has been used in face recognition [23,24] and emotion recognition [25,26] to address blurring and illumination challenges. It involves computing Fourier transform coefficients, decorrelation and quantization.

2D DFT is computed over a $m \times m$ neighborhood $\mathcal{N}_x$ at each pixel position $\mathbf{x}$ of image $f(\mathbf{x})$ given by

$$F(\mathbf{u},\mathbf{x}) = \sum_{\mathbf{y} \in \mathcal{N}_x} f(\mathbf{x} - \mathbf{y})e^{-j2\pi \mathbf{u}^T \mathbf{y}} = \mathbf{w}_{\mathbf{u}}^T \mathbf{f}_{\mathbf{x}} \qquad (9.3)$$

where $\mathbf{w_u}$ the basis vector of DFT at frequency $\mathbf{u}$ and $\mathbf{f_x}$ represents $m^2$ samples in $\mathcal{N_x}$. To ensure blur invariance, four low frequency coefficients corresponding to frequencies $\mathbf{u}_1 = [a, 0]^T$, $\mathbf{u}_2 = [0, a]^T$, $\mathbf{u}_3 = [a, a]^T$, and $\mathbf{u}_4 = [a, -a]^T$ are chosen where $a = 1/m$. This can be represented as

$$\mathbf{F_x} = \mathbf{W}\mathbf{f_x} \tag{9.4}$$

where $\mathbf{W} = [\text{Re}\{\mathbf{w_{u_1}}, \mathbf{w_{u_2}}, \mathbf{w_{u_3}}, \mathbf{w_{u_4}}\}, \text{Im}\{\mathbf{w_{u_1}}, \mathbf{w_{u_2}}, \mathbf{w_{u_3}}, \mathbf{w_{u_4}}\}]^T$.

Before quantizing these coefficients, they need to be decorrelated since neighboring pixels are highly correlated in natural images. Let the correlation coefficient between adjacent pixels be $\rho$ and covariance between two positions $\mathbf{x}_i$ and $\mathbf{x}_j$ be $\sigma_{ij} = \rho^{\|\mathbf{x}_i - \mathbf{x}_j\|}$. The covariance matrix of $\mathbf{f_x}$ can be represented as

$$C = \begin{bmatrix} 1 & \sigma_{1,2} & \cdots & \sigma_{1,m^2} \\ \sigma_{2,1} & 1 & \cdots & \sigma_{2,m^2} \\ \vdots & \vdots & \ddots & \vdots \\ \sigma_{m^2,1} & \sigma_{m^2,2} & \cdots & 1 \end{bmatrix} \tag{9.5}$$

The covariance matrix of $\mathbf{F_x}$ is given by

$$\mathbf{D} = \mathbf{W}\mathbf{C}\mathbf{W}^T \tag{9.6}$$

Decorrelation is achieved using a whitening transform

$$\mathbf{G_x} = \mathbf{V}^T \mathbf{F_x} \tag{9.7}$$

where $\mathbf{V}$ is obtained by singular value decomposition (SVD) of $\mathbf{D} = \mathbf{U}\Sigma\mathbf{V}^T$.

$\mathbf{G_x}$ is computed at all pixel positions and quantized using the sign of each component of $\mathbf{G_x}$. The quantized coefficients are represented as a 8-bit integer using binary coding. A histogram of these 8-bit codes is then computed for non-overlapping regions of an image and final LPQ descriptor is obtained by concatenating these 256-dimensional histograms from all regions.

### 9.2.3.5 Local color histograms

Local color histograms (LCHs) is a texture representation with remarkable discriminative power and computational simplicity. LCH quantizes color information using a 16-bin histogram. Only red and green color channels are used since red channel captures skin textures effectively and green channel contains luminance information. Each color channel intensity is quantized into 4 values spaced uniformly across 256 possible values. The combination of two color channels is thereby represented by a 16-bin histogram. LCH is computed for non-overlapping regions of an image and concatenated to provide the LCH feature descriptor.

### 9.2.3.6 Histogram of Gabor ordinal measures

Histogram of Gabor Ordinal Measures (HOGOMs) [27] uses a combination of Gabor filters [28] and Ordinal Measures [29] for texture representation. Gabor filters capture rich texture information by encoding local structure of an image at specific frequencies and scales while preserving spatial relations. Ordinal measures encode qualitative relationships between adjacent regions thereby representing local structures of different complexities. It can be considered as a generalization of LBP by expanding beyond adjacent pixels to adjacent regions. Hence, the combination of these two features provides a rich texture representation while being robust to noise. HOGOM feature extraction consists of Gabor filter convolution, computing Gabor Ordinal Measures, binary coding of Gabor Ordinal measures, and computing histogram of Gabor Ordinal Measures.

The Gabor magnitude images are computed by convolving an image $I(x, y)$ with a Gabor filter bank

$$\psi_{\mu,\nu}(z) = \frac{||k_{\mu,\nu}||^2}{\sigma^2} e^{\left(-\frac{||k_{\mu,\nu}||^2||x||^2}{2\sigma^2}\right)} \left[ e^{ik_{\mu\nu x}} - e^{-\sigma^2/2} \right] \tag{9.8}$$

where orientation $\mu \in \{0, \ldots, 7\}$ and scale $\nu \in \{0, \ldots, 4\}$.

The Gabor Ordinal Measures (GOMs) images are obtained by convolving the Gabor magnitude images with ordinal filters. Four ordinal filters are used – two lobes vertical, two lobes horizontal, three lobes vertical and three lobes horizontal. Hence, each ordinal filter produces 40 GOM images for eight orientations and five scales of Gabor wavelets. Since the ordinal codes are binary, given an ordinal filter at a certain scale, the GOM of all eight orientations at a pixel position are fused to produce a 8-bit representation. Hence, we get a codified GOM image for each ordinal filter at each scale. These codified GOM images are partitioned into non-overlapping regions to compute histograms of GOM.

All the features described above are histogram-based and are hence sparse. Further, features like HOGOM are high-dimensional. To reduce the overall size of the feature representations, Principal Component Analysis (PCA) is applied such that at least 95% of the variance is retained. A reduced size set of feature vectors are now available for soft biometric classification.

## 9.2.4 Classification

Support Vector Machines (SVMs) are used to perform the gender and ethnicity classification. The objective of SVM is to find an optimal hyperplane that separates samples of different classes with a maximum margin. Let $x_1, \ldots, x_N$ represent $N$ training samples in $d$ dimensions, i.e. $x_i \in R^d$. Let $y_1, \ldots, y_N$ represent the corresponding binary labels such that $y_i \in \{-1, 1\}$. SVM attempts to find the optimal hyperplane

$$f(x) = w^T \phi(\mathbf{x}) + b \tag{9.9}$$

where weight vector, $w = \sum_{i=1}^{N} \alpha_i y_i \phi(x_i)$, $\alpha_i$ denotes the Lagrange multipliers and $b$ is bias term. Training samples with $\alpha_i > 0$ are support vectors and lie on the margin of the hyperplane. $\phi(\mathbf{x})$ represents high-dimensional mapping used by the kernel function $k(\mathbf{x}, \mathbf{x}_i) = \phi(\mathbf{x}) \cdot \phi(\mathbf{x}_i)$. For a linear kernel, $\phi(\mathbf{x}) = \mathbf{x}$ and $k(\mathbf{x}, \mathbf{x}_i) = \mathbf{x}^T \mathbf{x}_i$. With non-linear kernels, the training samples are projected to a higher dimensional space and an optimal hyperplane is found in higher dimensional space. However, one does not have to compute $\phi(\mathbf{x})$ explicitly and can use the *kernel trick* to compute $k(\mathbf{x}, \mathbf{x}_i)$. Some non-linear kernels include:

- Polynomial: $k(\mathbf{x}, \mathbf{x}_i) = ((\mathbf{x}.\mathbf{x}_i) + 1)^p$
- Radial basis function (RBF): $k(\mathbf{x}, \mathbf{x}_i) = e^{-\gamma||\mathbf{x} - \mathbf{x}_i||^2}$, $\gamma > 0$

For gender and ethnicity classification, a one-vs-one multi-class classification approach is used. For every feature and $N$ classes, $N(N-1)/2$ binary SVM classifiers are used. The resulting classification is made based on which class has most number of binary classifications made towards it. The classification results of multiple feature representations are then fused using a simple voting scheme. The class with most votes is determined as the final classification decision.

## 9.3 Experiment results

### 9.3.1 Experiment protocol

The experimental design follows a fivefold cross validation scheme. In this scheme, the experiment is conducted five times with a different data partition each time. The data is split into fifths, ensuring that the data is equally split per class, and four sets are used for training and one set is used for testing. The five iterations of the experiment allows for all subsets of data to be used in the test set. The results of each experiment are reported as the average classification accuracy of the five iterations. For performance comparison purposes, a similar set of experiments are performed using local appearance features which have been extracted from the entire face. Three separate types of experiments are performed – gender only classification (two classes), ethnicity only classification (five classes) and gender and ethnicity classification (ten classes).

### 9.3.2 Gender classification

The performance of each of the local appearance feature representations is shown in Table 9.2. Of the six feature representations investigated, Local Binary Patterns (LBPs), Local Color Histograms (LCHs), and Local Phase Quantization (LPQ) performed well, suggesting that the differences between the periocular region appearance can be adequately captured in these local appearance feature representations. The remaining three feature representations, Local Salient Patterns (LSPs), Local Ternary Patterns (LTPs), and Histograms of Gabor Ordinal Measures (HOGOMs) did not perform nearly as well.

The small difference in gender classification performance between experiments involving only the periocular region and those involving the use of the entire face seems to support the assertion that the determination of an individual's gender can be made reasonably accurately using information only from periocular region of the face. Examination of the gender classification results indicate that a number of gender misclassification instances of the three best performing local representations are identical leading one to speculate that combining the results of the individual feature representations would not lead to a substantial improvement in overall gender classification performance. This is verified by an average of only one percent increase in gender classification performance when using the combination the three best performing feature representations. Further, we expect periocular region to outperform iris with respect to gender recognition [30,31].

*Table 9.2    Gender classification performance*

| Features | Left periocular (%) | Right periocular (%) | Face (%) |
|----------|---------------------|----------------------|----------|
| LBP | 88.86 | 89.48 | 95.19 |
| LCH | 86.61 | 86.39 | 93.82 |
| LPQ | 87.77 | 87.57 | 94.19 |
| LSP | 59.35 | 60.93 | 58.62 |
| LTP | 52.50 | 52.50 | 52.50 |
| HOGOM | 52.50 | 52.50 | 52.50 |

*Table 9.3    Ethnicity classification performance*

| Features | Left periocular (%) | Right periocular (%) | Face (%) |
|----------|---------------------|----------------------|----------|
| LBP | 69.27 | 69.04 | 78.38 |
| LCH | 64.44 | 64.52 | 73.90 |
| LPQ | 67.20 | 66.59 | 76.74 |
| LSP | 30.24 | 29.83 | 27.58 |
| LTP | 22.82 | 22.82 | 22.82 |
| HOGOM | 22.82 | 22.82 | 22.82 |

## 9.3.3    Ethnicity classification

Ethnicity classification experiments involve the classification of periocular and facial images correctly into one of the five classes of Black, Caucasian, Hispanic, Asian, and Native American. The results of these experiments are listed in Table 9.3. As with the gender classification experiments, LBP, LCH, and LPQ local feature representations perform best for ethnicity classification. However, this performance is considerably less that that achieved for gender classification. This is partially due to the increased number of possible classes compared to gender classification.

One of the classes in particular, the Hispanic category, was responsible for the largest number of ethnicity misclassifications. One possible explanation would be that the Hispanic ethnicity does not correspond directly to a specific "biogeographic region of origin" compared to the other ethnic groups. Individuals who may belong to the Hispanic ethnic group may have backgrounds (features) that are much more diverse. The ethnicity classification performance when using only the periocular region is not as close to that when using the entire face. This is probably due to the increased complexity of ethnic classification as compared to gender classification. Similarly as experienced during gender classification experiments, combining the performance of the three best performing representations leads to a marginal increase (approximately one percent) in ethnicity classification performance.

*Table 9.4  Gender and ethnicity classification performance*

| Features | Left periocular (%) | Right periocular (%) | Face (%) |
|---|---|---|---|
| LBP | 60.23 | 58.95 | 71.53 |
| LCH | 55.98 | 54.75 | 67.60 |
| LPQ | 55.00 | 56.39 | 69.12 |
| LSP | 16.65 | 15.45 | 14.16 |
| LTP | 11.42 | 11.42 | 11.42 |
| HOGOM | 11.42 | 11.42 | 11.42 |

## 9.3.4  Gender and ethnicity classification

In the final set of experiments, the periocular and facial images are to be classified into a total of 10 classes consisting of Black Male, Black Female, Caucasian Male, Caucasian Female, Hispanic Male, Hispanic Female, Asian Male, Asian Female, Native American Male, and Native American Female. The results of these experiments are listed in Table 9.4

As with the previous experiments, LBP, LCH, and LPQ local feature representations perform significantly better than the others. The performance difference between the three best performing local feature representations is much more pronounced compared to that observed in gender and ethnicity only classification experiments. The performance when using the entire face region was at least 11% higher than that when using only the periocular region. As observed in ethnicity only experiments, the Hispanic classes are responsible for a number of misclassifications. In addition, the Native American classes were also were difficult to accurately distinguish. It is suspected that the difficulty in the classification of the Native American classes was due to their lack of representation in the data set.

## 9.4  Summary

Although biometric technology has advanced tremendously over the past decade, high performance in challenging operational scenarios such as large-scale identification and identification using partial data has not been realized. The use of ancillary information from soft biometric classification has been proposed as a means to address these challenges. The face can reveal much information about an individual such as their identity, age, gender, ethnicity, and emotional state making it ideal for use in soft biometric classification tasks. However, there are situations in which the entire face is not visible. In these situations, there is a possibility that the periocular region can be used for soft biometric classification.

This chapter presents results which suggest that gender classification based upon local appearance-based representations of the periocular region is possible and comparable to performance achieved when using the entire face region. Ethnicity classification using the periocular region has shown to be a more challenging problem. The diversity of features within certain classes due the absence of a specific biogeographic region of origin makes classification based on ethnicity difficult. Future directions for exploration would include a detailed analysis of how certain periocular features are distributed within particular ethnic groups and perhaps the development of a new classification system in which classes are not based upon established definitions of ethnicity.

# References

[1]    J. L. Wayman. "Large-scale civilian biometric systems – issues and feasibilty," in *Proceedings of Card Tech/Secur Tech ID*, 1997.

[2]    Anil K. Jain, Sarat C. Dass, and Karthik Nandakumar. "Can soft biometric traits assist user recognition?" in *Defense and Security*, pp. 561–572. Bellingham, WA: International Society for Optics and Photonics, 2004.

[3]    Srinivas Gutta, Jeffrey R. J. Huang, P. Jonathon, and Harry Wechsler. "Mixture of experts for classification of gender, ethnic origin, and pose of human faces," *IEEE Transactions on Neural Networks*, 11(4):948–960, 2000.

[4]    Baback Moghaddam and Ming-Husan Yang. "Learning gender with support faces," *IEEE Transactions on Pattern Analysis and Machine Intelligence*, 24(5):707–711, 2002.

[5]    Koray Balci and Volkan Atalay. "PCA for gender estimation: which eigenvectors contribute?" in *Proceedings of the 16th IEEE International Conference on Pattern Recognition, 2002*, vol. 3, pp. 363–366, 2002.

[6]    Bo Wu, Haizhou Ai, and Chang Huang. "LUT-based adaboost for gender classification," in *Audio-and Video-Based Biometric Person Authentication*, pp. 104–110, 2003.

[7]    Satoshi Hosoi, Erina Takikawa, and Masato Kawade. "Ethnicity estimation with facial images," in *Proceedings of the Sixth IEEE International Conference on Automatic Face and Gesture Recognition, 2004*, pp. 195–200, 2004.

[8]    Chiraz BenAbdelkader and Paul Griffin. "A local region-based approach to gender classification from face images," in *IEEE Computer Society Conference on Computer vision and pattern recognition-workshops, 2005. CVPR Workshops*, pp. 52–52, 2005.

[9]    Agata Lapedriza, Manuel J Marin-Jimenez, and Jordi Vitria. "Gender recognition in non-controlled environments," in *18th IEEE International Conference on Pattern Recognition, 2006. ICPR 2006*, vol. 3, pp. 834–837, 2006.

[10]   Xiaoguang Lu, Hong Chen, and Anil K Jain. "Multimodal facial gender and ethnicity identification," in *International Conference on Biometrics*, pp. 554–561, 2006.

[11]   Zhiguang Yang, Ming Li, and Haizhou Ai. "An experimental study on automatic face gender classification," in *18th IEEE International Conference on Pattern Recognition, 2006. ICPR 2006*, vol. 3, pp. 1099–1102, 2006.

[12] Zhiguang Yang and Haizhou Ai. "Demographic classification with local binary patterns," in *International Conference on Biometrics*, pp. 464–473, 2007.

[13] Erno Mäkinen and Roope Raisamo. "Evaluation of gender classification methods with automatically detected and aligned faces," *IEEE Transactions on Pattern Analysis and Machine Intelligence*, 30(3):541–547, 2008.

[14] Ziyi Xu, Li Lu, and Pengfei Shi. "A hybrid approach to gender classification from face images," in *19th IEEE International Conference on Pattern Recognition, 2008. ICPR 2008*, pp. 1–4, 2008.

[15] Wei Gao and Haizhou Ai. "Face gender classification on consumer images in a multiethnic environment," in *International Conference on Biometrics*, pp. 169–178, 2009.

[16] Chuan Chin Teo, Han Foon Neo, and A. Beng Jin Teoh. "A study on partial face recognition of eye region," in *International Conference on Machine Vision, 2007. ICMV 2007*, pp. 46–49, Dec. 2007.

[17] Timo Ojala, Matti Pietikainen, and Topi Maenpaa. "Multiresolution gray-scale and rotation invariant texture classification with local binary patterns," *IEEE Transactions on Pattern Analysis and Machine Intelligence*, 24(7):971–987, 2002.

[18] Guillaume Heusch, Yann Rodriguez, and Sebastien Marcel. "Local binary patterns as an image preprocessing for face authentication," in *Seventh International Conference on Automatic Face and Gesture Recognition (FGR06)*, pp. 6–14, 2006.

[19] Timo Ahonen, Abdenour Hadid, and Matti Pietikainen. "Face description with local binary patterns: Application to face recognition," *IEEE Transactions on Pattern Analysis and Machine Intelligence*, 28(12):2037–2041, 2006.

[20] Xiaoyang Tan and Bill Triggs. "Enhanced local texture feature sets for face recognition under difficult lighting conditions," *IEEE Transactions on Image Processing*, 19(6):1635–1650, 2010.

[21] Zhenhua Chai, Zhenan Sun, Tieniu Tan, and Heydi Mendez-Vazquez. "Local salient patterns – a novel local descriptor for face recognition," in *2013 IEEE International Conference on Biometrics (ICB)*, pp. 1–6, 2013.

[22] Ville Ojansivu and Janne Heikkilä. "Blur insensitive texture classification using local phase quantization," in *Proceedings of the International Conference on Image and Signal Processing*, pp. 236–243, Jul. 2008.

[23] Chi Ho Chan, Josef Kittler, Norman Poh, Timo Ahonen, and Matti Pietikäinen. "(Multiscale) local phase quantisation histogram discriminant analysis with score normalisation for robust face recognition," in *2009 IEEE 12th International Conference on Computer Vision Workshops (ICCV Workshops)*, pp. 633–640, 2009.

[24] Masashi Nishiyama, Abdenour Hadid, Hidenori Takeshima, Jamie Shotton, Tatsuo Kozakaya, and Osamu Yamaguchi. "Facial deblur inference using subspace analysis for recognition of blurred faces," *IEEE Transactions on Pattern Analysis and Machine Intelligence*, 33(4):838–845, 2011.

[25] Songfan Yang and Bir Bhanu. "Facial expression recognition using emotion avatar image," in *2011 IEEE International Conference on Automatic Face & Gesture Recognition and Workshops (FG 2011)*, pp. 866–871, 2011.

[26] Abhinav Dhall, Akshay Asthana, Roland Goecke, and Tom Gedeon. "Emotion recognition using PHOG and LPQ features," in *2011 IEEE International Conference on Automatic Face & Gesture Recognition and Workshops (FG 2011)*, pp. 878–883, 2011.

[27] Zhenhua Chai, Ran He, Zhenan Sun, Tieniu Tan, and Heydi Mendez-Vazquez. "Histograms of Gabor ordinal measures for face representation and recognition," in *2012 Fifth IAPR IEEE International Conference on Biometrics (ICB)*, pp. 52–58, 2012.

[28] Ángel Serrano, Isaac Martín de Diego, Cristina Conde, and Enrique Cabello. "Recent advances in face biometrics with Gabor wavelets: a review," *Pattern Recognition Letters*, 31(5):372–381, 2010.

[29] Zhenan Sun and Tieniu Tan. "Ordinal measures for iris recognition," *IEEE Transactions on Pattern Analysis and Machine Intelligence*, 31(12):2211–2226, 2009.

[30] Vince Thomas, Nitesh V. Chawla, Kevin W. Bowyer, and Patrick J. Flynn. "Learning to predict gender from iris images," in *First IEEE International Conference on Biometrics: Theory, Applications, and Systems, 2007. BTAS 2007*, pp. 1–5, 2007.

[31] Denton Bobeldyk and Arun Ross. "Iris or periocular? Exploring sex prediction from near infrared ocular images," in *Proceedings of International Conference of the Biometrics Special Interest Group*, 2016.

*Chapter 10*

# Age predictive biometrics: predicting age from iris characteristics

*Márjory Da Costa-Abreu, Michael Fairhurst, and Meryem Erbilek*

As biometrics-based identity authentication systems have become more widely deployed, it has become evident that traditional identification and verification tasks are not the only application for such approaches. The prediction of individual, but non-unique, characteristics such as subject age is also an obvious option, since there are diverse situations in which information short of absolute identity is itself valuable. Physical ageing is an important issue for practical biometrics, since it is known that the associated physiological changes can impair performance for most modalities. Understanding the effects of ageing is necessary, therefore, both to optimise attainable performance but also to understand how to manage biometric templates, especially as the time elapsed between enrolment and use increases. Age prediction is relatively poorly represented in the literature. This chapter will explore applications of age prediction from iris biometrics and the implications for the underpinning computational structures.

## 10.1 Introduction

The field of biometrics, the identification of individuals from measurement of their physiological or behavioural characteristics, is now well established, and can provide practically viable solutions in many important application areas. However, for a number of years, there has been an increasing interest in exploiting the potential of biometric data to offer predictive capabilities for scenarios in which full identification of a specific individual is not the primary requirement, but where the aim is more generally the prediction of a 'soft' biometric marker, a piece of information which is characteristic of, but not unique to, an individual. A typical example of such an application might be the prediction of the age or gender of an individual. This type of prediction reveals a specific piece of information which is related to individual identity, but is nevertheless a characteristic which is common across a larger number of individuals. Despite the non-unique nature of the outcome of such a capacity for prediction, the process is nevertheless extremely valuable in a number of practical applications. This is perhaps most apparent in forensic analysis in criminal investigations, for example, or in providing security monitoring in electronic transactions, in subject profiling activities and a variety of other areas, including the assessment of entitlement to age-restricted goods and services [1].

Estimating a subject's age from biometric data is perhaps the most common and potentially most immediately valuable manifestation of this process. The literature shows that face [1,2], speech [3,4] and signature [5] for age prediction are the modalities which have received the most research attention in recent years. The age of an individual is particularly important personal information and is a characteristic which is commonly used in security transactions as well as in legal matters. The age of a subject can also have an impact on the overall design of a system as well as informing the planning of the enrolment and template update of a system user. Thus, age prediction of individuals which is based on available biometric data is increasingly becoming an important research topic in its own right [6]. Estimating the age of an individual may often be seen as an important factor in key applications such as forensics, medicine or in the support of criminal investigations [1].

In this context, age estimation uses the physical and/or behavioural characteristics embedded in an individual's biometric data, characteristics which will commonly vary over time and, as a result, provide the key to the required task. Reliable age estimation is typically a challenging task, and age progression has an effect not only on biometric systems which depend directly on behavioural characteristics such as signature, handwriting, keystroke dynamics, voice, etc., but also has an impact on biometric systems based on physiological features, such as iris, retina and fingerprint information, which are often regarded as being generally more stable over time. This is because the ageing process encompasses factors relating to health, environmental conditions, living and working environments, cosmetics and/or makeup usage, gender differences and so on. Increasingly, however, we see that a capability to predict age can be extremely influential in situations relating to networking, social interactions, security and an ability to regulate entitlement of access to age-restricted goods and services [7].

There is not a wide selection of studies regarding age estimation from iris. This may seem strange, since iris recognition is widely regarded as one of the most reliable biometrics and there has recently been an increasing interest in the development of iris recognition systems designed to capture eye images at a distance or while the subject is mobile (i.e. iris 'on the move'). Such scenarios broaden the scope and significantly enhance the potential usability of age prediction based on the iris modality for many important applications, including those noted above.

In this chapter we will specifically explore some of the challenges and advantages of performing age prediction from iris data, and will address questions about the principal factors which influence the likely success of iris-based age prediction. We will investigate experimentally how the availability of different types of features (such as might be dictated, e.g., by a specific technology or device configuration) will influence overall performance. In this way, we will take some important steps towards a better understanding of how to define an optimal mechanism to predict age from iris data. We will also broaden the base of our discussion by considering how we can enhance this predictive capability by integrating an iris-based system with an additional modality.

## 10.2    Background discussion and related studies

The importance of research concerning the estimation of age (or any other soft-biometric information) is not yet reflected in the number of studies reported in this area, which has not perhaps been as well explored as might be expected. We will first briefly consider some of the most relevant strands of work within the general area, and will include a limited discussion of some work to be found on bimodal studies since, as noted above, we will later report some relevant results which adopt a bimodal approach.

Considering prediction of soft-biometrics in general, studies in the literature show that the face modality has received the greatest attention (particularly for age and gender prediction) [2,8–15]. This is not really surprising since it is particularly natural and easy to acquire facial images in applications which involve, for example, criminal investigations or for networking purposes. There has also been a considerable degree of effort expended in investigating the prediction of gender and age from voice characteristics [3,16,17]. Some work can also be found which focuses on the estimation of soft-biometric information from gait [18,19].

If we turn to the estimation of age based on information derived from the hand-written signature modality, which is also particularly useful and relevant in criminal investigations, we find some relevant work presented in [5]. In this particular study, an analysis of how traditional classifiers behave, as well as how classifier fusion can support age prediction is presented. Three age bands of interest are adopted and the error rates for each individual band are analysed.

Physical ageing effects in signature biometrics have been investigated and explored in [20] to show quantitatively that it is very difficult categorically to identify clearly discernible trends when considering the relationship between error rate performance in signature biometrics as a function of age groupings within a user population.

The literature reports a variety of diverse studies concerning the predictive properties of the iris in relation to different demographic characteristics of individuals. For example, gender and ethnic group prediction from iris images is proposed in [21,22]. In [21], ethnic grouping prediction is carried out using only texture features of iris images, while in [22], gender prediction is carried out using both geometric and texture-based features.

Age prediction from iris characteristics is studied in [23] as already noted. This study proposes a classification technique which categorises each subject as belonging to either a 'young' or an 'old' age group as determined from the texture-based characteristics of the iris. The iris biometric data used in this study have been collected at the University of Notre Dame and are not generally publicly available. Biometric samples in the 'young' group are derived from 50 subjects (with 3 left and 3 right eye samples) whose ages lie between 22 and 25 years, while the 'old' group consists of 48 subjects (again with 3 left and 3 right eye samples) whose ages are more than 35 years. Hence, a total of 98 subjects with 6 samples from each are used for this experimental study. Initially, 630 features were computed from the segmented and normalised iris texture and classification then performed with the Random-Forest algorithm from the Weka software, using 300 trees. Experimental results have shown that the correct classification rate achieved using this method is 64.68%.

A particularly relevant earlier analysis of ageing issues in iris biometrics, which can be found in [24], shows that physical ageing effects with respect to the iris are primarily the result of the physiology of pupil dilation mechanisms, with pupil dilation responsiveness decreasing with age. Since pupil dilation is clearly related to the geometric appearance of the pupil and the iris, this study is valuable in the present context in suggesting that geometric features of the iris may also provide useful information for the task of predicting age from the iris image.

Also, in [6], the authors have investigated, analysed and documented the effects of different possible age-band assignments (the age categories which are to be used for the prediction process), in order to guide and enhance the management of age-related data and take a step towards the possibility of more objectively determining the optimal age-bands to adopt, in the sense of defining which set of age bands offer the greatest possibility of minimising the sensitivity of a system which relies on such information. According to the results presented, it is suggested that a structure which divides a test population into the three age bands defined by the boundaries '< 25', '25–60' and '> 60' is the partition which best reflects age-related trends and provides useful information to support both the analysis and practical management of age-related factors in iris-based biometric systems. In [25], the same authors extended the investigation of feature and classification impact in the age prediction task using these three age bands.

Hence, on the basis of the discussion presented, the techniques proposed in [25] explore the age prediction task using only five simple geometric features extracted from iris images. This approach not only reflects some fundamental iris properties, but provides the basis of a technique which is simpler and computationally both less expensive and faster than the requirements of computing texture-based features of the iris. In this approach, the age groups are defined to be based on three broad age groupings, which may be described as relating to the general categories 'young', 'middle aged' and 'older', which is a wider experimental study than that reported previously which evaluated only a two class problem [23].

As this brief overview of some of the relevant literature has demonstrated, even though it is possible to find some interesting and important work dealing with age estimation from iris information or iris combined with various different biometric modalities, studies focusing on this issue are not very extensive.

## 10.3 Predicting age in diverse scenarios

The basic approach to the processing of biometric data in our study of iris-based age prediction can be described as follows:

1. An eye image is captured in an *Acquisition step*. In our study all the data are taken from a database, and thus the details of this step will be dependent on the database used, but most of the available iris databases have been acquired in a standard 'office' environment. During the acquisition, spectacles are not allowed to be worn by subjects, although contact lenses are allowed. A limited number of samples are collected, traditionally acquiring four eye images per person (two from the left eye and two from the right).
2. There is also a *Segmentation step* that aims to localise the iris region within the overall acquired eye image. This step involves detection of the sclera/iris and pupil/iris boundaries. In the results to be presented here, each eye sample is segmented using a robust segmentation algorithm as described in [24]. Subsequently, the obtained iris and pupil parameters (which are specifically, the $x$-coordinate, $y$-coordinate and radius values) from the segmentation process are restored. In the case of the texture features, we have adopted the same features proposed and described in [22] for gender classification tasks.
3. After a segmentation stage, a *Normalisation step* is performed to transform the iris region into a fixed rectangular block, so that the iris region from the overall eye image is presented at the fixed size necessary for comparisons between samples. A technique [26] based on Daugman's rubber sheet model [27,28] is employed for this normalisation process, which produces a 2D array with horizontal dimensions of angular resolution and vertical dimensions of radial resolution. In this study the process produces unwrapped images of size $20 * 240$.
4. The *Feature Correlation step* is the next phase and aims to apply a correlation test to remove correlated geometric features to allow us to use only the more distinguishing and non-redundant features for the further processing in the proposed age estimation process of our experimental study. For the results presented here, the inter-feature correlations are evaluated by using Spearman's rank correlation approach [29] (a nonparametric-based estimate of correlation).
5. Finally, there is the *Prediction step* itself, which uses the data generated at the output of the previous step and performs the age prediction task according to the three age categories chosen.

The age-bands adopted generally in age prediction studies reported in the literature are found to vary considerably [3,16,30,31] and, to a large extent, depend on factors such as the biometric modality used. The choices made in this respect in different studies, often with no specified rationale, make inter-study comparisons and, indeed, an informed or objective choice of age bands, extremely difficult. The work reported in [6] has analysed this aspect of predictive biometrics and, as we noted earlier, has suggested a set of age-band assignments which can be considered appropriate to many studies. Accordingly, we have adopted these age bands in the experimental

investigation which is the focus of this chapter, and thus divide our database population into the three age bands defined by the boundaries $<$ 25 years, 25–60 years and
$>$ 60 years. This offers the benefits not only of adhering to the objectively determined
estimate of what might be considered best practice, but also turns out to allow us to
maintain a good representation of subjects in each of the age categories.

## 10.4  Experimental infrastructure for iris-based predictive biometrics

In the previous sections we have discussed some of the issues underpinning the experimental study which is at the heart of this chapter. In particular, we have briefly
described the overall processing structure adopted, and discussed the fundamental
issue of defining the age bands used in the prediction process. We now consider the
other principal components of the experimental study which we will present subsequently. Specifically, we will describe three key factors which are likely to influence
the performance available in the predictive task, namely the features which we extract
from the iris data available, the actual database of iris images which we adopt for
the study, and the range of classifiers which we use to execute the age estimations
themselves.

In the study presented here we are especially interested in two different feature
classes, and we will consider the predictive performance achievable with these two
feature classes in order to evaluate the practical impact of using either unimodal or
multimodal data in the age prediction process, and also to determine the extent to
which the classifier characteristics can compensate for non-enhancing performance
modulations arising from these different configurations. In particular, we wish to
explore the potential for adopting more powerful and intelligent decision-level fusion
structures.

Considering first the features which are chosen for the experimentation, we use
features which are related, on the one hand, to the geometry of the iris and, on the
other hand, features which are related to the texture of the iris.

In the case of the geometric feature extraction, five iris and pupil parameters (which are specifically: the $x$-coordinate ($x_i$ for the iris and $x_p$ for the pupil),
$y$-coordinate ($y_i$ for the iris and $y_p$ for the pupil) and radius values ($r_i$ for the iris
and $r_p$ for the pupil)) from the segmentation and the correlation processes (presented
in [25]) are determined and used to generate 12 geometric features, as shown in
Table 10.1.

For the texture features, the 1D Log-Gabor wavelets are used to encode
features [26]. Each row of the 2D normalised iris pattern corresponds to a circular ring on the iris region. These rows are divided into a number of 1D signals and
convolved with 1D Log-Gabor wavelets which outputs a template of size $20 * 480$
with both real and imaginary components. As in [22], we use only real components
(which corresponds to the array of complex numbers of size $20 * 240$ of the template).
The six texture features thereby extracted are specified in Table 10.2.

*Table 10.1 Iris GEOMETRICAL features*

| Number | Feature description | Formula |
|--------|---------------------|---------|
| GF1 | Scalar distance $x$ and pupil | $\mid p_x - i_x \mid$ |
| GF2 | Scalar distance $y$ and pupil | $\mid p_y - i_y \mid$ |
| GF3 | Scalar distance iris and pupil | $\mid GF1 - GF2 \mid$ |
| GF4 | Total area of the iris | $\pi * i_r^2$ |
| GF5 | Total area of the pupil | $\pi * p_r^2$ |
| GF6 | True area of the iris | $GF4 - GF5$ |
| GF7 | Area ratio | $\frac{GF4}{GF5}$ |
| GF8 | Dilation ratio | $i_r/p_r$ |
| GF9 | Iris circumference | $p_i * 2 * i_r$ |
| GF10 | Pupil circumference | $p_i * 2 * p_r$ |
| GF11 | Circumference ratio | $\frac{GF9}{GF10}$ |
| GF12 | Circumference difference | $GF9 - GF10$ |

*Table 10.2 Iris TEXTURE features*

| Number | Feature description |
|--------|---------------------|
| TF1 | Mean of the real components in row $x$ |
| TF2 | Standard deviation of the real components in row $x$ |
| TF3 | Variance of the real components in row $x$ |
| TF4 | Mean of the real components in col $y$ |
| TF5 | Standard deviation of the real components in col $y$ |
| TF6 | Variance of the real components in col $y$ |

The database used in our study is the Data Set 2 (DS2) of the BioSecure Multimodal Database (BMDB)[1] which was collected as part of an extensive multimodal database by 11 European institutions participating in the BioSecure Network of Excellence [32], and is a commercially available database. The multimodal nature of the database allowed us to use both the basic iris data, but also consider a second modality (in this case, the handwritten signature) from the same subjects, in order to evaluate both a unimodal and a multimodal configuration.

The 210 subjects providing the iris samples contained in this database are within the age range of 18–73 years. The iris samples of 10 subjects were found to be incorrectly labelled (some of the left eye samples labelled as right or right eye samples labelled as left). Hence, those subjects are removed from both the handwritten signature and the iris component datasets, thus reducing the available number of subjects to 200.

---

[1]Available from http://biosecure.it-sudparis.eu/AB/getdatabase.php

*Table 10.3   Number of subjects per age band*

| Sets | <25 | 25–60 | >60 |
|------|------|-------|-----|
| All | 70 | 115 | 15 |
| Train | 50 | 82 | 11 |
| Test | 20 | 33 | 4 |

We have used a leave-one-out [33] technique for the training of the classifiers. This process is straightforward for the more commonly encountered identity-based classification tasks, since the division of a population into the testing and the training sets is performed according to the identity labels of individuals. However, this process becomes a more challenging and subtle issue in the case of the age classification tasks, since the division of a population into the testing and the training sets cannot be performed only according to the age labels of individuals. Identity labels should also be taken into account and the same individuals' samples should not be included both in the testing and the training sets. If this is not taken into account, then the age classification task might act as if it were an identity classification task. Studies related to age classification using biometric characteristics reported in the literature do not always appear to address this issue.

In this study, we have therefore considered both the age and the identity labels of subjects while dividing the overall population into the testing and the training sets. Hence, we make sure that the same subjects' samples are only included in the testing or in the training set, making them person-disjoint sets. The available number of subjects in the testing and the training sets for each age band is shown in Table 10.3.

## 10.5   Experimental analysis of individual classification systems for iris-based predictive biometrics

One of the more difficult aspects of designing any classification task is making the best choice of classifier or, in the case of a typical multiclassifier approach, the base classifiers for the fusion method. The pool of base classifiers selected for the experimental study here is as follows:

- **Multi-Layer Perceptron (MLP)** [34]: MLP is a Perceptron-based neural network with multiple layers [35]. The output layer receives stimuli from the intermediate layer and generates a classification output. The intermediate layer extracts the features, their weights being a codification of the features presented in the input samples, and the intermediate layer allows the network to build its own representation of the problem. Here, the MLP is trained using the standard backpropagation algorithm to determine the weight values.

- **Optimised IREP (Incremental Reduced Error Pruning) (JRip)** [36]: The Decision Tree approach usually uses pruning techniques to decrease the error rates of a dataset with noise, one approach to which is the Reduced Error Pruning method. Specifically, we use Incremental Reduced Error Pruning (IREP). The IREP uses a 'divide and conquer' approach. This algorithm uses a set of rules which, one by one, are tested to check whether a rule matches, all samples related to that rule then being deleted. This process is repeated until there are no more samples or the algorithm returns an unacceptable error. Our implementation uses a delayed pruning approach to avoid unnecessary pruning, resulting in what is usually designated the JRip procedure.
- **Support Vector Machine (SVM)** [37]: This approach embodies a functionality very different from that of more traditional classification methods and, rather than aiming to minimise the empirical risk, aims to minimise the structural risk. In other words, the SVM tries to increase the performance when trained with known data based on the probability of a wrong classification of a new sample. It is based on an induction method which minimises the upper limit of the generalisation error related to uniform convergence, dividing the problem space using hyperplanes or surfaces, splitting the training samples into positive and negative groups and selecting the surface which keeps more samples.
- **J48 Decision Trees (DT)** [38]: This classifier uses a generalised 'divide and conquer' strategy, splitting a complex problem into a succession of smaller sub-problems, and forming a hierarchy of connected internal and external nodes. An internal node is a decision point determining, according to a logical test, the next node reached. When this is an external node, the test sample is assigned to the class associated with that node.
- **K-Nearest Neighbours (KNN)** [39]: This embodies one of the most simple learning methods. The training set is seen as composed of $n$-dimensional vectors and each element represents an $n$-dimensional space point. The classifier estimates the KKN of a test pattern in the whole training dataset based on an appropriate distance metric (Euclidian distance in the simplest case). The classifier checks the class labels of each selected neighbour and chooses the class that appears most in the set of labels.

We have chosen these representative classifiers in order to guarantee a high diversity among the individual classifier components, which is an essential prerequisite in the present context, especially when we investigate a multiclassifier approach.

The results obtained (shown in Figure 10.1) show that predictive accuracy across the five different individual classifiers ranges from approximately 50% to something approaching 75% in the best case. The classifier J48 (Decision tree) is well known to be able to differentiate this type of data.

What is surprising is the fact that the relatively simple KNN classifier produces almost comparable accuracy to more powerful classifiers, such as MLP and SVM, perhaps indicative of the specificities of this type of data.

When the different types of features are analysed, it is seen that the combination of both texture and geometric features always performs better than when adopting

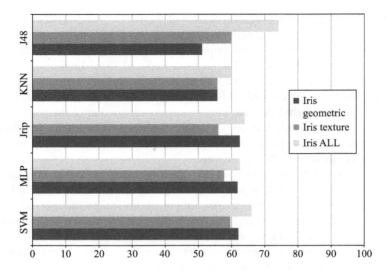

*Figure 10.1    Individual classifiers performing age prediction with different iris feature groupings*

either feature category alone. However, when only a single feature type is adopted, the geometric features tend to produce better results. However, the J48 classifier demonstrates a different trend, producing better results with the texture features.

The results overall show how classifier choice can have a significant impact on predictive performance, while feature choice is also a potentially important factor.

## 10.6    Experimental analysis of iris-based multiclassifier systems

The demands of computational systems which can deliver efficient and high performance for pattern recognition solutions have especially motivated the study of machine learning techniques in recent years [40–42]. Despite the fact that many classification algorithms appear to produce satisfactory performance in principal, they nevertheless sometimes fail to generate reliable results in real world tasks (when operating with very complex datasets, for instance) [43]. Thus, while optimising the performance attainable with any single particular classifier remains an important consideration, in order to make significant progress towards improving performance levels in more diverse situations, the idea of using the different (often complementary) characteristics of different classifiers within a single task domain has increasingly been considered as a more effective strategy, and hence the concept of multiclassifier systems (MCS) has become very important in recent years [44,45].

Multiclassifier systems can be divided into two categories: Parallel and Modular systems [46]. The first category can be described as defining a configuration where all the classifiers in the system perform the same recognition task, each classifier producing its own output for a specific input test sample. A single classifier then decides, based on all these outputs (which can only be given once), what is the overall output of the system. On the other hand, the second category defines a structure where each classifier or group of classifiers is responsible for the classification of a part of the problem. In the same way as in the Parallel category, in the Modular category, the base classifiers will produce their outputs and will pass them to a single classifier (again, only once) which will be responsible for deciding what is the overall output of the system.

Multiclassifier systems have been widely used in a range of pattern recognition problems such as speech recognition [47], writing recognition [48], face recognition [49,50], multimodal biometrics [51], protein classification [41] and many more applications.

We have therefore investigated the age prediction task using a multiclassifier approach, configured around two traditional combination techniques which can be described as follows.

- **Majority Voting (Vote)** [44]: This non-linear fusion-based method is one of the simplest and most intuitive methods for combining individual classifiers. It takes into account only the top outputs of the experts. The outputs of the classifiers are represented in a winner-take-all form (for each classifier, the output of the winner is 1 and the remaining outputs are 0) and the weights for all the experts are equal to 1. Once all the experts have made their decision, they 'vote' on their winner class and the class that has more votes is the overall winner of the system.

- **Sum-based fusion (Sum)** [52]: This is a linear fusion-based method that takes into account the confidence degree for each class of each classifier. In this sense, when an input pattern is presented to the base classifiers, the degrees of confidence for each class output are added to the other related outputs giving a score to that class. The winner class and hence the identity label of the system is the class with the highest score.

Comparing these MCS results with those for the individual classifiers presented in the previous section, it is seen that adopting a multiclassifier approach, with a traditional fusion technique such as the sum or vote configurations, improves (in almost all the cases) the prediction accuracy achieved, as might be expected.

A more surprising issue here is, perhaps, that the improvement achieved is relatively small. The predictive accuracy using the vote and sum combination configurations ranges from around 70% to just under 80%. In the case of the sum-based combination, the improvement is less than 25%. Considering that around a 75% accuracy can be achieved with the J48 (Decision Tree) classifier alone, it is arguable that in some applications the overhead of implementing the additional complexity of a multiclassifier arrangement is not easily justifiable.

These results demonstrate that, at least for a configuration based on traditional combination strategies, a further important consideration for a system designer is therefore to assess the benefits of moving to a multiclassifier platform in relation to the costs incurred.

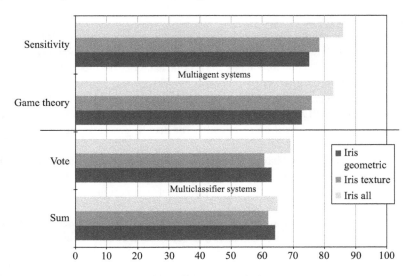

*Figure 10.2   Age prediction results for the combination of classifiers for the two traditional fusion techniques as well as for the multiagent systems*

## 10.7   Experimental analysis of multiagent intelligent systems for iris-based predictive biometrics

Matching the different component elements in biometric-based systems is always a very challenging task, but it is generally a good strategy to balance observed weaknesses in one step by enforcing compensation by means of another. Therefore, we address the limitations associated with the adoption of particularly simple features in some cases, by using decision-level fusion and exploiting a greater degree of 'intelligence' in the techniques we deploy.

When identification accuracy is the primary concern, a very common alternative approach, as we have seen, is to use a multiclassifier system approach in order to build a more accurate system [53]. This approach requires an effective method for combining information determined by each of the contributing classifiers (the base classifiers) in order to compute an overall classification decision for the system [54]. This is the approach considered in the previous section. Multiagent systems applied in classification scenarios provide a powerful alternative paradigm, offering the possibility of overcoming the difficulties involved in efficiently handling the combination problem, since they are structured to make their own decisions about the classification output of the system.

An intelligent agent is a software-based computer system that has autonomy, social ability, reactivity and pro-activeness [55]. Agents are entities which can communicate, cooperate and work together to reach a common goal, and they interact using negotiation protocols.

The main idea behind the functioning of an agent is that once an input pattern is provided, a *Controller module* passes the required information to a *Decision-making module*, which accesses a *classifier module* to produce its output. The *Classifier module* will produce a degree of confidence measure with respect to each enrolled class and this information will be used as a weight in order to assign a degree of importance to each output of the agent (each classifier agent will generate an individual degree of confidence to each possible output of the system and, therefore, each enrolled user in this scenario). The *Controller module* can decide to communicate with other agents in order to reach an agreed result. During the negotiation process, it might be necessary for the agent to change its opinion about its current output or to perform a new decision-making process. Also, an agent may decide to perform the decision-making process a further time, analysing other criteria or pattern features or, indeed, not changing the output at all.

The agents can also deal with other issues related with the system itself. For instance, the agents can decide which methods are more reliable when the system receives as input some specific modalities. The possibilities are many when more intelligent and autonomous elements are part of the biometric-based system. It is very important to have in mind the main benefit of this approach, which is an interactive and much more flexible process which differs from the traditional fusion approach by allowing the classifiers (here embedded in the agents) to change opinions, to discuss differences and, to take into account the other classifiers' (agents) opinions. Therefore, this is an extremely powerful approach which provides new ways of designing biometric-based systems.

We have therefore also experimentally investigated some multiagent negotiation techniques to support predictive processing, which can be described as follows.

- **The Game Theory-based negotiation** method has been used as a cooperation tool in multiagent systems. In game theory, the systematic description of the results can be carried out through the use of strategic games. A strategic game is a game in which a player chooses a plan of action only once and at the same time as his opponent. In order to help the players to make their decisions, a payoff matrix is used, in which each cell represents the payoff values which the players will have in a situation where these actions are chosen. The cell with the highest value is chosen [56]. Based on this approach, the game theory strategy has been adjusted to be able to implement a biometric classification task. It is important to remember that in such a scenario each agent labels an input sample based on all the enrolled users. Therefore, each agent will give an identity to this input sample.
- **The Sensitivity-based negotiation** method [57] is based on the idea that there is a decrease in the confidence level of the agents which is considered through the use of a sensitivity analysis during the testing phase of the system. This analysis can be achieved by excluding and/or varying the values of an input feature and analysing the variation in the performance of the classifier method. The main aim of this analysis is to investigate the sensitivity of each agent with respect to a certain feature and to use this information in the negotiation process.

This analysis is performed with respect to all features of the input patterns in the identity prediction classifier within the agents.

We have also implemented the two agent-based configurations referred to above when applied to the age prediction task, with the results shown in Figure 10.2. We can see that adopting the intelligent agent-based configuration increases the prediction accuracy quite substantially, returning an accuracy greater than 85% using the Sensitivity negotiation configuration, or 75% accuracy even when using only five texture-based features.

We also note that in all cases, the texture features performed better than when using only geometric features. This shows that the more 'intelligent' techniques offer the possibility of delivering the best predictive performance even when using a very small number of features, offering a more powerful configuration than the traditional fusion techniques.

## 10.8   Discussion

It is useful to briefly summarise what our series of experiments have so far established. The results obtained and documented in the preceding sections show that predictive accuracy across the five different individual classifiers ranges from approximately 50% to somewhat over 70% in the best case. It is seen that adopting a multiclassifier approach with a traditional fusion technique such as the sum or vote approaches improves this somewhat as might be expected. However, adopting the intelligent agent-based configuration increases the prediction accuracy substantially, returning an accuracy greater than 85% using the negotiation configuration. Obtaining this high level of predictive performance, especially the accuracy of just over 70% with only six texture features is a very positive outcome, which shows the benefits of making an appropriate and informed choice of processing configuration.

The performance levels we have been able to achieve – assigning each tested subject to one of three age groups (corresponding to 'younger', 'middle aged' and 'older' categories) in relation to prediction accuracy, even with such a small feature set, is seen to be comparable to that reported elsewhere for the prediction of only a two-class age determination problem, which also used a very much larger and more diverse feature set.

Two principal novel points emerge here. First, the work which we have summarised here includes the first reported results to show the reliable prediction of subject age from the iris characteristics of an individual on the basis of more than two possible age-related categories. Second, this study has provided some quantitative data to show the performance modulations which arise from different combinations of classifier type, feature characteristics and underlying processing strategy. Classifier choice, feature choice, single classifier platform adoption, possible multiclassifier arrangement and choice of underpinning processing strategy, are all factors which can be considered in developing a predictive structure which is well matched to a particular problem domain. This study is therefore valuable as a starting point for the design and implementation of practical predictive biometrics systems, not just for the prediction of age, but perhaps also for a wider range of important characteristics.

## 10.9 Experimental analysis of multimodal iris-based predictive systems

Finally, we turn to another option, which is to make use of more than one modality to improve the predictive capability of biometric data. For illustrative purposes here, we have adopted a two-modality approach, combining iris data with the alternative modality of information extracted from the handwritten signature. We have chosen these modalities partly to demonstrate how fundamentally different data sources can be beneficial, but also, more importantly, because of the availability of a dataset which contains biometric samples in both modalities captured from the same individuals.

Here, for the handwritten signature, the samples were collected using an A4-sized graphics tablet with a density of 500 lines per inch. Each user donated 30 genuine samples of his/her handwritten signature. For the iris modality, we used the dataset previously described. Thus, eye images were acquired in a standard 'office' environment managed by a supervisor and using the LG Iris Access EOU3000 system. During the acquisition, spectacles were not allowed to be worn by subjects, although contact lenses were allowed. Eight eye images (four left and four right) were acquired in two different sessions with a resolution of 640 * 480 pixels.

We are particularly interested in the impact which the complexity of the fundamentally different sets of features can have on the general prediction capabilities as well as balancing these differences against the option of using the more robust and intelligent decision-making structures described above. For the signature we consider both dynamic and static features, and for the iris we consider both textural and geometrical features (as before), thereby producing a range of features of differing types and of different complexity.

In the case of the signature, we can classify naturally two different types of features (static and dynamic). Static features in a signature context describe the geometry and general appearance of the sample. They can be extracted from both offline and online sample acquisition. Those used in our experiments are briefly described in Tables 10.4 and 10.5.

On the other hand, dynamic features are related to the interaction of the users with whichever device (s)he is using while executing the signing process. This type of feature is naturally collected in an online acquisition process, although some dynamic features can be inferred from offline samples. A brief description of the dynamic features used here can be seen in Table 10.5.

Our initial experiment is performed to test the accuracy of the proposed age prediction approach with respect to the different features and different individual classifiers for both the iris and the signature modality. The results are shown in

*Table 10.4   Signature STATIC features [58]*

| Feature | Description |
|---------|-------------|
| Signature width | *Width* of the image in mm |
| Signature height | *Height* of the image in mm |
| Height to width ratio | $\frac{height}{width}$ |
| Sum of horizontal coordinates | Ratio of *width* |
| Sum of vertical coordinates | Ratio of *height* |
| Horizontal centralness | Central horizontal point |
| Vertical centralness | Central vertical point |

*Table 10.5   Signature DYNAMIC features [58]*

| Feature | Description |
|---------|-------------|
| Execution time | Total time of the signature |
| Pen lift | Counts of the pen removal |
| Ave. hor. pen velocity in $X$ | Pen velocity in the $x$ |
| Ave. hor. pen velocity in $Y$ | Pen velocity in the $y$ |
| Vert. midpoint pen crossings | No. of times midline is crosses |
| Azimuth | Changing in the rotation with the $z$-axis |
| Altitude | Average angle toward the positive $z$-axis |

Figure 10.3. General observations on the results generated by the individual classifiers can be summarised as follows:

- Static features (signature) and geometric features (iris) always produce the classification scenario with the worst performance. This is expected because these types of features, by their inherent nature, are recognised as having limitations to the extent in which they identify an individual.
- Although these weaker features produce the weakest classification results, depending on the strength of the algorithm used, the difference between the performance when using only the weak features and the performance using only strong features is relatively small. This can be seen strongly when observing the performance of the SVM classifier for the iris features. This behaviour suggests that when not all the features are available, there is still a possible strategy available to improve performance.

Hence, following these observations, and in order to exploit and to investigate how iris and signature could be used effectively for age prediction, a second experiment is performed to define an optimal mechanism to predict age from both handwritten signature and iris biometric data. In order to analyse the impact of different types of features, five different scenarios for decision-level fusion are considered,

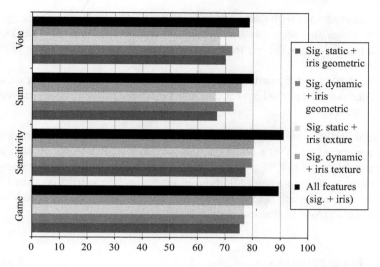

*Figure 10.3 Different combinations of features in the multimodal system*

and the results can be seen in Figure 10.3. General observations of the results can be summarised as follows:

- The first, and rather unsurprising result, is that all the fusion techniques produce better results than the individual classifiers. Also, all the negotiation techniques perform better than all the traditional fusion techniques, sometimes achieving an improvement of more than 10% in accuracy.
- The second, and also unsurprising result, is that when using all the features for both modalities, the performance is the best and, in some cases, there is a considerable difference compared with the other scenarios, as when the fusion technique is an agent-based solution. Since this is considered the most powerful, as it uses more information and more subtle manipulation of the data than most traditional fusion techniques, it is natural that it produces the best results, offering a very powerful processing structure for this type of task.
- Moreover, the more sophisticated the decision-level fusion technique, the bigger is the difference between using all the features and using only a smaller feature subset.
- But a very interesting scenario in this configuration is achieved when different sets of features for each modality are used. It might be expected that weaker features when combined produced worse results than when weak and strong features are combined, but that is not always the case.
- The second best result is achieved when using dynamic signature features and texture-based iris features.
- When observing the traditional fusion techniques (vote and sum), there is a slight variation in the observed performance rankings. In some cases, signature static

features combined with geometric iris features perform better or almost the same as static signature features combined with texture iris features. This can be taken as an indication that the iris features in general are less representative of the age of the subject.

- On the other hand, dynamic signature features, regardless of whether combined with geometric or texture iris features, can produce significant levels of predictive performance.

In summary, the multimodal (iris and signature) age prediction system is able to achieve accuracies up to 90% while the unimodal age prediction systems are able to achieve accuracies up to 85%. Also, all these results point to some further interesting observations, such as: signature is more effective for age prediction than iris and intelligent techniques can help to counter-balance the natural higher variability that can be expected from a behavioural modality.

## 10.10 Final remarks

This chapter has presented some possible approaches to age prediction based on the availability of iris images, either using different types of features in a unimodal scenario, or in a multimodal combination (here combined with signature samples by way of example).

This comparative study based on different feature sets, and different classification approaches, demonstrates how it is possible to provide the system designer with useful data to inform a choice of feature definition and classification approaches.

These experimental results are most encouraging, especially in a task domain which has not been widely investigated to date (especially for the iris modality and the combination of iris and a secondary modality).

Most importantly, this study demonstrates, in a very practical way, through a multi-stage experimental process, the significant potential for utilising conventional biometric data in a predictive capacity aimed at estimating diverse possible soft biometric characteristics of an individual, with important practical implications for a wide range of application domains.

## References

[1] G. Guo, Y. Fu, C.R. Dyer, and T.S. Huang. "Image-based human age estimation by manifold learning and locally adjusted robust regression," *IEEE Transactions on Image Processing*, 17(7):1178–1188, 2008.

[2] X. Geng, Z-H. Zhou, and K. Smith-Miles. "Automatic age estimation based on facial aging patterns," *IEEE Transactions on Pattern Analysis and Machine Intelligence*, 29(12):2234–2240, 2007.

[3] F. Metze, J. Ajmera, R. Englert, *et al.* "Comparison of four approaches to age and gender recognition for telephone applications," in *IEEE International*

Conference on Acoustics, Speech and Signal Processing, ICASSP 2007, vol. 4, pp. 1089–1092, 2007.

[4]   M. Nishimoto, Y. Azuma, Y. Miyamoto, T. X. Fujisawa, and N. Nagata. "Subjective age estimation using speech sounds: comparison with facial images," in *IEEE International Conference on Systems, Man, and Cybernetics*, SMC 2008, pp. 1900–1904, 2008.

[5]   M. C. Fairhurst and M. C. C. Abreu. "An investigation of predictive profiling from handwritten signature data," in *10th International Conference on Document Analysis and Recognition*, ICDAR 2009, pp. 1305–1309, 2009.

[6]   M. Erbilek and M. C. Fairhurst. "A methodological framework for investigating age factors on the performance of biometric systems," in *Multimedia and Security*, MM&Sec 2012, pp. 115–122, 2012.

[7]   M. C. Fairhurst and Da Costa-Abreu. "Using keystroke dynamics for gender identification in social network environment," in *The Fourth International Conference on Imaging for Crime Detection and Prevention*. Kingston University, 2011.

[8]   Y. Fu, G. Guo, and T.S. Huang. "Age synthesis and estimation via faces: a survey," *IEEE Transactions on Pattern Analysis and Machine Intelligence*, 32(11):1955–1976, 2010.

[9]   A. Gunay and V.V. Nabiyev. "Automatic age classification with LBP," in *The 23rd International Symposium on Computer and Information Sciences*, ISCIS 2008, pp. 1–4, 2008.

[10]   C.-H. Ju and Y.-H. Wang. "Automatic age estimation based on local feature of face image and regression," in *International Conference on Machine Learning and Cybernetics*, pp. 885–888, 2009.

[11]   C. Li, Q. Liu, J. Liu, and H. Lu. "Learning ordinal discriminative features for age estimation," in *IEEE Conference on Computer Vision and Pattern Recognition*, CVPR, pp. 2570–2577, 2012.

[12]   Z. Li, Y. Fu, and T.S. Huang. "A robust framework for multiview age estimation," in *IEEE Computer Society Conference on Computer Vision and Pattern Recognition Workshops*, CVPRW, pp. 9–16, 2010.

[13]   Y. Liang, L. Liu, Y. Xu, Y. Xiang, and B. Zou. "Multi-task GLOH feature selection for human age estimation," in *The 18th IEEE International Conference on Image Processing*, ICIP, pp. 565–568, 2011.

[14]   P. Turaga, S. Biswas, and R. Chellappa. "The role of geometry in age estimation," In *IEEE International Conference on Acoustics Speech and Signal Processing*, ICASSP, pp. 946–949, 2010.

[15]   C.-C. Wang, Y.-C. Su, C.-T. Hsu, C.-W. Lin, and H.Y.M. Liao. "Bayesian age estimation on face images," in *The 2009 IEEE International Conference on Multimedia and Expo*, ICME 2009, pp. 282–285, 2009.

[16]   J. Ajmera. "Effect of age and gender on LP smoothed spectral envelope," in *IEEE Odyssey 2006: The Speaker and Language Recognition Workshop*, San Juan, Puerto Rico, pp. 1–4, 2006.

[17]   M.H. Bahari and H. Van Hamme. "Speaker age estimation and gender detection based on supervised non-negative matrix factorization," in *IEEE Workshop on*

*Biometric Measurements and Systems for Security and Medical Applications*, BIOMS, pp. 1–6, 2011.

[18] J. Lu and Y-P Tan. "Gait-based human age estimation," in *IEEE International Conference on Acoustics Speech and Signal Processing*, ICASSP, pp. 1718–1721, 2010.

[19] Y. Makihara, M. Okumura, H. Iwama, and Y. Yagi. "Gait-based age estimation using a whole-generation gait database," in *International Joint Conference on Biometrics*, IJCB, pp. 1–6, 2011.

[20] M. Erbilek and M. C. Fairhurst. "Framework for managing ageing effects in signature biometrics," *IET Biometrics*, 1(2):136–147, 2012.

[21] X. Qiu, Z. Sun, and T. Tan. "Global texture analysis of iris images for ethnic classification," in D. Zhang and A.K. Jain, editors, *Advances in Biometrics, Lecture Notes in Computer Science*, vol. 3832, pp. 411–418, 2005.

[22] V. Thomas, N.V. Chawla, K.W. Bowyer, and P.J. Flynn. "Learning to predict gender from iris images," in *The First IEEE International Conference on Biometrics: Theory, Applications, and Systems*, BTAS 2007, pp. 1–5, 2007.

[23] A. Sgroi, K.W. Bowyer, and P.J. Flynn. "The prediction of old and young subjects from iris texture," in *IAPR International Conference on Biometrics*, ICB 2013, pp. 1–5, 2013.

[24] M. C. Fairhurst and M. Erbilek. "Analysis of physical ageing effects in iris biometrics," *IET Computer Vision*, 5(6):358–366, 2011. (Special issue on Future Trends in Biometric Processing.)

[25] M. Erbilek, M. C. Fairhurst, and M. Da Costa-Abreu. "Age prediction from iris biometrics," *IET Conference Proceedings*, pp. 1.07–1.07(1), 2013.

[26] L. Masek. "Recognition of human iris patterns for biometric identification," Bachelor of Engineering degree of the School of Computer Science and Software Engineering, The University of Western Australia, Crawley, WA, Australia, 2003.

[27] J.G. Daugman. "High confidence visual recognition of persons by a test of statistical independence," *IEEE Transactions on Pattern Analysis and Machine Intelligence*, 15:1148–1161, 1993.

[28] S.U. Maheswari, P. Anbalagan, and T. Priya. "Efficient iris recognition through improvement in iris segmentation algorithm," *International Journal on Graphics, Vision and Image Processing*, 8(2):29–35, 2008.

[29] C.E. Spearman. "The proof and measurement of association between two things," *The American Journal of Psychology*, 15(1):72–101, 1994.

[30] T. Miyoshi and M. Hyodo. "Aging effects on face images by varying vertical feature placement and transforming face shape," in *IEEE International Conference on Systems, Man and Cybernetics*, SMC 2006, vol. 2, pp. 1548–1553, 2006.

[31] S.K. Modi, S.J. Elliott, J. Whetsone, and Hakil Kim. "Impact of age groups on fingerprint recognition performance," in *IEEE Workshop on Automatic Identification Advanced Technologies*, WAIAT 2007, pp. 19–23, 2007.

[32] J. Ortega-Garcia, J. Fierrez, F. Alonso-Fernandez, *et al.* "The multiscenario multienvironment biosecure multimodal database (BMDB)," *IEEE*

*Transactions on Pattern Analysis and Machine Intelligence*, 32:1097–1111, 2010.

[33] F. Leisch, L.C. Jain, and K. Hornik. "Cross-validation with active pattern selection for neural-network classifiers," *IEEE Transactions on Neural Networks*, 9(1):35–41, 1998.

[34] S. Haykin. *Neural Networks: A Comprehensive Foundation*, 2nd edition, Prentice Hall PTR, Upper Saddle River, NJ, 1998.

[35] F. Rosenblatt. "The perception: a probabilistic model for information storage and organization in the brain," *Psychological Review*, 65(6):386–408, 1958.

[36] J. Furnkranz and G. Widmer. "Incremental reduced error pruning," in *Proceedings the 11th International Conference on Machine Learning*, ICML 1994, pp. 70–77, 1994.

[37] C. Nello and S.T. John. "An introduction to support vector machines and other kernel-based learning methods," *Robotics*, 18(6):687–689, 2000.

[38] J.R. Quinlan. *C4.5: Programs for Machine Learning*, Morgan Kaufmann Publishers Inc., San Francisco, CA, 1993.

[39] A. Arya. "An optimal algorithm for approximate nearest neighbors searching fixed dimensions," *Journal of ACM*, 45(6):891–923, 1998.

[40] F. M. Alkoot and J. Kittler. "Experimental evaluation of expert fusion strategies," *Pattern Recognition Letters*, 20(11–13):1361–1369, 1999.

[41] V.G. Bittencourt, M.C.C. Abreu, M.C.P. de Souto, and A.MdeP. Canuto. "An empirical comparison of individual machine learning techniques and ensemble approaches in protein structural class prediction," in *IEEE International Joint Conference on Neural Networks*, IJCNN 2005, vol. 1, pp. 527–531, 2005.

[42] J. Kittler, M. Hatef, R.P.W. Duin, and J. Matas. "On combining classifiers," *IEEE Transactions on Pattern Analysis and Machine Intelligence*, 20(3):226–239, 1998.

[43] G. Fumera and F. Roli. "A theoretical and experimental analysis of linear combiners for multiple classifier systems," *IEEE Transactions on Pattern Analysis and Machine Intelligence*, 27(6):942–956, 2005.

[44] L.I. Kuncheva. *Combining Pattern Classifiers: Methods and Algorithms*, Wiley-Interscience, New York, NY, 2004.

[45] L.I. Kuncheva and J.J. Rodriguez. "Classifier ensembles with a random linear oracle," *IEEE Transactions on Knowledge and Data Engineering*, 19(4):500–508, 2007.

[46] T.M. Mitchell. *Machine Learning*, McGraw-Hill, Inc., New York, NY, 1997.

[47] S. Mukhopadhyay, S. Peng, R. Raje, M. Palakal, and J. Mostafa. "Multi-agent information classification using dynamic acquaintance lists," *Journal of the American Society for Information Science and Technology*, 54(10):966–975, 2003.

[48] L. Heutte, A. Nosary, and T. Paquet. "A multiple agent architecture for handwritten text recognition," *Pattern Recognition Letter*, 37(4):665–674, 2004.

[49] J. Czyz, M. Sadeghi, J. Kittler, and L. Vandendorpe. "Decision fusion for face authentication," in *The First International Conference Biometric*

*Authentication, Lecture Notes in Computer Science*, vol. 3072, pp. 686–693, 2004.

[50] W.S. Lee and K.A. Sohn. "View influence analysis and optimization for multiview face recognition," *Journal on Image and Video Processing*, 2007(2):1–8, 2007.

[51] R. Tronci, G. Giacinto, and F. Roli. "Designing multiple biometric systems: measures of ensemble effectiveness," *Engineering Applications of Artificial Intelligence*, 22(1):66–78, 2009.

[52] J. Kittler and F. M. Alkoot. "Sum versus vote fusion in multiple classifier systems," *IEEE Transactions on Pattern Analysis and Machine Intelligence*, 25(1):110–115, 2003.

[53] J. Fierrez-Aguilar. *Adapted Fusion Schemes for Multimodal Biometric Authentication*. PhD thesis, Universidad Politecnica de Madrid, Madrid, Spain, 2006.

[54] A.M.P. Canuto, M.C.C. Abreu, A. Medeiros, F. Souza, M. F. Gomes Junior, and V.S. Bezerra. "Investigating the use of an agent-based multi-classifier system for classification tasks," in *The 11th International Conference on Neural Information Processing, Lecture Notes on Computer Science*, vol. 3316, pp. 854–859, 2004.

[55] M. Wooldridge. *An Introduction to Multi-Agent Systems*, 2nd edition, Wiley, New York, NY, 2009.

[56] N. Orsini, N. Rizzuto, and D. Nante. "Introduction to game-theory calculations," *Stata Journal*, 5(3):355–370(16), 2005.

[57] M.C.D.C. Abreu and M. C. Fairhurst. "Enhancing identity prediction using a novel approach to combining hard- and soft-biometric information," *IEEE Transactions on Systems, Man, and Cybernetics, Part C: Applications and Reviews*, 41(5):599–607, 2010.

[58] R.M. Guest. "Age dependency in handwritten dynamic signature verification systems," *Pattern Recognition Letter*, 27(10):1098–1104, 2006.

*Part IV*

**Security Aspects**

*Chapter 11*

# Presentation attack detection in iris recognition

*Javier Galbally and Marta Gomez-Barrero*

## 11.1 Introduction

Since the first pioneering works on automatic voice and face recognition over 40 years ago [1,2], steady and continuous progress has been made in the development of biometric technology [3]. Driven by the very appealing new security biometric paradigm *"forget about cards and passwords, you are your own key"*, researchers from many different fields, such as image processing, computer vision or pattern recognition, have applied the newest techniques in each of these areas to improve the performance of biometric systems. This path of technological evolution has permitted the use of biometrics in many diverse activities such as forensics, border and access control, surveillance or on-line commerce.

In this scenario of constant expansion, and as a consequence of its own natural progress, new concerns are arising regarding biometric technology different from the mere improvement of its recognition performance. Among these new issues and challenges that have emerged around biometrics, its resilience against external threats has lately drawn a significant level of attention.

Currently it is an accepted fact that, as the deployment of biometric systems keeps growing year after year in such different environments as airports, laptops or mobile phones, users are also becoming more familiar with this technology and, as a result, their security weaknesses are better known to the general public.

Attacks are no more restricted to a mere theoretical or academic sphere, but are starting to be carried out against real operational applications. The fairly easy hacking of the long anticipated new iPhone 5S fingerprint reader, just a day after it hit the shelves and using a regular and well-known type of fingerprint spoof [4], is only another example in the list of practical attacks and vulnerabilities of biometric systems that are being reported to the public. Nowadays we can find cases of attacks carried out by hacking groups attempting to get recognition [5–7], coming from real criminal cases [8,9], or even from live demonstrations at biometric and security specific conferences [10,11].

Among the different vulnerabilities analysed, intensive research efforts have been focused on the study of *spoofing attacks*, in modalities such as iris [12], fingerprint [13], face [14], voice [15], gait [16] or even palm vein recognition [17] and multimodal approaches [18].

In the specialized literature *biometric spoofing* [19,20] is generally understood as the ability to fool a biometric system into recognizing an illegitimate user as a genuine one by means of presenting to the sensor a synthetic forged version (i.e., artefact) of the original biometric trait (e.g., iris printed image, gummy finger or face mask) or trying to mimic the behaviour of genuine users (e.g., gait, signature). This way, *spoofing attacks* take advantage of the fact that our iris, fingerprints, face, voice or even our DNA are publicly available data. This is one of the well-known drawbacks of biometrics: *"biometric traits are not secrets"* [21,22].

Spoofing attacks fall within the larger category *presentation attacks*, defined in the latest draft of the ISO/IEC 30107 standard as *"presentation of an artefact or human characteristic to the biometric capture subsystem in a fashion that could interfere with the intended policy of the biometric system"* [23]. This wider group of attacks also includes the presentation to the acquisition device of human characteristics (and not only synthetic artefacts) such as dead fingers, mutilated traits, real living traits under coercion or a different living trait (i.e., *zero-effort* impostor attempts that try to take advantage of the False Acceptance Rate, FAR, of biometric systems) [24].

Although *presentation attacks* is the standard term, in practice, the vast majority of works related to this vulnerability are focused on the study of *spoofing* attacks. As a consequence, even if strictly speaking spoofing is a subtype of the more general presentation attacks, given its relevance, the term is widely used to refer to the overall threat. Hence, although not exactly synonyms, in the present chapter we will follow the common practice and use both terms indistinctly.

The spoofing-related works mentioned above and other analogue studies have clearly shown the need to propose and develop specific protection methods against these attacks. This way, researchers have focused on the design of specific countermeasures that enable biometric systems to detect fake samples and reject them, improving this way the robustness and security level of the systems. This has initiated a new research area known as biometric *presentation attack detection*.

A *presentation attack detection* method is usually accepted to be any technique that is able to automatically distinguish between real biometric traits presented to the sensor and access attempts carried out by means of presentation attacks [25]. Although this is the nomenclature advised by ISO standards [23], similarly to the case of the term *spoofing*, very often these approaches are also referred to in the specialized literature as *anti-spoofing*, *liveness detection* or *vitality detection* techniques. Following the same rationale as for spoofing and presentation attacks, in the present chapter we will use as synonyms the terms *presentation attack detection* and *anti-spoofing*.

From a general perspective, presentation attack detection techniques may be classified into one of three groups, depending on the part of the biometric system where they are integrated (see Figure 11.1):

- **Sensor-level techniques**. Usually referred in the literature as *hardware-based* techniques. These methods add some specific device to the sensor in order to detect particular properties of a living trait (e.g., specific reflection properties of the eye or pupil dynamics). In general, hardware-based approaches measure one of three characteristics, namely: (i) intrinsic properties of a living body,

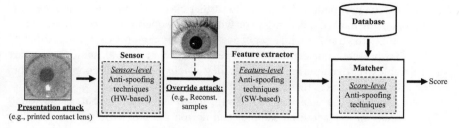

Figure 11.1 *General diagram of a biometric system specifying the modules where the three types of presentation attack detection techniques may be integrated (sensor-level, feature-level and score-level). Also displayed are the two different types of attacks for which presentation attack detection techniques may offer protection: presentation attacks and override attacks carried out for instance with synthetic or reconstructed samples*

including physical properties (e.g., density or elasticity), electrical properties (e.g., capacitance, resistance or permittivity), spectral properties (e.g., reflectance and absorbance at given wavelengths) or even visual properties (e.g., colour and opacity); (ii) involuntary signals of a living body, which can be attributed to the nervous system, being a good example the pupillary unrest (hippus); (iii) responses to external stimuli, also known as *challenge-response* methods, which require the user cooperation as they are based on detecting voluntary (behavioural) or involuntary (reflex reactions) responses to an external signal. Examples of such methods can be the contraction of the pupil after a lighting event (reflex), or the movement of the eye following a random path predetermined by the system (behavioural).

In the present chapter, multibiometric techniques are included in this category [26]. Multibiometric anti-spoofing is based on the hypothesis that the combination of different biometrics will increase the robustness to direct attacks, as, in theory, generating several fake traits is presumed to be more difficult than an individual trait. Such a strategy requires additional hardware acquisition devices, therefore these techniques may be included in the sensor-level group of presentation attack detection methods. Note that the above hypothesis (i.e., circumventing a multibiometric system implies breaking all unimodal modules) has already been shown to not hold for all cases as, depending on the selected configuration, bypassing just one of the unimodal subsystems is enough to gain access to the complete application [27,28].

- **Feature-level techniques**. Usually referred in the literature as *software-based* techniques. In this case the fake trait is detected once the sample has been acquired with a standard sensor, that is, features used to distinguish between real and fake traits are extracted from the biometric sample, and not the human body itself. These methods are installed after the sensor, usually operating as part of the feature extractor module. They can be further classified into *static* and *dynamic* anti-spoofing methods, depending on whether they operate with only one instance of the biometric trait, or with a sequence of samples captured over time. Although they may present some degradation in performance, in general, static features are preferable over dynamic techniques as they usually require less cooperation from the user, which makes them faster and less intrusive.

An appealing characteristic of software-based techniques is that, as they operate directly on the acquired sample (and not on the biometric trait itself), they are potentially capable of detecting other types of illegal break-in attempts not necessarily classified as presentation attacks. For instance, feature-level methods can protect the system against the injection of reconstructed or synthetic samples[1] into the communication channel between the sensor and the feature extractor as depicted in Figure 11.1 [29].

- **Score-level techniques**. Very recently, a third type of protection methods, which fall out of the traditional two-type classification software- and hardware-based, has started to be analysed in the field of fingerprint anti-spoofing. These protection techniques, much less common than the previous two categories, focus on the study of biometric systems at the *score-level*, in order to propose fusion strategies that increase their resistance against spoofing attempts. Due to their limited performance, they are designed as supplementary measures to the sensor-level and feature-level techniques presented above, and are usually integrated in the matcher. The scores to be combined may come from: (i) two or more unimodal biometric modules; (ii) unimodal biometric modules and anti-spoofing techniques; or (iii) only results from anti-spoofing modules. No work has yet been presented in this area for iris anti-spoofing and therefore will not be further treated in the present chapter.

A graphical diagram of the categorization proposed above is given in Figure 11.2. Although the present chapter will follow this three-group taxonomy, this is not a closed classification and some techniques may fall into one or more of these groups. Nonetheless, we believe that this classification can help to visualize the current scene in biometric presentation attack detection. The reader should also be aware that, even though this is a quite extended and accepted categorization, others are also possible.

It is also worth highlighting that the three types of anti-spoofing approaches presented here are not exclusive, and may be coupled in order to improve the overall security performance of the system. In fact, the two most deployed types of methods described above (hardware- and software-based) have certain advantages and drawbacks so that, in general, a combination of both would be the most desirable protection strategy to increase the security of biometric systems. As a coarse comparison, sensor-level schemes usually present a higher fake detection rate, while feature-level techniques are in general less expensive (as no extra device is needed), less intrusive and more user-friendly since their implementation is transparent to the user. As already

---

[1]Note the difference between a synthetic *artefact* (physical) used in spoofing attacks, and a synthetic *sample* (digital).

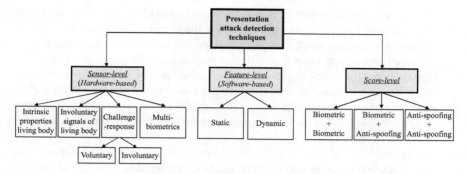

*Figure 11.2*   *General taxonomy of presentation attack detection methods considered in the present chapter with three main groups: sensor-level, feature-level and score-level techniques*

mentioned, score-level protection techniques present a much lower performance and are designed only as a support to the sensor- or feature-level protection measures.

Similarly to what has been recently published for the fingerprint and face modalities [30–32], the current chapter reviews in a comprehensive survey the presentation attack detection methods proposed in the iris modality, following the systematic categorization presented above. It also provides an overview of publicly available evaluation benchmarks for iris presentation attack detection approaches, with special attention to the international iris anti-spoofing competitions that have been organized up to date. Before the literature review of iris anti-spoofing methods, a brief summary of the most common iris presentation attacks is given. This initial section on spoofing can be useful to understand the rationale behind the design of some of the protection techniques later presented, and also to comprehend the structure of the evaluation anti-spoofing databases described in the final sections. The chapter concludes with an outline of the lessons learnt in these more than 10 years of intensive anti-spoofing research, and with a personal vision of the challenges to be faced and possible future research lines that may contribute to the general improvement of the security level offered by biometric systems against presentation attacks.

## 11.2   Presentation attacks in iris recognition

Whilst iris recognition is one of the most accurate biometric technologies, it is also a younger research field compared for instance to fingerprint or face, with the first pioneer research works dating of the early 1990s [33]. As a consequence, iris spoofing has also a somewhat shorter tradition than that of other long-studied modalities. Almost all iris presentation attacks reported in the literature follow one of three trends:

- **Photo attacks.** From a chronological point of view, these were the first attacks to be reported in the literature and they still remain popular, probably due to their great simplicity and, in many cases, high success rate [12,34,35]. They are carried out presenting to the scanner a photograph of the genuine iris. In the vast majority of cases this image is printed on paper (i.e., print-attacks), although it can also be displayed on the screen of a digital device such as a mobile phone or a tablet (i.e., digital-photo attacks). A slightly more evolved version of the basic print-attacks, which has also been considered in specialized works, consists of cutting out the pupil from the printout and placing it in front if the attacker's real eye. This way, countermeasures based on features extracted from this part of the eye lose much of their efficiency [36].

A more sophisticated variation of photo-attacks are *video-attacks*, which consist of the presentation to the scanner of an eye video replayed on a multimedia device such as a smart phone or a laptop. This way, cues related to the dynamics of the eye are not effective any more to detect such spoofing attempts. Although this type of attacks was mentioned in different iris-related works some years ago [37,38], only recently have they started to be systematically evaluated with the acquisition of the first video-based iris spoofing database [39,40].

Finally, it is also worth noting that recent works on iris image reconstruction have shown that a compromised iris template can be reversed engineered to produce an image with a very similar iriscode to the original sample [41,42]. These algorithms could also potentially lead to photo attacks using the reconstructed samples, although the vulnerability to such spoofing threat has not yet been rigorously assessed.

- **Contact-lens attacks**. These appeared as a further evolution of the classic photo-attacks. In this case, the pattern of a genuine iris is printed on a contact lens that the attacker wears during the fraudulent access attempt [43]. Such attacks are very difficult to be recognized even by human operators, and represent a real challenge for automatic protection methods as all the contextual and ancillary information of the iris corresponds to that of a living eye. In most cases, the impact analysis of this vulnerability has been carried out in the context of wider studies working on the development of appropriate presentation attack detection approaches for these artefacts [37,38,44,45].

- **Artificial-eye attacks**. These are far less common than the previous two types and have just started to be systematically studied [38,46]. Although some works may be found where very sophisticated spoofing artefacts are presented, such as the use of multilayered 3D artificial irises [47], in most cases these attacks are carried out with artificial eyes made of plastic or glass. Presentation attack detection methods based on the analysis of depth properties of the eye are more prone to be deceived by such 3D reproductions.

Some graphical examples of the attacks described above are shown in Figure 11.3. For real and fake iris images acquired during these attacks we refer the reader to Section 11.4, where more information about publicly available iris-spoofing databases may also be found.

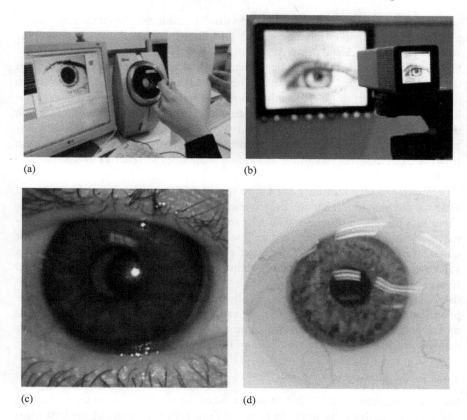

*Figure 11.3*   *(a) Photo-print attack (extracted from [34]), (b) Photo-digital attack (extracted from [35]), (c) contact-lens attack (extracted from [43]) and (d) artificial-iris attack (extracted from [47])*

## 11.3   Presentation attack detection in iris recognition

Although the first iris anti-spoofing works date back more than a decade [48], it has not been until the last three years that this technology has experimented a real revolution under the umbrella of the TABULA RASA European project focused on the study of spoofing attacks to biometric systems [49]. Another decisive factor for the development of new protection methods against direct attacks has been the acquisition and distribution of several public iris spoofing databases, that have made possible for researchers to focus on the design of efficient countermeasures and not on data acquisition issues [35,39,50,51]. Both factors have fostered the recent publication of multiple techniques in iris anti-spoofing.

In the next sections we review the works that have addressed so far the challenging problem of iris anti-spoofing. To this end, we follow the general categorization presented in Section 11.1 and depicted in Figure 11.2. It is difficult to clearly select one technique over the others, as their performance is highly dependent on the attacks and the use case considered. It is commonly observed that presentation attack detection methods fail to perform consistently across databases, experimenting a significant loss of accuracy when they are tested under different conditions to those for which they were designed. As such, usually the best results are achieved through the combination of several complementary algorithms or features, so that the final system benefits from the strengths of the individual techniques in order to overcome their weaknesses.

In order to give an initial overall perspective of the different methods studied so far in the literature related to presentation attack detection in iris-based systems, Table 11.1 presents a summary with relevant features of some of the most representative works referenced in this section. The table should be understood as a tool for quick reference and in no case as a strict comparative study, as most of the results shown in the last column have been obtained using proprietary databases designed to evaluate a specific method (see column "Database"). Moreover, in general, these databases are too small to obtain statistically meaningful results and are in most cases presented in their respective works only as a proof of concept. For further details on presentation attack detection evaluation and large publicly available iris spoofing datasets, the reader is referred to Section 11.4.

## 11.3.1    Sensor-level approaches

Daugman, regarded as the father of automatic iris recognition due to his pioneering and very successful early works in the field [33], presented some of the first ideas concerning sensor-level anti-spoofing countermeasures for iris biometrics. In some of his initial works he explored different aspects related to iris recognition, proposing some eye specific features that could be potentially used as hardware-based countermeasures against direct attacks [48,61]. Some of the characteristics mentioned in these early studies are the spectrographic properties of different parts of the eye (tissue, fat, blood, melanin pigment) and the four Purkinje reflections caused by each of the four optical surfaces comprised inside the eye.

A second group of possible anti-spoofing mechanisms highlighted in [61] are those based on behavioural eye features like the eye hippus (i.e., permanent oscillation that the eye pupil presents even under uniform lighting conditions) or the pupil response to a sudden lighting event (e.g., switching on a diode).

Although the works above do not contain any experimental validation of the proposed measures and, in the best cases, he just gave individual examples as valid proofs of concept, his proposals set the basis for many of the sensor-based iris anti-spoofing schemes that have been later developed in the literature and which are reviewed here.

As suggested in [48], the spectrographic properties of the eye tissue (e.g., fat or blood) can be used as a liveness cue in iris recognition. If the iris presented to the

*Table 11.1  Summary of the most relevant iris presentation attack detection techniques presented in Section 11.3. The column "subtype" corresponds to the algorithm subtype within each of the three main categories considered in the work (sensor-level, feature-level and score-level) as shown in the taxonomy in Figure 11.2. The column "attack" refers to the type of iris spoofing attacks considered in the work as defined in Section 11.2: photo, contact-lens or artificial-eye attacks. Sizes of databases that appear in column "database" are approximate and are given just as an indication of their order of magnitude (for the exact structure and size the reader should consult the corresponding reference). The same applies for results, as the ones shown in the table are an approximation of the different scenarios considered in each work. Works are ordered chronologically within the same subtype. For further information on iris spoofing public databases see Section 11.4.*

**Iris presentation attack detection techniques: General overview**

**Sensor-level techniques**

| Reference | Subtype | Features and methodology | Attack | Database | Error |
|---|---|---|---|---|---|
| 2013, [52] | Challenge-response | Pupil contraction after a lighting event | Photo | Proprietary, 12 identities, 322 samples | 0.2% |
| 2007, [53] | Intrinsic property | Difference in iris brightness between two LED lighting events | Photo | Proprietary, 1 subject, 5 samples | 0% |
| 2007, [54] | Intrinsic property | Reflectance difference between iris and sclera using multispectral imaging | Photo, contact-lens, artificial | Proprietary, 70 identities, 2800 samples | 0.2% |
| 2012, [46] | Intrinsic property | Conjunctival vessels detection using multispectral imaging | Photo, contact-lens, artificial | Proprietary, 100 identities, 2000 samples | 0.2% |
| 2014, [35] | Intrinsic property | Depth variation Light Field Camera | Photo | Public, GUC-LF-VIAr DB [35] 104 identities, 4847 samples | 1.1% |
| 2008, [55] | Invol. body signal | Four Purkinje reflections using NIR illumination | Photo, contact-lens, artificial | Proprietary, 30 identities, 500 samples | 0.3% |
| 2014, [56] | Invol. body signal | Pupil dynamics | Photo | Proprietary, 52 identities, 204 videos | 0% |

*(Continues)*

*Table 11.1* *(Continued)*

**Feature-level techniques**

| Reference | Subtype | Features and methodology | Attack | Database | Error |
|---|---|---|---|---|---|
| 2008, [44] | Static | Iris texture analysis using iris-texton + edge sharpness + co-occurrence matrix | Contact-lens | Proprietary, 30 identities, 600 samples | 3% |
| 2008, [57] | Static | Iris texture spectrum analysis using the FFT | Photo | Proprietary, 50 identities, 1,500 samples | 1% |
| 2010, [45] | Static | Iris texture analysis using Weighted LBP | Contact-lens | Proprietary, 72 identities, 1,400 samples | 0.5% |
| 2011, [38] | Static | Iris texture analysis using SIFT features and Hierarchical Visual Codebook (HVC) | Photo | Proprietary, 100 identities, 2,000 samples | 0.5% |
| 2014, [58] | Static | Iris texture analysis using image quality measures | Photo | Public, ATVS-FIr DB [34], 50 identities, 1,600 samples | 0.3% |
| 2015, [59] | Static | Iris texture analysis using deep learning | Photo | Public, ATVS-FIr DB [34] + LivDet-Iris DB [50] + MobBIOfake DB [51] | 0.9% |
| 2015, [60] | Dynamic | Iris texture analysis using multiscale binary statistical image features | Photo | Public, VSIA DB [60], 110 identities, 1,100 samples | 0% |

system is a glass eye, a photograph or dead tissue, spectrographic analyses could help to detect the spoofing attack. Following this approach, an anti-spoofing technique has been proposed using multispectral illumination to estimate the difference in the reflectance properties between the iris and the sclera at different wavelengths [62]. In a subsequent work, these measures were also complemented with the thickness of the corneoscleral limbus [54], and a comparison is established on a database of print, contact-lens and artificial-eye attacks. Based on this same multispectral principle, in [63], a novel sensor-level anti-spoofing method is proposed using NIR illumination at different wavebands and positions in order to detect the reflection properties of the different parts of the eye. Following a similar scheme to the spectrographic and reflectance signatures used in the previous works, in [46] an anti-spoofing technique was presented based on the specific characteristics of conjunctival vessels and iris textures that may be extracted from multispectral images. Also as part of this multispectral imaging research line, in two successive works [64,65] the authors present a novel iris recognition method using the grayscale sample resulting from the fusion of several images acquired at different wavelengths, and show that such an authentication approach is robust to attacks carried out with photos, contact lenses and artificial irises.

The four Purkinje reflections have also been exploited in the literature to develop iris liveness detection methods. These reflections are caused in a natural eye by the four optical surfaces that reflect light: the front and back surfaces of the cornea as well as the front and back surfaces of the lens. The position of the reflected light determines the position of the reflections. Therefore, a change in the location of the light source should even detect photographs displaying these reflections [55]. Varying positions of near-infrared light diodes used during image acquisition could also be used to analyse this property of the living eye as first suggested in [66]. However, as it was the case with the previous features, the efficiency of using such a characteristic against attack attempts where a real pupil is displayed (e.g., contact-lens attacks) is at least unclear.

Another intrinsic property of the human body that can be exploited for iris anti-spoofing purposes is the 3D nature of the eye. Based on this characteristic, a novel iris liveness detection method was proposed in [67] using a specific acquisition sensor with two NIR light sources attached to the sides of the camera. This way, the authors are able to acquire images where the 3D structure of the real irises is clearly visible thanks to the change of shadows. The system is tested against attacks carried out with printed images, contact lenses and artificial eyes.

Following a similar hypothesis to the previous work, that is, it should be possible to detect the 3D structure of the eye with respect to 2D surfaces used in photos attacks, the authors in [35] have developed a presentation attack detection method based on a Light Field Camera (LFC). The method, tested on a database comprising printed and digital-photo attacks from 104 unique iris patterns, measures the variation of focus between multiple depth images rendered by the LFC.

As mentioned above, involuntary signals of the body to measure liveness detection in iris recognition schemes. One of these interesting involuntary signals is the hippus, which is a pupillary diameter oscillation at about 0.5 Hz, occurring even under constant illumination. The coefficient of variation is at least 3%, although it declines with age. Several works have shown that this liveness cue can effectively be used to detect prosthetic eyes, high-resolution photographs or dead tissue [68,69]. However, as in the previous cases, its performance against video-based attacks or contact-lens attacks has not been tested yet.

Based on an analogous principle to the previous methods, the analysis of the pupil dynamics has also been studied as part of challenge–response approaches. In iris recognition, an involuntary reflex of the body that can be easily triggered by changing the illumination level is the variation of the pupil size. This size fluctuation in response to controlled external changes of light intensity has successfully been used for iris spoofing detection in the literature [52,56,68,70]. The pupillary light reflex does not only cause a change in the pupil size but also on the iris texture brightness, which was exploited in [53] to develop a liveness detection algorithm against print attacks. The two types of changes, pupil size and iris texture, were combined in [71] and tested against contact-lens attacks. In a more recent work [72], the pupil-dynamics metrics were combined with some of the previously proposed feature-level techniques based on the study of the iris texture, such as the multispectral reflectance analysis and the frequency spectrum analysis, and evaluated on a database of photo prints, contact lenses and artificial eyes showing very good results. Another interesting effect which can be observed when the pupil size changes due to a variation in illumination, and which could potentially be used for liveness detection, is the non-elastic distortion of the iris tissue known as iridal pattern deformation [73].

Although all the above-mentioned challenge–response schemes may be efficient even against contact-lens attacks or print-attacks where the pupil has been removed, they entail a high level of discomfort for the user due to the sudden lighting changes. Therefore, following this trend on the analysis of eye movement cues, other anti-spoofing techniques could be conceived based on voluntary responses of the users such as blinking, gazing at one specific point, looking to the sides, or moving the eyes up and down, so that the degree of unpleasantness is reduced. A preliminary study has been recently presented considering this line of research based on eye movement cues [74].

Following a similar dynamic-based perspective, a presentation attack detection method based on the Eulerian Video Magnification (EVM) was presented in [40] for video-based iris recognition systems. Since this approach requires an iris video, it is included within the sensor-level methods as standard iris scanners only acquire single iris images. The method has shown a remarkable performance even against video attacks.

Regarding multibiometric approaches, a recent study has proposed the combination of iris and electroencephalogram (EEG) recognition, not only to increase the system accuracy, but also as a protection method against presentation attacks [26]. Although it constitutes a first valuable research exercise, its usability in real applications is at least questionable.

## 11.3.2 Feature-level approaches

Regarding feature-level approaches, it was also Daugman who first proposed, mostly as a theoretical framework, some of the feature-level methods that have been later exploited in the field of presentation attack detection for iris biometrics [48,61].

One of the characteristics pointed out in [61] as a potential cue to detect fake and real irides were the retinal light reflections commonly known as the "red-eye effect". Essentially, light entering the eye is reflected back to the light source from the retina; this effect can be captured by a regular sensor, with no need for any additional hardware device, as long as the angle between light source, eye and camera is smaller than 2.5 degrees [75]. Although such anti-spoofing methods would be very efficient against regular print attacks or even artificial-eye attacks, its performance would be at least under question when dealing with contact lenses or iris printouts where the pupil has been cut out.

Daugman also indicated in his initial works that the printing process can leave detectable traces on spoofing artefacts, and that a simple 2D Fourier analysis of the acquired image can expose that unnatural behaviour. Pacut and Czajka [68], Czajka [76] and He *et al.* [57] have developed automated feature-level methods to analyse artificial frequencies in printed iris images. Following the same line, the Wavelet Transform has also been used, combined with a Support Vector Machine (SVM) classifier, as a way to extract discriminative features from the iris frequency spectrum in the task of detecting photo-attacks [37]. Another frequency-based approach was presented in [39], where the iris images are decomposed into Laplacian pyramids of various scales in order to analyse the frequency responses in different orientations. This constitutes the first iris anti-spoofing work that considers video attacks (i.e., the scanner is presented with a video of a real eye replayed in a digital device).

From those early works based on the spectral analysis of iris images, different image processing methods have been applied as an alternative to extract features that allow telling apart real and fake irides. For instance, in [77] four features based on the gray level values of the outer region of fake contact lenses are proposed for software-based spoofing detection. Similarly, in a subsequent work, gray level values of the iris texture are studied in order to characterize the visual primitives of the iris complementing them with measures related to the iris edge sharpness [44].

More recently, the use of colour cues from RGB images in the visible spectrum has also been analysed as a possible approach to detect fake irises taking advantage of the ancillary information that can be extracted from the periocular region [78]. Following this trend, a subsequent study has also compared the accuracy in fake iris detection of images captured in the near infrared (NIR) spectrum (i.e., grayscale pictures with only luminance information), and images acquired in the visible range that present also colour information [79].

One of the most active trends for presentation attack detection in iris systems is the use of local descriptors for the analysis of the iris texture. In [38,80], for example, iris texture primitives of real and fake images were modelled using a hierarchical visual codebook to represent the extracted Scale-Invariant Feature Transform (SIFT) features and tested against attacks carried out with printed photos, contact lenses and even plastic eyes. Also the very popular Local Binary Patterns (LBPs) have been considered for the texture representation of the iris image. This approach has been successfully applied to iris anti-spoofing in several works [45,81,82], where the efficiency of different configurations of LBPs has been evaluated against a number of known attacks (e.g., contact lenses, photo-attacks and artificial irises). Other texture-based features that have been studied include the use of Binary Statistical Image Features (BSIF) and Cepstral features that were fused for both iris and face presentation attack detection in [83]. A similar but more advanced version of this approach, using Multi-Scale Binary Statistical Image Features (M-BSIF) combined with SVM classifiers, was tested on different databases including a new large dataset comprising five types of presentation attacks [60]. Finally, a comprehensive study regarding local image descriptors for anti-spoofing purposes was presented in [84]. The work includes an analysis of the combination of multiple local features for the representation of iris texture including LBP, BSIF, SIFT or Local Phase Quantization (LPQ), tested on iris, face and fingerprint spoofing databases with very good results. A good comparative experimental study of several of the above-mentioned methods can also be found in the article reporting the results of the first Liveness Detection-Iris Competition (LivDet-Iris) held in 2013 [50] (further details about this competition may be found in Section 11.4).

Currently, one of the most promising trends in the wide field of image analysis and understanding is the use of deep neural networks for image classification. A pioneering work in this area has recently analysed the accuracy of two deep learning approaches applied to the problem of iris, face and fingerprint anti-spoofing reporting very good results against iris print-attacks [59]. However, the accuracy of these approaches against other type of presentation attacks still requires further attention.

The use of image quality assessment metrics has also been studied for iris anti-spoofing motivated by the assumption that a fake image captured in an attack attempt is expected to have different quality than a real sample acquired in the normal operation scenario for which the sensor was designed. Following this "quality-difference" hypothesis, several iris-specific quality metrics and general image quality metrics were studied in [85] and [58], respectively, to distinguish between real iris images and those acquired in photo-attacks.

Although in many cases not specifically designed for presentation attack detection purposes, a group of image processing methods that could be of great utility in this task are those developed for the detection of contact lenses [86–88]. Such algorithms can potentially play a major role as a pre-processing step in the protection against contact-lens attacks.

Many of the approaches mentioned above are starting to be also studied in the context of mobile iris applications, where sensor-level methods are more difficult to integrate due to the inherent hardware restrictions of this scenarios [89,90]. A recent independent competition organized to set the current state of the art of iris anti-spoofing in mobile environments has shown that this is a more challenging setting than the standard fixed scenario [51].

## 11.4 Evaluation of presentation attack detection methods

The present section gives an overview of the current publicly available anti-spoofing databases that may be used for the development and evaluation of new protection measures against direct attacks in the field of iris recognition.

In spite of the increasing interest in the study of vulnerabilities to direct attacks, the availability of spoofing databases is still scarce. This may be explained from both a technical and a legal point of view. (i) From a technical perspective, the acquisition of spoofing-related data presents an added challenge to the usual difficulties encountered in the acquisition of standard biometric databases (i.e., time-consuming, expensive, human resources needed, cooperation from the donors, etc.): the generation of a large amount of fake artefacts which are in many cases tedious and slow to generate on large scale (e.g., printed iris lenses or iris videos). (ii) The legal issues related to data protection are controversial and make the sharing and distribution of biometric databases among different research groups or industries very tedious and difficult. These legal restrictions have forced most laboratories working in the field of spoofing to acquire their own proprietary (and usually small) datasets on which to evaluate their protection methods, as may be seen in Table 11.2. Although these are very valuable efforts, they have a limited impact, since the results may not be compared or reproduced by other researchers.

Until very recently, compared to other traits largely analysed in the field of spoofing such as fingerprint and face, iris was still a step behind in terms of the organization of competitive liveness detection evaluations and also regarding the public availability of spoofing data. In that context of limited resources, several studies carried out in the field of iris security against presentation attacks have been

*Table 11.2   Comparative summary of the most relevant features corresponding to the publicly available iris presentation attacks databases described in Section 11.4*

| | Comparative summary: Public Iris Presentation Attacks DBs | | | | | | |
|---|---|---|---|---|---|---|---|
| | Overall info. (real/fake) | | Sensor info. | Attack (types) | | | |
| | # IDs | # Samples | # Sensors | Print | Digital | Video | Contact |
| ATVS-FIr DB [34] | 100/100 | 800/800 | 1 | ✓ | | | |
| LivDet-Iris DB [50] | 342/216 | 1,726/2,600 | 3 | ✓ | | | ✓ |
| IIITD-IS DB [82] | 202/202 | 4,848/4,848 | 2 | ✓ | | | |
| GUC-LF-VIAr DB [35] | 104/104 | 4,327/7,607 | 2 | ✓ | ✓ | | |
| MobBIOfake DB [51] | 110/110 | 550/2,750 | 1 | ✓ | | | |
| VSIA DB [60] | 110/110 | 550/2,750 | 1 | ✓ | ✓ | | |
| PAVID DB [39] | 100/100 | 800/800 | 1 | | | ✓ | |

performed using samples from previously acquired real datasets [38,72], so that in some cases real and fake users do not coincide [44,45,70].

In fact, until 2013, only one public iris spoofing database, the ATVS-FIr DB, was available [34]. In addition, its practical use was limited as it only related to one type of attack (i.e., print attacks without cutting out the pupil) acquired with one sensor. However, following the success of the LivDet competition series [91] organized in the fingerprint trait (held in 2009, 2011, 2013 and 2015), this situation of data scarcity was notably improved with the organization of the first Liveness Detection-Iris Competition of 2013 (LivDet-Iris 2013), that considered the submission of algorithms and systems [50]. The database used in the contest comprises three different subsets of print and contact-lens attacks and it is significantly larger than its predecessor. The second edition of the competition was held between 2015 and 2016 but upon the publication of the present chapter results had not been yet made public. Recently, another significant effort in the acquisition of iris spoofing data was published in [82] presenting the new IIITD Iris Spoofing public database with paper print attacks for over 200 identities. In the last two years, the Norwegian Information Security Laboratory[2] has also been very active in the field of presentation attack detection in iris recognition. As a result of their efforts, three new iris spoofing databases have been made available to researchers.

Table 11.2 presents a comparison of the most salient features of the iris spoofing public databases currently available. Examples of real and fake images that may be found in iris spoofing databases are shown in Figure 11.4 (extracted from the LivDet-Iris DB). In the following, all seven databases are briefly described in chronological order to their appearance.

## *ATVS-FIr DB*

The ATVS-FIr DB [34] is publicly available at the ATVS-Biometric Recognition Group website.[3]

The database comprises real and fake iris images (printed on paper) of 50 users randomly selected from the BioSec baseline corpus [92]. It follows the same structure as the original BioSec dataset, therefore, it comprises 50 users × 2 eyes × 4 images × 2 sessions =800 fake iris images and its corresponding original samples. The acquisition of both real and fake samples was carried out using the LG IrisAccess EOU3000 sensor with infrared illumination which captures bmp grayscale images of 640 × 480 pixels.

## *LivDet-Iris DB*

The first Liveness Detection-Iris Competition (LivDet-Iris) was held in 2013 [50]. The competition followed the approach initiated in the 2011 and 2013 Fingerprint LivDet evaluations, and two different categories, algorithms and systems, were opened for the

---

[2]http://www.nislab.no/
[3]http://atvs.ii.uam.es/

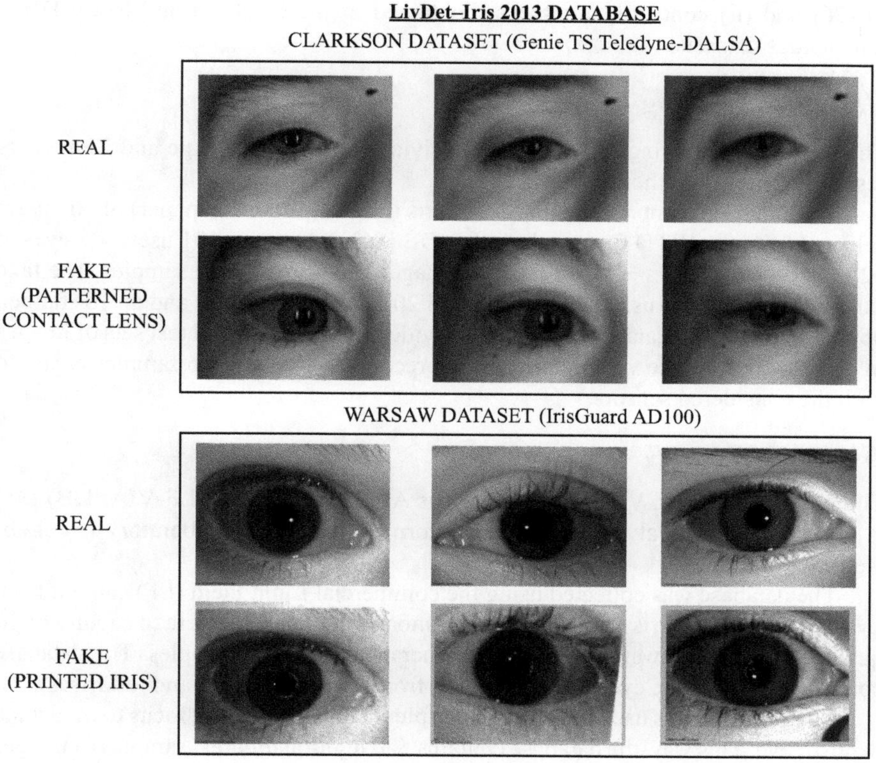

*Figure 11.4   Typical real iris images and fake samples from print and contact-lens attacks that may be found in the LivDet-Iris 2013 DB [50]*

participants. The LivDet-Iris 2013 DB used in the algorithms evaluation is distributed from the competition website.[4]

The train and test sets released are the same as the ones used in the LivDet-Iris 2013 competition so that future results achieved using these data may be directly compared to those obtained by the participants in the contest.

The database comprises over 4,000 samples acquired from around 500 different irises and is divided in three datasets captured at three different universities: University of Notre Dame, University of Warsaw and Clarkson University. Each dataset was captured with a different sensor: (i) IrisAccess LG4000 for the University of Notre Dame dataset; (ii) EyeGuard AD100 for the University of Warsaw dataset; and (iii) Genie TS from Teledyne-Dalsa, for the Clarkson University dataset.

Two different types of spoof attacks are considered in the database: (i) print attacks, that conform the University of Warsaw dataset, which is further described

---

[4]http://people.clarkson.edu/projects/biosal/iris/index.php

in [76] and (ii) contact-lens attacks, contained in the Clarkson and Notre Dame universities datasets.

## IIITD Iris Spoofing DB

The IIITD Iris Spoofing DB [82] is publicly available at the Image and Biometrics research group from the IIITD website.[5]

The database comprises real and fake iris images (printed on paper) of 101 users selected from the IIITD Contact Lens Iris DB [86]. It contains 101 users × 2 eyes × 12 images × 2 sensors $=4,848$ fake iris images and as many real samples. The fake samples are acquired using the Cogent CIS 202 dual eye scanner and an HP flatbed optical scanner. The database protocol is divided in three train and test sets of images in order to analyse the vulnerability of iris recognition systems to samples captured with the considered sensors.

## GUC-LF-VIAr DB

The GUC Light Field Visible Spectrum Iris Artefact DB (GUC-LF-VIAr DB) [35] is publicly available at the Norwegian Information Security Laboratory (NISLab) website.[6]

The database was collected using the commercial Light Field (LF) camera from Lytro to acquire real iris samples and the Canon 550D DSLR camera to capture high-quality samples that were later used to generate the artefact samples. The database comprises 104 unique eye patterns coming from 52 people (right and left irises).

The LF camera is used to acquire multiple images at different focus to model the depth of eye. This way, the database contains 520 high-quality real samples (5 images per iris) and 4,327 light field real samples. It also comprises 7,607 fake samples generated from five different types of attacks, which include both print attacks and digital attacks (i.e., the real high-resolution images are displayed in the screens of different mobile devices).

## MobBIOfake DB

The first Mobile Iris Liveness Detection Competition (MobILive) was held in 2014 [51]. The MobBIOfake DB used in the contest is distributed from the competition website.[7]

This is the first iris-spoofing database acquired in a mobile scenario. It was constructed from the set of iris images contained in the MobBIO Multimodal DB [93]. The MobBIO Multimodal DB comprises the biometric data from 105 volunteers. Each individual provided samples of face, iris and voice. The device selected for the samples acquisition was an Asus Transformer Pad TF 300T. Following the same structure, the MobBIOfake contains a subset of 800 iris images from MobBIO (8 real samples of

---

[5]https://research.iiitd.edu.in/groups/iab/resources.html
[6]http://www.nislab.no/biometrics_lab/guc_lf_viar_db
[7]http://mobilive2014.inescporto.pt/

100 different irides) and its corresponding fake copies, totalling 1,600 iris images. The fake samples were obtained from printed images of the original ones captured with the same handheld device and in similar conditions.

## VSIA DB

The GUC Visible Spectrum Iris Artefact DB (VSIA DB) [60] is publicly available at the Norwegian Information Security Laboratory (NISLab) website.[8]

This database is the subset of the GUC-LF-VIAr DB captured with the high-resolution Canon 550D DSLR camera. Therefore, it contains eye images captured from the right and left irides of 55 subjects (29 men and 26 women). For each iris, 5 real samples were acquired resulting in 550 real iris samples. As in the case of the GUC-LF-VIAr DB, the 2,750 fake samples are generated from five different types of photo attacks (both print and digital).

## PAVID DB

The Presentation Attack Video Iris DB (PAVID DB) [39] is publicly available at the Norwegian Information Security Laboratory (NISLab) website.[9]

This is the first iris spoofing database that contains replay video attacks (i.e., videos of a real eye are replayed in front of the iris acquisition scanner). The database comprises 152 unique eyes (right and left of 76 subjects) captured using two different smartphones: the Nokia Lumia 1020 and the iPhone 5S. For each of the eyes, two real and fake videos were acquired with each of the smartphones, totalling 608 real and fake samples.

## 11.5  Conclusion

The current biometric security context related to spoofing has promoted in the last 10 years a massive amount of research, which has flooded journals, conferences and media with new information, methods, algorithms and techniques regarding presentation attack detection approaches that intend to make this technology safer. This has been the case specially for some of the most deployed, popular and mature modalities such as the iris, which has also been shown to be one of the most exposed to spoofing. At the moment, the amount of new contributions and initiatives in the area of anti-spoofing is so overwhelming that it requires a significant condensation effort to keep track of all new information in order to form a clear picture of the state of the art as of today. The current chapter is an attempt to contribute to this difficult review task in the field of iris recognition.

When addressing the biometric spoofing issue we should always bear in mind that absolute security does not exist: given funding, willpower and the proper technology,

---

[8] http://www.nislab.no/biometrics_lab/vsia_db
[9] http://www.nislab.no/biometrics_lab/pavid_db

every security system can be compromised. However, there are a number of steps that can be taken in order to continue improving the security offered by biometric systems against presentation attacks, so that the effort and resources needed to break them, exceed the benefit gained by the attacker.

At the moment, one of the most important issues to be addressed in the field of biometric spoofing is the definition of a clear methodology to assess the "spoofability" of systems. This is not a straightforward problem, as there are new variables involved when the spoofing dimension is introduced. Although it is not yet generally deployed, an evaluation protocol which is gaining popularity for the assessment of biometric spoofing, defines two possible working scenarios [19]: *Licit scenario*, considered in classic recognition evaluations, it only takes into account genuine access attempts and zero-effort impostor access attempts. In this scenario performance is typically reported in terms of the FRR (False Rejection Rate, number of genuine access attempts wrongly rejected) and the FAR (False Acceptance Rate, number of zero-effort impostor access attempts wrongly accepted). *Spoofing scenario*, in which access attempts are either genuine or spoofing attacks. Although for this case there is still no agreed method for reporting results, two metrics which have been proposed and are starting to be used are the FRR (defined as in the licit scenario) and the SFAR (Spoofing False Acceptance Rate, corresponding to the number of spoofing attacks wrongly accepted).

Another shortcoming of most current anti-spoofing techniques, that should be analysed in the near future, is their lack of interoperability across databases. To date, the competitions organised have shown that top-ranked algorithms are able to achieve an accuracy close to 100%, however, their performance drops significantly when the testing dataset is changed. An interesting lesson may be learned from these results: There exists no clearly superior anti-spoofing technique. Selecting one particular protection method depends on the nature of the attack scenarios and acquisition conditions. Therefore, it is important to find complementary countermeasures and study the best fusion approaches in order to develop liveness-detection techniques that succeed at achieving a high performance over different spoofing data.

The anti-spoofing community should also consider engaging in new fundamental research regarding the biological dimension of biometric traits, in order to break with the current popular trend embraced by many of the latest research where some well known sets of features (e.g., LBP, LPQ, HOG or BSIF) are extracted from images in public databases and passed through a classifier. Although such a methodology is valid, in most cases it brings little new insight into the spoofing problem. A greater progress could potentially be obtained from new studies exploiting intrinsic biological differences between real and fake traits.

As a wrap up conclusion it may be stated that, although as shown in the present chapter, a great amount of work has been done in the field of iris spoofing detection and many advances have been reached, the attacking methodologies have also evolved and become more and more sophisticated. As a consequence, there are still big challenges to be faced in the protection against direct attacks, that will hopefully lead in the coming years to a whole new generation of more secure biometric systems.

# References

[1]  W. W. Bledsoe, "The model method in facial recognition," Panoramic Research Inc., Palo Alto, CA, Tech. Rep. PRI:15, 1964.

[2]  K. H. Davis, R. Biddulph, and S. Balashek, "Automatic recognition of spoken digits," *Journal of the Acoustical Society of America*, vol. 24, pp. 637–642, 1952.

[3]  A. K. Jain, "Biometric recognition," *Nature*, vol. 449, pp. 38–40, 2007.

[4]  The Guardian, "iPhone 5S fingerprint sensor hacked by Germany's chaos computer club," available on-line, 2013, http://www.theguardian.com/technology/2013/sep/22/apple-iphone-fingerprint-scanner-hacked.

[5]  The Register, "Get your German interior minister's fingerprint here," available on-line, 2008, http://www.theregister.co.uk/2008/03/30/german_interior_minister _fingerprint_appropriated/.

[6]  PRA Laboratory, "Fingerprint spoofing challenge," YouTube, 2013, http://www.youtube.com/watch?v=vr0FmvmWQmM.

[7]  Chaos Computer Club Berlin, "Hacking iPhone 5S touchid," YouTube, 2013, http://www.youtube.com/watch?v=HM8b8d8kSNQ.

[8]  Sky News, "Fake fingers fool hospital clock-in scanner," available on-line, 2013, http://news.sky.com/story/1063956/fake-fingers-fool-hospital-clock-in-scanner.

[9]  Tech Crunch, "Woman uses tape to trick biometric airport fingerprint scan," available on-line, 2009, http://techcrunch.com/2009/01/02/woman-uses-tape-to-trick-biometric-airport-fingerprint-scan/.

[10]  Tabula Rasa, "Tabula rasa spoofing challenge," available on-line, 2013, http://www.tabularasa-euproject.org/evaluations/tabula-rasa-spoofing-challenge-2013.

[11]  J. Galbally, "From the iriscode to the iris: a new vulnerability of iris recognition systems," in *Black Hat USA*, 2012.

[12]  T. Matsumoto, "Artificial irises: importance of vulnerability analysis," in *Proceedings of the Asian Biometrics Workshop (AWB)*, 2004.

[13]  J. Galbally, R. Cappelli, A. Lumini, *et al.*, "An evaluation of direct and indirect attacks using fake fingers generated from ISO templates," *Pattern Recognition Letters*, vol. 31, pp. 725–732, 2010.

[14]  J. Galbally and R. Satta, "3D and 2.5D face recognition spoofing using 3D printed models," *IET Biometrics*, vol. 5, no. 2, pp. 83–91, 2015.

[15]  F. Alegre, R. Vipperla, N. Evans, and B. Fauve, "On the vulnerability of automatic speaker recognition to spoofing attacks with artificial signals," in *Proceedings of the European Signal Processing Conference (EUSIPCO)*, 2012, pp. 36–40.

[16]  A. Hadid, M. Ghahramani, V. Kellokumpu, M. Pietikainen, J. Bustard, and M. Nixon, "Can gait biometrics be spoofed?" in *Proceedings of the IAPR International Conference on Pattern Recognition (ICPR)*, 2012, pp. 3280–3283.

[17]  P. Tome and S. Marcel, "On the vulnerability of palm vein recognition to spoofing attacks," in *Proceedings of the IEEE International Conference on Biometrics (ICB)*, 2015, pp. 319–325.

[18]  Z. Akhtar, G. Fumera, G. L. Marcialis, and F. Roli, "Evaluation of serial and parallel multibiometric systems under spoofing attacks," in *Proceedings of the IEEE International Conference on Biometrics: Theory, Applications and Systems (BTAS)*, 2012, pp. 283–288.

[19]  A. Hadid, N. Evans, S. Marcel, and J. Fierrez, "Biometrics systems under spoofing attack," *IEEE Signal Processing Magazine*, vol. 32, pp. 20–30, 2015.

[20]  Z. Akhtar, C. Micheloni, and G. L. Foresti, "Biometric liveness detection: challenges and research opportunities," *IEEE Security & Privacy Magazine*, vol. 13, pp. 63–72, 2015.

[21]  B. Schneier, "The uses and abuses of biometrics," *Communications of the ACM*, vol. 48, p. 136, 1999.

[22]  Y. Li, K. Xu, Q. Yan, Y. Li, and R. Deng, "Understanding OSN-based facial disclosure against face authentication systems," in *Proceedings of the ACM Asia Symposium on Information, Computer and Communications Security (ASIACCS)*, 2014, pp. 413–424.

[23]  ISO/IEC TC JTC1/SC 37, ISO/IEC 30107-1:2016, "Information technology – Biometric presentation attack detection – Part 1: Framework," International Standards Organization, 2016.

[24]  P. Johnson, R. Lazarick, E. Marasco, E. Newton, A. Ross, and S. Schuckers, "Biometric liveness detection: framework and metrics," in *Proceedings of the NIST International Biometric Performance Conference (IBPC)*, 2012.

[25]  S. Marcel, M. Nixon, and S. Z. Li, Eds., *Handbook of Biometric Anti-Spoofing*. Berlin: Springer, 2014.

[26]  B. Sabarigiri and D. Suganyadevi, "Counter measures against iris direct attacks using fake images and liveness detection based on electroencephalogram (EEG)," *World Applied Sciences Journal – Data Mining and Soft Computing Techniques*, vol. 29, pp. 93–98, 2014.

[27]  R. N. Rodrigues, N. Kamat, and V. Govindaraju, "Evaluation of biometric spoofing in a multimodal system," in *Proceedings of the IEEE International Conference on Biometrics: Theory, Applications and Systems (BTAS)*, 2010.

[28]  B. Biggio, Z. Akthar, G. Fumera, G. L. Marcialis, and F. Roli, "Security evaluation of biometric authentication systems under realistic spoofing attacks," *IET Biometrics*, vol. 1, pp. 11–24, 2012.

[29]  S. Shah and A. Ross, "Generating synthetic irises by feature agglomeration," in *Proceedings of the IEEE International Conference on Image Processing (ICIP)*, 2006, pp. 317–320.

[30]  C. Sousedik and C. Busch, "Presentation attack detection methods for fingerprint recognition systems: a survey," *IET Biometrics*, vol. 3, pp. 219–233, 2014.

[31]  E. Marasco and A. Ross, "A survey on anti-spoofing schemes for fingerprints," *ACM Computing Surveys*, vol. 47, pp. 1–36, 2014.

[32] J. Galbally, S. Marcel, and J. Fierrez, "Biometric anti-spoofing methods: a survey in face recognition," *IEEE Access*, vol. 2, pp. 1530–1552, 2014.

[33] J. Daugman, "High confidence visual recognition of persons by a test of statistical independence," *IEEE Transactions on Pattern Analysis and Machine Intelligence*, vol. 15, pp. 1148–1161, 1993.

[34] V. Ruiz-Albacete, P. Tome-Gonzalez, F. Alonso-Fernandez, J. Galbally, J. Fierrez, and J. Ortega-Garcia, "Direct attacks using fake images in iris verification," in *Proceedings of the COST 2101 Workshop on Biometrics and Identity Management (BioID)*, Springer LNCS, vol. 5372, 2008, pp. 181–190.

[35] R. Raghavendra and C. Busch, "Presentation attack detection on visible spectrum iris recognition by exploring inherent characteristics of light field camera," in *Proceedings of the IEEE International Joint Conference on Biometrics (IJCB)*, 2014.

[36] L. Thalheim and J. Krissler, "Body check: biometric access protection devices and their programs put to the test," *CT Magazine*, pp. 114–121, Nov. 2002.

[37] X. He, Y. Lu, and P. Shi, "A new fake iris detection method," in *Proceedings of the IAPR/IEEE International Conference on Biometrics (ICB)*. Springer Lecture Notes in Computer Science, vol. 5558, 2009, pp. 1132–1139.

[38] H. Zhang, Z. Sun, T. Tan, and J. Wang, "Learning hierarchical visual codebook for iris liveness detection," in *Proceedings of the IEEE International Joint Conference on Biometrics (IJCB)*, 2011.

[39] K. B. Raja, R. Raghavendra, and C. Busch, "Presentation attack detection using laplacian decomposed frequency response for visible spectrum and near-infra-red iris systems," in *Proceedings of IEEE International Conference on Biometrics: Theory and Applications (BTAS)*, 2015.

[40] K. B. Raja, R. Raghavendra, and C. Busch, "Video presentation attack detection in visible spectrum iris recognition using magnified phase information," *IEEE Transactions on Information Forensics and Security*, vol. 10, pp. 2048–2056, 2015.

[41] J. Galbally, A. Ross, M. Gomez-Barrero, J. Fierrez, and J. Ortega-Garcia, "Iris image reconstruction from binary templates: an efficient probabilistic approach based on genetic algorithms," *Computer Vision and Image Understanding*, vol. 117, pp. 1512–1525, 2013.

[42] S. Venugopalan and M. Savvides, "How to generate spoofed irises from an iris code template," *IEEE Transactions on Information Forensics and Security*, vol. 6, pp. 385–394, 2011.

[43] U. C. von Seelen, "Countermeasures against iris spoofing with contact lenses," in *Proceedings of the Biometrics Consortium Conference (BCC)*, 2005.

[44] Z. Wei, X. Qiu, Z. Sun, and T. Tan, "Counterfeit iris detection based on texture analysis," in *Proceedings of the IAPR International Conference on Pattern Recognition (ICPR)*, 2008.

[45] H. Zhang, Z. Sun, and T. Tan, "Contact lens detection based on weighted LBP," in *Proceedings of the IEEE International Conference on Pattern Recognition (ICPR)*, 2010, pp. 4279–4282.

[46]    R. Chen, X. Lin, and T. Ding, "Liveness detection for iris recognition using multispectral images," *Pattern Recognition Letters*, vol. 33, pp. 1513–1519, 2012.

[47]    A. Lefohn, B. Budge, P. Shirley, R. Caruso, and E. Reinhard, "An ocularist's approach to human iris synthesis," *IEEE Transactions on Computer Graphics and Applications*, vol. 23, pp. 70–75, 2003.

[48]    J. Daugman, *Biometrics. Personal Identification in a Networked Society*, Recognizing Persons by their Iris Patterns, Kluwer Academic Publishers, Dordrecht, 1999, pp. 103–121.

[49]    TABULA RASA, "Trusted biometrics under spoofing attacks," 2010, http://www.tabularasa-euproject.org/.

[50]    D. Yambay, J. S. Doyle, K. W. Boyer, A. Czajka, and S. Schuckers, "Livdet-iris 2013 – iris liveness detection competition 2013," in *Proceedings of the IEEE International Joint Conference on Biometrics (IJCB)*, 2014.

[51]    A. F. Sequeira, H. P. Oliveira, J. C. Monteiro, J. P. Monteiro, and J. S. Cardoso, "MobILive 2014 – Mobile Iris Liveness Detection Competition," in *Proceedings of the IEEE International Joint Conference on Biometrics (IJCB)*, 2014.

[52]    X. Huang, C. Ti, Q. Zhen Hou, A. Tokuta, and R. Yang, "An experimental study of pupil constriction for liveness detection," in *Proceedings of the IEEE Workshop on Applications of Computer Vision (WACV)*, 2013, pp. 252–258.

[53]    M. Kanematsu, H. Takano, and K. Nakamura, "Highly reliable liveness detection method for iris recognition," in *Proceedings of the SICE Annual Conference, International Conference on Instrumentation, Control and Information Technology (ICICIT)*, 2007, pp. 361–364.

[54]    S. J. Lee, K. R. Park, Y. J. Lee, K. Bae, and J. Kim, "Multifeature-based fake iris detection method," *Optical Engineering*, vol. 46, p. 127204, 2007.

[55]    E. C. Lee, Y. J. Yo, and K. R. Park, "Fake iris detection method using Purkinje images based on gaze position," *Optical Engineering*, vol. 47, p. 067204, 2008.

[56]    A. Czajka, "Pupil dynamics for iris liveness detection," *IEEE Transactions on Information Forensics and Security*, vol. 10, pp. 726–735, 2015.

[57]    X. He, Y. Lu, and P. Shi, "A fake iris detection method based on FFT and quality assessment," in *Proceedings of the IEEE Chinese Conference on Pattern Recognition (CCPR)*, 2008.

[58]    J. Galbally, S. Marcel, and J. Fierrez, "Image quality assessment for fake biometric detection: application to iris, fingerprint and face recognition," *IEEE Transactions on Image Processing*, vol. 23, pp. 710–724, 2014.

[59]    D. Menotti, G. Chiachia, A. Pinto, *et al.*, "Deep representations for iris, face, and fingerprint spoofing detection," *IEEE Transactions on Information Forensics and Security*, vol. 10, pp. 864–878, 2015.

[60]    R. Raghavendra and C. Busch, "Robust scheme for iris presentation attack detection using multiscale binarized statistical image features," *IEEE Transactions on Information Forensics and Security*, vol. 10, pp. 703–715, 2015.

[61] J. Daugman, "Iris recognition and anti-spoofing countermeasures," in *Proceedings of the International Biometrics Conference (IBC)*, 2004.

[62] S. J. Lee, K. R. Park, and J. Kim, "Robust fake iris detection based on variation of the reflectance ratio between the iris and the sclera," in *Proceedings of the Biometrics Symposium (BSym)*, 2006, pp. 66–71.

[63] Y. He, Y. Hou, Y. Li, and Y. Wang, "Liveness iris detection method based on the eye's optical features," in *Proceedings of the SPIE Optics and Photonics for Counterterrorism and Crime Fighting VI*, 2010, p. 78380R.

[64] J. H. Park and M. G. Kang, "Iris recognition against counterfeit attack using gradient based fusion of multi-spectral images," in *Proceedings of the International Workshop on Biometric Recognition Systems (IWBRS)*, Springer Lecture Notes in Computer Science, vol. 3781, 2005, pp. 150–156.

[65] J. H. Park and M.-G. Kang, "Multispectral iris authentication system against counterfeit attack using gradient-based image fusion," *Optical Engineering*, vol. 46, p. 117003, 2007.

[66] E. C. Lee, K. R. Park, and J. Kim, "Fake iris detection by using Purkinje image," in *Proceedings of the IAPR International Conference on Biometrics (ICB)*, 2006, pp. 397–403.

[67] E. C. Lee and K. R. Park, "Fake iris detection based on 3D structure of the iris pattern," *International Journal of Imaging Systems and Technology*, vol. 20, pp. 162–166, 2010.

[68] A. Pacut and A. Czajka, "Aliveness detection for iris biometrics," in *Proceedings of the IEEE International Carnahan Conference on Security Technology (ICCST)*, 2006, pp. 122–129.

[69] K. R. Park, "Robust fake iris detection," in *Proceedings of Articulated Motion and Deformable Objects (AMDO)*, Springer Lecture Notes in Computer Science, vol. 4069, 2006, pp. 10–18.

[70] R. Bodade and D. S. Talbar, "Dynamic iris localisation: a novel approach suitable for fake iris detection," *International Journal of Computer Information Systems and Industrial Management Applications*, vol. 2, pp. 163–173, 2010.

[71] N. B. Puhan, N. Sudha, and A. Suhas-Hegde, "A new iris liveness detection method against contact lens spoofing," in *Proceedings of the IEEE International Symposium on Consumer Electronics (ISCE)*, 2011.

[72] R. Bodade and S. Talbar, "Fake iris detection: a holistic approach," *International Journal of Computer Applications*, vol. 19, 2011.

[73] V. F. Pamplona, M. M. Oliveira, and G. V. G. Baranoski, "Photorealistic models for pupil light reflex and iridal pattern deformation," *ACM Transactions on Graphics*, vol. 28, pp. 106:1–106:12, 2009.

[74] O. Komogortsev and A. Karpov, "Liveness detection via oculomotor plant characteristics: attack of mechanical replicas," in *Proceedings of the International Conference of Biometrics (ICB)*, 2013.

[75] B. Toth, "Liveness detection for iris recognition," in *Proceedings of Biometrics and E-Authentication over Open Networks*, 2005.

[76] A. Czajka, "Database of iris printouts and its application: development of liveness detection method for iris recognition," in *Proceedings of the International*

Conference on Methods and Models in Automation and Robotics (MMAR), 2013, pp. 28–33.

[77]  X. He, S. An, and P. Shi, "Statistical texture analysis-based approach for fake iris detection using support vector machines," in *Proceedings of the IAPR International Conference on Biometrics (ICB)*.   Springer Lecture Notes in Computer Science, vol. 4642, 2007, pp. 540–546.

[78]  F. Alonso-Fernandez and J. Bigun, "Exploting periocular and RGB information in fake iris detection," in *Proceedings of the IEEE International Convention on Information and Communication Technology, Electronics and Microelectronics (MIPRO)*, 2014, pp. 1354–1359.

[79]  F. Alonso-Fernandez and J. Bigun, "Fake iris detection: a comparison between near-infrared and visible images," in *Proceedings of the IEEE International Conference on Signal-Image Technology and Internet-Based Systems (SITIS)*, 2014, pp. 546–553.

[80]  Z. Sun, H. Zhang, T. Tan, and J. Wang, "Iris image classification based on hierarchical visual codebook," *IEEE Transactions on Pattern Analysis and Machine Intelligence*, vol. 36, pp. 1120–1133, 2014.

[81]  Z. He, Z. Sun, T. Tan, and Z. Wei, "Efficient iris spoof detection via boosted local binary patterns," in *Proceedings of the IEEE International Conference on Biometrics (ICB)*, 2009.

[82]  P. Gupta, S. Behera, and M. V. V. Singh, "On iris spoofing using print attack," in *IEEE International Conference on Pattern Recognition (ICPR)*, 2014.

[83]  R. Raghavendra and C. Busch, "Presentation attack detection algorithm for face and iris biometrics," in *Proceedings of the IEEE European Signal Processing Conference (EUSIPCO)*, 2014, pp. 1387–1391.

[84]  D. Gragnaniello, G. Poggi, C. Sansone, and L. Verdoliva, "An investigation of local descriptors for biometric spoofing detection," *IEEE Transactions on Information Forensics and Security*, vol. 10, pp. 849–863, 2015.

[85]  J. Galbally, J. Ortiz-Lopez, J. Fierrez, and J. Ortega-Garcia, "Iris liveness detection based on quality related features," in *Proceedings of the IAPR International Conference on Biometrics (ICB)*, 2012, pp. 271–276.

[86]  D. Yadav, N. Kohli, J. S. Doyle, R. Singh, M. Vatsa, and K. W. Bowyer, "Unraveling the effect of textured contact lenses on iris recognition," *IEEE Transactions on Information Forensics and Security*, vol. 9, pp. 851–862, 2014.

[87]  K. W. Bowyer and J. S. Doyle, "Cosmetic contact lenses and iris recognition spoofing," *IEEE Computer*, vol. 47, pp. 96–98, 2014.

[88]  P. Silva, E. Luz, R. Baeta, H. Pedrini, A. X. Falcao, and D. Menotti, "An approach to iris contact lens detection based on deep image representations," in *Proceedings of the Conference on Graphics, Patterns and Images (SIBGRAPI)*, 2015.

[89]  A. F. Sequeira, J. Murari, and J. S. Cardoso, "Iris liveness detection methods in the mobile biometrics scenario," in *Proceedings of the IEEE International Joint Conference on Neural Networks (IJCNN)*, 2014, pp. 3002–3008.

[90]   A. F. Sequeira, J. Murari, and J. S. Cardoso, "Iris liveness detection methods in mobile applications," in *Proceedings of the IEEE International Conference on Computer Vision Theory and Applications (VISAPP)*, 2014, pp. 22–33.

[91]   LivDet, "LivDet: liveness detection competition series," online, September 2016, http://livdet.org/competitions.php.

[92]   J. Fierrez, J. Ortega-Garcia, D. Torre-Toledano, and J. Gonzalez-Rodriguez, "BioSec baseline corpus: a multimodal biometric database," *Pattern Recognition*, vol. 40, pp. 1389–1392, 2007.

[93]   A. F. Sequeira, J. C. Monteiro, A. Rebelo, and H. P. Oliveira, "MobBIO: a multimodal database captured with a portable handheld device," in *Proceedings of the IEEE International Conference on Computer Vision Theory and Applications (VISAPP)*, 2014, pp. 133–139.

*Chapter 12*

# Contact lens detection in iris images

*Jukka Komulainen, Abdenour Hadid,*
*and Matti Pietikäinen*

Iris texture provides the means for extremely accurate uni-modal person identification. However, the accuracy of iris-based biometric systems is sensitive to the presence of contact lenses in acquired sample images. This is especially true in the case of textured (cosmetic) contact lenses that can be effectively used to obscure the original iris texture of a subject and consequently to perform presentation attacks. Since also transparent contact lenses can degrade matching rates, automatic detection and classification of different contact lens types is needed in order to improve the robustness of iris-based biometric systems. This chapter introduces the problem of contact lens detection with particular focus on cosmetic contact lenses. The state of the art is analysed thoroughly and a case study on generalised textured contact lens detection is provided. The potential future research directions are also discussed.

## 12.1 Introduction

Among different biometric traits, iris is considered to be (one of) the most reliable and accurate biometric trait for person identification because iris patterns provide rich texture that is highly discriminative between individuals and stable during ageing of subjects [1]. Iris recognition is being increasingly deployed in large-scale applications requiring identity management, including border and access control, banking, mobile authentication and national identification programs, e.g., voter registration and social benefits. Probably the best example of a large-scale project was initiated by Unique Identification Authority of India[1] (UIDAI) that is implementing the scheme of providing a unique ID (AADHAAR number) for every Indian resident. So far the biometric samples of already over 1 billion citizens of 1.2 billion have been collected in the form of fingerprint, face and iris patterns.

In controlled environments, iris recognition is indeed extremely accurate but recent studies suggest that the iris texture is affected by covariates like pupil dilation [2] and sensor interoperability [3,4]. Presence of both textured and transparent soft

---

[1] http://uidai.gov.in

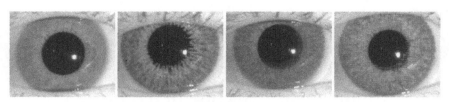

*Figure 12.1    Cropped iris images from the Notre Dame Contact Lens Detection 2013 (NDCLD'13) dataset [5] highlighting the variation in texture patterns between one genuine iris and three textured lens manufacturers, Cooper Vision, Johnson & Johnson and Ciba Vision, respectively. Reprinted with permission from [6] © 2014 IEEE*

contact lenses is another issue that may cause severe degradation in iris recognition performance in terms of false non-match rate (FNMR) [7–10]. The negative effect of contact lenses can be explained with general factors that can obscure the original iris texture like: (1) change in the optical properties of the eye, (2) presence of deliberate synthetic texture and (3) small movement of the contact lens on the surface of the eye between different acquisition times, e.g., enrolment and authentication [11].

Commercial iris recognition systems operate on eye images acquired using active near-infrared (NIR) illumination almost without an exception. The iris texture is rather hard to distinguish in conventional colour images, while rich texture can be observed in NIR images. Contact lenses with printed colour texture are designed for reshaping the appearance (colour and texture) of the iris tailored to wearer's preferences, e.g., transforming one's apparent eye colour from brown to blue, or from dark to light, thus they are also referred as cosmetic contact lenses. The change in colour and texture of the iris is due to a circular band on the contact lens containing pattern of pigment that is also visible in NIR wavelengths (see Figure 12.1). Typically, the observed iris texture is a mixture of the lens and original iris texture because the synthetic texture band (1) is not fully opaque when the original genuine iris texture is partly visible through it and (2) might not cover the whole iris region when a separate band of original iris texture is visible (see Figure 12.1) [12]. Still, it has been demonstrated that the presence of textured contact lenses yields to huge increase in FNMR when the gallery iris images of subjects with and without transparent contact lenses are compared with probe images of the same subjects with cosmetic lenses [7–10].

Textured contact lenses are indeed effective for obfuscating one's true biometric trait in order to avoid positive identification to one's previously enrolled identity because one is on a watchlist. This kind of presentation attack is a serious problem in real-world applications. For instance, iris recognition is utilised for checking if a person is on watch list of people who have been previously expelled from the United Arab Emirates (UAE) [13], thus cosmetic lenses could be potentially used to re-enter the country. In general, it is unlikely that cosmetic lenses would be used for other kinds of presentation attacks like targeted impersonation or creating a fake identity, e.g., extra social benefits. Soft contact lenses tend to move on the surface of the eye and can

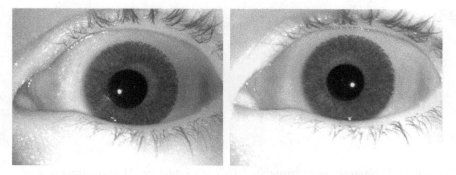

*Figure 12.2*    *Example of images of textured lenses from the NDCLD'13 benchmark dataset [5] illustrating the appearance variation on the same eye at different acquisition times due to small movement of the lens*

be put in differently, thus the iris texture of a subject wearing exactly the same cosmetic lens cannot be matched between different acquisition times (see Figure 12.2) [7–10]. However, soft contact lenses can be designed to maintain a specific orientation after eye blinking or any other movement by making it to be heavier at the bottom [8]. In theory, there exist no technical limitations of printing fully opaque cosmetic contact lenses with iris texture of a targeted person and wearing the customised lenses to successfully masquerade as someone else and thereby gaining illegitimate access and advantages [8,12,14]. However, there are no known real-world examples of successful presentation attack of this kind [12].

Textured contact lenses should be rejected before enrolment or during verification by labelling them as "failure to process", for instance [10]. The presence of cosmetic contact lenses is quite easy to reveal by manual visual inspection especially if the lens is not properly inserted or if the pure lens and original iris texture are otherwise separately observed [14]. However, detection of well-aligned cosmetic lenses of some printing patterns can be difficult using only the human eye. Furthermore, while manual inspection is feasible to perform, e.g., when enrolling a person to national ID system, the same does not hold for practical applications like automated border control. Automatic contact lens detection would be not only faster but potentially also more accurate than manual inspection [14]. The commercial iris recognition systems by IrisGuard are promoted to have a cosmetic contact lens detection feature[2] but no objective security evaluation on their robustness has been conducted [15].

Transparent (non-cosmetic) contact lenses can be considered to consist of two main categories based on the used material: rigid gas permeable (RGP) contact lenses and soft contact lenses [16]. The RGP lenses are well-known to degrade iris match quality because the whole lens fits within the iris region and thereby the boundary of

[2]http://irisguard.com/userfiles/file/Countermeasures.pdf

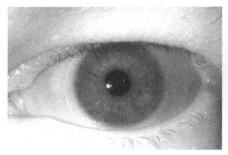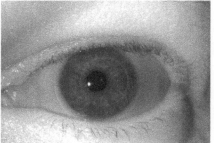

*Figure 12.3    Sample images from the NDCLD'13 dataset [5] illustrating typical*
*soft lens artefacts, e.g., circular lens boundary in the sclera region*
*and ring-like outline within the iris region*

the lens causes a severe circular artefact in the iris texture [8]. Daugman's iris recognition algorithm [1] is the basis for many iris-based biometric systems. Assuming that its improved version [17,18] is indeed able to detect and mask out these boundary artefacts, it seems likely that also many commercial iris recognition algorithms account for the presence of hard contact lenses [8]. In contrast, soft lenses, the most popular type of contacts, are much larger in diameter, thus it has been generally assumed for a long time that they do not substantially degrade the accuracy of iris recognition systems. However, the recent studies [7–10] have turned out this belief to be false.

Unlike cosmetic lenses, clear soft prescription lenses are not intentionally used for altering the original iris texture. Still, they must have some effect on the observed iris texture because they are designed to change the optical properties of the eye in order to correct eyesight [10]. For instance, the findings presented by Thompson *et al.* [19] imply that differences in iris curvature degrade matching ability of iris recognition systems. Some contact lenses can indeed yield to a ring-like artefact noticeable in the iris region (see Figure 12.3) [8]. As an example, toric lenses have additional curvature for correcting astigmatism in addition to "near-" or "far-sightedness curvature", which causes the circular outlines on the iris texture [8]. Occasionally, the lens boundary might be overlapping with the iris region like in the case of RGP lenses due to misplacement. Similarly to cosmetic lenses, clear lenses may contain also visible markings on them. For instance, toric lenses must maintain a specific orientation and need to be inserted correctly for proper vision correction, thus special visible markings are used to aid the insertion [8]. Some lenses may have even large visible artefacts, like a logo or numbers, on the iris region [8]. Also the alignment of transparent contact lenses can be different on the surface of the eye, thus the nature and location of the potential iris texture artefacts is likely to vary between different acquisition times.

One could try to locate the iris texture artefacts due to, e.g., lens boundaries or change in optical properties and mask them appropriately [8] or apply specific image correction techniques for the contact lens artefacts in order to increase the robustness of iris recognition to transparent lenses. The degradation in performance due to clear lenses can be potentially mitigated already without applying any sort of dedicated pre-processing before matching. The results by Baker *et al.* [8] suggest that false reject rates are generally lower for the case in which gallery images consist of subjects without (clear) contact lenses and probe images of subjects with contact lenses (none-soft) than rates for comparing contact lens images to contact lens images (soft-soft). Yadav *et al.* [10] conducted similar experiments but analysed the effect between two different iris sensors as well. Their intra-sensor and inter-sensors evaluations suggest that sensors can react very differently to soft contact lenses as the verification rates for soft-soft comparisons are very high in the intra-sensor scenario but drop dramatically in the inter-sensor case. In contrast, the use of natural iris in gallery images is not that sensitive to unknown sensors also when soft contact lenses are present in the probe images. The findings are somewhat consistent with the ones presented by Baker *et al.*. Therefore, it may be advisable to require enrolment without or, even better, with and without contact lenses to mitigate the effect of contact lenses at the time of verification or recognition regardless if the subject is wearing contact lenses or not [8].

There is indeed a need for a pre-processing stage consisting of robust automated detection and classification of both textured and transparent soft contact lenses in order to improve the robustness and reliability of iris recognition systems. This chapter gives an overview on the state of the art in contact lens detection with particular focus on software-based approaches and the problem of textured lenses. The remainder of the chapter is organised as follows. First, Section 12.2 introduces the prior works on hardware-based and software-based approaches for contact lens detection. A case study on generalised textured contact lens detection is presented in Section 12.3. In Section 12.4, the state of the art in software-based contact lens detection is thoroughly analysed. Finally, the conclusions and possible directions for future research are discussed in Section 12.5.

## 12.2 Literature review on contact lens detection

This section provides an overview on the prior works in contact lens detection. Since the presence of cosmetic contact lenses has been known to dramatically increase the FNMR of iris recognition systems for a long time and poses a severe security threat, the main research focus in the literature has been on textured contact lenses. Gradually, also the problem of transparent soft lenses has received more attention because the recent studies [7–10] have pointed out the benefits of detecting these types of lenses as well. Following the historical aspect and evolution of contact lens detection approaches, we introduce first methods developed particularly addressing the issue of cosmetic contact lenses, while transparent lenses are included into equation in the latter part of this section.

### 12.2.1 Textured contact lens detection

Iris-based biometric systems are prone to presentation attacks (traditionally referred to as spoofing) aiming at false positive or negative identification like solutions using any other biometric traits. Thus, dedicated countermeasures are needed in order to provide secure authentication solutions. Presentation attack detection (PAD) techniques can be broadly categorised into two groups based on the module in the biometric system pipeline in which they are integrated: hardware-based (sensor-level) and software-based (feature-level) methods [20]. Hardware-based methods introduce some custom sensor into the biometric system that is designed particularly for capturing the inherent differences between a valid living biometric trait and others. The measured characteristics can be divided into three groups: (1) intrinsic properties, e.g., reflectance at specific wavelengths [21–26], (2) involuntary signals, e.g., pupillary unrest [27] or (3) voluntary or involuntary responses to external stimuli (challenge-response), e.g., eye blink [28] or gaze [29], or pupillary light reflex [30–33]. Feature-level PAD techniques, on the other hand, are exploiting only the same data that is used for the actual biometric purposes, i.e., captured with the "standard" acquisition device. Counterfeit irises can be presented in many forms, including artificial glass and plastic eyes, photographs and videos, and printed textured contact lenses, for instance. In the following, we concentrate only on methods that have been proposed for detecting textured contact lenses. For further details on the advances in the field of iris PAD in general, interested readers are referred to surveys by Sun and Tan [34] and Galbally and Gomez-Barrero [20], for instance.

Cosmetic contact lenses have received significant attention in the research community in the context of presentation attack detection because they are easy to use in spoofing and are probably the most challenging ones to detect among different iris artefacts. For instance, the results of Iris Liveness Detection Competition 2013 (LivDet-Iris 2013) [35] demonstrate that cosmetic lenses are indeed much more difficult to detect compared to iris paper printouts. One reason for this is that the artefact is visible only within a very small part of the iris image, whereas usually the whole periocular region corresponds to the artefact in the case of a print attack. Furthermore, a cosmetic contact lens might have a transparent region in the vicinity of the pupil boundary (see Figures 12.1 and 12.2), thus the real pupil and its pupillary response, i.e., pupil dilation and contraction due to light stimulus is visible even through the lens. Therefore, PAD methods relying solely on eye blink detection [28], gaze estimation [29], Purkinje images [36,37], pupillary unrest [27] or pupillary light reflex [30–33] cannot detect the presence of printed contact lenses.

Few works [31–33] proposed to solve the failure modes of biological approaches utilising stimulated pupillary light reflex. The printed iris texture is not capable of being dilated and contracted like rubber band model along hippus movement unlike genuine iris patterns. Thus, measurement of iris texture deformation within the vicinity of pupil boundary can be exploited for detecting the presence of textured contact lenses.

Daugman suggested in one of his pioneer works on iris recognition [22] that multi-spectral reflectance analysis of eye region could be used for presentation attack detection. Consequently, multi-spectral measurements at specific wavelengths have been applied in cosmetic contact lens detection by exploiting the light absorption ratio between the iris and the sclera [23,24] and specific characteristics of conjunctival vessels and iris textures [21]. In addition, a greyscale image resulted from gradient-based fusion of multiple images acquired at different wavelengths do not present clear iris texture in the case of cosmetic contact lenses, unlike genuine iris images [25,26].

Natural iris can be considered roughly as a planar surface, whereas a textured contact lens is more curved as it is lying on the eye surface. Two proof-of-concept studies have suggested that 3D shape analysis of the observed iris can be indeed utilised for cosmetic contact lens detection. Connell *et al.* [38] applied structured light projection to measure the curvature of the observed iris, while Hughes and Bowyer [14] used a stereo camera set-up for recovering the 3D structure of the iris surface for cosmetic contact lens detection.

While these kinds of hardware-based solutions may provide efficient and generalised means for presentation attack detection, they can be also rather impractical due to unconventional imaging solutions that are not always possible to deploy or increased system complexity, or usability issues, e.g., time-consuming data acquisition, additional interaction demands or unpleasant sudden active lighting stimuli. Furthermore, these kinds of techniques have been usually evaluated on limited datasets just to demonstrate a proof of concept, like in [14,25,38], or, in the worst case, have not been experimentally validated at all, like in [30,33].

In the ideal case, textured contact lens detection would be performed entirely in software, i.e., only by further processing the NIR images acquired with "standard" iris sensors. Software-based approaches have been the most popular technique in cosmetic contact lens detection. Unsurprisingly, Daugman has been a pioneer in this field as well. In [39], he demonstrated that Fourier domain can be used well for describing periodic "dot-matrix style" cosmetic lens printing iris patterns (see Figure 12.1). However, defocus blur or the newer lens types with multiple layers of printing smooth the Fourier response when no evident peaks cannot be found.

As seen in Figure 12.1, each printing process leaves its own characteristic signature (or artefacts) that can be detected by analysing the evident textural differences between genuine iris and fake one. In general, genuine iris texture is rather smooth, while the printed texture patterns of contact lenses are somewhat coarse despite the more advanced printing techniques using, e.g., multiple layers. Consequently, by far the most common and promising approach has been to apply different local descriptors for cosmetic contact lens detection, including grey level co-occurrence matrix (GLCM)-based features [40], a combination of GLCM features and iris-textons [41], multi-resolution local binary patterns (LBP) [42,43], weighted-LBP [44], scale-invariant feature transform (SIFT)-based hierarchical visual codebook [45] and binarised statistical image features (BSIF) [6,15]. Most of these works are reporting excellent detection rates very close to 100% in controlled scenarios but novel printed lens texture patterns (not seen during training) and sensor interoperability can yield to dramatic decrease in system performance [5,44]. In-depth analysis on these issues will be provided in Sections 12.3 and 12.4.

## 12.2.2 Classification of contact lens type

The research on the problem of soft contact lenses is just in its infancy because their negative effect on the accuracy of iris-based biometric systems has not been discovered until recently. To the best of our knowledge, only one work has proposed a sensor-level

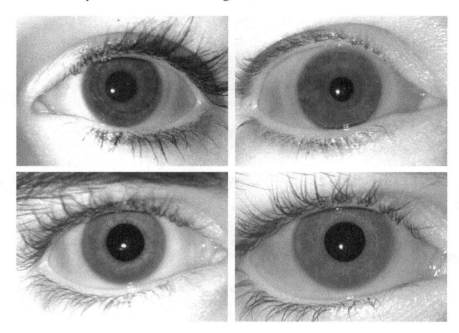

*Figure 12.4    Sample images from the NDCLD'13 dataset [5] illustrating the
difficulty of lens boundary detection, i.e., thin or virtually invisible
boundary outlines in sclera region due to specular reflections or
otherwise inconsistent illumination, defocus blur and lens placement*

approach to the problem. Kywe *et al.* [46] noticed that the variations in temperature
on the surface of the eye due to evaporation of water during blinking is different
for eyes with and without contact lenses. However, the proposed approach requires
a thermal camera and its detection accuracy is highly dependent on environmental
conditions like temperature and humidity. Thus, so far even hardware-based methods
have not been able to provide generalised solutions for robust detection of transparent
contact lenses.

Software-based detection of (non-cosmetic) soft contact lenses in NIR iris images
in general is far more difficult compared to the case of textured lenses. The appearance
differences between no lens and soft lens iris images are very subtle and highly
dependent on the input image quality, for instance. Even by the human eye it is
generally hard to tell if there is a soft contact lens present on the observed image or
not unless the lens boundary is visible in the sclera region (see Figure 12.3). Erdogan
and Ross [16] proposed a method for detecting non-cosmetic contact lenses in NIR
ocular images by checking if the lens boundary can be found in the vicinity of the
segmented limbic boundary. They reported moderate overall classification accuracies
between 66.8% and 76.0% because inconsistent illumination and defocus blur lead
to unsuccessful detections. As illustrated in Figure 12.4, the lens boundary can be

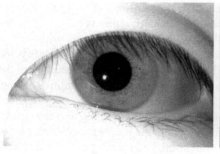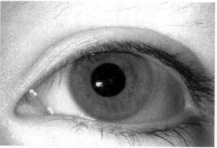

*Figure 12.5   No lens vs soft lens images of different eyes from the NDCLD'13 benchmark dataset [5] demonstrating the appearance similarity of no lens and soft lens iris images in the right sclera region*

indeed hard to describe because its appearance can have significant variations or the outline is virtually invisible due to acquisition conditions, e.g., specular reflections or otherwise inconsistent illumination or defocus blur, or the alignment of the lens. Another issue is that the similar ring-like artefacts can be observed in iris images of a subject without a soft lens as seen in Figure 12.5.

Ideally, contact lens detection could be seen as a three-class problem of categorising acquired iris images as no lens, clear lens or cosmetic lens because the classification of contact lens type is important in addition to detection [5,9,10,42]. Since the pioneer works in automated classification of contact lens type released public benchmarks the Notre Dame Contact Lens Detection 2013 [5,10] dataset and IIITD Contact Lens Iris Database [9,10] consisting of NIR iris images of all the three classes, an increasing trend has been to propose methods for distinguishing cosmetic and transparent soft contact lenses from natural irises [11,47,48]. In general, these kinds of approaches are fundamentally the same as the algorithms proposed in the context of presentation attack detection as they are both based on extracting some local feature description from the given iris image. The main differences are that the sclera region is also utilised (lens boundary detection) and three-class classification is performed instead of binary decisions. The state-of-the-art overall accuracy across the three classes varies between 83% and 93% depending on the used iris sensor. While the detection rates for textured lenses are again almost 100%, the rates for no lens and non-cosmetic soft lenses are generally significantly lower and vary from 79% to 96%, and from 76% to 84%, respectively [47]. However, again the overall performances drop in inter-sensor evaluations to 75% due to unsatisfying the classification rates for natural iris and non-cosmetic soft contact lens images [11].

## 12.3   Case study: generalised software-based textured contact lens detection

As seen in Section 12.2, many software-based approaches for detecting textured contact lenses have been indeed proposed in the literature. Since most of the existing works are reporting astonishing correct classification results of almost 100% on proprietary (e.g., [44]) and publicly available benchmark datasets (e.g., [45]), the cosmetic lens detection appeared to be a solved problem. However, without few exceptions, the effect of two important factors, namely, sensor interoperability and previously unseen cosmetic lens patterns, has been overlooked when validating the robustness of the proposed algorithms. Thus, their generalisation capabilities beyond laboratory conditions can be questioned. For instance, Zhang *et al.* [44] reported overall classification rates between 97% and 99% when the training and test sets of iris images were captured with the same sensors, while the performance decreased to between 83% and 88% under cross-sensor evaluation, i.e., training and test sets were captured with different iris sensors. Later, Doyle *et al.* [5] demonstrated that even more dramatic drop in the detection accuracy may be observed if a cosmetic lens with previously unseen printed texture is introduced to a lens detection algorithm.

In practice, biometric systems are installed in open environments, thus the assumption of full prior knowledge of the used sensors and different cosmetic lens patterns, that will be confronted in operation, is far from reasonable. Thus, there is a need for generalised textured lens detection algorithms that can operate under unpredictable conditions. Possible directions towards more robust and generalized software-based solutions include: (1) designing novel feature representations having milder assumptions about the cosmetic contact lens patterns, (2) augmenting training data of counterfeit iris patterns online like databases of anti-virus software [45], (3) using a combination of several (lens-specific) methods and (4) modelling variability in the mixture of genuine and fake iris feature representations.

One important question is also whether the cosmetic contact lens detection (or PAD in general) should be considered as two-class or one-class problem. While huge amount of genuine iris data is available for training natural iris model, the same does not hold for counterfeit irises because novel brands with previously unseen cosmetic lens patterns will be eventually experienced in operation. Since it is practically impossible to cover every existing printing pattern in training datasets, ideally, the problem can be solved using one-class classifiers for modelling the variations of the only fully known class (genuine). This kind of approach has already shown to be promising direction in speaker verification PAD [49].

In any case, all aforementioned main aspects towards generalised textured lens detection can be considered in both, fundamentally different, principles of one-class and two-class modelling. For instance, in feature design, one needs to figure out how to emphasise and capture the variations characteristic to only genuine iris texture or the differences between genuine iris texture and fake one without exploiting the prior knowledge of known printing signatures too much. Furthermore, where an up-to-date database of known lens printing patterns is useful when upgrading two-class models, it is also necessary for tuning the hyperparameters of the one-class model.

Eventually, the success in real-world applications is mostly dependent on how robust the applied feature representation is. The varying nature of different lens printing techniques makes the problem indeed difficult to solve. First, the market of textured contact lenses is growing, thus the number of texture patterns increases at the same time. Already the three example images seen in Figure 12.1 demonstrate

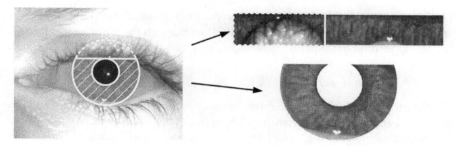

*Figure 12.6*  *Comparison of between traditional polar and the proposed geometric normalisation techniques highlighting the distorted texture patterns of Ciba Vision lenses and occlusion in polar domain. The dashed lines represent the omitted area because of possible occlusion due to eyelashes and eyelids. Reprinted with permission from [6] © 2014 IEEE*

how dissimilar the printing signatures of various suppliers can be. Generalised texture representation of various (unseen) printing techniques itself is a challenging problem but the semi-transparent nature of cosmetic contact lenses makes the problem even less trivial as the natural iris texture may be at least partly observed. In the following, we introduce our work [6] on designing a more generalised (natural) iris texture representation for detecting the presence of cosmetic contact lenses.

## 12.3.1  Pre-processing

Iris image pre-processing, including localisation, geometric normalisation and occlusion handling due to eyelashes and eyelids, is as important part of a counterfeit iris detection pipeline as the actual texture feature extraction. Traditionally iris images are normalised into polar coordinate system also when detecting counterfeit irises [10,40–43,45]. However, the geometric transformation causes severe distortion on the regular lens texture patterns. In addition, valuable details are probably lost due to interpolation when mapping the ring-shaped iris region into rectangular image. As seen in Figure 12.6, the printing signatures are far less evident in the resulting polar coordinate image as opposed to the original Cartesian domain, hence probably also harder to describe. While the polar coordinate system is convenient for finding distinctive features across different individuals and matching purposes, it might be unsuitable for creating generalised feature representation for cosmetic contact lens detection.

In order to preserve the regularity of the lens patterns, we compute the iris texture description in the original image space by normalising the square bounding box of the limbic boundary into $N \times N$ pixels. We omit all pixels belonging to the pupil and sclera region using the pupillary boundary and the limbic boundary to avoiding the effect of irrelevant information that is not part of iris texture (see Figure 12.6).

Furthermore, we take the possible occlusion due to eyelashes and eyelids into account and further refine our region of interest (ROI) to focus on the lower part of the iris, i.e., the upper limit of the pupillary boundary acts also as the upper limit for the ROI.

## 12.3.2    Texture description

For describing the inherent textural differences between natural and synthetic iris texture, we adopted binarised statistical image features (BSIF) [50] because they have shown potential tolerance to image degradations appearing in practice, e.g., rotation and blur. More importantly, the BSIF filters are derived from statistics of natural images, thus they are probably suitable for emphasising the textural properties of those characteristic to natural iris images and not synthetic ones.

Many local descriptors, such as LBP [51], compute the statistics of labels for pixels in local neighbourhoods by first convolving the image with a set of linear filters and then quantising the filter responses. The bits in the resulting code string correspond to binarised responses of different filters. Kannala and Rahtu [50] proposed to learn the filters by utilising statistics of natural images instead of using manually predefined heuristic code constructions.

Given an image patch $X$ of size $l \times l$ pixels and a linear filter $W_i$ of the same size, the filter response $s_i$ is obtained by:

$$s_i = \sum_{u,v} W_i(u, v)X(u, v) = w_i^T x, \qquad (12.1)$$

where vector notation is introduced in the latter stage, i.e., the vectors $w$ and $x$ contain the pixels of $W_i$ and $X$. The binarised $b_i$ feature is then obtained by setting $b_i = 1$ if $s_i > 0$ and $b_i = 0$ otherwise. Given $n$ linear filters the bit strings for all image patches of size $l \times l$, surrounding each pixel of an image, can be computed conveniently by $n$ convolutions. Like in the case of LBP, the properties of the iris texture are represented using histograms of BSIF values extracted over normalised and masked iris image.

The filters $W_i$ are obtained using independent component analysis (ICA) by maximising the statistical independence of $s_i$. In general, this approach has shown to produce good features for image processing [52]. Furthermore, the independence of $s_i$ provides justification for the independent quantisation of the elements of the response vector $s$ [50]. Thus, costly vector quantisation, used, e.g., in [45], is not needed in this case for obtaining a discrete texton vocabulary.

BSIF descriptor has two parameters: the filter size $l$ and the length $n$ of the bit string (i.e., number of learned filters). We used the set of filters[3] provided by the authors of [50]. The filters $W_i$ with different choices of parameter values were learned using 50,000 patches that were randomly sampled from a set of 13 natural images provided in [52]. The BSIF filters used in the proposed iris description can be seen in Figure 12.7.

---

[3] http://www.ee.oulu.fi/~jkannala/bsif/bsif.html

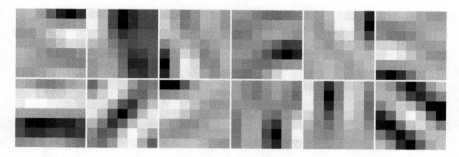

*Figure 12.7   BSIF filters size of 7 × 7 learned from natural images. Reprinted with permission from [6] © 2014 IEEE*

## 12.3.3   Experimental analysis

Next, we conduct extensive set of experiments in order to compare the performance of different feature representations for textured contact lens detection, including iris image pre-processing and texture features. The main purpose of our experimental analysis is to evaluate the generalisation capability of the different algorithms under two varying conditions, cosmetic lenses with previously unseen texture patterns and unknown iris sensors. For performance evaluation, we considered the extended version of the Notre Dame Contact Lens Detection 2013 (NDCLD'13) benchmark dataset [5] as it provides the means for evaluating the effect of both of these variables separately, unlike many other databases.

### 12.3.3.1   Experimental set-up

The NDCLD'13 database consists of iris images of subjects without contact lenses, with soft and textured contact lenses, acquired under near-infrared illumination. The database can be used for evaluating methods that try to solve a three-class problem of categorising iris images as no lens, clear lens or cosmetic lens [42]. In the following, we focus on two-class problem considering iris images without contact lenses and with clear lenses as genuine samples, whereas the iris images with cosmetic lenses are regarded as artefact samples.

The updated NDCLD'13 database contains three subsets of iris images. The first subset (ND I) consists of a training set of 3,000 samples and a test set of 1,200 samples acquired using an LG 4000 iris camera. Second subset (ND II) contains 600 for training images and 300 images for testing acquired with an IrisGuard AD100 iris sensor. Both datasets are balanced across the three different categories of iris images. Moreover, the iris images are divided into 10 subject-disjoint and class-balanced folds for training and tuning the algorithms. Although the number of iris images in training and test sets between ND I and ND II is a bit unbalanced, the consistent composition enables fair cross-sensor experiments.

The third subset (ND III) is an extended set of ND I because it contains constant set of no lens and soft lens images from ND I but, more importantly, 1,427

Table 12.1    Composition of the different subsets of the NDCLD'13
database. Reprinted with permission from [6] © 2014 IEEE

| Subset | Train | | Test | |
|---|---|---|---|---|
| | **Genuine** | **Fake** | **Genuine** | **Fake** |
| ND I | 2,000 | 1,000 | 800 | 400 |
| ND II | 400 | 200 | 200 | 100 |
| Cooper | 2,160 | 2,160 | 667 | 667 |
| J & J | 1,955 | 1,955 | 872 | 872 |
| Ciba | 1,539 | 1,539 | 1,288 | 1,288 |

additional iris images with cosmetic lenses acquired with the LG4000 camera.
ND III can be used for testing how well a textured contact lens detection algorithm is
able to perform when an unseen cosmetic texture pattern is shown because it provides
leave-one-out protocol across the three types of cosmetic contact lenses by Cooper
Vision,[4] Johnson & Johnson[5] and Ciba Vision[6] (667, 872 and 1,288 counterfeit iris
images, respectively). Table 12.1 summarises the composition of different subsets in
the NDCLD'13 database.

Since the focus of our experiments is on exploring the generalisation capabilities of textured lens detection algorithms, we utilise the segmentation information
included in the NDCLD'13 dataset that provides centre and radius for circles defining the pupillary boundary and the limbic boundary. The average radius of the limbic
boundary in the provided segmentation data is bit over 130 pixels, thus the bounding
box of the proposed Cartesian iris image normalisation is set to $255 \times 255$ pixels. For
training and tuning the classifiers, we follow the predefined folds in the ND I and ND
II. However, due to the lack of fixed folds in the ND III, cross-validation is applied
during the leave-one-out lens test.

LBP texture feature based algorithms have shown to be effective in cosmetic
contact lens detection [42–44], thus we adopted multi-resolution LBP [42,43] using
uniform patterns (denoted by $LBP^{u2}$) [51] as a baseline descriptor for comparing the
benefits of the proposed pre-processing approach and the BSIF-based iris texture
description. In our experiments, the multi-resolution representation was extracted
by applying $LBP^{u2}_{P,R}$ operator with eight sampling points ($P = 8$) at multiple scales
(radii $R$) and concatenating the resulting LBP histograms into a single feature vector.
For fair comparison, the best-performing combination of different $LBP^{u2}_{8,R}$ operators and best-performing BSIF descriptor are selected when comparing the different
features in general.

---

[4]Expressions Colours: http://coopervision.com/contact-lenses/expressions-color-contacts
[5]ACUVUE2 Colours: http://www.acuvue.com/products-acuvue-2-colours
[6]FreshLook Colourblends: http://www.freshlookcontacts.com

*Table 12.2   Mean CCR and EER (in %) across different lens types using different geometric normalisation approaches, i.e., polar and Cartesian coordinate systems and ROI processing. Reprinted with permission from [6] © 2014 IEEE*

| Pre-processing | LBP | | BSIF | |
|---|---|---|---|---|
| | **CCR** | **EER** | **CCR** | **EER** |
| Polar | 87.16 | 6.30 | 93.23 | 2.05 |
| Polar with ROI [41] | 91.55 | 4.95 | 94.40 | 1.11 |
| Proposed | **94.01** | **3.11** | **98.42** | **0.33** |

### 12.3.3.2   Leave-one-out lens validation

Since the algorithms are likely to fail in detecting cosmetic lenses with textured patterns not present in training material, we begin our experiments with the leave-one-out protocol of the ND III. Following the protocol used in [5], the models are trained on images of two of the three lens manufacturers while third one is left out for evaluating the generalisation capability.

First, we want to find out the actual benefit that we gain by extracting the features in the original image domain with the proposed pre-processing method instead of applying traditional normalisation into polar coordinate transform. For the sake of simplicity, a linear support vector machine (SVM) is utilised for classifying both feature representations. The mean correct classification rate (CCR) and equal error rate (EER) across different lens folds are shown in Table 12.2. Since the eyelids and eyelashes often cause severe occlusion over the iris texture, it is not surprising that the use of ROI already in polar domain leads to increase in performance for both features. More importantly, the results support our hypotheses as even more significant performance enhancement is obtained when the proposed pre-processing approach is used. Thus, we utilise it in the following experiments.

In Table 12.3, we can see the mean performance across different lens folds when different SVM classification schemes are applied. The use of non-linear SVM leads to best results for every feature representation, especially for multi-resolution LBP as it is able to reach the same performance level as with BSIF descriptors. However, it is also important to note that the BSIF features perform extremely well even with linear classifier and the performance does not gain too much when utilising non-linear SVM, which shows that the BSIF-based iris texture description is indeed suitable for detecting cosmetic contact lenses.

We also included one-class SVM for comparison and tried to approach the problem by modelling only the genuine iris texture inspired by the work by Alegre *et al.* [49]. The use of one-class SVM leads to satisfactory results but both of the two-class SVMs performed better because the false rejection rate of one-class SVM was much higher. Although better generalisation capability was achieved with proper

*Table 12.3    Mean CCR and EER (in %) across different lens types using different SVM classifiers. Reprinted with permission from [6] © 2014 IEEE*

| Feature | Linear | | RBF | | One-class | |
|---|---|---|---|---|---|---|
| | CCR | EER | CCR | EER | CCR | EER |
| BSIF 6b 5 × 5 | 98.23 | 0.95 | **98.31** | **0.90** | 92.88 | 5.52 |
| BSIF 9b 5 × 5 | **97.65** | 0.61 | **97.65** | **0.37** | 94.86 | 4.18 |
| BSIF 12b 7 × 7 | 98.42 | 0.33 | **98.47** | **0.11** | 95.28 | 2.09 |
| $LBP^{u2}_{8,[1-3]}$ | 94.01 | 3.11 | **97.67** | **0.69** | 94.00 | 5.18 |

*Table 12.4    Lens specific CCR and EER (in %) of the leave-one-out lens test. Reprinted with permission from [6] © 2014 IEEE*

| Feature | Cooper | | J & J | | Ciba | | Mean |
|---|---|---|---|---|---|---|---|
| | CCR | EER | CCR | EER | CCR | EER | CCR |
| BSIF 6b 5 × 5 | 99.55 | 0.60 | 96.73 | 1.49 | 98.64 | 0.62 | 98.31 |
| BSIF 9b 5 × 5 | 98.28 | 0.30 | 94.67 | 0.80 | **100.00** | **0.00** | 97.65 |
| BSIF 12b 7 × 7 | **99.78** | **0.00** | 95.76 | **0.34** | 99.88 | **0.00** | **98.47** |
| $LBP^{u2}_{8,[1-3]}$ | 99.33 | 0.60 | **97.25** | 0.46 | 96.43 | 1.01 | 97.67 |

feature design in our experiments, one-class approach should not be forgotten in future studies because of the limited number of different textured printing signatures and (genuine) training samples in the leave-one-out lens validation.

The final results of the leave-one-out lens validation and lens-specific breakdown for the best features and non-linear SVM can be seen in Table 12.4. Like in [5], we found out that the Cooper Vision lens appears to be the easiest one to detect probably because its printing texture is somewhere in between Ciba Vision and Johnson & Johnson. The results suggest that certain lens types can be more useful in training and tuning generalised models, like Johnson & Johnson seems to be in the case of the BSIF based description. It is also worth mentioning that the CCR and EER do not always go hand in hand due to the cross-validation used during training and tuning the lens detection models. A proper pre-defined validation set could be probably used for tuning the operating point in order to avoid this kind of overfitting.

### 12.3.3.3    Device independent validation

Another important property in practical applications is device independent performance. Next, we perform intra-sensor and inter-sensor experiments using ND I and ND II in the NDCLD'13 database. In other words, we train the algorithms on the training set of ND I and evaluate the models on the test sets of ND I and ND II,

*Table 12.5   CCR and EER (in %) of the cross-sensor validation. Reprinted with permission from [6] © 2014 IEEE*

| Feature | Trained on LG4000 | | | | Trained on AD100 | | | |
|---|---|---|---|---|---|---|---|---|
| | Intra-sensor | | Inter-sensor | | Intra-sensor | | Inter-sensor | |
| | CCR | EER | CCR | EER | CCR | EER | CCR | EER |
| BSIF 6b | 98.75 | 1.25 | 89.00 | 5.00 | 97.67 | 2.00 | 97.67 | 2.50 |
| BSIF 9b | 99.33 | 0.50 | **93.00** | 3.00 | 98.67 | **1.00** | 98.67 | 1.50 |
| BSIF 12b | **99.42** | **0.00** | 92.33 | **1.00** | **99.00** | **1.00** | **99.33** | **0.75** |
| LBP$_{8,[1-3]}^{u2}$ | 98.67 | 1.25 | 85.33 | 16.00 | 96.33 | 4.00 | 90.75 | 9.25 |

and vice versa. The results in Table 12.5 depict that all texture descriptions using the proposed pre-processing perform extremely well in the intra-sensor test. However, BSIF outperforms multi-resolution LBP in the inter-sensor validation, especially in terms of EER when even 6-bit version of BSIF performs reasonably well. The better performance of BSIF might be due to its potential tolerance to image degradations, e.g., blur [50].

## 12.4   Discussion

The experimental results from Section 12.3 show that it is indeed possible to perform more generalised textured contact lens detection across unknown iris sensor data and novel lens printing texture patterns with carefully designed iris texture representation. Further works by other researchers have been reporting similar findings highlighting the importance of proper pre-processing and feature spaces in the context of generalised textured contact lens detection [15] and the three-class problem of categorising iris images as no lens, clear lens or cosmetic lens [11,47,48].

### 12.4.1   Further work on generalised textured contact lens detection

Doyle and Bowyer [15] extended their prior work [5] and released the Notre Dame Contact Lens Detection 2015 (NDCLD'15) dataset that introduces additional sensor and two additional lens brands (printing patterns), Clearlab[7] and United Contact Lens,[8] that were not included in the extended version of the NDCLD'13 database [5]. Their experiments on the more comprehensive dataset provided additional evidence verifying our preliminary results on the robustness of BSIF based iris description extracted from the original Cartesian image domain. First, the CCR drops only from 100% in the intra-sensor case to just over 95% in the inter-sensor case, while the CCR

---

[7] Eyedia Clear Colour Elements: http://www.clearlabusa.com/eyedia-clear-color.php
[8] Cool Eyes Opaque: http://www.unitedcontactlens.com/contacts/opaque-lenses.html

regains back to almost 100% when the data of the different sensors is included into the training set. Second, the average CCR across all leave-$n$-out novel lens experiments, where $n = \{1, 2, 3, 4\}$ and models are trained $5 - n$ and evaluated on $n$ lens types, were 97.65%, 95.97%, 92.59% and 85.69%, respectively. It is also worth mentioning that BSIF outperformed LBP in all of their experiments and generalised better in every test protocol.

These results suggest that: (1) reasonably good interoperability between different iris sensors can be achieved in textured contact lens detection but additional models are (probably) required for novel iris sensors in order to achieve extremely high detection rates and (2) cosmetic lens detection algorithms can be robust to a previously unseen printed lens texture patterns and the more lens brands are introduced to the training set, the better generalisation capabilities can be obtained. When the detection algorithms are trained with iris images of the target sensor and no previously unseen lens type is confronted in operation, textured contact lens detection seems to be a solved problem. In less restricted conditions, sensor-specific factors, e.g., how their different NIR wavelengths interact with the pigments used in the textured lenses [15], and the different printing signatures types may degrade the detection rates but still reasonable generalisation capability may be achieved if substantial number of known lens types are used in training.

## 12.4.2    The role of pre-processing in contact lens detection

Pre-processing, e.g., iris image segmentation and normalisation, is an important factor in contact lens detection. The contact lens detection algorithm pipelines, i.e., pre-processing, feature extraction and classification, have been usually evaluated as a whole and compared with other approaches having significant differences especially in the used pre-processing techniques and feature descriptors. Therefore, based on the literature, it is very hard to tell what is the actual effect of different pre-processing strategies.

Intuitively, textured contact lenses are the more challenging to detect than other types of presentation attacks because the artefact is visible only within a very small part of the iris image. Thus, while periocular region might be useful in, e.g., printed iris image detection, it can be considered to be irrelevant for detecting cosmetic lens or even distracting. However, recent studies [15,47] demonstrated that accurate segmentation and geometric normalisation of iris region might not be required for robust and generalised textured contact lens detection when operating in the original Cartesian image domain.

Doyle and Bowyer [15] experimented with three different pre-processing strategies for extracting the BSIF description using: (1) the entire given iris image (whole image), (2) average iris location for the dataset to guess the ROI (best guess) and (3) accurate automatic segmentation for the given iris image (known segmentation). In the last two cases, radius of the estimated or segmented ROI is increased by 30 pixels to include sclera region where the contact lens boundary might be present. The pixels not belonging to the ROI were masked out and no geometric normalisation, e.g., resizing, was applied on the used iris region. Interestingly, the CCR was only slightly decreased when best guess estimate of ROI was utilised instead of exact segmentation in inter-sensor validation, while the performance of all three strategies was roughly the same in intra-sensor evaluation. Gragnaniello *et al.* [47] used three different ROIs, whole iris image, segmented iris region and segmented region containing both iris and sclera for computing dense scale-invariant local descriptors (SID) and bag-of-features (BoF) paradigm-based iris representation. Also their experiments showed that cosmetic contact lenses can be detected with high accuracy using all three pre-processing techniques.

While accurate segmentation and geometric normalisation might not be important for robust textured lens detection, the same does not hold for the three-class problem because the differences between no lens and soft lens iris images are not that evident (as seen in Figure 12.5) compared with textured lenses (at least in the case of familiar printing patterns). Gragnaniello *et al.* [47] found out that the specifically segmented sclera region plays a key role in accurate contact lens detection especially for no lens and soft lens classes, while the whole image and iris region based feature representations fail to capture crucial fine details. Therefore, they approached the three-class problem by utilising accurate sclera and iris segmentation and avoiding any kind of geometric normalisation on the iris images that might distort the subtle texture variations in sclera and iris regions due to the presence of a contact lens. The proposed method extracting SID and BoF encoding from the combined sclera and iris regions obtained the state-of-the-art overall CCR of 88.04% across the three classes in intra-sensor tests. Unfortunately, the generalisation capabilities of the proposed pre-processing across different sensors remain unexplored because no inter-sensor results were reported.

Silva *et al.* [48] conducted a preliminary study on the effectiveness of deep convolutional neural networks (CNN) in contact lens detection. Based on their experiments, even CNNs require approximation of ROI containing iris and possible visible contact lens boundary in order to reach comparable performance with the state of the art as overall CCRs of 82.165% and 70.57% across the three classes in intra-sensor tests and 76.67% and 42.30% in inter-sensor tests were obtained by processing localised iris region and entire given iris images, respectively. However, these preliminary experiments cannot be regarded as the final word on the effectiveness of deep neural networks in contact lens detection. Further attention is likely required in order to find out their full potential.

The idea of using minimal pre-processing, e.g., operating on "raw iris images" or approximate iris region, or avoiding any geometric normalisation in order to preserve all valuable fine details, is reasonable when operating on "classic-still" iris images captured in controlled and cooperative conditions. For instance, the iris cameras, like LG4000 and IrisGuard AD100, aim at placing the iris (or the pupil) in the centre of the acquired iris image and acquire an image only when the built-in proximity sensor tells that there is a subject within a certain distance from the camera. Therefore, the variation in the position and the size of the iris in the captured images is not that significant. However, this kind of approaches are not likely to be suitable for operating, e.g., on iris images taken at a distance using an iris on the move (IOM) system that captures NIR face video of subjects while they are walking through a portal at a distance of 3 m away from the camera.

Robustness and generalisation of software-based contact lens detection in less controlled conditions can be probably improved by combining different (complementary) pre-processing strategies. Instead of operating on a single Cartesian image, Raghavendra *et al.* [11] proposed to combine the iris representations computed over three different images: (1) whole eye image resized into $120 \times 120$ pixels, (2) $300 \times 50$ strip image cropped from the original iris image starting from the pupil centre, i.e., roughly representing the iris, pupil and sclera regions and (3) "traditional" unrolled iris image resized into $512 \times 64$. While the ensemble of BSIF descriptions extracted at multiple scales was not able beat the state of the art [47] on three-class problem in intra-sensor evaluation (overall CCR of 81.46% vs 88.04), very promising state-of-the-art performance of 74.73% was obtained in inter-sensor tests. Again, BSIF based iris image description obtained very high accuracy in textured contact lens detection in both intra-sensor and inter-sensor tests.

In general, the textured contact lenses can be detected quite well while there seems to be confusion between natural irises and non-cosmetic soft contact lenses. This suggests that alternatively the three-class iris image classification problem could be simplified into a cascade of two binary classification tasks: (1) considering iris images with and without clear contact lenses as genuine samples and the one with cosmetic lenses as counterfeit samples, and then (2) discriminating images with transparent lenses from natural images. In this manner, the algorithms could be tuned to distinguish the fine differences between the two more difficult classes.

## 12.4.3 On the evaluation of contact lens detection algorithms

The current publicly available benchmark databases have been a very important kick-off for finding out best practices for contact lens detection. The existing databases have been and are still useful for developing and evaluating contact lens detection algorithms, especially for transparent soft lenses. However, the almost perfect CCRs for textured contact lens detection in "homogeneous" training and test conditions indicate that even more challenging configurations are needed. On the other hand, LivDet-Iris 2013 competition [35] showed that already these kinds of experimental set-ups can be made challenging if the actual test data is kept inaccessible during algorithm development and third-party evaluation is deployed in order to simulate real-world operating conditions. The resulting performances were far from satisfactory even for the winning method, which suggest that the laboratory results reported in scientific papers might be indeed overly optimistic estimate on the true generalisation capabilities of the existing textured contact lens detection methods. It is worth highlighting that any later comparison to these competition results should be treated with caution because it is impossible to reproduce the "blind" evaluation conditions anymore and, consequently, to achieve a fair comparison.

Another issue in textured contact lens detection is that while the training and test folds are subject-disjoint, the individual folds contain usually different genuine and fake subjects, e.g., in [41,44,53]. Intuitively, the arrangement of not including subjects with and without (cosmetic) lenses can be justified in the training phase as it would prevent methods from learning the subject features instead of lens properties [10].

However, this kind of configuration in the test set might lead to biased results because presentation attack detection methods are well-known for their subject-dependent performance [54,55]. Furthermore, on this type of database it is impossible to conduct user-specific studies, e.g., to analyse the impact of contact lens detection algorithms and according follow-up procedures on the recognition accuracy. When subjects for the different classes are not the same, the acquisition conditions, e.g., used iris sensor [30,45], may be different [56]. This potential flaw can be prevented by following exactly the same data collection protocol for all classes, like that conducted, e.g., in [5,10,15]. It is worth mentioning that the different benchmark datasets, like [10,15], enable different kinds of contact lens detection studies but the effect of important factors, e.g., users and novel lenses, cannot be isolated and analysed together.

The general evaluation protocols and performance metrics could be also improved. The benchmark datasets contain usually separate folds only for training and testing which may cause bias due to "data peeking". While independent (third-party) evaluations are impossible to arrange without collective evaluations, like LivDet-Iris 2013 [35], the use of pre-defined training, development and test sets would mitigate the effect of tuning the methods on the test data. Unambiguous evaluation protocols would also allow fairer and direct comparison between different studies. The specific validation set would also help to improve and standardise performance metrics that have been not been corresponding to real-world operating scenarios, e.g., with specific operating points, or otherwise very informative. For instance, the detection accuracy for each class in the three-class problem is not that meaningful if the confusion between the different classes [5,42], especially in the case of no lens and transparent lens, is not analysed, e.g., in [10,11,47,48]. As the generalisation capabilities of contact lens detection algorithms in unknown conditions, e.g., novel sensors and lenses, have shown to be a real issue, the inter-test protocols should be followed when provided in a dataset.

## 12.5   Conclusions

The presence of a contact lens in the acquired iris image degrades the performance of iris recognition systems. This is particularly true for textured contact lenses that are designed to alter the original iris texture of the wearer but also for clear soft lenses due to the change in optical properties of the eye and varying alignment. Automatic detection of both types of contact lenses would be beneficial for detecting iris presentation attacks (cosmetic lenses) and engaging adaptation algorithms, e.g., distortion correction or masking, that could increase the accuracy of iris recognition systems against transparent lenses.

Since the effect of clear soft prescription lenses on recognition accuracy has been understated until recently, the research focus has been on textured contact lens detection and many hardware- and software-based approaches (and combination of both) have been proposed in the literature. While hardware-based solutions provide efficient and generalised means for presentation attack detection, they can be also rather impractical due to additional interaction or unconventional imaging requirements, and unpleasant active lighting. Furthermore, these kinds of techniques have been usually evaluated just to demonstrate a proof of concept or, in the worst case, have not been experimentally validated at all.

Software-based approaches exploiting only the data captured with "standard" iris sensors would be an inexpensive and attractive solution for contact lens detection. Under known operating conditions, cosmetic contact lens detection can be considered as a solved problem but sensor interoperability and previously unseen printed lens texture patterns can cause dramatic degradation in detection accuracy. The recent studies on generalised contact lens detection have demonstrated, however, that with careful feature design and comprehensive training sets containing target sensors and multiple contact lens types/brands reasonable, or even rather high, performance can be maintained in challenging operating conditions. The issues with generalised contact lens detection have been recognised but there is still room for future work. Since there are virtually no technical limitations to fabricate cosmetic contact lenses suitable for targeted impersonation, it would be interesting to see if new approaches are needed for discriminating original natural iris texture from replica of it.

The research on the problem of soft prescription lenses, on the other hand, is just in its infancy and only a few approaches have been introduced. Detection of clear soft contact lenses is far more difficult compared to the problem of textured lenses because the appearance differences between no lens and soft lens iris images are very subtle and highly dependent on the input image quality, for instance. The reported results have not been satisfactory so far and even hardware-based methods have not been able to provide generalised solutions yet. Thus, robust soft contact lens detection is still an open research topic.

Compared with "classic-still" NIR iris images, distant iris acquisition would be more practical, e.g., in watchlist applications but there are no studies that perform contact lens detection from images captured with iris on the move (IOM) systems. While NIR iris sensors are emerging in mobile devices, already the standard high-quality cameras embedded in smartphones facilitate iris recognition [57]. Detection of print [58] and video [59] based presentation attacks from conventional RGB iris images has received already some attention but textured contact lenses have not been included in these preliminary works. Even though there are still unresolved issues when operating on "classic-still" NIR iris images, contact lens detection in iris images captured in less restricted acquisition conditions and application scenarios would be an interesting and important research topic.

## Acknowledgement

The financial support of Academy of Finland is gratefully acknowledged.

## References

[1]  Daugman, J. "High confidence visual recognition of persons by a test of statistical independence," *IEEE Transactions on Pattern Analysis and Machine Intelligence* **15**(11), 1148–1161 (1993).

[2] Hollingsworth, K., Bowyer, K.W., Flynn, P.J. "Pupil dilation degrades iris biometric performance," *Computer Vision and Image Understanding* **113**(1), 150–157 (2009).

[3] Arora, S.S., Vatsa, M., Singh, R., Jain, A.K. "On iris camera interoperability," in: *IEEE International Conference on Biometrics: Theory, Applications and Systems (BTAS)*, pp. 346–352 (2012).

[4] Connaughton, R., Sgroi, A., Bowyer, K.W., Flynn, P.J. "A cross-sensor evaluation of three commercial iris cameras for iris biometrics," in: *Conference on Computer Vision and Pattern Recognition Workshops (CVPRW)*, pp. 90–97 (2011).

[5] Doyle, J.S., Bowyer, K.W., Flynn, P.J. "Variation in accuracy of textured contact lens detection based on sensor and lens pattern," in: *IEEE International Conference on Biometrics: Theory, Applications and Systems (BTAS)*, pp. 1–8 (2013).

[6] Komulainen, J., Hadid, A., Pietikäinen, M. "Generalized textured contact lens detection by extracting BSIF description from Cartesian iris images," in: *IEEE International Joint Conference on Biometrics (IJCB)*, pp. 1–7 (2014).

[7] Baker, S.E., Hentz, A., Bowyer, K.W., Flynn, P.J. "Contact lenses: Handle with care for iris recognition," in: *IEEE International Conference on Biometrics: Theory, Applications and Systems (BTAS)*, pp. 190–197 (2009).

[8] Baker, S.E., Hentz, A., Bowyer, K.W., Flynn, P.J. "Degradation of iris recognition performance due to non-cosmetic prescription contact lenses," *Computer Vision and Image Understanding* **114**(9), 1030–1044 (2010).

[9] Kohli, N., Yadav, D., Vatsa, M., Singh, R. "Revisiting iris recognition with color cosmetic contact lenses," in: *IAPR International Conference on Biometrics (ICB)*, pp. 1–7 (2013).

[10] Yadav, D., Kohli, N., Doyle, J.S., Singh, R., Vatsa, M., Bowyer, K.W. "Unraveling the effect of textured contact lenses on iris recognition," *IEEE Transactions on Information Forensics and Security* **9**(5), 851–862 (2014).

[11] Raghavendra, R., Raja, K.B., Busch, C. "Ensemble of statistically independent filters for robust contact lens detection in iris images," in: *ACM Indian Conference on Computer Vision Graphics and Image Processing (ICVGIP)*, pp. 24:1–24:7 (2014).

[12] Bowyer, K.W., Doyle, J.S. "Cosmetic contact lenses and iris recognition spoofing," *Computer* **47**(5), 96–98 (2014).

[13] Al-Raisi, A.N., Al-Khouri, A.M. "Iris recognition and the challenge of homeland and border control security in UAE," *Telematics and Informatics* **25**(2), 117–132 (2008).

[14] Hughes, K., Bowyer, K.W. "Detection of contact-lens-based iris biometric spoofs using stereo imaging," *Hawaii International Conference on System Sciences (HICSS)*, pp. 1763–1772 (2013).

[15] Doyle, J.S., Bowyer, K.W. "Robust detection of textured contact lenses in iris recognition using BSIF," *IEEE Access* **3**, 1672–1683 (2015).

[16] Erdogan, G., Ross, A. "Automatic detection of non-cosmetic soft contact lenses in ocular images," in: *SPIE Biometric and Surveillance Technology for Human and Activity Identification X* (2013).

[17]   Daugman, J. "The importance of being random: statistical principles of iris recognition," *Pattern Recognition* **36**(2), 279–291 (2003).
[18]   Daugman, J. "How iris recognition works," *IEEE Transactions on Circuits and Systems for Video Technology* **14**(1), 21–30 (2004).
[19]   Thompson, J., Flynn, P.J., Bowyer, K.W., Santos-Villalobos, H. "Effects of iris surface curvature on iris recognition," in: *IEEE International Conference on Biometrics: Theory, Applications and Systems (BTAS)*, pp. 1–8 (2013).
[20]   Galbally, J., Gomez-Barrero, M. "A review of iris anti-spoofing," in: *International Workshop on Biometrics and Forensics (IWBF)*, pp. 1–6 (2016).
[21]   Chen, R., Lin, X., Ding, T. "Liveness detection for iris recognition using multispectral images," *Pattern Recognition Letters* **33**(12), 1513–1519 (2012).
[22]   Daugman, J. "Recognizing persons by their iris patterns," in: A.K. Jain, R. Bolle, S. Pankanti (eds.). *Biometrics: Personal Identification in Networked Society*, pp. 103–121 (1996).
[23]   Lee, S.J., Park, K.R., Kim, J. "Robust fake iris detection based on variation of the reflectance ratio between the iris and the sclera," in: *Biometrics Symposium: Special Session on Research at the Biometric Consortium Conference*, pp. 1–6 (2006).
[24]   Lee, S.J., Park, K.R., Lee, Y.J., Bae, K., Kim, J. "Multifeature-based fake iris detection method," *Optical Engineering* **46**(12), 127204–127204-10 (2007).
[25]   Park, J.H., Kang, M.G. "Iris recognition against counterfeit attack using gradient based fusion of multi-spectral images," in: *Advances in Biometric Person Authentication: International Workshop on Biometric Recognition Systems (IWBRS)*, pp. 150–156 (2005).
[26]   Park, J.H., Kang, M.G. "Multispectral iris authentication system against counterfeit attack using gradient-based image fusion," *Optical Engineering* **46**(11), 117003–117003-14 (2007).
[27]   Czajka, A. "Pupil dynamics for iris liveness detection," *IEEE Transactions on Information Forensics and Security* **10**(4), 726–735 (2015).
[28]   Pan, G., Wu, Z., Sun, L. "Liveness detection for face recognition," in: *Recent Advances in Face Recognition*, pp. 109–124. In-Teh (2008).
[29]   Rigas, I., Komogortsev, O.V. "Eye movement-driven defense against iris print-attacks," *Pattern Recognition Letters* **68**(2), 316–326 (2015). Special Issue on Soft Biometrics.
[30]   Bodade, R., Talbar, S. "Fake iris detection: A holistic approach," *International Journal of Computer Applications* **19**(2), 1–7 (2011).
[31]   Huang, X., Ti, C., Hou, Q., Tokuta, A., Yang, R. "An experimental study of pupil constriction for liveness detection," in: *IEEE Workshop on Applications of Computer Vision (WACV)*, pp. 252–258 (2013).
[32]   Park, K.R. "Robust fake iris detection," in: Articulated Motion and Deformable Objects, *Lecture Notes in Computer Science*, vol. 4069, pp. 10–18 (2006).
[33]   Puhan, N.B., Sudha, N., Suhas Hegde, A. "A new iris liveness detection method against contact lens spoofing," in: *IEEE International Symposium on Consumer Electronics (ISCE)*, pp. 71–74 (2011).

[34] Sun, Z., Tan, T. "Iris anti-spoofing," in: S. Marcel, S.M. Nixon, Z.S. Li (eds.). *Handbook of Biometric Anti-Spoofing: Trusted Biometrics under Spoofing Attacks*, chap. 6, pp. 103–123. Springer, London (2014).

[35] Yambay, D., Doyle, J.S., Bowyer, K.W., Czajka, A., Schuckers, S. "Livdet-iris 2013 – iris liveness detection competition 2013," in: *IEEE International Joint Conference on Biometrics (IJCB)*, pp. 1–8 (2014).

[36] Lee, E.C., Ko, Y.J., Park, K.R. "Fake iris detection method using Purkinje images based on gaze position," *Optical Engineering* **47**(6), 067204–067204-16 (2008).

[37] Lee, E.C., Park, K.R., Kim, J. "Fake iris detection by using Purkinje image," in: *International Conference on Biometrics (ICB)*, pp. 397–403 (2006).

[38] Connell, J., Ratha, N., Gentile, J., Bolle, R. "Fake iris detection using structured light," in: *IEEE International Conference on Acoustics, Speech and Signal Processing (ICASSP)*, pp. 8692–8696 (2013).

[39] Daugman, J. "Demodulation by complex-valued wavelets for stochastic pattern recognition," *International Journal of Wavelets, Multi-resolution and Information Processing* **1**, 1–17 (2003).

[40] He, X., An, S., Shi, P. "Statistical texture analysis-based approach for fake iris detection using support vector machines," in: *International Conference on Biometrics (ICB)*, pp. 540–546 (2007).

[41] Wei, Z., Qiu, X., Sun, Z., Tan, T. "Counterfeit iris detection based on texture analysis," in: *IEEE International Conference on Pattern Recognition (ICPR)*, pp. 1–4 (2008).

[42] Doyle, J.S., Bowyer, K.W., Flynn, P.J. "Automated classification of contact lens type in iris images," in: *IAPR International Conference on Biometrics (ICB)*, pp. 1–6 (2013).

[43] He, Z., Sun, Z., Tan, T., Wei, Z. "Efficient iris spoof detection via boosted local binary patterns," in: *Advances in Biometrics*, pp. 1080–1090 (2009).

[44] Zhang, H., Sun, Z., Tan, T. "Contact lens detection based on weighted LBP," in: *International Conference on Pattern Recognition (ICPR)*, pp. 4279–4282 (2010).

[45] Sun, Z., Zhang, H., Tan, T., Wang, J. "Iris image classification based on hierarchical visual codebook," *IEEE Transactions on Pattern Analysis and Machine Intelligence* **36**, 1120–1133 (2013).

[46] Kywe, W.W., Yoshida, M., Murakami, K. "Contact lens extraction by using thermo-vision," in: *International Conference on Pattern Recognition (ICPR)*, pp. 570–573 (2006).

[47] Gragnaniello, D., Poggi, G., Sansone, C., Verdoliva, L. "Using iris and sclera for detection and classification of contact lenses," *Pattern Recognition Letters* **82**(Part 2), 251–257 (2015).

[48] Silva, P., Luz, E., Baeta, R., Pedrini, H., Falcao, A.X., Menotti, D. "An approach to iris contact lens detection based on deep image representations," in: *SIBGRAPI – Conference on Graphics, Patterns and Images*, pp. 157–164 (2015).

[49]  Alegre, F., Amehraye, A., Evans, N. "A one-class classification approach to generalised speaker verification spoofing countermeasures using local binary patterns," in: *IEEE International Conference on Biometrics: Theory, Applications and Systems (BTAS)*, pp. 1–8 (2013).

[50]  Kannala, J., Rahtu, E. "BSIF: Binarized statistical image features," in: *International Conference on Pattern Recognition (ICPR)*, pp. 1363–1366 (2012).

[51]  Ojala, T., Pietikäinen, M., Mäenpää, T. "Multiresolution gray-scale and rotation invariant texture classification with local binary patterns," *IEEE Transactions on Pattern Analysis and Machine Intelligence* **24**, 971–987 (2002).

[52]  Hyvärinen, A., Hurri, J., Hoyer, P.O. *Natural Image Statistics: A Probabilistic Approach to Early Computational Vision*. Springer, London (2009).

[53]  Bodade, R., Talbar, S. "Dynamic iris localisation: A novel approach suitable for fake iris detection," in: *International Conference on Ultra Modern Telecommunications Workshops*, pp. 1–5 (2009).

[54]  Chingovska, I., Anjos, A. "On the use of client identity information for face anti-spoofing," *IEEE Transactions on Information Forensics and Security* **10**(4), 787–796 (2015).

[55]  Yang, J., Yi, D., Li, S.Z. "Person-specific face anti-spoofing with subject domain adaptation," *IEEE Transactions on Information Forensics and Security* **10**(4), 797–809 (2015).

[56]  Galbally, J., Marcel, S., Fiérrez, J. "Biometric antispoofing methods: A survey in face recognition," *IEEE Access* **2**, 1530–1552 (2014).

[57]  De Marsico, M., Galdi, C., Nappi, M., Riccio, D. "Firme: face and iris recognition for mobile engagement," *Image Vision Computing* **32**(12), 1161–1172 (2014).

[58]  Gragnaniello, D., Sansone, C., Verdoliva, L. "Iris liveness detection for mobile devices based on local descriptors, *Pattern Recognition Letters* **57**, 81–87 (2015).

[59]  Raja, K.B., Raghavendra, R., Busch, C. "Video presentation attack detection in visible spectrum iris recognition using magnified phase information," *IEEE Transactions on Information Forensics and Security* **10**(10), 2048–2056 (2015).

## Chapter 13

# Software attacks on iris recognition systems

### Marta Gomez-Barrero and Javier Galbally

## 13.1  Introduction

Since the first works on biometric verification [1,2], increasingly accurate and time-efficient biometric recognition systems have been proposed over the years [3–6]. These works have allowed a wider deployment of biometric authentication techniques, including identity management using biometrics within the e-passport [7]. Biometric technologies have also become more prevalent on personal computers and smartphones with more biometric-enabled functions [8].

Unfortunately, with this wide deployment of biometric systems, new concerns have arisen regarding its vulnerabilities to external attacks. More specifically, in 2001, Ratha *et al.* identified and classified in a biometric recognition system eight possible points of attack [9]. These vulnerable points can be broadly divided into presentation attacks (those directed to the sensor) and software attacks (directed to the inner modules of the system), as depicted in Figure 13.1.

- **Presentation attacks**. Also known as *spoofing* or *direct attacks*, defined in the latest draft of the ISO/IEC 30107 standard as "presentation of an artefact or human characteristic to the biometric capture subsystem in a fashion that could interfere with the intended policy of the biometric system" [10]. In practice, the vast majority of presentation attacks studied in the literature are *spoofing* attacks. Biometric spoofing is generally understood as the ability to fool a biometric system into recognizing an illegitimate subject as a genuine one by means of presenting to the sensor a synthetic forged version (i.e., artefact) of the original biometric characteristic (e.g., printed iris image, gummy finger or face mask). Therefore, these attacks require no knowledge for the attacker of the inner parts of the system (matching algorithm used, feature extraction method, template format, etc.). In order to fabricate the synthetic characteristics, methods such as the ones described in Section 13.3 can be used to generate a *digital* representation, which can be then transformed into the appropriate *physical* object presented to the sensor.
- **Software attacks**. Also known as *indirect attacks*. While for presentation attacks the intruder needs no knowledge about the inner modules of the system, this knowledge is a main requisite here, together with access to some of the system

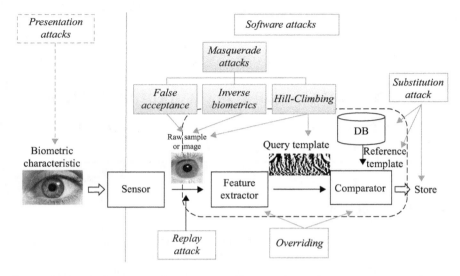

*Figure 13.1*    *Architecture of an automated biometric verification system, with different possible attacks and the points to which they are directed. Software attacks, right of the vertical line and the focus of the present chapter, are those directed to one of the inner modules of the system. The specific types of software attacks analysed in this chapter are highlighted with shaded boxes*

components (e.g., database, feature extractor or comparator) or one of the inner communication channels. In fact, one of the main differences among the different types of software attacks described in the literature is the amount of knowledge required about the target of the attack (i.e., the biometric system). As a consequence, this level of knowledge is one of the key parameters that needs to be assessed in order to evaluate the risk posed by these attacks, which have been traditionally divided into four groups:

–    *Masquerade attacks*, in which a *digital* synthetic biometric sample or a synthetic template is injected in the system. Examples of these attacks include the *hill-climbing* attacks analysed in Section 13.2 or the so-called *false acceptance attacks*, which take advantage of the False Acceptance Rate of the system by presenting a sufficiently large number of synthetic biometric samples until one is positively matched to the stored reference. Such samples can be synthesized using the methods described in Section 13.3.1. Additionally, *inverse biometrics* methods, such as those presented in Section 13.3.2, aim at the reconstruction of synthetic biometric samples which are positively matched to a particular stored reference.

–    *Replay attacks* where an intercepted template is presented to the feature extractor.

- *Overriding* one of the inner modules of the system (e.g., feature extractor or comparator) using a Trojan Horse, so that it produces a pre-selected score or query template, which matches the input biometric sample presented to the sensor by the attacker.

- *Substitution attack* on the database or the communication channel with the comparator, in which the reference template is substituted by a pre-selected reference template, in order to match the impostor query template, thereby allowing an impostor positive claim of identity. The final similarity score can be analogously substituted by a matching score above the verification threshold.

Since that initial classification of external attacks on biometric systems, many efforts have been directed to this particular area within the biometric community. Such works include not only new attacks or countermeasures [11–16] but also new standards on this relevant topic. In 2009, the ISO/IEC JTC 1/SC 27 [17] on IT Security techniques published the ISO/IEC International Standard 19792 on security evaluation of biometrics [18], which includes a clause on vulnerability assessment of biometric systems. More recently, some EU FP7 projects such as Tabula Rasa [19] or BEAT [20] have been focused on this topic. More specifically, within the latter, a free online platform[1] has been developed to test not only the performance of biometric systems but also the robustness to external attacks. The relevance of this area of research is also highlighted by the inclusion in international conferences of specific tracks on vulnerability evaluation of biometric systems [21,22].

As highlighted in Figure 13.1 with solid lines and shaded blocks, among all indirect or software attacks, we focus in this chapter on those attacks that require a complex algorithmic process in order to succeed: *masquerade attacks*. A summary of all the attacks is presented in Tables 13.1–13.3. We first introduce hill-climbing based attacks in Section 13.2. Then, synthetic generation of iris samples is summarized in Section 13.3. In addition, software attacks on biometric template schemes are described in Section 13.4, and countermeasures to all the aforementioned attacks are analysed in Section 13.5.

## 13.2   Hill-climbing based attacks

Most software attacks are based on some hill-climbing technique, for which a general diagram is depicted in Figure 13.2. These approaches start by synthetically generating an initial pool of templates or images (see Figure 13.2, left). Each of them is used in an attempt to access the system. If none of them succeeds, the templates are iteratively changed, according to a particular modification scheme, so that at each iteration a higher similarity score between the synthetic and the attacked template is obtained. The attack will be successful when one of the synthetic templates or samples is positively matched to the stored reference template. In Figure 13.2 (right), an example

---

[1] https://www.beat-eu.org/platform/

*Table 13.1    Summary of key software attacks to iris recognition systems, where 'iter'*
*stands for number of iterations and 'MS' for matching score*

| Type | Reference | Acceptance rate | Efficiency | Database |
|------|-----------|-----------------|------------|----------|
| Hill-climbing | [13] | 100% ($>0.9$ MS) | 9,800 iter. ($>10^7$ comp.) | CASIAv3 Interval (249 subjects) |
| | [23] | 62% (0.01% FMR) | 9,074 comp. | BioSecure (210 subjects) |
| Inv. biometrics | [24] | >95% (0.1% FMR) | – – | NIST ICE 2005 (132 subjects) |
| | [15] | 94% – % (0.01% FMR) | – – | BioSecure (210 subjects) |

*Table 13.2    Summary of key fully synthetic iris generation algorithms*

| Approach | Reference | Technique | Real textures? |
|----------|-----------|-----------|----------------|
| Anatomic | [65] | Sum of layers | Yes |
| | [43,66] | 3D fibres + external effects | No |
| Computer | [39] | PCA + Super-resolution | Yes |
| | [41,42] | MRF model + Features | Yes |
| | [44] | MRF model + Patch-sampling | Yes |

*Table 13.3    Summary of key software attacks to iris BTP systems*

| Type | Target system | Reference | Sub-type |
|------|---------------|-----------|----------|
| Inversion | Fuzzy vault | [81] | Chaff points |
| | Fuzzy commitment | [84] | EECs |
| | | [85] | EECs |
| | Bloom filters | [86] | Average codeword |
| Record multiplicity | Fuzzy vault | [87] | Correlation + cross-matching |
| | Bloom filters | [88] | Cross-matching |

of the evolution of the scores across the iterations of a successful hill-climbing attack
is depicted. As may be observed, with each iteration the score increases, meaning that
we are closer to our objective: reaching a score higher than the verification threshold $\delta$,
depicted with a horizontal dashed line. However, the modification scheme may not
work for all cases, thus leading to solutions (i.e., synthetic templates/images) very
different from the reference template, and which are hence not positively matched to
it and might lead to infinite loops. In order to avoid those situations, a limit on the
number of iterations is usually established.

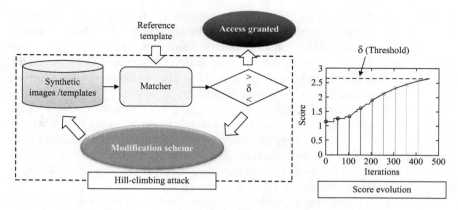

Figure 13.2   Diagram of a generic hill-climbing attack (left) with an example of the evolution of the scores (right). The threshold δ for which access to the system is granted is depicted with a horizontal dashed line

As depicted in Figure 13.1, depending on the target to optimize, or, in other words, on what they generate, hill-climbing attacks can be broadly divided into two groups:

- *Hill-climbing attacks on synthetic samples*, which assume that the attacker has (i) access to the communication channel between the sensor and the feature extractor and (ii) access to the similarity score.
- *Hill-climbing attacks on synthetic templates*, which require from the attacker (i) knowledge of the template format, (ii) access to the communication channel between the feature extractor and the comparator and (iii) access to the similarity score.

Following this paradigm, in the last two decades hill-climbing attacks have been devised for numerous biometric characteristics, including iris [13,23], face [11,12,25–27], fingerprint [14,28], EEGs [29], signature [30–32], speaker verification [33] or multimodal schemes [16,34].

Most of those works present algorithms designed ad hoc for a particular biometric characteristic. However, some generic modification schemes have been also proposed, namely: (i) a Bayesian approach applied to on-line signature [30] and face [26], (ii) the Uphill simplex algorithm applied to on-line signature [31] and face [27] and (iii) the SPSA algorithm for EEG [29] and multi-biometric systems [16]. Given their generality, they could also be applied to iris systems. However, in the following we will focus on two different algorithms which have been proposed for the particular case of iris-based schemes: (i) a method to reconstruct iris textures [13] and (ii) a method to reconstruct iriscodes [23].

Rathgeb and Uhl proposed in [13] a *hill-climbing attack on synthetic samples*. In spite of the high dependency of the positive verification of iris textures on the feature extraction algorithm, the authors describe a general approach to synthetic iris textures generation based on a hill-climbing algorithm, where no knowledge about the extracted features is required. This method builds upon the fact that, in spite of their differences, most iris recognition algorithms share a common characteristic: they tend to average pixels in a block-wise manner. Taking advantage of this concept, the hill-climbing algorithm can be sped up finding the size of such blocks and applying the modification scheme in a block-wise manner.

In that work, the initial synthetic iris texture is set to either a plain texture where all pixels are set to 128, or an initial eigeniris texture averaged out of five randomly chosen textures. The hill-climbing modification scheme updates the iris texture from the top left to the right bottom iteratively increasing or decreasing the pixel values of complete blocks by a fixed value, retaining changes only if the similarity score with respect to the stored reference template is improved.

Using this hill-climbing technique, the authors show that, after 9,800 iterations, which would require at least $9,800 \times 1,024 \approx 10^7$ comparisons, an iris texture can be recovered, yielding a similarity score of at least 0.9 with respect to the stored reference. This would lead to a positive identification in most current iris recognition schemes.

A different approach to generate iriscodes is presented by Gomez-Barrero *et al.* in [23], where a *hill-climbing attack on the templates* is proposed. Given the binary nature of the iriscodes [35], in this case a genetic algorithm is used as modification scheme. Analogously to the previous work, no knowledge about the feature extraction method is assumed. In this case, no real iris samples or templates are used for the initialization step: an initial random population of iriscodes, comprising 30 binary vectors, is generated. Then, the genetic algorithm is used to optimize those iriscodes until their similarity score with respect to the stored reference falls within the intra-class variability (i.e., the matching threshold is exceeded). At each iteration, the two iriscodes yielding the highest similarity score are kept unaltered for the next generation. Among the remaining individuals, stochastic universal sampling is used to choose the parents, which are combined to generate two children per pair following a scattered crossover method. Finally, some bits are randomly flipped in the mutation phase.

The authors evaluate the success chances of this attack for different operating points. In contrast to the previous scheme, not all iriscodes are successfully recovered. For instance, for a False Match Rate (FMR) of 0.1%, the attacker succeeds in 91% of the cases. On the other hand, for FMR = 0.01%, this rate is reduced to 62%, which still represents a threat to the security of the system. The main advantage of this scheme relies on its probabilistic nature (i.e., not only one iriscode but several can be generated to positively match the stored reference) and its efficiency: only around 8,000 comparisons are needed to recover the synthetic iriscode.

To sum up, these two approaches offer some differences:

- **Type of approach.** While [13] proposes a hill-climbing attack on the samples (i.e., aims at generating iris textures), [23] describes a hill-climbing attack on the templates (i.e., iriscodes). In addition, whereas each iteration in [13] is deterministic, [23] follows a probabilistic approach for computing the updated iriscodes.

- **Knowledge required**. Whereas both attacks require access to the score, in [13] the impostor needs access to channel between the sensor and the feature extractor, and in [23] the attack requires knowledge of the template format and access to the communication channel between the feature extractor and the comparator.
- **Number of iriscodes**. In addition, while, at each iteration, [13] updates a single iriscode, in a block-wise manner, the second scheme [23] generates 28 new iriscodes, generated in a bit-wise basis, thereby allowing a fast convergence.

## 13.3 Synthetic iris generation

As depicted in Figure 13.1, other types of attack to biometric systems are those carried out with either *physical* synthetic samples, presented to the sensor (i.e., presentation attacks) or *digital* synthetic samples (i.e., templates), injected in the communication channel between the sensor and the feature extractor.

While manually produced *physical* biometric characteristics such as fingerprints, signatures or forged handwriting have been a point of concern for experts from a forensic point of view since the early 1920s [36,37], the *automatic* generation of *digital* synthetic samples has not been considered until the wider development of automatic biometric recognition systems in the last two decades [25,38,39]. As a consequence, together with other applications (e.g., pseudo-identities or completely synthetic databases generation), the use of synthetically generated samples for vulnerability assessment of biometric systems has awakened a growing interest in the biometric community over the last decade for different characteristics, such as voice [38], fingerprints [40], iris [39,41–44], handwriting [45], face [46] or signature [47–49]. Depending on the final purpose, different types of synthetic data might be created:

- *Duplicated samples*: in these methods the generation algorithm starts from one or more real samples of a given person and, through different transformations, produces different synthetic (or duplicated) samples corresponding to the same subject. This approach has been applied to signature [50,51], handwriting [52,53] or face synthesis [46,54,55], being type of algorithms useful to artificially increase the amount of already acquired biometric data [46,56], and thereby posing no threat to biometric systems.
- *Synthetic individuals*: in this case, some kind of a priori knowledge about a certain biometric characteristic (e.g., iris structure, minutiae distribution, and signature length) is used to create a model of the biometric characteristic for a population of subjects. New synthetic individuals can then be generated sampling the constructed model. In a subsequent stage of the algorithm, multiple samples of the synthetic subjects can be generated by any of the procedures for creating duplicated samples. Different model-based algorithms have been presented in the literature to generate synthetic individuals for biometric characteristics such as iris [39,42,43], fingerprint [40], speech [57,58] or signature [49]. The synthetic samples produced with these methods can be used to either generate fully synthetic

databases, thereby avoiding legal constraints derived from sharing real biometric data, or to launch *false acceptance attacks* (see Figure 13.1 and Section 13.3.1).

* *Inverse biometrics*: in these methods, given a genuine template, the aim is the reconstruction of a synthetic biometric sample, which matches the stored biometric reference according to a particular biometric recognition system. In other words, it denotes a reverse engineering process which reconstructs synthetic samples from the information conveyed in real biometric templates, and which has already been applied to iris [15,24], fingerprint [59–61], handshape [62] and face [25,26,63]. Those reconstructed samples can be presented to the system to gain fraudulent access to a particular application, hence launching a so-called *inverse biometrics attack* (Figure 13.1 and Section 13.3.2).

In this chapter we will focus on the last two types of synthetic data, since duplicated samples pose no additional threat to the enrolled subject – the potential attacker is already in possession of real biometric data belonging to the same subject. On the other hand, the generation of fully synthetic individuals can pose severe security issues in biometric systems, since those synthetic samples can be used in false acceptance attacks. Similarly, inverse biometric methods do result in a potential violation of the subject's privacy: given only a theoretically secure representation of the subject's biometric characteristics (template), sensitive biometric information can be obtained. In addition, contrary to the common belief that templates do not comprise enough information in order to reconstruct the original sample from them [64], synthetic samples can be generated and used to impersonate a particular subject, launching masquerade attacks. In the following, we will focus on these two types of synthetic samples generation for the particular case of iris.

## 13.3.1   *False acceptance attacks based on fully synthetic iris*

As mentioned before, fully synthetic irides can be used to launch false acceptance attacks: a high enough number of iris textures can be generated and presented to the system until one of them is positively matched to the stored reference due to the intrinsic False Acceptance Rate of the system. Since the synthesized textures represent, according to some quality or visual metrics, instances of iris samples, the success chances of the attack will be considerably higher than those of a brute force attack, in which random textures would be presented to the system. In addition, even if both attacks were successful, a lower number of attempts could be expected if synthetic iris images were used. These synthetic irides can be additionally used to increase the success chances of hill-climbing or inverse biometrics attacks (see Section 13.3.2) by using them for initialization instead of average or random textures, not needing any real images to carry out the attack.

In spite of their advantages for eventual impostors, generating realistic textures is in most cases a challenging task. In the particular case of iris, given its complex texture pattern and randomness, models for the generation of realistic synthetic iris textures are hard to obtain. In spite of such difficulties, several approaches, summarized in Table 13.2, have been proposed in the literature in the last two decades.

The first approach to iris synthesis described in the literature by Lefohn *et al.* [65] is based on the knowledge of artificial eye makers (ocularists). In order to make realistic looking eyes, given the high complexity of the real layers that compose them, ocularists paint eyes using 30–70 different layers, with layers of clear coat between them. To mimic their procedure, Lefohn *et al.* present in [65] an incremental method, where new texture layers are added one at a time to the model, rendering intermediate results. The first layer contains the most dominant colours found in the eye, and is assumed to be opaque. Each subsequent semitransparent layer consists of one or two of the following components: (i) the stroma, comprising dots or radial smears, (ii) the collarette, containing radial spokes or dots, (iii) the sphincter, being a coloured ring lying close to the pupil, (iv) the limbus, fuzzy and darker than the rest of the eye and (v) the pupil. All those components will be chosen from a predefined database and added in a free order to the synthetic texture, only trying to alternate high-frequency, fairly opaque layers with more transparent cloudy ones to add a sense of depth to the model. Finally, in order to model the geometry of the eye, the authors propose a two-dimensional model which is later converted to a 3D object using cones.

Similarly, based on extensive anatomical studies of the iris, Zuo and Schmid proposed in [66] a different synthesis process based on a combination of radial fibres, a raised collarette and a semitransparent layer covering the fibres with irregular edges. Therefore, their algorithm starts with the generation of continuous 3D fibres in cylindrical coordinates, which are then projected into a two-dimensional image. The pupil and iris boundaries are modelled with B-spline functions to take into account deviations from a perfect circular shape. In order to include the effect of the raised collaret, its area is brightened. In addition, a semi-transparent top layer is imitated blurring the image, and its irregular edges modelled with cosine functions. Finally, a bumpy pattern is incorporated in the last layer and an additional smoothed Gaussian noise layer included to make the texture more realistic. To create eye images, eyelids are included with a random degree of opening and eyelashes randomly generated in the last step. As reported by the authors, the whole process is based on several thousands of random parameters, which are selected from a uniform distribution over intervals that ensure a final appearance close to a real iris.

In a subsequent work, Zuo *et al.* [43] added a few external effects to the ideal synthesized iris in order to mimic real iris databases. For instance, shot noise is added following a Poisson model, motion blur is approximated with a two-dimensional filter and defocus blur is modelled as a two-dimensional circular averaging filter. Contrast and brightness are also adjusted with four additional parameters. Eyelids are enhanced by including two stripes of shadow along their inner side. To add specular reflections, a sphere is generated in order to simulate the shape of the eyeball, which is then lit based on a Phong lighting model [67]. The lighting pattern is finally projected onto the two-dimensional image. A database comprising 10,000 different synthetic irides generated with this method, and 15 samples per iris, is freely available online.[2]

[2]http://www.clarkson.edu/citer/research/collections/wvu_syn_iris_model.html

Other approaches are based on a more computer-oriented perspective. For instance, Cui *et al.* presented in [39] a method based on clustering iris textures, and then generating different samples for each class. Therefore, the method comprises two steps. First, Principal Component Analysis (PCA) is used to classify real iris images and extract the corresponding eigeniris components. The accuracy of the PCA classifier is tested on the CASIA iris database, and an optimum performance is obtained for 75 components. Then, synthetic images are generated varying the coefficients of each component. In order to obtain realistic textures, those coefficients are kept within the boundaries found within the CASIA database. Finally, to enhance the quality of the synthesized images, a super-resolution method is used [68], where high-resolution patches are inferred from low-resolution areas using a bank of circular symmetric filters.

A second set of computer-based methods is based on Markov Random Field (MRF) models. Inspired by the texture synthesis method described in [69], Makthal and Ross proposed in [41] a deterministic and fast iris texture synthesis method based on the MRF model. These models had been previously used to synthesize a wide variety of textures [70], including facial images [71]. In this particular approach, they are used to describe the probability distribution of the intensity of each pixel value in a given neighbourhood. More specifically, the authors show that realistic iris textures can be synthesized using only three primitives (i.e., radial furrows, crypts or limbus) extracted from real iris images. Randomness in this case is added with the assignment of different weights to each primitive, being those weights treated as probability measures over the whole iris texture. More specifically, irides are synthesized starting from a white noise image, and successively selecting and updating pixels in a neighbourhood basis: the new pixel value will be that of the primitive image that minimizes the Euclidean norm between the neighbourhood in the synthetic image and all the neighbourhoods in the primitive image. The algorithm thus requires no sampling of probability distributions, thereby allowing a fast synthesis, which is further enhanced using tree-structured vector quantization for the neighbourhoods comparison.

Building upon this last scheme, an enhanced iris synthesis was later developed by Shah and Ross [42]. A background texture is first generated based on MRF models and a single primitive element. Then, to achieve a more realistic appearance, features of the iris such as the radial and concentric furrows, collarette and crypts are added to the synthetic images. Radial furrows are generated by constructing straight lines radially from the centre of the pupil, and perturbing them at random points. Periodic spline curves are used to interpolate the perturbed coordinates, and lineal integral convolution to impart texture to the radial furrows. The collarette is then added as a circle with a radius 20–30 pixels larger than the pupil. Again, the circle is randomly perturbed and interpolated to recreate the zig-zag form of the collarette, and only furrows present within the collarette curve are retained. Then, concentric furrows are added in the ciliary region and merged with the background. Finally, a random number of circles are generated on the periphery of the collarette to represent the crypts. As with other elements, circles are randomly perturbed and interpolated to provide them with a realistic appearance, and then lower intensity values are assigned to the pixels inside the crypt. In order to blend the intensity of these new features with the background image, a smoothing filter is applied. A database comprising 1,000 different irides synthesized with this method, including seven samples per iris, is freely available online.[3]

Even if most of the aforementioned methods are able to yield very realistic irides, they also share a common drawback: their high complexity. Inspired by a previous work on efficient texture synthesis using patch-based sampling [72], Wei *et al.* proposed a new and more efficient approach to iris synthesis in [44], where texture patches of a real input sample are used to generate the new synthetic texture. One by one, patches are selected and added to the synthetic texture at hand. To avoid mismatching features across patch boundaries, MRF models are used to find patches within the input image with boundaries as similar as possible to the last patch added to the synthetic image. This step also provides the stochastic characteristics of iris patterns. Once the image has been fully synthesized, the authors proposed to generate similar textures applying non-linear deformations to the pupil, Gaussian filters to achieve defocus effects, linear shifts to rotate the eye or adding noise.

The main similarities and differences of these synthetic iris generation algorithms can be summarized as follows:

- **Type of approach**. All the aforementioned works are based on iterative algorithms. However, while the first approaches [43,65,66] are based on anatomical studies and try to mimic the iris structure step by step, the remaining algorithms [39,41,42,44] are based on global models of the iris texture.
- **Knowledge required**. Most of these methods [39,41,42,44,65] require access to real iris databases in order to either synthesize the global iris texture or particular elements such as crypts or furrows. In contrast, the method described in [43,66] requires no such knowledge.

## 13.3.2   Inverse biometrics attacks

As pointed out in the beginning of Section 13.3, in addition to launching false acceptance attacks, impostors may try to reconstruct synthetic samples that match a particular subject's template in order to impersonate them. This can be done using inverse engineering methods to recover the biometric data from the stored reference template in what is known as inverse biometrics attacks. One of the first works that addressed the problem posed by these attacks was carried out in [59]. The work proposed a general scheme for the reconstruction of biometric samples, consisting of four successive steps, where only access and knowledge of the templates format stored in the database is required. In the first step, the attacker needs to gain access to one or more biometric templates. Then, he or she needs to understand the structure of the template (i.e., what information is stored, which format is used, etc.). After fully determining the structure of the templates, in the third step a reverse engineering process is carried out in order to reconstruct one or more *digital* biometric samples.

---

[3] http://www.clarkson.edu/citer/research/collections/wvu_syn_iris_texture.html

Finally, the eventual attacker could also reconstruct *physical* artefacts from the digital synthetic samples.

The most challenging step is the third one, that is, devising a method for reconstructing digital samples given only the stored templates. In the particular case of iris verification, three different methods have been proposed: (i) the hill-climbing algorithm proposed by Rathgeb and Uhl in 2010 [13] can be also considered an inverse biometrics attack, as it tries to reconstruct synthetic iris textures, (ii) a deterministic approach proposed in [24] and (iii) a probabilistic scheme presented in [15]. Since the work by Rathgeb and Uhl has already been described in Section 13.2, in the following we will focus on reviewing the two specific works that have addressed to date the challenging problem of iris reconstruction [15,24].

In the work described in [24], Venugopalan and Savvides take advantage of the prior knowledge of the feature extraction scheme used by the recognition system (i.e., functions defining the filters used during feature extraction) in order to reverse engineer the iriscode. Then, real images are used to impart a more realistic appearance to the synthetic iris patterns. The feature extraction scheme used is based on the method outlined by Daugman [73], which computes the convolution between normalized iris images and Gabor functions. The knowledge of this extraction scheme is used to generate user-specific discriminating patterns from their respective iriscodes. These patterns may be estimated based on the original Gabor filter used for feature extraction or based on actual iris patches from an exemplar image that resembles this Gabor filter.

On the other hand, the challenging reverse engineering problem of reconstructing an iris pattern from its iriscode was solved by Galbally *et al.* [15] using a probabilistic approach based on genetic algorithms. The approach needs to have access to a matching score which is not necessarily that of the system being attacked. This way the reconstruction approach is independent of the comparator or feature extractor being used. However, the algorithm will be most effective if the development comparator and the final test comparator use a similar feature extraction strategy. This is usually the case, as most iris recognition methods are based on the principles first introduced by Daugman [74]. In addition, the authors showed that the algorithm can successfully bypass black-box commercial systems with unknown feature-extraction algorithms.

Regarding the particular choice for genetic algorithms among other optimization strategies, it is based on the unknown nature of the search space. Although previous work in [13] partially supports the assumption of smoothness/continuity of the similarity score function with respect to the iris textures, so far this fact has not been proven. Consequently, whereas the efficiency of classical stochastic gradient descent methods would be at least unclear, genetic search algorithms generally obtain better results in such scenarios: by simultaneously searching for multiple solutions in the solution space, they are more likely to avoid potential minima or even plateaus in the search space.

Therefore, the main differences between the two iris reconstruction approaches described in [15,24] may be summarized as follows:

- **Type of approach**. The method proposed in [24] is a deterministic one, which means that, given an iriscode and a fixed set of parameter values, the resulting reconstructed synthetic pattern is always the same. On the other hand, the algorithm described in [15] is probabilistic. As such, given an iriscode and a fixed set of parameters it is capable of producing different iris-like patterns with very similar iriscodes.

- **Knowledge required**. In the development stage, the method proposed in [24] requires knowledge of the feature-extraction scheme being used by the recognition system (i.e., type of filters). On the other hand, the technique presented in [15] needs to have access in the development stage to the output score of an iris comparator (which can be a publicly available one) to reconstruct the image.
- **Images required**. In order to generate realistic iris-like patterns, both algorithms rely on information extracted from a pool of real iris samples which can be obtained, for instance, from one of the many publicly available iris databases.

## 13.4   Software attacks on biometric template protection schemes

The aforementioned works, especially the hill-climbing and inverse biometrics attacks, have shown that, contrary to the traditional belief that the extracted templates did not reveal enough information to reconstruct the underlying biometric data, it is indeed possible to recover synthetic biometric samples which are identified as genuine subjects by the systems. To tackle this severe issue, numerous biometric template protection (BTP) schemes have been developed in the last decade to protect the security and privacy of the subjects. However, as we will see in this section, these systems are still vulnerable to some new types of attacks.

In the first place, hill-climbing and inverse biometrics attacks can be extended to biometric template protection schemes. For instance, Adler adapts the hill-climbing algorithm proposed in [11] to attack quantized face recognition schemes, to attack a biometric cryptosystem in [75]. It is shown in the article that the hill-climbing approach yields synthetic images that disclose the concealed secret within the protected template, even if the synthetic images do not resemble the images used at enrolment.

Similarly, since template protection schemes consist of additional transformations of the input samples or unprotected templates extracted from them, attacks based on presenting them with fully synthetic iris images (see Section 13.3.1 for more details) would have the same impact on these systems if no additional countermeasures are incorporated to the system.

Furthermore, attacking methodologies can be developed to exploit new vulnerabilities of BTP schemes, as we will see in the following. In 2007, Scheirer and Boult emphasized the lack of research on attacks against BTP schemes [76], due to security proofs that assume unrealistic situations, and proposed some guidelines as to what attacks might be possible on general BTPs. They classify the attacks in three main groups, namely:

- *Attacks via record multiplicity*, in which an attacker in possession of several protected templates belonging to the same subject but protected with different secret keys can combine them to extract more biometric information. Within this group, different attacks might be implemented:

—   On the one hand, in a *correlation attack*, given two templates which conceal the same biometric instance but protected with different keys, an eventual attacker might try to correlate them and extract the unprotected template or reconstruct the biometric characteristic.

—   On the other hand, a *cross-matching attack* will try to determine whether two protected templates, generated with different secret keys, conceal the same biometric instance.

• *Stolen key-inversion attacks*, in which the attacker, in possession of the secret key of the system, tries to reconstruct the biometric characteristic. These attacks would be analogous to the *inverse biometrics* attacks described for unprotected biometric systems.

• *Blended substitution attacks*, in which the subject and the attacker's data can be combined in a single template to allow both of them to be positively identified.

In subsequent works the authors include more theoretical details for particular implementations to attack biometric fuzzy vault schemes [77] and other biometric cryptosystems [78]. Similarly, an extensive analysis is carried out by Ignatenko and Willems in [79] on the security of fuzzy commitment schemes, where it is shown that such schemes leak information on both the secret key and the biometric data.

In the next sections, particular implementations for each kind of attack will be described, and a summary of their main characteristics is provided in Table 13.3.

## 13.4.1   *Stolen key-inversion attacks*

In the last decade, several attacks on BTP schemes have been proposed exploiting the non-randomness of the biometric data or the protected template construction. Since most biometric cryptosystems are based on the fuzzy vault [77] and fuzzy commitment [80] schemes, the majority of the attacks have been directed to those systems.

In general, fuzzy vault schemes hide real biometric data or features by iteratively adding synthetic or chaff points from a secret polynomial. Therefore, in order to recover the biometric characteristic, a methodology for discriminating between real and chaff points should first be devised. In 2006, Chang *et al.* proposed such a methodology [81] for minutiae-based templates, which can be easily extended to fuzzy vault schemes based on iris [82,83]. In order to succeed, the authors exploit the fact that the process of adding chaff points is not memoryless and lacks the necessary randomness to fully conceal the real biometric data, since the selection of a new point depends on the location of the previously selected points. Hence, statistical properties of the points that arrive early (i.e., real and first chaff points) may be different from the latecomers (i.e., remaining chaff points). In particular, the authors observed that the latecomers tend to have more nearby points. This fact can be used to identify a high number of chaff points (i.e., latecomers) and thus isolate the real points. This would reveal the original unprotected template to the attacker, who could recover the iris image using one of the inverse biometrics methods described in Section 13.3.2.

In 2008, Stoianov *et al.* suggested in [84] that not only the non-random nature of biometric data can be exploited to attack fuzzy commitment schemes [89,90], but also the Error Correcting Codes (EECs) used in those schemes may leak important information. For instance, some EECs can be run in a soft decoding mode, always returning a list of the nearest codewords. If a relatively small database of images is run against helper data with soft decoding, a histogram can be generated with the codewords obtained for each chunk of the images. The bin with the maximum value yields the most probable codeword used in the commitment process, and thus reveals the concealed iriscode.

A similar approach based on soft decoding is also used by Rathgeb and Uhl in [85] to retrieve secret keys in fuzzy commitment schemes. By analysing the statistics of the decoded codewords with histograms, they show that correct keys are retrieved after presenting on average 251 impostor iriscodes to the system proposed in [89]. In addition, only 3 impostor templates are required to correctly identify 33 correct codewords within the system described in [91].

A second attack is outlined in [84], using syndrome decoding (i.e., a minimum distance decoding using a reduced lookup table, which is allowed by the linearity of the code). The authors state that this will reduce the problem of finding the underlying biometric characteristic to solving a set of linear equations, even if no empirical results are provided.

More recently, Bringer *et al.* proposed in [86] an inversion attack on the iris BTP based on Bloom filters first presented in [92]. The reconstruction approach for iriscodes given their corresponding Bloom filter templates is based on recovering the average word from each Bloom filter and repeating it in a block-wise manner. They showed that such an approach was able to produce iriscodes which, even if not visually realistic, were authenticated by the original unprotected iris system.

Aside from the particularities of each BTP scheme attacked, these algorithms differ in the knowledge they require to be carried out: whereas the works [81,86] assume access to the stored templates, [84,85] require access to a real database and the ability to use the corresponding ECC in soft decoding mode.

## 13.4.2 Attacks via record multiplicity

A different type of attack arises when the attacker is able to compromise two or more templates enrolled in different systems, and hence protected with different secret keys. Those templates, even if they conceal the same biometric instance, should not be related to each other, in order to prevent an eventual impostor from gaining additional information (*correlation attack*) or tracking a particular subject (*cross-matching attack*). In spite of the importance of this property, very few efforts have been directed to analysing this kind of attacks.

In 2008, Kholmatov and Yanikoglu implemented a correlation attack on a finger-print fuzzy vault scheme [87], which is straightforward to extend to iris-based fuzzy vaults. In their attack, the authors use exhaustive matching to align vaults, taking into account all possible rotations and translations between the vaults. Afterwards the number of matching genuine and chaff points is computed and the secret polynomial recovered in 59% of the cases. This can hence be considered as well as an instance of stolen key-inversion attacks. In addition, they proposed a cross-matching attack based on the same alignment principle.

A different cross-matching attack was proposed by Hermans *et al.* in [88] against iris BTP schemes based on Bloom filters [92]. This attack takes advantage of the linearity of the XOR operation proposed in [92] to achieve unlinkable templates, and hence prevent cross-matching attacks. Being a linear operation, different secret keys produce protected templates with identical Hamming weights, which can thus be used to correctly link templates to a single subject in over 96% of the cases.

Also on the same iris BTP [92], and based on the same principle as [93], Bringer *et al.* proposed in [86] a different cross-matching attack. Using exhaustive search on the secret key space, the authors show that almost with the same error rate as under a normal verification attempt, the attack allows to determine whether the protected templates conceal the same iriscode.

In this case, all the aforementioned attacks [87,88,93] require access to the templates stored in the database to succeed.

## 13.5　Countermeasures to software attacks

After having unveiled a number of different attacking strategies to undermine the security of biometric systems, some efforts have also been directed to the development of efficient countermeasures that minimize the effects of the aforementioned attacks. For instance, the BioAPI Consortium [94] recommends that biometric systems output only quantized similarity scores. Using quantization steps as big as possible without affecting the recognition accuracy, hill-climbing attacks such as the ones described in Section 13.2 (that require feedback on whether or not the score is increasing in each step of the algorithm) can be prevented. However, it has been shown that some hill-climbing attacks are still robust to this countermeasure [11,34]. As an alternative, non-uniform quantization is also evaluated by Maiorana *et al.* [16] as a possible countermeasure for limiting the effectiveness of the considered attacks in terms of both acceptance rate and average number of required attempts. Based on the Lloyd-Max quantizer [95], a fixed number of quantization levels is chosen, determining the intervals so that the mean-square error (MSE) between the original and quantized distributions is minimized. In particular, genuine scores distributions over the training sets are used for the estimation. The authors highlight that one of the main advantages of this method is not only its higher efficiency when compared to uniform quantization, but also its capabilities to adapt to different attacking scenarios. While hill-climbing attacks are especially relevant at low FMR operating points, false acceptance attacks launched with synthetic images are preferable when the systems works at a higher FMR. Therefore, a finer quantization should be chosen for the appropriate range of similarity scores values.

Nevertheless, score quantization is not adequate to prevent attacks carried out with synthetic images. In this context, some software approaches to liveness detection can be also extended to detect the use of such images, generated with any of the methods described in Section 13.3. For instance, inspired in previous works on image quality assessment for image manipulation detection [96,97], Galbally *et al.* employ 25 different image quality measures in [98], including the peak image to noise ratio, normalized cross-correlation or high–low frequency indices. Several biometric characteristics are analysed in the article. In the particular case of iris, synthetic images generated with the method proposed in [42] and real images are successfully classified in about 98% of the cases.

In contrast to those general methods, specific countermeasures have also been developed to deal with particular attacks. For instance, a two-step approach is considered by Galbally *et al.* in [99] to detect synthetic iris images such as those generated with the inverse biometrics method described in [15]. First of all, edge detection is applied to the image in order to ensure that no image with a homogeneous background is accepted into the system. In a real image, eyelashes generate a high number of edges outside the iris boundaries. Therefore, if no edges are detected outside that boundary, we may conclude that we are dealing with a synthetic image with a plain background. In a subsequent step, to detect iris textures embedded into eye images, including other elements such as eyelids and eyelashes, the power spectrum is analysed. Due to the characteristics of the inverse biometrics algorithm, the resulting synthetic images are formed by blocks with sharp edges. This configuration results in an abnormal amount of high-frequency energy compared to real irides, which have a much smoother surface. As a consequence, this property may be used to distinguish between real and synthetic irides by computing the ratio between the total low-frequency power and that found in the high frequencies. Given the block-wise nature of the hill-climbing algorithm proposed in [13], this countermeasure could be also effective to detect those reconstructed images.

As pointed out in Section 13.4, biometric template protection approaches have been devised to prevent inverse biometrics attacks in general, and not specific methodologies. However, some of them are still susceptible to specific software attacks developed to exploit particular vulnerabilities of a given BTP approach.

To improve the robustness of BTP schemes, [84] proposed increasing the size of EEC blocks or the application of randomizing transformations to increase in turn the randomness of the biometric data. Another widely suggested approach towards increasing that randomness has been the inclusion of inner permutations of biometric templates. However, Bringer *et al.* showed in [93] that it is still possible to both determine whether two iriscodes protected with different permutations conceal the same instance, and also recover the original iriscodes (without permutation) of the database with some residual errors.

On the other hand, Gomez-Barrero *et al.* presented in [100] an enhanced Bloom filters based BTP, by adding a structure-preserving random re-arrangement of the features to the initial scheme [92], in order to deal with the attacks proposed in [86,88]. Given the information loss due to the Bloom filter transformation, this technique has been shown to improve the robustness of the original system to the aforementioned attacks.

Other approaches to counterfeit these external attacks include the use of Homomorphic Encryption for biometric recognition [101,102] and secure Hamming Distance computation [103–105].

## 13.6 Conclusions

The wide deployment of biometric systems over the last few years has motivated an increased amount of research on both vulnerabilities of biometric schemes to software attacks and the development of appropriate countermeasures. The focus is not limited any more to decreasing error rates or verification time, but it is slowly moving towards the analysis of the security and privacy granted by biometric systems.

The works summarized in this chapter have shown that even the biometric characteristics considered the most secure due to its inherent stochastic nature, such as the iris, are vulnerable to a non-negligible number of external attacks. In an attempt to reduce such vulnerabilities, some countermeasures have been proposed. However, as new countermeasures and systems are developed, new attacking methodologies are also devised. As a consequence, still further research and development efforts are required on the field of protection measures in order to provide the necessary security assurance levels to the subjects. To reach such levels, a careful evaluation of the risks posed by software attacks needs to be carried out, analysing the following aspects for each attack:

- Success rate of the attack: the higher it is, the more dangerous the attack.
- Time required for the attack to be successful. In some cases (e.g., hill-climbing attacks or some inverse biometrics methods), it can be measured in terms of the number of scores accessed before the desired result is obtained.
- Knowledge needed for the attack to be carried out: access to which communication channel or module? Access to the score? Knowledge about the feature extraction method or template format?
- Type of equipment required to carry out the attack. For instance, as pointed out in [65], ocularists utilize very specialized equipment to fabricate prosthetic eyes. However, a simpler, computer-based method can imitate their results with a high enough accuracy to fool unprotected iris verification systems.
- Likelihood that an attacker can have sufficient access to the system to carry out the attack.
- Likelihood that an attacker, if necessary, can access real databases comprising enough textures or samples.

All these features have to be carefully assessed to determine how dangerous a particular software attack is. Similarly, if a countermeasure is introduced, the impact of the protection measure on the above-mentioned parameters should be evaluated in order to determine the improvement of the overall security of the system.

## References

[1]  K. Davis, R. Biddulph, and S. Balashek, "Automatic recognition of spoken digits," *The Journal of the Acoustical Society of America*, vol. 24, no. 6, pp. 637–642, 1952.

[2] W. Bledsoe, "The model method in facial recognition," Panoramic Research Inc., Palo Alto, CA, Technical Report, Technical Report PRI: 15, Technical Report, 1964.

[3] W. Zhao, R. Chellappa, P. J. Phillips, and A. Rosenfeld, "Face recognition: A literature survey," *ACM Computing Surveys*, vol. 35, no. 4, pp. 399–458, 2003.

[4] A. Jain, R. Bolle, and S. Pankanti, *Biometrics: Personal Identification in Networked Society*. Berlin: Springer Science & Business Media, 2006, vol. 479.

[5] A. Kong, D. Zhang, and M. Kamel, "A survey of palmprint recognition," *Pattern Recognition*, vol. 42, no. 7, pp. 1408–1418, 2009.

[6] K. W. Bowyer, K. P. Hollingsworth, and P. J. Flynn, "A survey of iris biometrics research: 2008–2010," in *Handbook of Iris Recognition*. Berlin: Springer, 2013, pp. 15–54.

[7] ICAO, "ICAO document 9303, part 1, volume 2: Machine readable passports – Specifications for electronically enabled passports with biometric identification capability," 2006.

[8] "Precise BioMatch Mobile." [Online]. Available: http://precisebiometrics. com/fingerprint-technology/precise-biomatch-mobile/

[9] N. Ratha, J. H. Connell, and R. M. Bolle, "An analysis of minutiae matching strength," in *Proceedings of the IAPR on Audio- and Video-Based Person Authentication (AVBPA)*. Berlin: Springer LNCS-2091, 2001, pp. 223–228.

[10] ISO/IEC JTC1 SC37 Biometrics, *ISO/IEC 30107:2016, Biometric Presentation Attack Detection*, International Organization for Standardization, 2016.

[11] A. Adler, "Images can be regenerated from quantized biometric match score data," in *Proceedings of the Canadian Conference on Electrical and Computer Engineering (CCECE)*, 2004, pp. 469–472.

[12] J. Galbally, C. McCool, J. Fierrez, S. Marcel, and J. Ortega-Garcia, "Hill-climbing attack to an eigenface-based face verification system," in *Proceedings of the IEEE International Conference on Biometrics, Identity and Security (BIdS)*, 2009.

[13] C. Rathgeb and A. Uhl, "Attacking iris recognition: An efficient hill-climbing technique," in *Proceedings of the International Conference on Pattern Recognition (ICPR)*, 2010, pp. 1217–1220.

[14] M. Martinez-Diaz, J. Fierrez, J. Galbally, and J. Ortega-Garcia, "An evaluation of indirect attacks and countermeasures in fingerprint verification systems," *Pattern Recognition Letters*, vol. 32, pp. 1643–1651, 2011.

[15] J. Galbally, A. Ross, M. Gomez-Barrero, J. Fierrez, and J. Ortega-Garcia, "Iris image reconstruction from binary templates: An efficient probabilistic approach based on genetic algorithms," *Computer Vision and Image Understanding*, vol. 117, no. 10, pp. 1512–1525, 2013.

[16] E. Maiorana, G. E. Hine, and P. Campisi, "Hill-climbing attacks on multibiometrics recognition systems," *IEEE Transactions on Information Forensics and Security*, vol. 10, no. 5, pp. 900–915, 2015.

[17]   ISO/IEC JTC 1/SC 27, "IT security techniques," 2009, http://www.jtc1.org/sc27/.

[18]   ISO/IEC JTC1 SC27 IT Security Techniques, "ISO/IEC 19792:2009, information technology – Security techniques – Security evaluation of biometrics," International Organization for Standardization, 2009.

[19]   TABULA RASA, "Trusted biometrics under spoofing attacks," 2010. [Online]. Available: http://www.tabularasa-euproject.org/

[20]   BEAT, "Biometrics evaluation and testing," 2012. [Online]. Available: http://www.beat-eu.org/

[21]   M. Tistarelli, W.-Y. Yau, V. Areekul, V. Bhagavatula, M. Nixon, and S. Li, Eds., *Proceedings of the Eighth International Conference on Biometrics (ICB)*. Piscataway, NJ: IEEE, 2015.

[22]   K. Bowyer, A. Ross, R. Beveridge, P. Flynn, and M. Pantic, Eds., *Proceedings of Seventh International Conference Biometrics: Theory, advances and systems (BTAS)*. Piscataway, NJ: IEEE, 2015.

[23]   M. Gomez-Barrero, J. Galbally, P. Tome-Gonzalez, and J. Fierrez, "On the vulnerability of iris-based systems to software attacks based on genetic algorithms," in *Proceedings of the Iberoamerican Conference on Pattern Recognition (CIARP)*, 2012, pp. 114–121.

[24]   S. Venugopalan and M. Savvides, "How to generate spoofed irises from an iris code template," *IEEE Transactions on Information Forensics and Security*, vol. 6, no. 2, pp. 385–395, June 2011.

[25]   A. Adler, "Sample images can be independently restored from face recognition templates," in *Proceedings of the Canadian Conference on Electrical and Computer Engineering (CCECE)*, vol. 2, 2003, pp. 1163–1166.

[26]   J. Galbally, C. McCool, J. Fierrez, and S. Marcel, "On the vulnerability of face verification systems to hill-climbing attacks," *Pattern Recognition*, vol. 43, pp. 1027–1038, 2010.

[27]   M. Gomez-Barrero, J. Galbally, J. Fierrez, and J. Ortega-Garcia, "Face verification put to test: A hill-climbing attack based on the uphill-simplex algorithm," in *Proceedings of the International Conference on Biometrics (ICB)*, 2012, pp. 40–45.

[28]   U. Uludag and A. Jain, "Attacks on biometric systems: a case study in fingerprints," in *Proceedings of the SPIE Seganography and Watermarking of Multimedia Contents VI*, vol. 5306, no. 4, 2004, pp. 622–633.

[29]   E. Maiorana, G. E. Hine, D. La Rocca, and P. Campisi, "On the vulnerability of an EEG-based biometric system to hill-climbing attacks algorithms' comparison and possible countermeasures," in *Proceedings of the International Conference on Biometrics: Theory, Applications and Systems (BTAS)*, 2013, pp. 1–6.

[30]   J. Galbally, J. Fierrez, and J. Ortega-Garcia, "Bayesian hill-climbing attack and its application to signature verification," in *Proceedings of the IAPR International Conference on Biometrics (ICB)*. Berlin: Springer Lecture Notes in Computer Science, vol. 4642, 2007, pp. 386–395.

[31] M. Gomez-Barrero, J. Galbally, J. Fierrez, and J. Ortega-Garcia, "Hill-climbing attack based on the uphill simplex algorithm and its application to signature verification," in *Proceedings of the European Workshop on Biometrics and Identity Management (BioID)*. Lecture Notes in Computer Science, vol. 6583, 2011, pp. 83–94.

[32] E. Maiorana, G. E. Hine, and P. Campisi, "Hill-climbing attack: Parametric optimization and possible countermeasures. An application to on-line signature recognition," in *Proceedings of the International Conference on Biometrics (ICB)*, 2013, pp. 1–6.

[33] M. Gomez-Barrero, J. Gonzalez-Dominguez, J. Galbally, and J. Gonzalez-Rodriguez, "Security evaluation of i-vector based speaker verification systems against hill-climbing attacks," in *Proceedings the Interspeech*, 2013.

[34] M. Gomez-Barrero, J. Galbally, and J. Fierrez, "Efficient software attack to multimodal biometric systems and its application to face and iris fusion," *Pattern Recognition Letters*, vol. 36, pp. 243–253, 2014.

[35] J. Daugman, "How iris recognition works," *IEEE Transactions on Circuits and Systems for Video Technology*, vol. 14, no. 1, pp. 21–30, 2004.

[36] A. Wehde and J. N. Beffel, *Finger-Prints Can Be Forged*. Chicago, IL: The Tremonia Publishing Co., 1924.

[37] A. S. Osborn, *Questioned Documents*. Albany, NY: Boyd Printing Co., 1929.

[38] T. Dutoit, *An Introduction to Text-to-Speech Synthesis*. Dordrecht: Kluwer Academic Publishers, 2001.

[39] J. Cui, Y. Wang, J. Huang, T. Tan, and Z. Sun, "An iris image synthesis method based on PCA and super-resolution," in *Proceedings of the IAPR International Conference on Pattern Recognition (ICPR)*, 2004, pp. 471–474.

[40] R. Cappelli, "Synthetic fingerprint generation," in *Handbook of Fingerprint Recognition*. Berlin: Springer, 2003, pp. 203–231.

[41] S. Makthal and A. Ross, "Synthesis of iris images using Markov random fields," in *Proceedings of the European Signal Processing Conference (EUSIPCO)*, 2005.

[42] S. Shah and A. Ross, "Generating synthetic irises by feature agglomeration," in *Proceedings of the IEEE International Conference on Image Processing (ICIP)*, 2006, pp. 317–320.

[43] J. Zuo, N. A. Schmid, and X. Chen, "On generation and analysis of synthetic iris images," *IEEE Transactions on Information Forensics and Security*, vol. 2, pp. 77–90, 2007.

[44] Z. Wei, T. Tan, and Z. Sun, "Synthesis of large realistic iris databases using patch-based sampling," in *Proceedings of the IAPR International Conference on Pattern Recognition (ICPR)*, 2008, pp. 1–4.

[45] A. Lin and L. Wang, "Style-preserving English handwriting synthesis," *Pattern Recognition*, vol. 40, pp. 2097–2109, 2007.

[46] N. Poh, S. Marcel, and S. Bengio, "Improving face authentication using virtual samples," in *Proceedings of the IEEE International Conference on Acoustics, Speech and Signal Processing (ICASSP)*, 2003.

[47]   D. V. Popel, "Signature analysis, verification and synthesis inpervasive environments," in *Synthesis and Analysis in Biometrics*. Singapore: World Scientific, 2007, pp. 31–63.

[48]   J. Galbally, R. Plamondon, J. Fierrez, and J. Ortega-Garcia, "Synthetic on-line signature generation. Part I: Methodology and algorithms," *Pattern Recognition*, vol. 45, pp. 2610–2621, 2012.

[49]   J. Galbally, J. Fierrez, J. Ortega-Garcia, and R. Plamondon, "Synthetic on-line signature generation. Part II: Experimental validation," *Pattern Recognition*, vol. 45, pp. 2622–2632, 2012.

[50]   M. E. Munich and P. Perona, "Visual identification by signature tracking," *IEEE Transactions on Pattern Analysis and Machine Intelligence*, vol. 25, no. 2, pp. 200–217, 2003.

[51]   C. Oliveira, C. A. Kaestner, F. Bortolozzi, and R. Sabourin, "Generation of signatures by deformations," in *Proceedings of the IAPR International Conference on Advances in Document Image Analysis (ICADIA)*. Springer Lecture Notes in Computer Science, vol. 1339, 1997, pp. 283–298.

[52]   M. Mori, A. Suzuki, A. Shio, and S. Ohtsuka, "Generating new samples from handwritten numerals based on point correspondence," in *Proceedings of the IAPR International Workshop on Frontiers in Handwriting Recognition (IWFHR)*, 2000, pp. 281–290.

[53]   J. Wang, C. Wu, Y.-Q. Xu, H.-Y. Shum, and L. Ji, "Learning-based cursive handwriting synthesis," in *Proceedings of the IAPR International Workshop on Frontiers of Handwriting Recognition (IWFHR)*, 2002, pp. 157–162.

[54]   H. Wang and L. Zhang, "Linear generalization probe samples for face recognition," *Pattern Recognition Letters*, vol. 25, p. 829–840, 2004.

[55]   H. R. Wilson, G. Loffler, and F. Wilkinson, "Synthetic faces, face cubes, and the geometry of face space," *Vision Research*, vol. 42, no. 34, pp. 2909–2923, 2002.

[56]   C. Rabasse, R. M. Guest, and M. C. Fairhurst, "A method for the synthesis of dynamic biometric signature data," in *Proceedings of the International Conference on Document Analysis and Recognition (ICDAR)*, vol. 1, 2007, pp. 168–172.

[57]   D. H. Klatt, "Software for a cascade/parallel formant synthesizer," *Journal Acoustic Society of America*, vol. 67, pp. 971–995, 1980.

[58]   N. B. Pinto, D. G. Childers, and A. L. Lalwani, "Formant speech synthesis: Improving production quality," *IEEE Transactions on Acoustics, Speech and Signal Processing*, vol. 37, pp. 1870–1887, 1989.

[59]   C. J. Hill, "Risk of masquerade arising from the storage of biometrics," Master's thesis, Australian National University, Canberra, Australia, 2001.

[60]   R. Cappelli, D. Maio, A. Lumini, and D. Maltoni, "Fingerprint image reconstruction from standard templates," *IEEE Transactions on Pattern Analysis and Machine Intelligence*, vol. 29, pp. 1489–1503, Sep. 2007.

[61]   A. Ross, J. Shah, and A. K. Jain, "From template to image: reconstructing fingerprints from minutiae points," *IEEE Transactions on Pattern Analysis and Machine Intelligence*, vol. 29, pp. 544–560, 2007.

[62] M. Gomez-Barrero, J. Galbally, M. A. Ferrer, A. Morales, J. Fierrez, and J. Ortega-Garcia, "A novel hand reconstruction approach and its application to vulnerability assessment," *Information Sciences*, vol. 268, pp. 103–121, 2014.

[63] P. Mohanty, S. Sarkar, and R. Kasturi, "From scores to face templates: A model-based approach," *IEEE Transactions on Pattern Analysis and Machine Intelligence*, vol. 29, pp. 2065–2078, 2007.

[64] International Biometric Group, "Generating images from templates," White paper, 2002.

[65] A. Lefohn, B. Budge, P. Shirley, R. Caruso, and E. Reinhard, "An ocularist's approach to human iris synthesis," *Computer Graphics and Applications, IEEE*, vol. 23, no. 6, pp. 70–75, Nov.–Dec. 2003.

[66] J. Zuo and N. A. Schmid, "A model based, anatomy based method for synthesizing iris images," in *Proceedings of the International Conference on Biometrics (ICB)*, 2005, pp. 428–435.

[67] B. T. Phong, "Illumination for computer generated pictures," *Communications of the ACM*, vol. 18, no. 6, pp. 311–317, 1975.

[68] J. Huang, L. Ma, T. Tan, and Y. Wang, "Learning based resolution enhancement of iris images." in *Proceedings of the British Machine Vision Conference (BMVC)*, 2003, pp. 1–10.

[69] L.-Y. Wei and M. Levoy, "Fast texture synthesis using tree-structured vector quantization," in *Proceedings of the Annual Conference on Computer Graphics and Interactive Techniques (SIGGRAPH)*, 2000, pp. 479–488.

[70] S. Z. Li, *Markov Random Field Modeling in Image Analysis*. Berlin: Springer Science & Business Media, 2009.

[71] S. C. Dass, A. K. Jain, and X. Lu, "Face detection and synthesis using Markov random field models," in *Proceedings of the International Conference on Pattern Recognition (ICPR)*, vol. 4, 2002, pp. 201–204.

[72] L. Liang, C. Liu, Y.-Q. Xu, B. Guo, and H.-Y. Shum, "Real-time texture synthesis by patch-based sampling," *ACM Transactions on Graphics (ToG)*, vol. 20, no. 3, pp. 127–150, 2001.

[73] J. Daugman, "Probing the uniqueness and randomness of iris codes: Results from 200 billion iris pair comparisons," *Proceedings of the IEEE*, vol. 94, pp. 1927–1935, 2006.

[74] J. Daugman, "How iris recognition works," in *Proceedings of the IEEE International Conference on Image Processing (ICIP)*, 2002, pp. I.33–I.36.

[75] A. Adler, "Vulnerabilities in biometric encryption systems," in *Proceedings of the IAPR Audio- and Video-Based Biometric Person Authentication (AVBPA)*. Springer Lecture Notes in Computer Science, vol. 3546, 2005, pp. 1100–1109.

[76] W. J. Scheirer and T. E. Boult, "Cracking fuzzy vaults and biometric encryption," in *Proceedings of the Biometrics Symposium*, 2007, pp. 1–6.

[77] A. Juels and M. Sudan, "A fuzzy vault scheme," *Designs, Codes and Cryptography*, vol. 38, no. 2, pp. 237–257, 2006.

[78]   C. Soutar, D. Roberge, A. Stoianov, R. Gilroy, and B. V. Kumar, "Biometric encryption using image processing," in *Proceedings of the Photonics West Electronic Imaging*, 1998, pp. 178–188.

[79]   T. Ignatenko and F. Willems, "Information leakage in fuzzy commitment schemes," *IEEE Transactions on Information Forensics and Security*, vol. 2, no. 5, pp. 337–348, 2010.

[80]   A. Juels and M. Wattenberg, "A fuzzy commitment scheme," *ACM Conference on Computer and Communications Security*, pp. 28–36, 1999.

[81]   E.-C. Chang, R. Shen, and F. W. Teo, "Finding the original point set hidden among chaff," in *Proceedings of the ACM Symposium on Information, Computer and Communications Security*, 2006, pp. 182–188.

[82]   Y. J. Lee, K. Bae, S. J. Lee, K. R. Park, and J. Kim, "Biometric key binding: Fuzzy vault based on iris images," in *Proceedings of the International Conference on Biometrics (ICB)*. Berlin: Springer, 2007, pp. 800–808.

[83]   M. Fouad, A. El Saddik, J. Zhao, and E. Petriu, "A fuzzy vault implementation for securing revocable iris templates," in *Proceedings of the IEEE International Systems Conference (SysCon)*, 2011, pp. 491–494.

[84]   A. Stoianov, T. Kevenaar, and M. Van der Veen, "Security issues of biometric encryption," in *Proceedings of the IEEE Toronto International Conference on Science and Technology for Humanity, TIC-STH*, 2009.

[85]   C. Rathgeb and A. Uhl, "Statistical attack against iris-biometric fuzzy commitment schemes," in *International Conference on Computer Vision and Pattern Recognition Workshops (CVPRW)*, 2011, pp. 23–30.

[86]   J. Bringer, C. Morel, and C. Rathgeb, "Security analysis of bloom filter-based iris biometric template protection," in *Proceedings of the International Conference on Biometrics (ICB)*, 2015, pp. 527–534.

[87]   A. Kholmatov and B. Yanikoglu, "Realization of correlation attack against the fuzzy vault scheme," in *Proceedings of SPIE*, vol. 6819, 2008, pp. 68190O–68190O-7.

[88]   J. Hermans, B. Mennink, and R. Peeters, "When a bloom filter is a doom filter: Security assessment of a novel iris biometric," in *Proceedings of the International Conference of the Biometrics Special Interest Group (BIOSIG)*, 2014.

[89]   F. Hao, R. Anderson, and J. Daugman, "Combining crypto with biometrics effectively," *IEEE Transactions on Computers*, vol. 55, no. 9, pp. 1081–1088, 2006.

[90]   J. Bringer, H. Chabanne, G. Cohen, B. Kindarji, and G. Zémor, "Optimal iris fuzzy sketches," in *Proceedings of the IEEE International Conference on Biometrics: Theory, Applications, and Systems (BTAS)*, 2007, pp. 1–6.

[91]   J. Bringer, H. Chabanne, G. Cohen, B. Kindarji, and G. Zémor, "Theoretical and practical boundaries of binary secure sketches," *IEEE Transactions on Information Forensics and Security*, vol. 3, no. 4, pp. 673–683, 2008.

[92]   C. Rathgeb, F. Breitinger, and C. Busch, "Alignment-free cancelable iris biometric templates based on adaptive bloom filters," in *Proceedings of the International Conference on Biometrics (ICB)*, 2013, pp. 1–8.

[93] J. Bringer, H. Chabanne, and C. Morel, "Shuffling is not sufficient: Security analysis of cancelable iriscodes based on a secret permutation," in *Proceedings of the IEEE International Joint Conference on Biometrics (IJCB)*, 2014, pp. 1–8.

[94] BioAPI Consortium, "BioAPI specification (version 1.1)," March 2001, www.bioapi.org/Downloads/BioAPI

[95] S. P. Lloyd, "Least squares quantization in PCM," *IEEE Transactions on Information Theory*, vol. 28, no. 2, pp. 129–137, 1982.

[96] S. Bayram, İ. Avcıbaş, B. Sankur, and N. Memon, "Image manipulation detection," *Journal of Electronic Imaging*, vol. 15, no. 4, pp. 041102–041102-17, 2006.

[97] M. C. Stamm and K. Liu, "Forensic detection of image manipulation using statistical intrinsic fingerprints," *IEEE Transactions on Information Forensics and Security*, vol. 5, no. 3, pp. 492–506, 2010.

[98] J. Galbally, S. Marcel, and J. Fierrez, "Image quality assessment for fake biometric detection: Application to iris, fingerprint and face recognition," *IEEE Transactions on Image Processing*, vol. 23, no. 2, pp. 710–724, 2014.

[99] J. Galbally, M. Gomez-Barrero, A. Ross, J. Fierrez, and J. Ortega-Garcia, "Securing iris recognition systems against masquerade attacks," in *Proceedings of the SPIE Biometric and Surveillance Technology for Human and Activity Identification X, BSTHAI*, vol. 8712, 2013.

[100] M. Gomez-Barrero, C. Rahtgeb, J. Galbally, C. Busch, and J. Fierrez, "Unlinkable and irreversible biometric template protection based on bloom filters," *Information Sciences*, vol. 370–371, pp. 18–32, 2016.

[101] M. Barni, G. Droandi, and R. Lazzeretti, "Privacy protection in biometric-based recognition systems: A marriage between cryptography and signal processing," *IEEE Signal Processing Magazine*, vol. 32, no. 5, pp. 66–76, 2015.

[102] M. Gomez-Barrero, J. Galbally, E. Maiorana, P. Campisi, and J. Fierrez, "Implementation of fixed-length template protection based on homomorphic encryption with application to signature biometrics," in *Proceedings of the IEEE Conference on Computer Vision and Pattern Recognition Workshops (CVPRW)*, 2016.

[103] C. Karabat, M. S. Kiraz, H. Erdogan, and E. Savas, "THRIVE: Threshold homomorphic encryption based secure and privacy preserving biometric verification system," *EURASIP Journal on Advances in Signal Processing*, vol. 2015, no. 1, pp. 1–18, 2015.

[104] M. Osadchy, B. Pinkas, A. Jarrous, and B. Moskovich, "SCiFI: A system for secure face identification," in *Proceedings of the IEEE Symposium on Security and Privacy*, 2010, pp. 239–254.

[105] J. Bringer, H. Chabanne, M. Favre, A. Patey, T. Schneider, and M. Zohner, "GSHADE: Faster privacy-preserving distance computation and biometric identification," in *Proceedings of the ACM Workshop on Information Hiding and Multimedia Security*, 2014, pp. 187–198.

*Part V*

**Privacy Protection and Forensics**

*Chapter 14*

# Iris biometric template protection

*Christian Rathgeb, Johannes Wagner,*
*and Christoph Busch*

## 14.1 Introduction

Iris recognition technologies [1,2] are already deployed in numerous large-scale nation-wide projects in order to provide robust and reliable biometric recognition of individuals. In order to safeguard individuals' privacy and biometric systems' security biometric reference data, i.e., templates, need to be protected, especially if these are stored in centralized repositories. Unprotected storage of biometric templates poses serious privacy threats, e.g., identity theft, cross-matching, or limited re-newability [3,4]. More recently, different researchers have shown that, when an attacker has full knowledge of the employed feature extraction, iris-codes can be utilized in order to reconstruct images of subjects' iris textures [5,6]. Artificial attack presentation instruments (e.g., printouts, electronic displays) showing such reconstructed images can be presented to an iris recognition system in order to successfully launch presentation attacks [7].

It is well-known that biometric samples, e.g., iris images, underlie a natural intra-class variance. In an iris recognition system such variance can be caused by several factors, e.g., pupil dilation, partial closure of eyelids, or change in gaze angle. Example variations are depicted in Figure 14.1. In order to tolerate a certain variance, generic biometric systems compare obtained (dis)similarity scores (between a probe and a reference template) against an adequate decision threshold yielding acceptance or rejection. However, this vital step of every biometric recognition system prevents from a secure application of conventional cryptographic techniques. The desirable "avalanche effect" property of cryptographic algorithms, i.e., a small change in either the key or the plaintext should cause a drastic change in the ciphertext, obviously obviates a comparison of protected templates in the encrypted domain resulting in two major drawbacks [3]: on the one hand, an encrypted template needs to be decrypted before each authentication attempt, i.e., an attacker can glean the biometric template by simply launching an authentication attempt; on the other hand, an encrypted template will be secure only as long as the corresponding decryption key is unknown to the attacker, i.e., the problem is shifted yielding two-factor authentication.

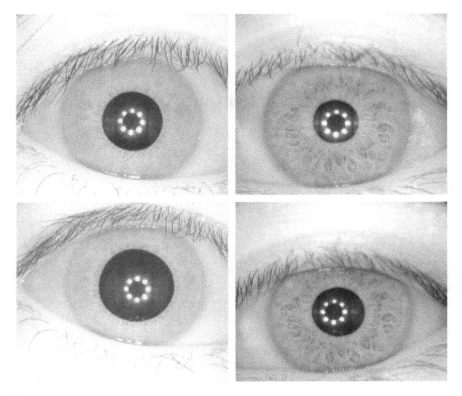

*Figure 14.1    Examples for intra-class variations between pairs of NIR iris images.
Images taken from CASIA-v4-Interval iris database [8]*

Technologies of *biometric template protection* [4] offer solutions to privacy preserving biometric authentication, improving the public confidence and acceptance of biometric systems. Moreover, such technologies will enable biometric systems to become compliant to the recently issued EU regulation 2016/679 (General Data Protection Regulation) [9]. The two major requirements of biometric template protection are defined in the ISO/IEC IS 24745 [10]: (1) *irreversibility*: knowledge of a protected template cannot be exploited to reconstruct a biometric sample which is equal or close (within a small margin of error) to an original captured sample of the same source; (2) *unlinkability*: different versions of protected biometric templates can be generated based on the same biometric data (re-newability), while protected templates should not allow cross-matching.

Based on the standardized architecture biometric template protection schemes convert an unprotected biometric template to a pseudonymous identifier and auxiliary data [10], which allow for a privacy-preserving authentication. Apart from satisfying the above properties, an ideal biometric template protection scheme shall not cause a

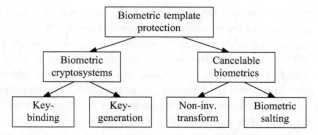

*Figure 14.2    Generic categorization of (iris) biometric template protection schemes*

decrease in biometric performance, i.e., recognition accuracy, with respect to the corresponding unprotected system [3]. However, due to several reasons the vast majority of proposed approaches have been found to suffer from a significant drop in biometric performance. Nevertheless, the rich random structure of the iris, and hence its resistance to false matches, constitutes one of the most powerful biometric characteristics, which might still enable the construction of biometric template protection schemes operating at practical acceptable biometric performance.

This chapter provides a summary of iris biometric template protection schemes. A review and discussion of state-of-the-art approaches is presented, based on which an outlook to future prospects is given. In addition, we provide a detailed insight into the construction of two prominent iris biometric template protection schemes in particular, the *fuzzy vault scheme* and the *bin-combo scheme*. Both schemes are evaluated and compared on a publicly available iris database with respect to provided privacy protection and biometric performance.

The remainder of this chapter is organized as follows: an overview of proposed iris biometric template protection schemes is given in Section 14.2. Section 14.3 provides practical insight into the implementation of different types of biometric template protection schemes. Experimental evaluations are presented in Section 14.4. Finally, a summary is given and future research directions are discussed in Section 14.5.

## 14.2    Iris template protection schemes

As shown in Figure 14.2, (iris) biometric template protection schemes are commonly categorized as *biometric cryptosystems*, also referred to as helper data schemes, and *cancelable biometrics*, also referred to as feature transformation approaches. Biometric cryptosystems are designed to securely bind a digital key to a biometric or generate a digital key from a biometric signal [11]. Cancelable biometrics consist of intentional, repeatable distortions of biometric signals based on transforms that provide a comparison of biometric templates in the transformed domain [12,13].

Biometric cryptosystems can be further divided into the groups of *key-binding* and *key-generation* schemes [4]. Within a key-binding scheme helper data is obtained by binding a chosen key to a biometric template. As a result of the binding process a fusion of the secret key and the biometric template is stored, which ideally does neither reveal any information about the key nor about the original biometric template. Applying an appropriate key retrieval algorithm, the key is reconstructed from the stored helper data during authentication. An update of the key usually requires re-enrolment in order to generate new helper data. In a key-generation scheme the helper data is derived from the biometric template so that the key is generated from the helper data and a given biometric template. While the storage of helper data is not obligatory the majority of proposed key-generation schemes do store helper data, which allow for an update of keys. Key-generation schemes in which helper data are applied are also referred to as "fuzzy extractors" [14]. A fuzzy extractor should reliably extract a uniformly random string, i.e., key, from a biometric template while helper data is used to reconstruct that string from another biometric template. Hence, biometric cryptosystems either retrieve/generate a key or return a failure message.

Non-invertible transforms are either applied in image domain or to the biometric template. In order to provide re-newable templates, parameters of the applied transforms are modified. Non-invertible transforms are designed in a way that potential impostors are not able to reconstruct the original biometric image or template even if these are compromised. However, applying non-invertible transforms mostly implies a loss in recognition accuracy, which might be caused by information loss or infeasible feature alignment at comparison [4]. Biometric salting usually denotes transforms which are selected to be invertible. Hence, the parameters of an applied transform have to be kept secret, since impostors may be able to recover the original biometric template when transform parameters are compromised. Generic approaches to biometric salting maintain the biometric performance of the underlying unprotected biometric systems. In case subject-specific transforms are applied, the parameters of the transform have to be presented at each authentication. It is important to note that the majority of cancelable biometric systems employ the biometric comparator of the original unprotected system, i.e., (dis)similarity scores are obtained, in contrast to biometric cryptosystems.

## 14.2.1   Iris-biometric cryptosystems

The first iris-based key binding scheme, in which the *fuzzy commitment scheme* is employed, was presented in [15]. The fuzzy commitment scheme by Juels and Wattenberg [16] represents a cryptographic primitive, which combines techniques from the area of error correcting codes and cryptography. At key-binding a pre-chosen key is prepared with error correcting codes and bound to a binary biometric feature vector of the same length by XORing both resulting in a difference vector. In addition, a hash of the key is stored together with the difference vector forming the commitment. During key retrieval another binary probe feature vector is XORed with the stored difference vector and error correction decoding is applied. If the presented feature vector is sufficiently "close" to the stored one, then the correct key is returned, which can be tested using the stored hash. In [15] iris-codes are bound to keys prepared with Hadamard and Reed–Solomon error correction codes. In order to provide a more efficient error correction decoding in an iris-based fuzzy commitment scheme, two-dimensional iterative min-sum decoding was introduced in [17]. In [18] a systematic approach to the construction of iris-based fuzzy commitment schemes is proposed, where error correction codes are chosen according to score distributions of genuine *HD*-based comparisons. A so-called context-based reliable component selection used to extract keys from iris-codes which are then bound to Hadamard codewords is presented in [19]. Further, diverse techniques to improve the performance and security of (iris) fuzzy commitment schemes have been proposed, e.g., [20–23].

In [24,25] diverse attacks have been suggested, which utilize the fact that error correction codes underlie distinct structures. Statistical attacks based on so-called error correction code histograms have been successfully conducted against iris-based fuzzy commitment schemes in [26]. In [27] it was found that fuzzy commitment schemes leak information in bound keys and non-uniform templates. Suggestions to prevent from information leakage in fuzzy commitment schemes have been proposed in [28]. In addition, attacks via record multiplicity could be applied to decode stored commitments [29,30]. In [20] a bit-permutation process was introduced to prevent from cross-matching attacks. It is, however, important to note that even if one uses record-specific bit-permutation processes, cross-matching may still be possible for records protecting very similar feature vectors and then even enables reversibility attacks from record multiplicity [31]. Considering proposed attacks, the security provided by iris-based fuzzy commitment schemes is rather doubtable.

An implementation of an iris-based *fuzzy vault scheme* was presented in [32,33]. The fuzzy vault scheme [34,35] by Juels and Sudan enables a protection and error-tolerant verification with feature sets. In the key-binding process a chosen key serves as polynomial based on which the set-based (integer-valued) biometric feature vector is encoded. The resulting genuine points are dispersed among a set of chaff points not lying on the graph of the polynomial. At the time of authentication a given feature vector is intersected with the vault resulting in a set of unlocking pairs. If the unlocking pairs are mostly genuine it is possible to distinguish the genuine pairs from the chaff pairs so that the polynomial, and hence the key, can be reconstructed by tolerating errors within certain limits of an error correction code. This approach offers order invariance, meaning that the sets used to lock and unlock the vault are unordered. It is due to this property that led researchers to consider the fuzzy vault scheme as a promising tool for protecting fingerprint minutiae sets [36,37]. In [33] salient feature points are extracted from iris textures. Based on several enrolment samples independent component analysis and *k*-means clustering are employed to extract coefficients from most stable parts of the iris textures which are locked in a vault. In order to prevent from attacks via record multiplicity a iris fuzzy vault which is hardened with an additional password is presented in [38,39]. In this scheme image enhancement techniques are applied to extract iris fibres from which minutiae-like coordinates are extracted. In [40] a multi-biometric fuzzy vault based on fingerprint and iris is proposed. Secret keys, which are used to generate an iris-based fuzzy commitment, are directly encoded, fused with fingerprint data at feature level, and locked in the fuzzy vault scheme. Further proposals of iris-based fuzzy vaults have been presented, e.g., [41–43]. It is important to note that the vast majority of iris-based

*Table 14.1*    Overview of most relevant proposals of iris biometric cryptosystems based on key-binding schemes

| Method | Scheme | Database | FNMR | FMR | Key size |
|---|---|---|---|---|---|
| Hao *et al.* [15] | Fuzzy commitment | 700 subjects | 0.47% | 0% | 140 bit |
| Bringer *et al.* [17] | Fuzzy commitment | ICE 2005 | 5.62% | 0% | 40 bit |
| Lee *et al.* [33] | Fuzzy vault | CASIAv3 | ~20% | 0% | 128 bit |
| Nandakumar and Jain [40] | Multi-biometric fuzzy vault | MSU-DBI, CASIAv1 | ~2% | 0.01% | 208 bit |

key-binding scheme resemble either the fuzzy commitment scheme or the fuzzy vault scheme [4]. Table 14.1 lists reported biometric performance and key sizes of most relevant approaches to iris biometric cryptosystems.

A conceptual proposal of an iris-based key-generation scheme has been presented in [44,45]. In the *private template scheme* the iris biometric feature vector itself serves as a key. During enrolment several iris-codes are combined using a majority decoder and error correction check bits are generated for the resulting bit string, which represent the helper data. At authentication faulty bits of a given iris-codes should be corrected using the stored helper data. Experimental results are omitted and it is commonly expected that the proposed system reveals poor performance due to the fact that the authors restrict to the assumption that only 10% of bits change among different genuine iris-codes. Moreover, the direct use of biometric features as key contradicts with the requirement of re-newability and unlinkability.

Different types of *quantization schemes* have been proposed, which coarsely quantize biometric feature vector elements based on a segmentation of the feature space. Obtained intervals are encoded with bit strings whereas an alteration of the encoding order should grant unlinkability. Given an iris-based feature vector, bits are obtained from mapping feature elements to according intervals, which should result in a stable (biometric) key. For instance, in [46] a context-based reliable component selection is applied to construct intervals for the most reliable features of an iris-code. Similar approaches have been presented in [47,48]. Generally speaking, key-generation schemes cause a more severe drop in biometric performance compared to key-binding schemes [4].

## 14.2.2    Cancelable iris biometrics

Ratha *et al.* [13] were the first to introduce the concept of cancelable biometrics. In their work the authors apply image-based block permutations and surface-folding in order to obtain cancelable biometric templates from face and fingerprint images. In further work [49] the authors propose different techniques to generate cancelable iris biometrics based on non-invertible transforms, which are applied in image and feature domains. In order to preserve a computational efficient alignment of resulting iris-codes based on circular bit-shifting, iris textures, and iris-codes are obscured in a row-wise manner, which means adjacency of pixels and bits is maintained along

*Table 14.2*   *Overview of most relevant proposals of cancelable iris biometrics based on non-invertible transforms*

| Method | Scheme | Database | FNMR | FMR |
|---|---|---|---|---|
| Zuo *et al.* [49] | Gray-combo, | MMU1 | ~4% | 0.1% |
| | bin-combo | | ~4% | 0.1% |
| Hämmerle-Uhl *et al.* [50] | Block re-mapping, | CASIAv3 | ~5% | 0.01% |
| | image warping | | ~2% | 0.01% |
| Ouda *et al.* [51] | BioEncoding | CASIAv3 | ~6% | 0.01% |
| Rathgeb *et al.* [52] | Bloom filter | CASIAv3 | 2.1% | 0.01% |

$x$-axis in image and feature domain, respectively. In [50] block re-mapping and image warping have been applied to normalized iris textures. For both types of transforms a proper alignment of resulting iris-codes is infeasible causing a significant decrease of biometric performance [4]. In [51] several enrolment templates are processed to obtain a vector of consistent bits. Re-newability is provided by encoding the iris-code according to a subject-specific random seed. In case subject-specific transforms are applied in order to achieve cancelable biometrics, these transforms have to be considered compromised during inter-class comparisons [53]. Subject-specific secrets, be it transform parameters, random numbers, or any kind of passwords are easily compromised, i.e., performance evaluations have to be performed under the "stolen-secret scenario", where each impostor is in possession of valid secrets. In [54] cancelable iris templates are achieved by applying sector random projection to iris images. Recognition performance is only maintained, if subject-specific random matrices are applied. In [55] non-invertible iris-codes are computed by thresholding inner products of the feature vector with randomly generated vectors. The random vectors are created by using a per-subject secret and a pseudo random number generator. Several normalized iris textures are multiplied with a random kernel in [56] to create concealed feature vectors. In [52] a cancelable iris biometric system is presented, in which iris-codes bits are transformed to an non-invertible Bloom filter-based representation. The scheme is extended to a multi-iris system in [57], where cancelable templates of the left and right iris are combined to further improve privacy protection. Most relevant approaches to iris-based non-invertible transforms are summarized in Table 14.2.

In [58] two approaches to biometric salting have been presented. In the first scheme iris-codes are XORed with a randomly generated application-specific bit stream. In order to assure that resulting iris-codes sufficiently differ from original ones it is suggested retaining at least half of the bits in the randomly generated bit stream as one. In the second scheme an application specific permutation is employed to scramble bits of an iris-code (identical permutation is also applied to their bit masks). In [49] the first scheme of [58] is evaluated, which is referred to as bin-salt. Since a reference template should be permanently protected, the random bit stream will have to be XORed with the probe iris-code at various shifting positions in order to achieve an appropriate alignment of iris-codes (the same holds for the permutation scheme). Further, a salting approach in the image domain is proposed in [49]. In this scheme a random noise image is combined pixel-wise with an iris texture using either addition or multiplication. A salting approach based on key-dependent wavelet transforms, which are applied in the feature extraction stage of an iris recognition system, is presented in [59].

Focusing on provided security, diverse attacks might be applied to the above summarized cancelable iris recognition system. For instance, cross-matching attacks might be feasible if non-invertible transforms are applied in image domain, e.g., block re-mapping or image warping [50]. Such attacks could be performed based on visual inspection or automatically using image descriptors such as SIFT. Cross-matching attacks could be applied to salting approaches which are based on bit permutations, e.g., [58]. If a single iris-code is protected using two different permutations, these could be linked simply via an estimation of the Hamming weights of protected iris-codes. Finally, it is important to note that, in contrast to biometric cryptosystems, cancelable biometrics are vulnerable to hill-climbing attacks [60]. The key idea behind hill-climbing is to consecutively modify the biometric probe, e.g., an iris image, which is presented to the biometric recognition system in order to access a distinct account. The attacker observes the internal comparison score returned by the system at the time of each authentication attempt and retains changes in the input, which improves the score. The process of changing the biometric input is repeated until successful authentication is achieved.

## 14.3 Implementation of iris template protection schemes

Traditional iris recognition schemes extract two-dimensional binary feature vectors, i.e., iris-codes, from iris images [2], cf. Figure 14.3(d) and (e). Hence, in the following template protection schemes we assume to be presented with a two-dimensional binary feature vector $\mathbf{I}$ of size $W \times H$ bits, where columns and rows are denoted as $\mathbf{c}_i$, $i = 0, \ldots, W - 1$ and $\mathbf{r}_j, j = 0, \ldots, H - 1$, respectively.

### 14.3.1 Iris fuzzy vault

As previously described, the fuzzy vault scheme [34] operates on unordered integer sets. Hence, a feature-type transformation [61] from binary to integer representation needs to be performed. For this purpose, $\mathbf{I}$ is vertically divided into $B$ blocks of size $w \times h$ bits, where each block and hence each column, is associated with a block index $b = 0, \ldots, B - 1$. Subsequently, a transform $f$ is applied to each column $\mathbf{c}_i \in \{0, 1\}^h$, $h \leq H, i = 0, \ldots, w - 1$, of a block with index $b$,

$$f(\mathbf{c}_i, b) = \text{int}(\mathbf{c}_i) \cdot \log_2 B + b, \tag{14.1}$$

where int denotes the conversion of a binary vector to its integer representation. In case $h < H$, columns consist of the $h$ upper most bits, i.e., features originating from outer iris bands, which are expected to contain less discriminative information, are ignored. Each block yields a vector $\mathbf{V}_b$ of feature values, $\mathbf{V}_b = (v_0, \ldots, v_{w-1})$, with

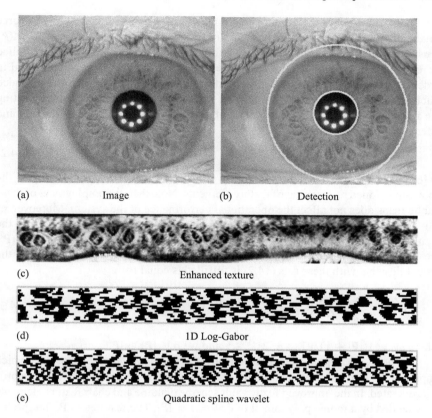

(a)        Image          (b)        Detection

(c)                    Enhanced texture

(d)                    1D Log-Gabor

(e)                  Quadratic spline wavelet

*Figure 14.3    Iris recognition processing chain: (a) iris image, (b) iris detection,*
*(c) normalized pre-processed iris texture, and (d) and (e) iris-codes*
*of applied feature extractor. Image taken from CASIA-v4-Interval*
*iris database [8]*

$v_i = f(\mathbf{c}_i, b)$. For each vector $\mathbf{V}_b$, $\mathbf{P}_b$ denotes the sets of unique vector elements, $\mathbf{P}_b = \{v_0, \ldots, v_l\}_{\neq}$, $l \leq w$. The set-based representation conceals the ordering of columns within each block, in addition, multiple entries are discarded, which improves privacy protection. The final set of feature values $\mathbf{P}$ is estimated as the union of all sets of unique vector elements of all blocks, $\mathbf{P} = \bigcup_{b=0}^{B-1} \mathbf{P}_b$, where $|\mathbf{P}| = \sum_{b=0}^{B-1} |\mathbf{P}_b|$, since each block index $b$ is unique.

The proposed representation is alignment-free to a certain extent, since equal columns within certain blocks are mapped to identical feature values. Hence, self-propagating errors caused by an inappropriate alignment of iris-codes are eliminated (radial neighbourhoods persist). The rotation-compensating property of the proposed system comes at the cost of location information of iris-code columns. However, as will be shown in experiments, the proposed representation maintains distinctiveness if $w$ and $h$ are chosen appropriately.

In the first step of the binding process, a secret polynomial $\kappa \in F[X]$ of degree smaller than $k$ is chosen, and the hash $SHA(\kappa)$ is stored. Given a feature set $\mathbf{P}$, $SHA(\kappa)$ is used as seed for a record-specific but public bijection $\sigma : \mathbf{F} \to \mathbf{F}$, which is applied to re-map the elements of $\mathbf{P}$, $\hat{\mathbf{P}} = \sigma(\mathbf{P}) = \{\sigma(v)|v \in \mathbf{P}\}$. This step is performed in order to prevent from the attack proposed in [62], which is based on the extended Euclidean algorithm. Let $V(X)$ and $W(X)$ be two related vault records protecting the feature sets $\mathbf{P}$ and $\mathbf{P}'$, respectively, unlinkability can be attacked efficiently and effectively, provided that

$$|\mathbf{P} \cap \mathbf{P}'| \geq (\max(|\mathbf{P}|, |\mathbf{P}'|) + k)/2. \tag{14.2}$$

In [62] it is also shown that the probability of (14.2) can be destroyed by applying the above-mentioned re-mapping of feature elements. Note that, the employment of these public maps does not affect the operational performance of the system. Moreover, as $\sigma$ is generated based on $SHA(\kappa)$, no additional data storage is required. Due to the assumed randomness of two maps $\sigma$ and $\sigma'$, the corresponding sets $\sigma(\mathbf{P})$ and $\sigma'(\mathbf{P}')$ are random and, based on the definition of the *hyper-geometric distribution*, the probability that with these sets (14.2) is fulfilled is equal to

$$1 - \binom{\rho}{|\mathbf{P}|}^{-1} \sum_{j=0}^{\omega_0 - 1} \binom{|\mathbf{P}'|}{j}\binom{\rho - |\mathbf{P}'|}{|\mathbf{P}| - j}, \tag{14.3}$$

where $\omega_0 = \lceil(|\mathbf{P}| + k)/2\rceil$, $\rho = 2^{h + \log_2 B}$, and w.l.o.g. $|\mathbf{P}| \geq |\mathbf{P}'|$.

The next step is performed based on the improved fuzzy vault scheme [14], which improves the original construction in a way that significantly more compact records are generated. In the improved fuzzy vault scheme genuine and chaff feature elements are encoded by a monic polynomial of degree $t = |\hat{\mathbf{P}}|$. The features in $\hat{\mathbf{P}}$, interpreted as elements of a finite field $\mathbf{F}$ where $|\mathbf{F}| = \rho$, are bound to the secret polynomial $\kappa$ by computing $V(X) = \kappa(X) + \prod_{v \in \hat{\mathbf{P}}} (X - v)$, such that the pair $(V(X), SHA(\kappa))$ builds our final record.

All elements of $\hat{\mathbf{P}}$ represent columns consisting of exactly $h + \log_2 B$ bits. The size of the vault increases with $t$ and is upper bounded by the width of processed iris-codes, i.e., $|V(X)| \leq W(h + \log_2 B)$ bits. Further, the size of the vault, which might leak information about the protected iris-code, can be obscured by appending an additional zero-bit to each element in $\hat{\mathbf{P}}$ and adding random values to $\hat{\mathbf{P}}$, for which this additional bit is set to 1, until a desired maximum vault size is reached.

It is important to note that the correlation attack [63,64], which is a special linkage attack, cannot be applied due to the fact that the improved fuzzy vault scheme encodes a maximal number of chaff points. In such a way no correlation can be exploited between feature sets protected by two (or more) related instances of the applied fuzzy vault records.

At the time of authentication, an unlocking set $\mathbf{U} \subset \mathbf{F} \times \mathbf{F}$ of size $u$ is built containing $\omega$ pairs being interpolated by the polynomial $\kappa$. For decoding a Guruswami–Sudan algorithm [65] is used. This class of algorithms can potentially recover $\kappa$ from $\mathbf{U}$ provided that $\omega > \sqrt{u \cdot (k - 1)}$. The correctness of $\kappa$ is finally verified using the hash value $SHA(\kappa)$.

## 14.3.2   Bin-combo for iris-codes

The aforementioned bin-combo scheme [49] comprises two main steps. In the first step a shifting vector $\mathbf{s}_j \in [0, W), j = 0, \ldots, H - 1$ is employed in order to shift each row $\mathbf{r}_j \in \{0, 1\}^W$ of $\mathbf{I}$. Let $g(\mathbf{r}_j, s)$ denote a row shifted by $s$ bits. Each row $\mathbf{r}_j$ is shifted according to the defined shifting vector resulting in a shifted row $\hat{\mathbf{r}}_j$,

$$\hat{\mathbf{r}}_j = g(\mathbf{r}_j, \mathbf{s}_j). \tag{14.4}$$

The shifting process is performed circularly, similar to the shifting in the alignment process of a generic iris recognition scheme. For a shift of $s$ bits, $\mathbf{r}_{jl}$, i.e., bit at position $l$, is shifted to position $l + \mathbf{s}_j \mod W$. Note that, a left shift of $s$ bits is equal to a right shift of $W - s$.

This circular shifting procedure does not obscure the neighbourhood of iris-code bits within each row. Hence, if a pair of iris-codes is shifted by the same shifting vector, the dissimilarity score between them will be the same as in the case where no shifting is applied. In other words, proper iris-code alignment can still be performed using a generic iris biometric comparator, which compensates for eye rotations by estimating Hamming distance-based scores at various shifting positions.

In the second step randomly chosen pairs of rows are combined by XORing them. We draw $H/2$ pairs without returning from the total number of $H$ rows. The $k$-th row, $k = 0, \ldots, H - 1$, of the bin-combo cancelable template $\mathbf{B}$ of size $w \times H/2$ is estimated as

$$\mathbf{B}_k = \hat{\mathbf{r}}_p \oplus \hat{\mathbf{r}}_q, \tag{14.5}$$

where $p$ and $q$ are two randomly drawn row indexes (without returning), $p, q \in \{0, \ldots, H - 1\}$. The size of the key $\kappa$ (in bits), which defines the shifting vector as well as the pairing of shifted rows is estimated as,

$$|\kappa| = H \times (\log_2 (W) + \log_2 (H)). \tag{14.6}$$

That is, $|\kappa|$ bits are necessary to describe one out of $H \times W \times H!/2^{H/2}$ possible combinations of shifting and pairing of rows. In [49] it is noted that a single row can conceivably be used in more than one combination.

## 14.4   Experimental evaluations

Experiments are carried out on all left eye images of the CASIAv4-Interval iris database [8], which consists of good-quality NIR iris images of size $320 \times 280$ pixels. Note that, low intra-class variability at high inter-class variability is considered a fundamental premise for biometric template protection schemes, which can only be achieved in case biometric traits are acquired under favourable environmental conditions. For a number of 198 subjects and 1,332 images we perform all 4,454 genuine comparisons and 19,503 impostor comparisons using the first image of each subject. Sample iris images of the employed database are depicted in Figure 14.1.

At pre-processing the iris of a given sample image is detected, un-wrapped to an enhanced rectangular texture of $512 \times 64$ pixel, shown in Figure 14.3(a)–(c) applying the weighted adaptive Hough algorithm proposed in [66] and contrast enhancement. In the feature extraction stage two different iris recognition algorithms are employed where normalized enhanced iris textures are divided into stripes to obtain 10 one-dimensional signals, each one averaged from the pixels of 5 adjacent rows (the upper $512 \times 50$ rows are analysed). The first feature extraction method follows the Daugman-like 1D Log Gabor feature extraction algorithm of Masek [67] (LG) and the second follows the algorithm proposed by Ma *et al.* [68] (QSW) based on a quadratic spline wavelet. Both feature extraction techniques generate iris-codes of $512 \times 20 = 10,240$ bit. Sample iris-codes generated by both feature extraction methods are shown in Figure 14.3(d) and (e). For further details on the employed feature extraction algorithms the reader is referred to [69].

Biometric performance is evaluated in terms of genuine match rate (GMR), false non-match rate (FNMR), false match rate (FMR), as well as equal error rate (EER) [70]. In addition, the security level provided by both template protection schemes will be analysed.

## 14.4.1   Performance evaluation

The unprotected baseline systems, which perform a Hamming distance-based comparison of iris-codes at $\pm 8$ shifting positions, obtain EERs of 0.817% and 0.783% for the LG and the QSW feature extraction method, respectively. At an FMR of 0.01% LG feature extractor achieves a GMRs of 98.26% and the QSW feature extractor achieves a GMRs of 98.92%. In Figure 14.6 corresponding receiver operation characteristic (ROC) curves are plotted.

Any fuzzy vault can be attacked by a brute-force attack, where the attacker repeatedly samples $k$ points from the vault and tries to interpolate the secret polynomial from these. In contrast, in a false-accept attack biometric authentications are simulated employing features of randomly chosen (real) iris-codes. The success probability of this attack is equal to the FMR a system is operated at. We report the security of the presented iris-based fuzzy vault in terms of false-accept security (FAS) [71]. Estimating high security levels assumes sharp estimations of FMRs when they are close to zero. However, the FMR can only be estimated down to the magnitude of $1/N$, where $N$ is the number of impostor verifications performed in the evaluation. Since the verification of our implementation is probabilistic as soon as the unlocking set contains more than $k$ points, a heuristic estimation of FMRs that are much smaller than $1/N$ is possible. For each single impostor verification, we compute the success probability based on the size of the unlocking set and the number of correct points contained, and, finally, we estimate the FMR as the mean over all verifications [71].

For both feature extractors obtained GMRs and FMRs as well as the resulting FAS for different parameter settings are listed in Table 14.3. We observe that biometric performance of the iris-based fuzzy vaults slightly drops compared to the corresponding baseline systems. At FMRs of approximately 0.01%, for both feature extraction techniques, iris-based vaults achieve GMRs above 93%. We further observe a clear trade-off between biometric performance and FAS. At practical recognition accuracy provided FAS values tend to be only slightly above 12 bits. For higher security levels

*Table 14.3* *Biometric performance in terms of GMR and FMR (in %) and privacy protection in terms of FAS (in bits) for different parameter settings of the iris-based fuzzy vault using both feature extractors*

| | System | | | 1D Log-Gabor | | | Quadratic spline wavelet | | |
|---|---|---|---|---|---|---|---|---|---|
| *h* | *w* | *k* | *n* | GMR | FMR | FAS | GMR | FMR | FAS |
| 8 | 16 | 1 | $2^{13}$ | 99.74 | 66.081 | 4.257 | 99.75 | 72.740 | 4.322 |
| 8 | 16 | 2 | $2^{13}$ | 97.12 | 0.3354 | 8.466 | 97.17 | 0.2762 | 8.608 |
| 8 | 16 | 3 | $2^{13}$ | 94.14 | 0.0115 | 12.62 | 94.36 | 0.0025 | 12.85 |
| 8 | 16 | 4 | $2^{13}$ | 90.58 | 0.0012 | 16.74 | 91.17 | 0 | 17.07 |
| 8 | 16 | 6 | $2^{13}$ | 82.37 | 0 | 24.85 | 84.02 | 0 | 25.38 |
| 8 | 16 | 8 | $2^{13}$ | 73.86 | 0 | 32.78 | 75.89 | 0 | 33.54 |
| 8 | 32 | 1 | $2^{12}$ | 99.95 | 98.469 | 3.397 | 99.97 | 99.226 | 3.439 |
| 8 | 32 | 2 | $2^{12}$ | 99.26 | 24.483 | 6.769 | 99.44 | 29.712 | 6.856 |
| 8 | 32 | 3 | $2^{12}$ | 98.29 | 3.9015 | 10.11 | 98.64 | 4.8949 | 10.25 |
| 8 | 32 | 4 | $2^{12}$ | 96.88 | 0.5847 | 13.43 | 97.38 | 0.7492 | 13.62 |
| 8 | 32 | 6 | $2^{12}$ | 92.33 | 0.0102 | 20.00 | 94.01 | 0.0167 | 20.30 |
| 8 | 32 | 8 | $2^{12}$ | 85.63 | 0 | 26.45 | 88.65 | 0 | 26.88 |
| 9 | 16 | 1 | $2^{14}$ | 99.17 | 25.167 | 4.928 | 99.20 | 25.987 | 5.045 |
| 9 | 16 | 2 | $2^{14}$ | 93.12 | 0.0115 | 9.781 | 93.27 | 0 | 10.03 |
| 9 | 16 | 3 | $2^{14}$ | 86.76 | 0 | 14.56 | 87.55 | 0 | 14.96 |
| 9 | 16 | 4 | $2^{14}$ | 80.06 | 0 | 19.28 | 81.56 | 0 | 19.84 |
| 9 | 16 | 6 | $2^{14}$ | 68.21 | 0 | 28.52 | 69.36 | 0 | 29.44 |
| 9 | 16 | 8 | $2^{14}$ | 57.31 | 0 | 37.48 | 58.11 | 0 | 38.80 |
| 9 | 32 | 1 | $2^{13}$ | 99.87 | 81.765 | 4.067 | 99.91 | 85.163 | 4.153 |
| 9 | 32 | 2 | $2^{13}$ | 97.29 | 1.3005 | 8.093 | 97.86 | 1.1681 | 8.270 |
| 9 | 32 | 3 | $2^{13}$ | 93.82 | 0.0462 | 12.07 | 94.81 | 0.0334 | 12.35 |
| 9 | 32 | 4 | $2^{13}$ | 89.12 | 0 | 16.02 | 91.06 | 0.0025 | 16.39 |
| 9 | 32 | 6 | $2^{13}$ | 78.63 | 0 | 23.78 | 81.41 | 0 | 24.38 |
| 9 | 32 | 8 | $2^{13}$ | 67.98 | 0 | 31.36 | 71.12 | 0 | 32.20 |
| 10 | 16 | 1 | $2^{15}$ | 98.29 | 5.2123 | 5.617 | 98.25 | 4.2626 | 5.784 |
| 10 | 16 | 2 | $2^{15}$ | 85.21 | 0.0012 | 11.11 | 85.34 | 0 | 11.47 |
| 10 | 16 | 3 | $2^{15}$ | 74.77 | 0 | 16.51 | 75.27 | 0 | 17.08 |
| 10 | 16 | 4 | $2^{15}$ | 65.46 | 0 | 21.82 | 65.80 | 0 | 22.62 |
| 10 | 16 | 6 | $2^{15}$ | 50.44 | 0 | 32.17 | 50.00 | 0 | 33.47 |
| 10 | 16 | 8 | $2^{15}$ | 39.42 | 0 | 42.16 | 38.52 | 0 | 44.01 |
| 10 | 32 | 1 | $2^{14}$ | 99.58 | 37.674 | 4.757 | 99.55 | 39.335 | 4.883 |
| 10 | 32 | 2 | $2^{14}$ | 91.77 | 0.0231 | 9.450 | 92.54 | 0.0089 | 9.714 |
| 10 | 32 | 3 | $2^{14}$ | 83.18 | 0.0012 | 14.08 | 84.89 | 0 | 14.49 |
| 10 | 32 | 4 | $2^{14}$ | 74.87 | 0 | 18.65 | 76.43 | 0 | 19.21 |
| 10 | 32 | 6 | $2^{14}$ | 59.61 | 0 | 27.60 | 60.80 | 0 | 28.50 |
| 10 | 32 | 8 | $2^{14}$ | 47.25 | 0 | 36.27 | 48.32 | 0 | 37.55 |

we observe a significant drop in biometric performance. The interrelation between FNMR, FMR and privacy protection, i.e., degree of applied polynomial, for different parameter settings for both feature extractors is shown in Figure 14.4.

For the cancelable iris biometric systems a sample of protected iris-codes for both feature extraction algorithms are shown in Figure 14.5. Note that, due to the XORing of pairs of shifted iris-code rows, the resulting cancelable iris-codes are of size $W \times H/2$. For the bin-combo scheme a significant decrease in biometric performance is observed, similar effects have been reported in [49]. For estimating the biometric performance for ten different randomly chosen keys average EERs of 4.411% and 4.403% are obtained for the LG and QSW feature extractor, respectively. At an FMR of 0.01% an average GMR of 90.13% is achieved for using the LG method and an average GMR of 89.42% is achieved for using the QSW method. A comparison of ROC curves between baseline systems and corresponding cancelable systems is depicted in Figure 14.6. The relatively large key space of the scheme, $|\kappa| \simeq 280$ bit is expected to prevent from brute force attacks.

In order to assess whether presented bin-combo schemes meet the unlinkability requirement, three types of score distributions are analysed: (1) *Same image*: scores computed from iris-codes extracted from a single iris image of the same subject using different keys; (2) *Same iris*: scores computed from iris-codes extracted from different iris images of the same subject using different keys. (3) *Different iris*: scores yielded by iris-codes generated from iris images of different subjects using different keys. In Figure 14.7 obtained score distributions are illustrated for cancelable systems using both feature extraction techniques. As can be seen, the three types of score distributions are almost identical. Hence, for a potential impostor it is not feasible to link protected templates of a single subject if these are transformed with different keys.

## 14.4.2    Discussion

Protecting iris-codes in a fuzzy vault scheme yields only a slight drop in biometric performance. It is important to note that the presented adaptation to the improved fuzzy vault scheme [14] achieves unlinkability. The correlation attack in [63,64], which is a special cross-matching attack, cannot be applied due to the fact that a maximal number of chaff points is encoded. While obtained results coincide with previously proposed schemes, e.g., [33,40], a maximum security of 40 bits may be considered inadequate from the cryptographic point of view.

Given the large number of possible keys, the bin-combo scheme might provide a higher level of security. However, on the one hand, for the proposed transform (combination of shifting and pairing of iris-code rows), similar keys are expected to yield similar protected templates, which reduces the overall key space. On the other hand, bits in iris-codes are not mutually independent such that protected iris-codes underlie a distinct structure, see Figure 14.5, which might be exploited by a potential attacker. In contrast to the fuzzy vault scheme a more severe drop in recognition accuracy is observed, which might be avoided by using iris-code rows more than once, as suggested in [49]. However, a repeated use of rows might evolve security implications requiring further analysis.

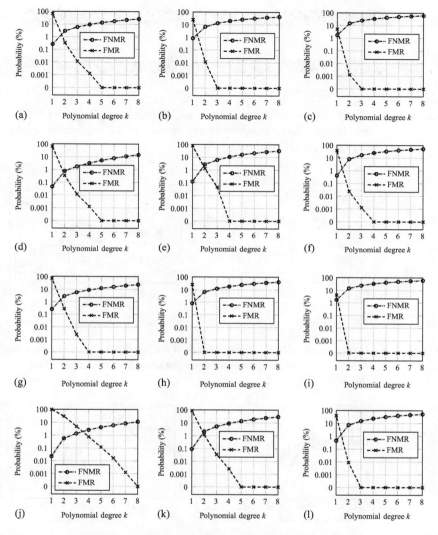

*Figure 14.4 Biometric performance in terms of FNMR and FMR (in %) in relation to the polynomial degree k for different parameter settings of the iris-based fuzzy vault using both feature extractors: (a) LG: w = 16, h = 8, (b) LG: w = 16, h = 9, (c) LG: w = 16, h = 10, (d) LG: w = 32, h = 8, (e) LG: w = 32, h = 9, (f) LG: w = 32, h = 10, (g) QSW: w = 16, h = 8, (h) QSW: w = 16, h = 9, (i) QSW: w = 16, h = 10, (j) QSW: w = 32, h = 8, (k) QSW: w = 32, h = 9, and (l) QSW: w = 32, h = 10*

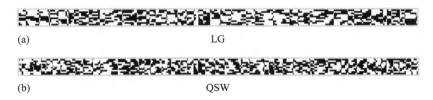

Figure 14.5   *Protected iris-codes for a single randomly generated shifting (LG) and permutation parameter setting applied to the original iris-codes of Figure 14.3(d) and (e) (QSW)*

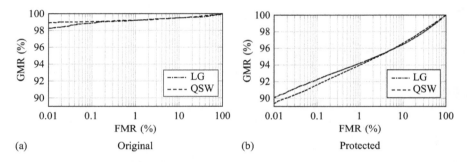

Figure 14.6   *Comparison of ROC curves obtained for (a) the original systems and (b) the cancelable systems for both feature extraction algorithms*

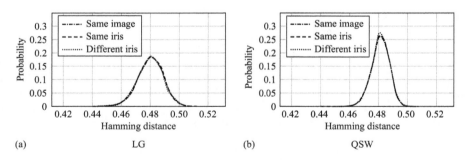

Figure 14.7   *Score distributions of which the overlap indicates the degree of unlinkability for the cancelable systems for both feature extraction algorithms (LG and QSW)*

## 14.5   Summary and research directions

Most concerns against the common use of biometrics arise from the storage and misuse of biometric data. Technologies of biometric template protection address this issue offering solutions to privacy-preserving biometric authentication. In past decades numerous template protection techniques have been proposed in the literature with the objective of ensuring requirements of biometric information protection defined in ISO/IEC IS 24745 [10]. While some proposed techniques are theoretically sound, these seldom guarantee the desired irreversibility and unlinkability properties without significantly degrading biometric performance [3]. One fundamental challenge regarding template protection represents the issue of feature alignment. Biometric signals are obscured within template protection schemes, i.e., alignment of obscured templates is highly non-trivial [4]. This motivates the construction of feature extraction methods which are capable of generating discriminative alignment-free feature vector representations. Further, template protection schemes are designed to operate on a certain form of input, e.g., point-sets or binary features. Hence, incompatibility issues arise when the type of intended biometric features does not match the acceptable input type of a template protection system [61]. Ideally, biometric features can be transformed to the required feature-type in an appropriate manner. However, in most cases information loss is inevitable in case feature type transformations are applied. Numerous template protection schemes, in particular biometric cryptosystems, require complex (computationally demanding) authentication procedures, which makes them unusable for biometric identification systems. Within cancelable biometric systems local neighbourhoods of feature elements might be obscured, which also causes a loss of information. Finally, result reporting remains an issue since unrealistic preconditions distort performance rates. Moreover, a lack of standardized metrics to measure the irreversibility and the unlinkability provided by a template protection scheme hampers a fair comparison of proposed approaches [3]. These issues will be addressed in the forthcoming ISO/IEC IS 30136 [72] on "Performance testing of template protection schemes", which is currently under development. However, it is important to note that it is impossible to guarantee that a potential attacker cannot come up with a better (system-specific) attack than what is known to the system designers [3].

Focusing on iris recognition, proposed template protection systems have been found to achieve practical recognition accuracy, even if these schemes cause a significant drop in biometric performance, see Section 14.2. Depending on the design of an iris-based template protection system, an appropriate alignment of iris biometric feature vectors can be achieved, e.g., in [15,49]. Nevertheless, it is generally conceded that there exists a clear trade-off between biometric performance and privacy protection, which is highlighted by conducted experiments, see Section 14.4. In the iris-based fuzzy vault the security level is rather low at acceptable recognition accuracy. However, it is worth noting that biometric cryptosystems eliminate the key management problem, which represents a major issue in practical cryptosystems. In contrast, the presented cancelable iris biometric system is expected to provide a higher level of security, while the observed drop in recognition accuracy is more severe. Future research might focus on the above mentioned open issues. Furthermore, multi-biometric template protection schemes, e.g., based on left and right iris [57], might improve both, recognition accuracy and provided security, compared to the use of a single biometric instance. With respect to template protection feature level has been identified as the only suitable level of fusion. Since score and decision level fusion would require a separate storage of protected templates, which would make the system vulnerable to parallelized attacks. Finally, the constant and tremendous increase in computational power might even make homomorphic encryption methods applicable in the upcoming future [73].

# References

[1]   M. J. Burge and K. Bowyer. *Handbook of Iris Recognition* (2nd edition). Springer-Verlag Inc., New York, 2016.

[2]   J. Daugman. "How iris recognition works," *IEEE Transactions on Circuits and Systems for Video Technology*, 14(1):21–30, 2004.

[3]   K. Nandakumar and A. K. Jain. "Biometric template protection: Bridging the performance gap between theory and practice," *IEEE Signal Processing Magazine – Special Issue on Biometric Security and Privacy*, pp. 1–12, 2015.

[4]   C. Rathgeb and A. Uhl. "A survey on biometric cryptosystems and cancelable biometrics," *EURASIP Journal on Information Security*, 2011(3):1–25, 2011.

[5]   J. Galbally, A. Ross, M. Gomez-Barrero, J. Fiérrez, and J. Ortega-Garcia. "Iris image reconstruction from binary templates: An efficient probabilistic approach based on genetic algorithms," *Computer Vision and Image Understanding*, 117(10):1512–1525, 2013.

[6]   S. Venugopalan and M. Savvides. "How to generate spoofed irises from an iris code template," *IEEE Transactions on Information Forensics and Security*, 6(2):385–395, 2011.

[7]   S. Marcel, M. Nixon, and S. Z. Li. *Handbook of Biometric Anti-Spoofing*. Springer-Verlag Inc., New York, 2014.

[8]   Chinese Academy of Sciences, Institute of Automation. CASIA Iris Image Database V4.0 – Interval, 2016.

[9]   European Parliament. Regulation (EU) 2016/679 of the European Parliament and of the Council on the protection of natural persons with regard to the processing of personal data and on the free movement of such data, and repealing Directive 95/46/EC (General Data Protection Regulation). April 2016.

[10]  ISO/IEC JTC1 SC27 Security Techniques. *ISO/IEC 24745:2011. Information Technology – Security Techniques – Biometric Information Protection*. International Organization for Standardization, 2011.

[11]  U. Uludag, S. Pankanti, S. Prabhakar, and A. K. Jain. "Biometric cryptosystems: issues and challenges," *Proceedings of the IEEE*, 92(6):948–960, 2004.

[12]  V. M. Patel, N. K. Ratha, and R. Chellappa. "Cancelable biometrics: A review," *IEEE Signal Processing Magazine*, 32(5):54–65, 2015.

[13]  N. Ratha, J. Connell, and R. Bolle. "Enhancing security and privacy in biometrics-based authentication systems," *IBM Systems Journal*, 40(3): 614–634, 2001.

[14]  Y. Dodis, R. Ostrovsky, L. Reyzin, and A. Smith. "Fuzzy extractors: How to generate strong keys from biometrics and other noisy data," *SIAM Journal of Computation*, 38(1):97–139, 2008.

[15]  F. Hao, R. Anderson, and J. Daugman. "Combining cryptography with biometrics effectively," *IEEE Transactions on Computers*, 55(9):1081–1088, 2006.

[16]  A. Juels and M. Wattenberg. "A fuzzy commitment scheme," in *Proceedings of the Sixth ACM Conference on Computer and Communications Security*, pp. 28–36, 1999.

[17]  J. Bringer, H. Chabanne, G. Cohen, B. Kindarji, and G. Zemor. "Theoretical and practical boundaries of binary secure sketches," *IEEE Transactions on Information Forensics and Security*, 3(4):673–683, 2008.

[18]  C. Rathgeb and A. Uhl. "Systematic construction of iris-based fuzzy commitment schemes," in *Proceedings of the International Conference on Biometrics (ICB)*, pp. 947–956, 2009.

[19]  C. Rathgeb and A. Uhl. "Context-based biometric key-generation for iris," *IET Computer Vision*, 5(6):389–397, 2011.

[20]  E. R. C. Kelkboom, J. Breebaart, T. A. M. Kevenaar, I. Buhan, and R. N. J. Veldhuis. "Preventing the decodability attack based cross-matching in a fuzzy commitment scheme," *IEEE Transactions on Information Forensics and Security*, 6(1):107–121, 2011.

[21]  E. Maiorana, P. Campisi, and A. Neri. "Iris template protection using a digital modulation paradigm," in *Proceedings of the International Conference on Acoustics, Speech and Signal Processing (ICASSP)*, pp. 3759–3763, 2014.

[22]  C. Rathgeb, A. Uhl, and P. Wild. "Reliability-balanced feature level fusion for fuzzy commitment scheme," in *Proceedings of the International Joint Conference on Biometrics (IJCB)*, pp. 1–7, 2011.

[23]  L. Zhang, Z. Sun, T. Tan, and S. Hu. "Robust biometric key extraction based on iris cryptosystem," in *Proceedings of the International Conference on Biometrics (ICB)*, pp. 1060–1070, 2009.

[24]  A. Cavoukian and A. Stoianov. "Biometric encryption: The new breed of untraceable biometrics," in *Biometrics: Fundamentals, Theory, and Systems*. Wiley, New York, NY, 2009.

[25]  A. Stoianov, T. Kevenaar, and M. van der Veen. "Security issues of biometric encryption," in *Proceedings of the Toronto International Conference on Science and Technology for Humanity*, pp. 34–39, 2009.

[26]  C. Rathgeb and A. Uhl. "Statistical attack against fuzzy commitment scheme," *IET Biometrics*, 1(2):94–104, 2012.

[27]  T. Ignatenko and F. M. J. Willems. "Information leakage in fuzzy commitment schemes," *IEEE Transactions on Information Forensics and Security*, 5(2):337–348, 2010.

[28]  P. Failla, Y. Sutcu, and M. Barni. "eSketch: A privacy-preserving fuzzy commitment scheme for authentication using encrypted biometrics," in *Proceedings of the 12th Workshop on Multimedia and Security*, pp. 241–246, 2010.

[29]  I. Buhan-Dulman, J. G. Merchan, and E. Kelkboom. "Efficient strategies for playing the indistinguishability game for fuzzy sketches," in *Proceedings of IEEE Workshop on Information Forensics and Security (WIFS)*, 2010.

[30]   K. Simoens, P. Tuyls, and B. Preneel. "Privacy weaknesses in biometric sketches," in *Proceedings of the IEEE Symposium on Security and Privacy*, pp. 188–203. Piscataway, NJ: IEEE, 2009.

[31]   B. Tams. "Decodability attack against the fuzzy commitment scheme with public feature transforms," *CoRR*, abs/1406.1154, 2014.

[32]   Y. J. Lee, K. Bae, S. J. Lee, K. R. Park, and J. Kim. "Biometric key binding: Fuzzy vault based on iris images," in *Proceedings of the Second International Conference on Biometrics*, pp. 800–808, 2007.

[33]   Y. J. Lee, K. R. Park, S. J. Lee, K. Bae, and J. Kim. "A new method for generating an invariant iris private key based on the fuzzy vault system," *Transactions on Systems, Man, and Cybernetics, Part B: Cybernetics*, 38(5):1302–1313, 2008.

[34]   A. Juels and M. Sudan. "A fuzzy vault scheme," in *Proceedings of the IEEE International Symposium on Information Theory*, p. 408, 2002.

[35]   A. Juels and M. Sudan. "A fuzzy vault scheme," *Design Codes Cryptography*, 38(2):237–257, 2006.

[36]   T. C. Clancy, N. Kiyavash, and D. J. Lin. "Secure smartcard-based fingerprint authentication," in *Proceedings of the ACM SIGMM Workshop on Biometrics Methods and Applications (WBMA)*, pp. 45–52, 2003.

[37]   K. Nandakumar, A. K. Jain, and S. Pankanti. "Fingerprint-based fuzzy vault: Implementation and performance," *IEEE Transactions on Information Forensics and Security*, 2(4):744–757, 2007.

[38]   E. Reddy and I. Babu. "Performance of iris based hard fuzzy vault," *IJCSNS International Journal of Computer Science and Network Security*, 8(1): 297–304, 2008.

[39]   E. Reddy and I. Ramesh Babu. "Authentication using fuzzy vault based on iris textures," in *Second Asia International Conference on Modeling Simulation AICMS'08*, pp. 361–368, 2008.

[40]   K. Nandakumar and A. Jain. "Multibiometric template security using fuzzy vault," in *Second International Conference on Biometrics: Theory, Applications and Systems (BTAS'08)*, pp. 1–6, 2008.

[41]   M. Fouad, A. El Saddik, J. Zhao, and E. Petriu. "A fuzzy vault implementation for securing revocable iris templates," in *International Systems Conference (SysCon'11)*, pp. 491–494, 2011.

[42]   R. Á. Mariño, F. H. Álvarez, and L. H. Encinas. "A crypto-biometric scheme based on iris-templates with fuzzy extractors," *Information Sciences*, 195:91–102, 2012.

[43]   S. Sowkarthika and N. Radha. "Securing iris and fingerprint templates using fuzzy vault and symmetric algorithm," in *International Conference on Intelligent Systems and Control (ISCO)*, pp. 189–193, 2013.

[44]   G. Davida, Y. Frankel, and B. Matt. "On enabling secure applications through off-line biometric identification," in *Proceedings of the IEEE Symposium on Security and Privacy*, pp. 148–157, 1998.

[45]   G. Davida, Y. Frankel, and B. Matt. "On the relation of error correction and cryptography to an off line biometric based identification scheme," in *Proceedings of the Workshop on Coding and Cryptography*, pp. 129–138, 1999.

[46]   C. Rathgeb and A. Uhl. "Privacy preserving key generation for iris biometrics," in *Proceedings of the International Conference on Communications and Multimedia Security (CMS)*, pages 191–200, 2010.

[47]   C. Rathgeb and A. Uhl. "An iris-based interval-mapping scheme for biometric key generation," in *Proceedings of the Sixth International Symposium on Image and Signal Processing and Analysis*, pp. 511–516. Piscataway, NJ: IEEE, 2009.

[48]   C. Rathgeb and A. Uhl. "Iris-biometric hash generation for biometric database indexing," in *Proceedings of the International Conference on Pattern Recognition (ICPR)*, pp. 2848–2851. Piscataway, NJ: IEEE, 2010.

[49]   J. Zuo, N. K. Ratha, and J. H. Connel. "Cancelable iris biometric," *Proceedings of the 19th International Conference on Pattern Recognition*, pp. 1–4, 2008.

[50]   J. Hämmerle-Uhl, E. Pschernig, and A. Uhl. "Cancelable iris biometrics using block re-mapping and image warping," in *Proceedings of the 12th International Information Security Conference*, pp. 135–142, 2009.

[51]   O. Ouda, N. Tsumura, and T. Nakaguchi. "Tokenless cancelable biometrics scheme for protecting iris codes," in *Proceedings of the International Conference on Pattern Recognition (ICPR)*, pp. 882–885, 2010.

[52]   C. Rathgeb, F. Breitinger, and C. Busch. "Alignment-free cancelable iris biometric templates based on adaptive bloom filters," in *Proceedings of the Sixth IAPR International Conference on Biometrics (ICB'13)*, pp. 1–8, 2013.

[53]   A. Kong, K.-H. Cheunga, D. Zhanga, M. Kamelb, and J. Youa. "An analysis of BioHashing and its variants," *Pattern Recognition*, 39(7):1359–1368, 2006.

[54]   J. K. Pillai, V. M. Patel, R. Chellappa, and N. K. Ratha. "Secure and robust iris recognition using random projections and sparse representations," *IEEE Transactions on Pattern Analysis and Machine Intelligence*, 33(9):1877–1893, 2011.

[55]   S. C. Chong, A. T. B. Jin, and D. N. C. Ling. "High security iris verification system based on random secret integration," *Computer Vision and Image Understanding*, 102(2):169–177, 2006.

[56]   S. C. Chong, A. T. B. Jin, and D. N. C. Ling. "Iris authentication using privatized advanced correlation filter," in *Proceedings of the International Conference on Biometrics (ICB)*, pp. 382–388, 2006.

[57]   C. Rathgeb and C. Busch. "Cancelable multi-biometrics: Mixing iris-codes based on adaptive bloom filters," *Elsevier Computers and Security*, 42:1–12, 2014.

[58]   M. Braithwaite, U. von Seelen, J. Cambier, *et al.* "Applications-specific biometric template," in *Proceedings of the IEEE Workshop on Automatic Identification Advanced Technologies*, pp. 167–171, 2002.

[59]   J. Hämmerle-Uhl, E. Pschernig, and A. Uhl. "Cancelable iris-templates using key-dependent wavelet transforms," in *International Conference on Biometrics (ICB)*, pp. 1–8, 2013.

[60]   C. Rathgeb and A. Uhl. "Attacking iris recognition: An efficient hill-climbing technique," in *International Conference on Pattern Recognition (ICPR)*, pp. 1217–1220, 2010.

[61]  M.-H. Lim, A. B. J. Teoh, and J. Kim. "Biometric feature-type transformation: Making templates compatible for template protection," *IEEE Signal Processing Magazine*, 32(5), 2015.

[62]  J. Merkle and B. Tams. Security of the improved fuzzy vault scheme in the presence of record multiplicity (full version). *CoRR*, http://arxiv.org/abs/1312.5225abs/1312.5225, 2013.

[63]  A. Kholmatov and B. Yanikoglu. "Realization of correlation attack against the fuzzy vault scheme," in *Proceedings of the SPIE*, vol. 6819, 2008.

[64]  W. J. Scheirer and T. E. Boult. "Cracking fuzzy vaults and biometric encryption," in *Proceedings of the Biometrics Symposium*, pp. 1–6, 2007.

[65]  V. Guruswami and M. Sudan. "Improved decoding of Reed–Solomon and algebraic-geometric codes," *IEEE Transactions on Information Forensics and Security*, 45:1757–1767, 1998.

[66]  A. Uhl and P. Wild. "Weighted adaptive hough and ellipsopolar transforms for real-time iris segmentation," in *Proceedings of the Fifth International Conference on Biometrics*, pp. 1–8, 2012.

[67]  L. Masek. "Recognition of human iris patterns for biometric identification," Master's thesis, University of Western Australia, Crawley, Australia, 2003.

[68]  L. Ma, T. Tan, Y. Wang, and D. Zhang. "Efficient iris recognition by characterizing key local variations," *IEEE Transactions on Image Processing*, 13(6):739–750, 2004.

[69]  C. Rathgeb, A. Uhl, and P. Wild. *Iris Biometrics: From Segmentation to Template Security*. Number 59 in Advances in Information Security. Berlin: Springer, 2012.

[70]  ISO/IEC JTC1 SC37 Biometrics. *ISO/IEC 19795-1:2006. Information Technology – Biometric Performance Testing and Reporting – Part 1: Principles and Framework*. International Organization for Standardization and International Electrotechnical Committee, 2006.

[71]  B. Tams, P. Mihăilescu, and A. Munk. "Security considerations in minutiae-based fuzzy vaults," *IEEE Transactions on Information Forensics and Security*, 10(5):985–998, 2015.

[72]  ISO/IEC JTC1 SC27 Security Techniques. *ISO/IEC CD 30136. Information Technology – Security Techniques – Performance testing of template protection schemes*. International Organization for Standardization, 2016.

[73]  J. Bringer, H. Chabanne, and A. Patey. "Privacy-preserving biometric identification using secure multiparty computation: An overview and recent trends," *IEEE Signal Processing Magazine*, 30(2):42–52, 2013.

*Chapter 15*

# Privacy-preserving distance computation for IrisCodes

*Julien Bringer, Hervé Chabanne,*
*and Constance Morel*

In this chapter, we describe how to perform a secure and efficient IrisCode-based identification. In this use case, the first party has an IrisCode, the second party has one or several IrisCodes and they would like to discover whether the first party's IrisCode is close to at least one IrisCode belonging to the second party without revealing any information about their own IrisCodes to the opposing party. Secure Two-Party Computation (S2PC) protocols are dedicated to this use case because they enable two parties to jointly evaluate a function over their inputs while preserving the privacy of their inputs. In this chapter, we explain how to efficiently use S2PC protocols for secure iris-based identification.

## 15.1 Introduction

As biometric data are sensitive, their use leads to specific security and privacy issues. Two methods are widely applied to protect computations on biometric data: biometric template transformation such as biohashing [1] or cancelable biometric [2,3] and biometric cryptosystems such as fuzzy commitment [4], fuzzy vault [5] or homomorphic encryption. Biometric template transformation consists in transforming the biometric template (the features extracted from the biometric trait) using parameters derived from a password or a key. Only the transformed template is stored, and the comparison is performed directly in the transformed domain. Unfortunately, the security of these systems is often overestimated and many protection schemes have been proved non-secure [6–9]. Due to the vulnerability or the moderate level of protection of biometric template transformation systems and the need to rely on strong cryptographic properties for some applications, we focus on a biometric cryptosystem called S2PC [10] in this chapter. In S2PC protocols, two parties input, respectively, $X$ and $Y$ and they would like to securely and efficiently evaluate a function over these inputs $f(X, Y)$. At the end of the protocol, they obtain the function evaluation without sharing the value of their input with the opposing party.

When working with S2PC protocols, it is paramount to define some security models. Firstly, S2PC protocols have to respect two requirements: *privacy* and *correctness*. A protocol is private if the parties learn their outputs and nothing else. A protocol is correct if the parties obtain the correct outputs. In addition, two main adversarial behaviors are considered in S2PC protocols: *semi-honest* and *malicious* adversaries. Semi-honest adversaries follow the protocol but try to obtain as much information as possible from their view of the protocol: inputs, outputs and all messages they received during the execution. This adversarial behavior is also called honest but curious or passive. Malicious adversaries also aim to obtain as much information as possible and they may follow their own strategy and deviate from the protocol, e.g. modify the sent messages/inputs, abort when they want. This adversarial behavior is also called active.

In this chapter, we focus on a privacy IrisCodes distance computation use case involving two parties: a client $\mathscr{C}$ and a server $\mathscr{S}$. The client $\mathscr{C}$ inputs one IrisCode called the search IrisCode $IC^{sea}$ and the server $\mathscr{S}$ inputs $N$ IrisCodes called the reference IrisCodes $IC_1^{ref}, \ldots, IC_N^{ref}$. They would like to securely and efficiently evaluate whether the search IrisCode is close to one of the reference IrisCodes. To do that, they securely evaluate the distance between the search template and each reference templates and compare these distances to a threshold. If at least one distance is below the threshold, it is an acceptance, otherwise it is a reject. As the biometric data are sensitive, the client $\mathscr{C}$ and the server $\mathscr{S}$ do not want to leak some information about their IrisCodes to the opposing party. If the server $\mathscr{S}$ stores a unique reference template ($N = 1$), it is a biometric authentication scenario, otherwise it is a biometric identification. Figure 15.1 sums up this use case.

In most of this chapter, the IrisCodes are extracted from iris images following the Daugman's approach [11]. Firstly, the iris boundaries are detected and the iris is split into small parts. Then a vector of 256 bytes is obtained by extracting the information of each part of the iris with Gabor filters. Another vector of 256 bytes called the mask is also created. This mask corresponds to reliable bits positions of the iris vector. For instance, a mask bit set to 0 means an erasure due to eyelids, blurs, etc. To sum up, an IrisCode contains two $\ell$-bit vectors $IC = (I, M)$ with $\ell = 2,048$ where $I$ is the iris vector and $M$ is the mask vector. To compare two IrisCodes, a normalized Hamming distance (NHD) is computed $NHD((I^{sea}, M^{sea}), (I^{ref}, M^{ref})) = \frac{HW((I^{sea} \oplus I^{ref}) \wedge (M^{sea} \wedge M^{ref}))}{HW(M^{sea} \wedge M^{ref})}$ where for any bit vector $a$, $HW(a)$ denotes the Hamming weight of $a$. In order to deal with iris orientation variation, the normalized Hamming distances between the search IrisCode $IC^{sea}$ and some rotations of the reference IrisCode $IC^{ref}$ (e.g. $\pm 3$ radii) are computed and then the lowest score is kept to compare it with the threshold $t$. For simplicity in the sequel of this chapter, we will work on pre-aligned IrisCodes. If the IrisCodes are not pre-aligned, we have to increase the database size from $N$ to $7N$ by adding the rotations of the reference IrisCodes into the database.

**Notations.** Given a positive integer $m$, the notation $[x]_m$ denotes the smallest non-negative integer which is equal to $x$ modulo $m$. In addition, $[1 \ldots n]$ represents the integer set $\{1, 2, \ldots, n\}$. Given two vectors $a$ and $b$, $a \parallel b$ denotes the concatenation of the vector $a$ with the vector $b$.

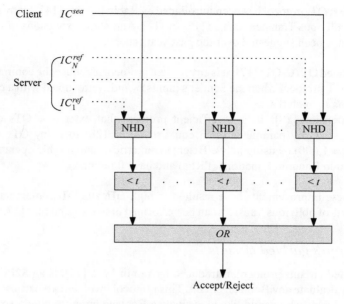

*Figure 15.1   IrisCode identification*

*Figure 15.2   The Oblivious Transfer protocol*

## 15.2   Secure distance computation in the semi-honest model

In this section, we present Yao's garbled circuit [12] and GSHADE [13] which are efficient S2PC protocols widely used in related scientific projects on biometric identification. We then describe how to combine these two protocols in order to obtain a secure and efficient protocol for IrisCodes identification in the semi-honest model.

### 15.2.1   Oblivious transfer

Oblivious Transfer (OT) [14] is the basis of many S2PC protocols such as Yao's garbled circuit and GSHADE protocols. OT protocol is a two-party protocol where one party (the Receiver) inputs a bit $b$ and the other party (the Sender) inputs two strings $x_0$ and $x_1$. At the end of the protocol, the Receiver obtains $x_b$ but learns no information about $x_{1-b}$ and the Sender obtains nothing (i.e. has no clue about $b$). Figure 15.2 sums up this protocol.

The first OT protocol has been introduced by Rabin in 1981 [14]. Many protocols exist for Oblivious Transfer such as [15] or [16]. And some extensions of Oblivious Transfer have been suggested over the past years such as:

- Correlated OT (C-OT) [17]: it is an efficient protocol to reduce the communication of the OT protocol when the Sender's inputs $x_0$ and $x_1$ are linked with a correlated function $f_\Delta$ such that $x_1 = f_\Delta(x_0)$.
- OT extension [18]: it is an efficient protocol that extends $\kappa$ OTs on $\kappa$ bits ($\kappa$ is a security parameter often equal to 80 or 128) to many OTs on $\ell$ bits (e.g. $\ell = 1{,}000$) by using only efficient symmetric cryptographic operations such as Pseudo Random Functions (PRFs) and hash functions.

All these improvements have resulted in high-efficient OT implementations. A few millions of oblivious transfers can be performed in one second (see [17, Table 4]).

## 15.2.2 Yao's garbled circuits

Yao's garbled circuits protocol, introduced by Yao in 1986 [12], is an S2PC protocol to securely evaluate any Boolean circuit. This protocol involves two parties: a Builder and an Evaluator who would like to evaluate a Boolean circuit securely so that their inputs remain secret. Firstly the Builder "garbles" the circuit by replacing all Boolean values in the circuit by random strings called garbled values. For security reasons, he keeps this mapping secret. Then he sends to the Evaluator the garbled circuit and the garbled inputs corresponding to his own inputs. However, since the Evaluator has only the garbled inputs corresponding to the Evaluator's inputs and these garbled values are random, the Evaluator cannot infer from the garbled inputs any information about the Builder's inputs. Then the Evaluator performs some OTs with the Builder in order to obtain the garbled inputs corresponding to his own inputs. Note that the OT's properties prevent the Builder from learning about the Evaluator's inputs. Finally the Evaluator evaluates the garbled circuit with the garbled inputs to obtain a garbled output. During the evaluation of the garbled circuit, the wires mapped with random values prevent the Evaluator from learning about the Builder's inputs and intermediate values. The end of the protocol depends on which party obtains the output. If the Evaluator obtains the output, the Builder has to send to the Evaluator the output mapping and then the Evaluator can deduce the plaintext output from his garbled output. If the Builder obtains the output, the Evaluator has to send to the Builder the garbled output and then the Builder deduces the plaintext output from the output mapping. See [19] for more details on this protocol. Figure 15.3 sums up this protocol. The Builder is the party who inputs $X$ and the Evaluator is the party who inputs $Y$.

Many extensions have been suggested during the last decade in order to dramatically improve the efficiency of this protocol such as the free XOR optimization [12], the point-and-permute technique [20], the gabled row reduction [15,21] and the pipelined circuit evaluation [22]. With the free XOR optimization, only the no XOR gates (e.g. AND and OR gates) impact the complexity of the Yao's garbled circuit

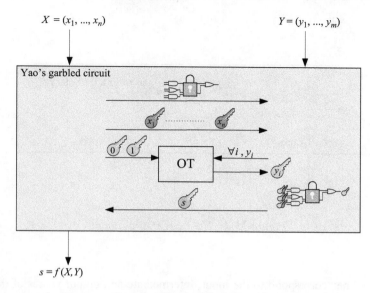

$X = (x_1, ..., x_n)$

$Y = (y_1, ..., y_m)$

$s = f(X, Y)$

*Figure 15.3 The Yao's garbled circuit protocol*

protocol. The Yao's garbled circuit protocol is secure against a malicious Evaluator and a semi-honest Builder.

## 15.2.3 GSHADE in the semi-honest model

GSHADE [13] is a protocol based only on Oblivious Transfers which allows to efficiently and securely evaluate following functions $f(X, Y) = f_X(X) + f_Y(Y) + \sum_{i=1}^{n} f_i(x_i, Y)$ where $X = (x_1, \ldots, x_n) \in \{0, 1\}^n$ is the first party's input and $Y$ is the second party's input. Many usual metrics can be evaluated with GSHADE such as the Hamming distance, the Euclidean distance and the Mahalanobis distance. For instance, to compute a Hamming distance, the functions $f_X, f_Y$ and $f_i$ are defined as followed $f_X(X) = f_Y(Y) = 0$ and $f_i(x_i, Y) = x_i \oplus y_i$.

During a GSHADE protocol, all computations are performed into $\mathbb{Z}_m$ with $m$ an integer such that the codomain of $f$ is included into $\mathbb{Z}_m$. For each first party's input bit $x_i$, an Oblivious Transfer is performed. The first party is the Receiver and inputs $x_i$. The second party is the Sender and inputs $([r_i + f_i(0, Y)]_m, [r_i + f_i(1, Y)]_m)$ where $r_i$ is randomly picked into $\mathbb{Z}_m$. At the end of the Oblivious Transfer, the first party outputs $t_i = [r_i + f_i(x_i, Y)]_m$ and the second party outputs nothing. After the $n$ Oblivious Transfers (one per input bit of the first party), the first party computes $T = [f_X(X) + \sum_{i=1}^{n} t_i]_m$ and the second party computes $R = [-f_Y(Y) + \sum_{i=1}^{n} r_i]_m$. Note that $[T - R]_m$ is equal to $f(X, Y)$. Consequently at the end of the protocol, the parties obtain a sharing of $f(X, Y)$. Figure 15.4 sums up the GSHADE protocol.

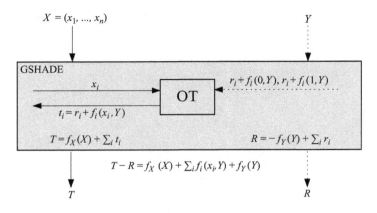

*Figure 15.4   The GSHADE protocol*

The plain lines correspond to the input, intermediate and output values of the first party and the dotted lines to the ones of the second party.

The complexity of the GSHADE protocol is equal to the complexity of performing $n$ OTs. Since the OT protocol is highly efficient, the GSHADE protocol is also highly efficient. In addition, GSHADE can be efficiently extended in order to simultaneously evaluate $N$ functions $f(X, Y_1), \ldots, f(X, Y_N)$ where the first party's input $X$ is the same for all the function evaluations and $Y_1, \ldots, Y_N$ are the second party's inputs. To do that, the number $n$ of OTs remains unchanged and the different masked intermediate values are concatenated in the OTs. In each OT, the Sender inputs $([r_{i,1} + f_i(0, Y_1)]_m \parallel \ldots \parallel [r_{i,N} + f_i(0, Y_N)]_m, [r_{i,1} + f_i(1, Y_1)]_m \parallel \ldots \parallel [r_{i,N} + f_i(1, Y_N)]_m)$ where $r_{i,1}, \ldots, r_{i,N}$ are randomly picked into $\mathbb{Z}_m$ and the Receiver inputs $x_i$ to obtain $t_i = ([r_{i,1} + f_i(x_i, Y_1)]_m \parallel \ldots \parallel [r_{i,N} + f_i(x_i, Y_N)]_m)$. Then the Receiver extracts for $j$ in $[1 \ldots N]$, $t_{i,j} = [r_{i,j} + f_i(x_i, Y_j)]_m$ from $t_{i,j}$. Finally the first party computes and outputs $T_j = [f_X(X) + \sum_{i=1}^{n} t_{i,j}]_m$ for $j$ in $[1 \ldots N]$ and the second party computes and outputs $R_j = [-f_Y(Y_j) + \sum_{i=1}^{n} r_{i,j}]_m$ for $j$ in $[1 \ldots N]$. Noted that for each $j$ in $[1 \ldots N]$, $T_j - R_j = f(X, Y_j)$. This extension is called 1-vs-$N$ GSHADE extension. Thanks to OT's properties, the GSHADE protocol and its extensions are secure against semi-honest adversaries.

## 15.2.4   *Privacy-preserving distance computation for IrisCodes in the semi-honest model*

### 15.2.4.1   Authentication in the semi-honest setting

In this section, we will explain in details how to combine the GSHADE protocol and the Yao's garbled circuit protocol in order to obtain a secure and efficient IrisCodes authentication in the semi-honest setting. As explained in [13, Section 4.2], we can securely and efficiently evaluate with the GSHADE protocol the numerator and denominator of the normalized Hamming distance. To do that, we need to adopt the

Table 15.1  Definition of $f_i^{num}$ for computing the numerator of the normalized Hamming distance $(i \in [1 \ldots \ell])$

| $I^{ref}[i]$ | 0 | | 1 | |
|---|---|---|---|---|
| $M^{ref}[i]$ | 0 | 1 | 0 | 1 |
| $f_i^{num}(I^{sea}[i]=0, IC^{ref})$ | 0 | 0 | – | 1 |
| $f_i^{num}(I^{sea}[i]=1, IC^{ref})$ | 0 | 1 | – | 0 |
| $f_{i+\ell}^{num}(M^{sea}[i]=0, IC^{ref})$ | 0 | 0 | – | -1 |
| $f_{i+\ell}^{num}(M^{sea}[i]=0, IC^{ref})$ | 0 | 0 | – | 0 |

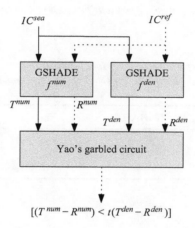

$$[(T^{num} - R^{num}) < t(T^{den} - R^{den})]$$

*Figure 15.5  Secure IrisCodes authentication in the semi-honest setting*

following convention, if a mask bit is set to 0, the corresponding iris bit is also set to 0. This convention will not modify the result of the normalized Hamming distance and will enable to build $4\ell$ functions $f_i$ to evaluate with GSHADE the NHD numerator and the denominator between two IrisCodes $IC^{sea} = (I^{sea}, M^{sea})$ and $IC^{ref} = (I^{ref}, M^{ref})$. The functions $f_i$ are defined as following: for $i$ in $[1 \ldots \ell]$, $f_i^{den}(I^{sea}[i], IC^{ref}) = 0$, $f_{i+\ell}^{den}(M^{sea}[i], IC^{ref}) = M^{sea}[i] \cdot M^{ref}[i]$ and $f_i^{num}$ is defined as in Table 15.1 where the $i$th bit coordinate of the vectors $I^{sea}, I^{ref}, M^{sea}$ and $M^{ref}$ is denoted with brackets.

At the end of the GSHADE protocol, each party obtains a sharing of the NHD numerator and denominator: one party obtains $T^{num}$ and $T^{den}$ and the second party obtains $R^{num}$ and $R^{den}$ such that $(T^{num} - R^{num})$ is equal to the numerator of the NHD and $(T^{den} - R^{den})$ is equal to the denominator. Consequently no party has information about the numerator and the denominator of the NHD. Finally, a Yao's garbled circuit is performed to obtain the authentication result (accept or reject). This Boolean circuit inputs $T^{num}, T^{den}, R^{num}, R^{den}$ and a threshold $t$ and outputs 1 if $(T^{num} - R^{num}) < t(T^{den} - R^{den})$ and 0 otherwise. Figure 15.5 sums up this protocol.

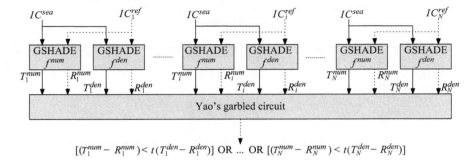

*Figure 15.6    Secure IrisCodes identification in the semi-honest setting*

*Table 15.2    Performance of an iris-based authentication (N = 1)*
*and identification (N > 1) in the semi-honest model*

|  | **GSHADE** | **Yao** | **Overall** |
|---|---|---|---|
| $N = 1$ | 0.053 s | 0.051 s | 0.104 s |
| $N = 500$ | 0.679 s | 0.151 s | 0.830 s |
| $N = 1,000$ | 1.223 s | 0.258 s | 1.481 s |

The plain lines correspond to the input, intermediate and output values of the client $\mathscr{C}$ and the dotted lines to the ones of the server $\mathscr{S}$.

### 15.2.4.2    Identification in the semi-honest setting

With the GSHADE extension, we can efficiently and securely evaluate the $N$ numerators $NHD^{num}(IC^{sea}, IC_1^{ref}), \ldots, NHD^{num}(IC^{sea}, IC_N^{ref})$ and the $N$ denominators $NHD^{den}(IC^{sea}, IC_1^{ref}), \ldots, NHD^{den}(IC^{sea}, IC_N^{ref})$. After GSHADE execution, each party obtains a sharing of the NHD numerator and denominator: for each $i$ in $[1 \ldots N]$, the first party obtains $T_i^{num}$ and $T_i^{den}$ and the second party obtains $R_i^{num}$ and $R_i^{den}$ such that $NHD(IC^{sea}, IC_i^{ref}) = \frac{T_i^{num} - R_i^{num}}{T_i^{den} - R_i^{den}}$. Finally, a Yao's garbled circuit is performed to obtain the identification result (accept or reject). The Boolean circuit within the Yao's garbled circuit protocol outputs 1 if there exists $i$ in $[1 \ldots N]$ such that $(T_i^{num} - R_i^{num}) < t(T_i^{den} - R_i^{den})$ and 0 otherwise. Figure 15.6 sums up these steps.

### 15.2.4.3    Performance

As shown in Table 15.2, this protocol which combines GSHADE and Yao's garbled circuit is highly efficient and thus can be foreseen for loosely constrained real-world applications. These experiments were done on two identical standard computers (Ubuntu 12.04, Intel Core i5 with 5GB RAM) connected via LAN.

## 15.3 Secure distance computation in the malicious model

In real scenarios, the adversaries cannot always be considered as semi-honest. Consequently, it is important to protect S2PC protocol against malicious adversaries.

### 15.3.1 Yao's garbled circuits in the malicious setting

The Yao's garbled circuit protocol is secure against a semi-honest Builder and a malicious Evaluator. There are two main protocols to secure Yao's garbled circuit against a malicious Builder: the cut-and-choose construction [23] and the DualEx technique [24].

#### 15.3.1.1 The cut-and-choose construction

The cut-and-choose construction [23] ensures that the garbled circuit is constructed correctly and thus secures the Yao's garbled circuit protocol against a malicious Builder. In this construction, the Builder creates $s$ garbled circuits instead of one and he sends them to the Evaluator. Then the Evaluator randomly chooses half of the garbled circuits and asks to the Builder to open them in order to verify that they are correct. Then the Evaluator evaluates the remaining circuits (unchecked circuits). If at least one checked circuits is incorrect or if all the evaluated circuits do not output the same result, then the Builder is caught cheating. For a successful cheating, all the checked circuits must be correct and all the evaluated circuits must be incorrect and must output the same wrong output. Consequently a malicious Builder has a low probability to cheat without being caught.

Initial cut-and-choose construction [23] requires $17s$ circuits to achieve security $2^{-s}$. Many articles have suggested some optimizations of this technique to reduce to $3s$ circuits [25] and then $s$ circuits [26] for security $2^{-s}$. Unfortunately these techniques are not efficient for real-world applications due to the number of garbled circuits to create, check and evaluate.

#### 15.3.1.2 The DualEx technique

The DualEx construction [24] is based on the work of Mohassel and Franklin [27, Section 4.1]. The idea is to perform two semi-honest Yao's garbled circuit executions by swapping the parties' roles between each execution and then to perform a secure equality test to check that the two Yao's garbled circuit executions obtain the same outputs. In more details, the first party builds a garbled circuit $GC_1$ and sends it to the second party. At the same time, the second party creates a garbled circuit $GC_2$ and sends it to the first party. Then the first party evaluates $GC_2$ to obtain the garbled output $G_2(out_2)$ where $out_2$ is the plaintext output of the evaluation of the Boolean circuit and $G_2$ is a mapping held by the garbled circuit Builder (here the second party). At the same time, the second party evaluates $GC_1$ to obtain the garbled output $G_1(out_1)$ where $out_1$ is the plaintext output of the evaluation of the Boolean circuit and $G_1$ is a mapping held by the garbled circuit Builder (here the first party). Since the outputs are garbled, no party has the plaintext output. Finally, the both parties perform an inexpensive secure equality test, to verify that the two garbled outputs correspond to the same output. If this is the case, the plaintext output is revealed. Otherwise,

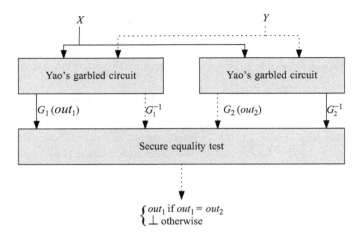

*Figure 15.7   The DualEx technique*

nothing is revealed. Figure 15.7 sums up this technique. The plain lines correspond to the input, intermediate and output values of the client $\mathscr{C}$ and the dotted lines to the ones of the server $\mathscr{S}$.

This protocol has similar running times than the semi-honest Yao's garbled circuit protocol because the two executions of the semi-honest Yao's garbled circuit protocols are performed in parallel. This protocol leaks at most one bit. In fact, a malicious adversary can partition the opposing party input set into two subsets and can create a circuit that outputs incorrect garbled values on one of the subset. Since the malicious party learns when the equality test fails, he will learn in which subset the opposing party input belongs. For large enough input set, the one-bit information leakage is negligible.

### 15.3.2   GSHADE in the malicious setting

Bringer *et al.* [28] introduced a method to secure GSHADE against malicious adversaries. This construction is also based on the idea of Mohassel and Franklin [27, Section 4.1] with some extra modifications in order to be fully secure and correct. One of the modifications is a small reduction of the function domain into the functions $f(X, Y) = \sum_i f_i(x_i, Y) = \sum_i f_i(y_i, X)$ where $X = (x_1, \ldots, x_n) \in \{0, 1\}^n$ is the input of the first party $P_1$ and $Y = (y_1, \ldots, y_n) \in \{0, 1\}^n$ is the input of the second party $P_2$. In more details, all computations are performed into $\mathbb{Z}_m$ with $m$ a prime number such that the codomain of $f$ is included into $\mathbb{Z}_m$. Each party $P_j$ picks randomly $n$ integers $r_{1,j}, \ldots, r_{n,j}$ from $\mathbb{Z}_m$ and one integer $a_j$ from $\mathbb{Z}_m^*$. In the first GSHADE execution instance, the two parties perform $n$ Oblivious Transfers where $P_1$ inputs $(a_1(r_{i,1} + f_i(0, X)), a_1(r_{i,1} + f_i(1, X)))$ for $i \in [1 \ldots n]$ and $P_2$ inputs $y_i$ to obtain $t_{i,1} = a_1(r_{i,1} + f_i(y_i, X))$. After the $n$ Oblivious Transfers, $P_1$ outputs $a_1$ and $R_1 = a_1 \sum_i r_{i,1}$ and $P_2$ outputs $T_1 = \sum_i t_{i,1}$. At the same time, a second GSHADE

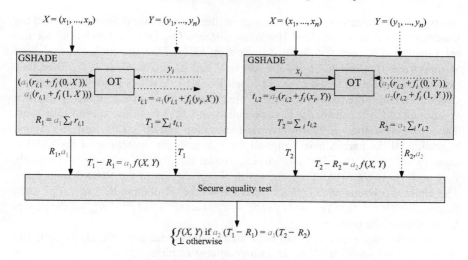

*Figure 15.8   The GSHADE$_{malicious}$ protocol*

execution is performed but the roles of the two parties have been swapped: $P_1$ outputs $T_2 = \sum_i t_{i,2}$ and $P_2$ outputs $a_2$ and $R_2 = a_2 \sum_i r_{i,2}$ where $t_{i,2} = a_2(r_{i,2} + f_i(x_i, Y))$ for $i \in [1 \ldots n]$. Finally a secure equality test is performed to check that the two GSHADE executions output the same value. In this equality test, $P_1$ inputs $a_1, R_1, T_2$ and $P_2$ inputs $T_1, R_2, a_2$. The equality test outputs 1 if $a_2(T_1 - R_1) = a_1(T_2 - R_2)$ and 0 otherwise. This equality test is secure if the two parties learn only the output and nothing about the inputs of the opposing party. If the equality test outputs 0, the parties abort and the protocol outputs nothing because one party has cheated. Otherwise the parties have a sharing of the $f(X, Y)$ because $P_1$ knows $a_1, R_1, T_2$ and $P_2$ knows $T_1, R_2, a_2$ such that $f(X, Y) = a_1^{-1}(T_1 - R_1) = a_2^{-1}(T_2 - R_2)$. Figure 15.8 sums up this protocol. The plain lines correspond to the input, intermediate and output values of the client $\mathscr{C}$ and the dotted lines to the ones of the server $\mathscr{S}$.

Since the two internal GSHADE executions are independent, their executions can be overlapped by using two independent threads working in parallel. Consequently, GSHADE$_{malicious}$ has only a low extra cost compared to GSHADE in the semi-honest setting. The GSHADE$_{malicious}$ protocol requires two parties with a similar role but in the GSHADE 1-vs-N extension, one party has one input and the second party has $N$ inputs and thus the parties have not a similar role. Consequently GSHADE$_{malicious}$ is not compatible with the 1-vs-N GSHADE extension.

## 15.3.3   Privacy-preserving distance computation for IrisCodes in the malicious model

### 15.3.3.1   Authentication in the malicious setting

In this section following the idea of GSHADE$_{malicious}$, we combine GSHADE and Yao's garbled circuit in dual execution in order to securely and efficiently perform an iris-based authentication in the malicious setting. To evaluate the numerator and the denominator of the normalized Hamming distance (NHD) with GSHADE, we use the functions $f_i^{num}$ and $f_i^{den}$ defined in Section 15.2.4.1. In this protocol, the client $\mathscr{C}$ inputs one IrisCode $IC^{sea}$ and the server $\mathscr{S}$ inputs one IrisCode $IC^{ref}$. Then they perform four internal GSHADEs. The first two GSHADEs evaluate the numerator of the NHD with the parties swapping roles between the two executions. The client $\mathscr{C}$ outputs $a_1^{num}$, $T_1^{num}$ from the first execution and $R_2^{num}$ from the second execution. The server $\mathscr{S}$ outputs $R_1^{num}$ from the first execution and $a_2^{num}$, $T_2^{num}$ from the second execution. If the parties have followed the protocol, the numerator of the NHD is equal to $(a_1^{num})^{-1}(T_1^{num} - R_1^{num})$ which is also equal to $(a_2^{num})^{-1}(T_2^{num} - R_2^{num})$. Similarly they perform two GSHADE executions to obtain sharing of the NHD denominator. The client $\mathscr{C}$ obtains $a_3^{den}$, $T_3^{den}$, $R_4^{den}$ and the server $\mathscr{S}$ obtains $R_3^{den}$, $a_4^{den}$, $T_4^{den}$ such that $NHD^{den}(IC^{sea}, IC^{ref}) = (a_3^{den})^{-1}(T_3^{den} - R_3^{den}) = (a_4^{den})^{-1}(T_4^{den} - R_4^{den})$ if the parties have followed the protocol.

Then the parties perform Yao in DualEx to obtain the authentication result. The Boolean circuit within the Yao execution outputs two bits:

- the first bit $c$ means whether the GSHADE executions are correct. If $(a_1^{num})^{-1}(T_1^{num} - R_1^{num}) = (a_2^{num})^{-1}(T_2^{num} - R_2^{num})$ and $(a_3^{den})^{-1}(T_3^{den} - R_3^{den}) = (a_4^{den})^{-1}(T_4^{den} - R_4^{den})$, then $c$ is equal to 1, otherwise $c$ is equal to 0.
- the second bit $s$ is the authentication result. If $c$ is equal to 0 (one party has cheated during the GSHADE executions), by convention $s$ is set to 0 to reveal no information. Otherwise, $s$ is equal to the authentication result: if $(a_1^{num})^{-1}(T_1^{num} - R_1^{num}) < t \cdot (a_3^{den})^{-1}(T_3^{den} - R_3^{den})$, $s$ is equal to 1 and otherwise $s$ is equal to 0.

Consequently, this Yao in DualEx contains two steps: the verification that the parties have followed the protocol during the previous GSHADE executions and the evaluation of the authentication result from the numerator and the denominator of the NHD. From the first (resp. second) Yao execution, the client $\mathscr{C}$ (resp. the server $\mathscr{S}$) outputs the garbled values $G_1(c_1)$ and $G_1(s_1)$ (resp. $G_2(c_2)$ and $G_2(s_2)$) and the server $\mathscr{S}$ (resp. the client $\mathscr{C}$) outputs the mapping to obtain the plaintext value from a garbled value noted $G_1^{-1}$ (resp. $G_2^{-1}$). Finally a secure equality test is performed to verify that no party cheats during the Yao execution in DualEx. This security equality test checks that $c_1$ is equal to $c_2$ and $s_1$ is equal to $s_2$. If the equality test succeeds, the protocol outputs two bits $c = c_1 = c_2$ and $s = s_1 = s_2$ and otherwise the protocol outputs nothing. Figure 15.9 sums up this protocol. The plain lines correspond to the input, intermediate and output values of the client $\mathscr{C}$ and the dotted lines to the ones of the server $\mathscr{S}$.

### 15.3.3.2  Identification in the malicious setting

As in the authentication in the malicious setting, we combine GSHADE and Yao's garbled circuit in dual execution in order to securely and efficiently evaluate the identification result. The GSHADE protocol in dual execution evaluate the $N$ numerators and denominators of the NHD between the search IrisCode and the $N$ reference

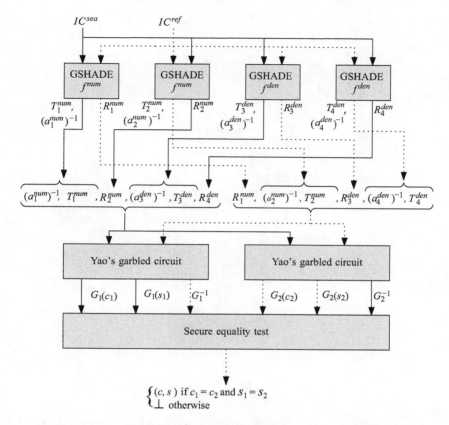

*Figure 15.9   Secure IrisCodes authentication in the malicious setting*

IrisCodes. Unfortunately, the 1-vs-$N$ GSHADE extension in the semi-honest setting is not compatible with the GSHADE$_{malicious}$ protocol. Consequently, for each NHD to evaluate, four GSHADEs are still performed: two for the numerator of the NHD and two for the denominator of the NHD. At the end of these GSHADE executions, parties obtain sharing of the $N$ numerators and denominators of the NHD. For each $i$ in $[1 \ldots N]$, the client $\mathscr{C}$ obtains $T_{i,1}, a_{i,1}^{-1}, R_{i,2}, T_{i,3}, a_{i,3}^{-1}, R_{i,4}$ and the server $\mathscr{S}$ obtains $R_{i,1}^{num}, T_{i,2}^{num}, (a_{i,2}{}^{num})^{-1}, R_{i,3}^{den}, T_{i,4}^{den}, (a_{i,4}^{den})^{-1}$ such that $NHD^{num}(IC^{sea}, IC_i^{ref}) = (a_{i,1}^{num})^{-1}(T_{i,1}^{num} - R_{i,1}^{num}) = (a_{i,2}^{num})^{-1}(T_{i,2}^{num} - R_{i,2}^{num})$ and $NHD^{den}(IC^{sea}, IC_i^{ref}) = (a_{i,3}^{den})^{-1}(T_{i,3}^{den} - R_{i,3}^{den}) = (a_{i,4}^{den})^{-1}(T_{i,4}^{den} - R_{i,4}^{den})$ if the parties have followed the protocol.

Then the parties perform Yao in DualEx to obtain the identification result. The Boolean circuit within the Yao execution outputs two bits:

- the first bit $c$ means whether the GSHADE executions are correct. If for each $i$ in $[1 \ldots N]$, $(a_{i,1}^{num})^{-1}(T_{i,1}^{num} - R_{i,1}^{num})$ is equal to $(a_{i,2}^{num})^{-1}(T_{i,2}^{num} - R_{i,2}^{num})$

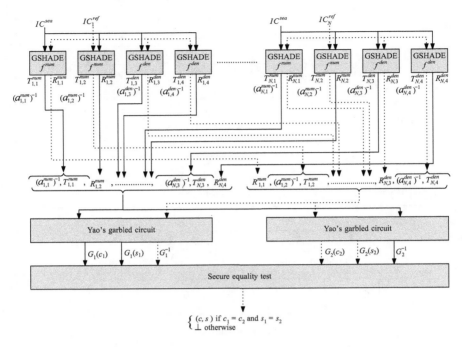

*Figure 15.10    Secure IrisCodes identification in the malicious setting*

and $(a_{i,3}^{den})^{-1}(T_{i,3}^{den} - R_{i,3}^{den})$ is equal to $(a_{i,4}^{den})^{-1}(T_{i,4}^{den} - R_{i,4}^{den})$ then $c$ is equal to 1, otherwise $c$ is equal to 0.

- the second bit $s$ is the identification result. If $c$ is equal to 0 (one party has cheated during the GSHADE executions), by convention $s$ is set to 0. Otherwise, $s$ is equal to the identification result: if at least one $(a_{i,1}^{num})^{-1}(T_{i,1}^{num} - R_{i,1}^{num}) < t \cdot (a_{i,3}^{den})^{-1}(T_{i,3}^{den} - R_{i,3}^{den})$ for $i$ in $[1 \ldots N]$, $s$ is equal to 1 and otherwise $s$ is equal to 0.

Finally a secure equality test is performed to check that no party cheats during the Yao in DualEx. This equality test checks that $c_1$ is equal to $c_2$ and $s_1$ is equal to $s_2$. If the test succeeds, the protocol outputs two bits $c = c_1 = c_2$ and $s = s_1 = s_2$. Figure 15.10 sums up this protocol. The plain lines correspond to the input, intermediate and output values of the client $\mathscr{C}$ and the dotted lines to the ones of the server $\mathscr{S}$.

### 15.3.3.3   Performance

Table 15.3 presents the running times of these protocols which combine GSHADE and Yao's garbled circuit in dual execution. These experiments were done on two identical standard computers (Ubuntu 12.04, Intel Core i5 with 5GB RAM) connected via LAN and only two threads per computer are used. These protocols could be easily

*Table 15.3*  *Performance of an iris-based authentication (N = 1) and*
           *identification (N > 1) in the malicious model*

|         | GSHADE | Yao | Overall |
|---------|--------|-----|---------|
| $N = 1$    | 0.075 s | 0.07 s | 0.145 s |
| $N = 500$  | 37.5 s  | 3.952 s | 41.452 s |
| $N = 1000$ | 75 s    | 7.974 s | 82.9740 s |

*Table 15.4*  *Performance of an iris-based authentication (N = 1) and identification*
           *(N > 1) on Masek and Ma et al. templates*

|         | Semi-honest setting | | | Malicious setting | | |
|---------|--------|-----|---------|--------|-----|---------|
|         | GSHADE | Yao | Overall | GSHADE | Yao | Overall |
| $N = 1$    | 0.064 s | 0.087 s | 0.151 s | 0.076 s | 0.069 s | 0.145 s |
| $N = 500$  | 1.021 s | 0.109 s | 1.130 s | 38 s | 3.495 s | 41.495 s |
| $N = 1000$ | 1.974 s | 0.184 s | 2.158 s | 76 s | 6.516 s | 82.516 s |

parallelized and thus they can be foreseen for real-world applications with a pipelined implementation executing on several threads.

## 15.4  Application to other iris representations

IrisCode is nowadays the most used algorithm for iris recognition. Masek [29] and Ma *et al.* [30] are two other widely used algorithms for iris recognition. These two feature extraction algorithms transform an iris image into a vector of 10 240 bits. With Masek or Ma *et al.* algorithm, the comparison between two iris templates requires a simple Hamming distance and not an NHD. Compared to IrisCode templates, the distance is easier to compute but the template size is higher. Table 15.4 presents the running times on these new templates for authentication and identification in the semi-honest and malicious settings. These experiments were done on two identical standard computers (Ubuntu 12.04, Intel Core i5 with 5GB RAM) connected via LAN.

In the malicious setting, we obtain similar performances compared to ones with IrisCodes (see Table 15.3). On the contrary, in the semi-honest setting, the IrisCode representation obtains the best performances (see Table 15.2).

## 15.5  Conclusion

In this chapter, we presented efficient protocols for the iris authentication and iden-tification in the semi-honest and malicious setting. In the semi-honest setting, the authentication and the identification protocols are based on the combination of GSHADE and Yao's garbled circuit protocols. In the malicious setting, a dual execution of the semi-honest protocols followed by a secure equality test is suggested and some extra modifications within the GSHADE protocol are performed in order to obtain fully secure protocols. The experiments of these protocols prove their efficiency and their usability in real-world applications with realistic adversary models.

## Acknowledgments

This work is carried out under the funding of the Research Council of Norway (Grant No. IKTPLUSS 248030/O70, SWAN project).

## References

[1]　A. B. J. Teoh, A. Goh, and D. C. L. Ngo. "Random multispace quantization as an analytic mechanism for biohashing of biometric and random identity inputs," *IEEE Transactions on Pattern Analysis and Machine Intelligence*, vol. 28, no. 12, pp. 1892–1901, 2006.

[2]　V. M. Patel, N. K. Ratha, and R. Chellappa. "Cancelable biometrics: A review," *IEEE Signal Processing Magazine*, vol. 32, no. 5, pp. 54–65, 2015.

[3]　N. K. Ratha, J. H. Connell, and R. M. Bolle. "Enhancing security and privacy in biometrics-based authentication systems," *IBM Systems Journal*, vol. 40, no. 3, pp. 614–634, 2001.

[4]　A. Juels and M. Wattenberg. "A fuzzy commitment scheme," in *CCS'99, Sixth Conference on Computer and Communications Security*, 1999.

[5]　A. Juels and M. Sudan. "A fuzzy vault scheme," in *CCS 2001, Eighth Conference on Computer and Communications Security*, 2001.

[6]　J. Bringer, H. Chabanne, and C. Morel. "Shuffling is not sufficient: Security analysis of cancelable IrisCodes based on a secret permutation," in *IJCB 2014, Second International Joint Conference on Biometrics*, 2014.

[7]　J. Bringer, C. Morel, and C. Rathgeb. "Security analysis of bloom filter-based iris biometric template protection," in *ICB 2015, Eighth International Conference on Biometrics*, 2015.

[8]　M. Gomez-Barrero, J. Galbally, P. Tome-Gonzalez, and J. Fiérrez. "On the vulnerability of iris-based systems to a software attack based on a genetic algorithm," in *CIARP 2012, 17th Iberoamerican Congress on Pattern Recognition*, 2012.

[9]　A. Nagar, K. Nandakumar, and A. K. Jain. "Biometric template transformation: A security analysis," in *SPIE 2010, Media Forensics and Security II, Part of the IS&T-SPIE Electronic Imaging Symposium*, 2010.

[10]　C. Hazay and Y. Lindell. "*Efficient Secure Two-Party Protocols – Techniques and Constructions*," Information Security and Cryptography. Springer, Berlin, 2010.

[11]  J. Daugman. "How iris recognition works," in *ICIP 2002, Ninth International Conference on Image Processing*, 2002.

[12]  A. C.-C. Yao. "How to generate and exchange secrets (extended abstract)," in *FOCS 1986, 27th Annual Symposium on Foundations of Computer Science*, 1986.

[13]  J. Bringer, H. Chabanne, M. Favre, A. Patey, T. Schneider, and M. Zohner. "GSHADE: Faster privacy-preserving distance computation and biometric identification," in *IH&MMSec'14, Second Information Hiding and Multimedia Security Workshop*, 2014.

[14]  M. O. Rabin. "How to exchange secrets with oblivious transfer," *Harvard University Technical Report*, 1981.

[15]  M. Naor and B. Pinkas. "Efficient oblivious transfer protocols," in *SODA 2001, 12th Annual Symposium on Discrete Algorithms*, 2001.

[16]  T. Chou and C. Orlandi. "The simplest protocol for oblivious transfer," in *LATINCRYPT 2015, Fourth International Conference on Cryptology and Information Security in Latin America*, 2015.

[17]  G. Asharov, Y. Lindell, T. Schneider, and M. Zohner. "More efficient oblivious transfer and extensions for faster secure computation," in *CCS'13, 20th Conference on Computer and Communications Security*, 2013.

[18]  Y. Ishai, J. Kilian, K. Nissim, and E. Petrank. "Extending oblivious transfers efficiently," in *CRYPTO 2003, 23rd Annual International Cryptology Conference*, 2003.

[19]  P. Snyder. "Yao's garbled circuits: Recent directions and implementations," 2014.

[20]  M. Naor, B. Pinkas, and R. Sumner. "Privacy preserving auctions and mechanism design," in *EC'99, First Conference on Electronic Commerce*, 1999.

[21]  B. Pinkas, T. Schneider, N. P. Smart, and S. C. Williams. "Secure two-party computation is practical," in *ASIACRYPT 2009, 15th International Conference on the Theory and Application of Cryptology and Information Security*, 2009.

[22]  Y. Huang, D. Evans, J. Katz, and L. Malka. "Faster secure two-party computation using garbled circuits," in *USENIX Security'11, 20th USENIX Security Symposium*, 2011.

[23]  Y. Lindell and B. Pinkas. "An efficient protocol for secure two-party computation in the presence of malicious adversaries," in *EUROCRYPT 2007, 26th Annual International Conference on the Theory and Applications of Cryptographic Techniques*, 2007.

[24]  Y. Huang, J. Katz, and D. Evans. "Quid-pro-quo-tocols: Strengthening semi-honest protocols with dual execution," in *SP 2012, 33rd Symposium on Security and Privacy*, 2012.

[25]  Y. Lindell and B. Pinkas. "Secure two-party computation via cut-and-choose oblivious transfer," *Journal of Cryptology*, vol. 25, no. 4, pp. 680–722, 2012.

[26]  Y. Lindell. "Fast cut-and-choose based protocols for malicious and covert adversaries," in *CRYPTO 2013, 33rd International Cryptology Conference*, 2013.

[27]    P. Mohassel and M. K. Franklin. "Efficiency tradeoffs for malicious two-party computation," in *PKC 2006, Ninth International Conference on Theory and Practice of Public-Key Cryptography*, 2006.

[28]    J. Bringer, H. Chabanne, O. El Omri, and C. Morel. "Boosting GSHADE capabilities: New applications and security in malicious setting," in *SACMAT 2016, 21st Symposium on Access Control Models and Technologies*, 2016.

[29]    L. Masek. "Recognition of human iris patterns for biometric identification," Master's thesis, University of Western Australia, Crawley, Australia, 2003.

[30]    L. Ma, T. Tan, Y. Wang, and D. Zhang. "Efficient iris recognition by characterizing key local variations," *IEEE Transactions on Image Processing*, vol. 13, no. 6, pp. 739–750, 2004.

*Chapter 16*

# Identifying iris sensors from iris images

*Luca Debiasi, Christof Kauba, and Andreas Uhl*

The base component of iris sensors deployed in practical applications is a digital image sensor, mostly supported by a near infra-red (NIR) light source to improve the iris recognition results [1]. These sensors acquire digital images, which are then further processed and inserted into a biometric system's processing chain.

The authenticity and integrity of the acquired iris images play an important role for the overall security of a biometric system. Ratha *et al.* [2] identified eight stages in a generic biometric system where attacks may occur. Figure 16.1 shows an insertion and presentation attack on an exemplary biometric system. An insertion attack bypasses the biometric sensor by inserting data (biometric sample) into the transmission from the sensor to the feature extractor. This transmission is the most relevant point for an attack on the integrity and authenticity of the acquired iris images, where the iris image inserted during the attack could be acquired with another sensor off-site, even without the knowledge of a genuine user, or a manipulated image to spoof the biometric recognition system. In contrast to the insertion attack, in the case of the presentation attack a forged or fake biometric trait, i.e., an artificially manufactured fake fingerprint or a print of an iris image, is presented to the genuine sensor installed in the biometric system. The presentation of a forged biometric trait can usually be detected by deploying different liveness detection systems.

Encryption and other classical authentication techniques like digital signatures or data-hiding have been suggested to secure the previously mentioned transmission channel by verifying the senders (i.e., sensor and feature extractor) authenticity, as well as the integrity of the entire authentication mechanism. The proposed approaches can be divided into active and passive-blind approaches.

Active methods consist of data hiding approaches [3,4] and the digital signature approaches [5–8]. Höller *et al.* [9] describe the pros and cons of these active methods as follows:

- **Classical digital signatures** work by adding additional data to verify the original data, whereas watermarks become an integral part of the sample data, and moreover, spatial locations of eventual tampering can be identified [10].
- **Fragile watermarks** (as proposed for these tasks in e.g., [11–13]) cannot provide any form of robustness against channel errors and unintentional signal processing "attacks" like compression, which is the same as with classical digital signatures.

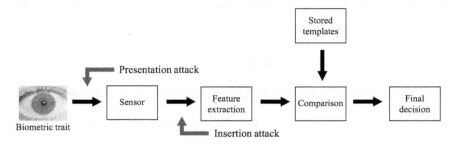

*Figure 16.1   Exemplary biometric system and point of insertion and presentation attacks*

- **Semi-fragile watermarks** have been designed to differentiate between allowed signal processing operations and malicious attacks and have also been suggested for employment in biometric systems [14–17].

Höller *et al.* [9] also mention that a general drawback of watermarks is the representation of additional data which is inserted into the sample data, where an impact on recognition accuracy may be expected. In fact, literature reports on corresponding effects in case of iris recognition [18], speech recognition [19], and fingerprint recognition [20].

Passive-blind approaches, in contrast to active methods, do not need any prior information about the image. As stated in [21], passive-blind approaches are mostly based on the fact that forgeries can bring specific detectable changes into the image (e.g., statistical changes). In high-quality forgeries, these changes cannot be found by visual inspection.

The field of digital image forensics deals with still images and analyzing traces in still image data. Two major tasks in this field are establishing an image's origin and its integrity. In contrast to digital watermarking as authenticity technique, as mentioned in [22], digital image forensics do not require any active embedding step at the time of creation or publication. Evidence is extracted merely from structural analysis of image files and statistical analysis of the image data (i.e. the two-dimensional array of pixel intensities).

To determine an image's origin several approaches have been proposed exploiting hardware- and software-related artifacts. Investigated hardware-related artifacts cover optical defects, like chromatic aberrations [23] or lens distortions [24], or sensor artifacts, like sensor defects [25] and noise. Software artifacts are introduced during the processing of the images in the cameras and can be unveiled using statistical features [26] or by analyzing the common image processing pipeline of the images [27].

The photo-response non-uniformity (PRNU) of imaging sensors, as described in [28,29], is an intrinsic property of all digital imaging sensors due to slight variations among individual pixels in their ability to convert photons to electrons. Consequently, every sensor casts a weak noise-like pattern onto every image it takes. This pattern, which plays the role of a "sensor fingerprint," is essentially an unintentional stochastic spread-spectrum watermark that survives processing, such as lossy compression or filtering. This fingerprint can be estimated from images taken by the camera and later detected in a given image to establish image origin and integrity.

Even though the PRNU is stochastic in nature, it is a relatively stable component of the sensor over its life span, providing a unique sensor fingerprint with the following important properties [29]:

1. **Dimensionality**: The fingerprint is stochastic in nature and has a large information content, which makes it unique to each sensor.
2. **Universality**: All imaging sensors exhibit PRNU.
3. **Generality**: The fingerprint is present in every picture independently of the camera optics, camera settings, or scene content, with the exception of completely dark images.
4. **Stability**: It is stable in time (except for aging-related sensor defects) and under a wide range of environmental conditions (temperature, humidity, etc.).
5. **Robustness**: It survives lossy compression, filtering, gamma correction, and many other typical processing procedures.

Slight variations of individual pixels during the conversion of photons to electrons in digital image sensors are the source of the PRNU, thus it is considered an intrinsic property which is present in all digital imaging sensors. Every digital image sensor adds this weak, noise-like pattern into every image acquired with it. The sensor identification can be performed at different levels, as described by Bartlow *et al.* [30]: Technology, brand, model, unit. Due to the datasets evaluated in this work we focus on the model level, which corresponds to a differentiation according to model and brand.

The PRNU can also be used for the verification of an image's integrity. The integrity is compromised if an image has been geometrically transformed (e.g., cropped, rotated, turned, and flipped) or if parts of the image have been tampered (e.g., deleted, copied, replaced, and altered). These manipulations lead to changes in the PRNU which can be detected as shown in [31,32].

In the context of biometric systems security the PRNU fingerprint of a sensor can be used to ensure the integrity and authenticity of images acquired with a biometric sensor. Höller *et al.* [9] propose a suitable passive approach to secure the transmission channel between the sensor and the feature extractor, making use of sensor fingerprints based on a sensor's PRNU [31]. Besides image integrity, this technique can also provide authenticity by identifying the source sensor uniquely and important properties as required in a biometric scenario have been demonstrated: suitability to manage large datasets [33,34], robustness against common signal processing operations like compression and malicious signal processing [35,36], and finally methodology to reveal forged PRNU fingerprints has been established [37].

To ensure the authenticity of the biometric sensor, first the discriminative power of the biometric sensors has to be evaluated, as it has been done in [9,30] using the PRNU.

The results from Höller *et al.* [9], where the discriminative power of five iris sensors from the *CASIA-Iris V4* database, have been evaluated to show high variations. Other work by Kalka *et al.* [38] regarding the differentiability of iris sensor showed varying results, while studies conducted on fingerprint sensors by Bartlow *et al.* [30] showed more satisfactory results. In order for PRNU fingerprints being useful as an authentication measure for biometric systems, the sources of the poor differentiation results have to be determined. Some possible explanations are given in [38,9] and consist of the highly correlated data of biometric datasets, saturated pixels and the use of multiple sensors of the same model. An additional caveat for the PRNU extraction is the image content. Since the PRNU covers the high frequency components of an image, it is contaminated with other high frequency components within the images, such as edges. Li [39] proposed an approach for attenuating the influence of details from scenes on the PRNU so as to improve the device identification rate of the identifier. Moreover the PRNU fingerprint can be extracted from images of a biometric sensor and injected into forged images, as described by Goljan *et al.* [40]. Using several images captured by the sensor deployed in the biometric system a suitable PRNU fingerprint can be generated by the attacker. This attack can only be detected with a triangle test [9], which requires additional genuine images acquired under controlled conditions.

To overcome the reported problems with the PRNU extraction for some sensors and the injection attack, there exist other approaches like the one by El-Naggar and Ross [41], who proposed a passive approach tailored to iris recognition. At first the ocular image is segmented to get the iris region, then the iris texture is unwrapped, followed by a normalization step to get a normalized iris image. Only the inner half of this normalized iris image is used and further split into a set of overlapping blocks. For each block 50 Gabor and 68 statistical features are extracted to form a 118 dimensional feature vector representing the iris image. These feature vectors are then classified using a three-layer artificial neural network. They were able to achieve accuracies of 80%–85%. We propose a similar approach which follows their evaluation methodology but uses different features and an SVM classifier. In this chapter we evaluate this approach and a PRNU-based one, trying to identify the iris data set which an iris image belongs to. If the correct data set can be determined for a given iris image, these approaches could be used to secure an iris recognition system against insertion attacks.

The chapter is organized as follows: In Section 16.1 we describe the two different approaches and the examined iris data sets are listed in Section 16.2. The experimental setup and the results are illustrated in Sections 16.3 and 16.4, respectively. In Section 16.5 we discuss how the previously examined techniques can be used in a practical application. Finally Section 16.6 concludes the chapter.

## 16.1   Techniques for sensor identification/dataset classification

In this section we present two different techniques that allow to infer which dataset an iris image originates from. The first technique, PRNU-based Sensor Identification (PSI), does this by identifying the sensor used to acquire the image. The second technique, Iris Texture Classification (ITC), makes use of the iris texture and its inherent features to classify the iris images according to the source sensor. Both techniques are presented in detail in the following section.

### 16.1.1  PRNU-based sensor identification (PSI)

A digital image sensor consists of lots of small photosensitive, usually rectangular detectors that capture the incident light and generate an electric signal. These detectors are commonly known as pixels. The image acquired with the sensor is constructed by the aggregate of all pixels. Due to imperfections in the manufacturing and the inhomogeneity of the manufacturing material, silicon, the efficiency of each pixel of converting photons to electrons varies slightly. According to Fridrich [29], the raw output of a sensor with $w \times h$ pixels can be modeled as:

$$Y = I + I \circ K + \tau D + C + \Theta$$

$$\text{with} \quad Y, I, K, D, C, \Theta \in \mathbb{R}^{w \times h}; \tau \in \mathbb{R} \tag{16.1}$$

where $Y$ is the sensor output (image). $I$ represents the incoming light, $I \circ K$ the photoresponse non-uniformity PRNU, $\tau D$ the dark current (with $\tau$ being a multiplicative factor representing exposure settings, sensor temperature, etc.). The matrix $C$ is a light-independent offset and $\Theta$ some modeling noise, which is a collection of all other noise sources mostly random in nature (e.g., readout noise, shot noise or photonic noise, and quantization noise). Since all pixels are independent and all operations element-wise, the matrix-elements $y_{x,y} \in Y$ are denoted as $y \in Y$ for simplicity reasons. The same applies to $i \in I$, $k \in K$, $d \in D$, $c \in C$ and $\theta \in \Theta$.

The extraction of the PRNU noise residuals is performed as indicated by Fridrich in [42]. For each image $I$ the noise residual $W_I$ is estimated:

$$W_I = I - F(I) \tag{16.2}$$

where $F$ is a denoising function filtering out the sensor pattern noise. We used two different denoising techniques to extract the PRNU from the images: The wavelet-based denoising filter as described in Appendix A of [43] and the BM3D filter proposed in [44], which is reported to produce better and more consistent results in filtering out the PRNU in [45]. The extracted PRNU noise residual is then normalized in respect to the $L_2$-norm because its embedding strength is varying between different sensors as explained by [9]. As additional post processing steps a zero mean operation is applied to each extracted PRNU noise residual to suppress artifacts with regular grid structure.

To reduce the PRNU contamination effect from scene details, we apply an image content attenuating PRNU enhancement technique (Model 3 in [39]), subsequently denoted as *ELi*.

Estimating a sensor's PRNU from a single image is usually not sufficient, because that specific image may contain various kinds of disturbances as modeled by $\Theta$ in (16.1). Thus multiple images from the same sensor are averaged to isolate the systematic components of all images and suppress these random noise components, as shown in Figure 16.2. This averaged noise is denoted as PRNU fingerprint or

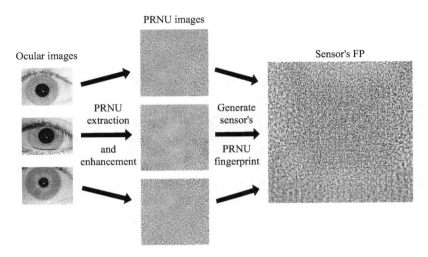

*Figure 16.2    PRNU noise residual extraction and fingerprint generation with multiple iris images of the same sensor*

reference pattern noise (RPN) in literature. The PRNU fingerprint $\hat{K}$ of a sensor is then estimated using a maximum likelihood estimator for images $I_i$ with $i = 1, \ldots, N$.

$$\hat{K} = \frac{\sum_{i=1}^{N} W_I^i I^i}{\sum_{i=1}^{N} (I^i)^2} \tag{16.3}$$

The PRNU fingerprint is enhanced using a Wiener filter applied in the DFT domain to suppress periodic artifacts as described in [46].

To determine if an image has been acquired with a specific sensor, the presence of a sensor's PRNU fingerprint in the questioned image has to be detected. Since images acquired with iris sensors are usually not geometrically transformed, this can be done by means of calculating the normalized Cross Correlation (NCC):

$$\rho_{[J,\hat{K}]} = NCC(W_J, J\hat{K}) \tag{16.4}$$

where $\rho$ indicates the correlation between the PRNU residual $W_j$ of the image $J$ and the fingerprint $\hat{K}$ weighted by the image content of $J$.

An alternative correlation measure to detect the presence of a PRNU fingerprint $\hat{K}$ in an Image $I$ is the Peak Correlation Energy (PCE), proposed by Fridrich in [46]. Fridrich notes that with the PCE the detection threshold will not vary as much as for NCC detector with varying signal length, different cameras and their on-board image processing. It is applied like the NCC detector:

$$\rho_{[J,\hat{K}]} = PCE(W_J, J\hat{K}) \tag{16.5}$$

A schematic illustration for the detection of the correct PRNU fingerprint in a questioned image is given in Figure 16.3.

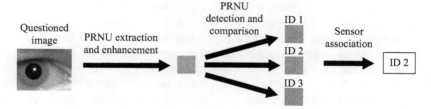

Figure 16.3 *PRNU noise residual extraction and identification of corresponding sensor*

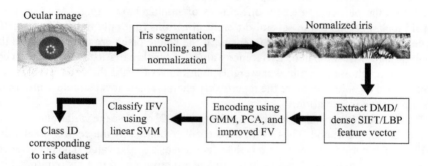

Figure 16.4 *Flowchart of the Iris Texture Classification (ITC) approach*

## 16.1.2 Iris texture classification (ITC)

The input for the Iris Texture Classification (ITC) approach are the preprocessed, segmented, unrolled, and normalized iris images originating from various different iris datasets and the output of the classifier is a prediction of the iris sensor used to capture the image or the dataset where the input iris image belongs to, respectively (Figure 16.4). As the ITC is SVM based, a training phase is needed prior to the use of the classifier, similar to generating a PRNU fingerprint for the PSI approach. In the following the three chosen feature extraction methods, namely DenseSIFT, DMD and LBP, are briefly explained. Then the classification approach using a GMM, Fisher Vector encoding and an SVM classifier is described.

### 16.1.2.1 Feature extraction

**DenseSIFT:** Is a variant of SIFT. SIFT, the scale invariant feature transform, is a general purpose feature extraction technique used in object recognition proposed by Lowe [47]. It is invariant to image scale and rotation and robust against various affine distortions, addition of noise, illumination changes and changes of the viewpoint. SIFT locates extrema in the scale-space, localizes keypoints, determines their dominant orientation, and finally constructs a local descriptor for the keypoint based on a region around it. Fei-Fei *et al.* [48] proposed to use the local SIFT descriptors on a predefined grid defined across the whole image instead of localizing their positions according to scale space extrema. This approach is known as dense SIFT. A 128-dimensional SIFT feature vector is extracted each 3 pixels in 5 different scales ($2^0$, $2^{-1/2}$, $2^{-1}$, $2^{-3/2}$, $2^{-2}$). The spatial bins of the SIFT feature descriptor histogram consist of four bins in $x$, four bins in $y$, and eight orientation bins. vl_feat's (http://www.vlfeat.org) implementation of DenseSIFT is utilized.

**DMD:** Dense Micro-block difference is a local feature extraction and texture classification technique proposed by Mehta and Egiazarian [49]. It captures the local structure from image patches ($9 \times 9$ to $15 \times 15$ pixels) at high scales. Instead of the pixels, small blocks of the image which capture the micro-structure are processed. Therefore the pairwise intensity differences of smaller blocks (e.g., $2 \times 2$ or $3 \times 3$ pixel blocks) calculated in several different directions (not only radial-like in LBP) in combination with the average intensity of the whole patch are used to encode the local structure of the patch. Difference values of block pairs located near the center of the patch are given higher weights than blocks toward the patch boundaries. This should be able to capture the repetitively characteristic local structure providing discriminative information.

**LBP:** The local binary patterns proposed by Ojala [50] observe the variations of pixels in a local neighborhood. These variations are thresholded by the central pixel value to obtain a binary decision, which is then encoded as a scalar value. The occurrences of each scalar value for all pixels in the image are represented in a histogram, which forms the extracted feature vector.

### 16.1.2.2  Feature encoding

We utilize the Improved Fisher Vector Encoding (IFV) scheme [51] in the same way as it is done in [49,52]. IFV is usually used in object recognition. Fisher vector encoding starts by extracting local SIFT descriptors densely (DenseSIFT) and at multiple scales to get a feature vector $f$. We not only use DenseSIFT features but also DMD and LBP ones as input for the next steps. The feature vector $f$ is then soft-quantized using a Gaussian Mixture Model (GMM) with $K$ modes where the Gaussian covariance matrices are assumed to be diagonal. The local descriptors present in $f$ are first decorrelated and then dimensionality reduced (optional) by PCA. So far this describes the standard Fisher Vector encoding [53]. The IFV now adds signed square rooting and $l^2$ normalization as described in [51].

### 16.1.2.3  Classification

A linear SVM is then used to classify the IFV encoded features. We experimented with different types of kernels $K(x', x'')$ (linear, Hellinger, exponential) and the linear kernel lead to the most promising results. The input data to the SVM (IFV encoded feature vectors) is normalized such that $K(x', x'') = 1$ which usually improves the performance. The SVM is trained using a standard nonlinear SVM solver on a subset of the unrolled, normalized iris images which is subsequently not used for the testing (evaluation) step.

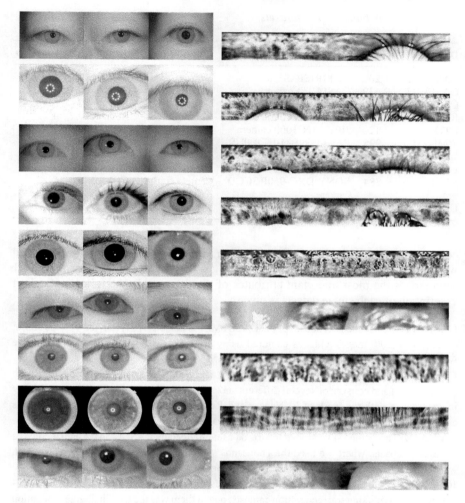

*Figure 16.5    Ocular image and normalized iris image samples from different datasets, from top to bottom: CASIA V2, CASIA V3, CASIA V4, ICE2005, IITD, MMU2, UBIRIS, UPOL, UTIRIS*

## 16.2 Datasets

To enable a meaningful comparison with the previous work of El-Naggar and Ross [41] we attempted to use the same iris datasets they originally used and extend the number of datasets. Unfortunately we were not able to acquire the MGBC and the WVU iris dataset. Thus we use the remaining six datasets they used plus three additional iris dataset which are described in the following. All of them are publicly available,

*Table 16.1   Attributes of iris datasets*

| Dataset | # IMG | Sensor | Illumination | Resolution | Class ID |
|---------|-------|--------|--------------|------------|----------|
| CASIA V2 | 1,200 | OKI IRISPASS-h | Near infrared | 480 × 640 | 1 |
| CASIA V3 | 2,639 | CASIA Iris camera | Near infrared | 320 × 280 | 2 |
| CASIA V4 | 20,000 | IrisKing IKEMB-100 | Near infrared | 640 × 480 | 3 |
| ICE2005 | 2,953 | LG EOU 2200 iris camera | Near infrared | 480 × 640 | 4 |
| IITD | 1,120 | JIRIS, JPC1000 camera | Near infrared | 240 × 320 | 5 |
| MMU2 | 995 | Panasonic BM-ET100US Authenticam | Near infrared | 320 × 238 | 6 |
| UBIRIS | 1,877 | Nikon E5700 | Natural lighting | 200 × 150 | 7 |
| UPOL | 384 | SONY DXC-950P 3CCD camera | Camera flash | 768 × 576 | 8 |
| UTIRIS | 793 | ISG Lightwise LW | Near infrared | 1,000 × 776 | 9 |

common datasets which have been utilized in many different iris recognition related works. Figure 16.5 shows some example images for each of the datasets. Table 16.1 summarizes the most important attributes of the datasets. In the following a short description of each single dataset is given:

**CASIA V2:** We use the first subset of the CASIA V2 iris database [54] (device 1). This subset consists of 1,200 images and was captured using an OKI Irispass-h sensor by the Chinese Academy of Sciences Institute of Automation (CASIA).

**CASIA V3:** was again captured by the Chinese Academy of Sciences Institute of Automation (CASIA) [54] and consists of several different subsets. We used the CASIA V3 Interval subset in accordance with the work of El-Naggar and Ross. This subset consists of 2,639 images captured with a self-developed close-up iris camera.

**CASIA V4:** This is the V4 version of the iris dataset provided by CASIA [54]. Again it consists of several subsets, where we used the Thousands subset. It consists of 20,000 images which were collected using an IrisKing IKEMB-100 camera.

**ICE2005:** NIST, the National Institute of Standards and Technology in the United States conducted a series of biometric recognition contests, one of them was the Iris Challenge Evaluation (ICE) in 2005. The ICE2005 [55] images were captured at the University of Notre Dame with an LG EOU 2200 iris camera and consists of 2,953 images.

**IITD:** The IIT Delhi Iris Database [56] consists of 1,120 images and was acquired by the Biometrics Research Laboratory in the Indian Institute of Technology Delhi (IITD) in 2007. The images were captures with an JIRIS JPC1000 digital CMOS iris camera.

**MMU2:** The MMU V2 iris database [57] consists of 995 iris images. These images are collected using a Panasonic BM-ET100US Authenticam.

**UBIRIS:** The Noisy Visible Wavelength Iris Image Database UBIRIS V1 [58] consists of 1,877 images collected in 2004. The images were captured with a Nikon E5700 digital camera in two sessions.

**UPOL:** The Univerzita Palackého v Olomouci iris dataset [59] consists of 384 images. The irises were scanned by TOPCON TRC50IA optical device connected with SONY DXC-950P 3CCD camera.

**UTIRIS:** University of Tehran IRIS (UTIRIS) image dataset [60] consists of two different sessions, one captured using visible wavelength illumination and the other one using near-infrared illumination. We only used the near infrared subset, which consists of 793 images. The infrared images were captured with an ISG Lightwise LW iris camera.

## 16.3 Experimental setup

We follow the same test methodology as El-Naggar and Ross [41]. For the Iris Texture Classification (ITC) approach as described in Section 16.2 each dataset is randomly split into two distinct subsets, a training and a testing one. UPOL is the iris dataset containing the least images, 384 images only, thus it is split 50:50 into 192 training and 192 testing images. Consequently, for all other images also 192 training and 192 testing images are chosen for the corresponding subsets. The first step in the processing chain is the preprocessing of the ocular images, including iris segmentation and iris unrolling. The unrolled iris patches are then normalized and all having a size $512 \times 64$ pixels. This is done utilizing the USIT (University of Salzburg Iris Toolkit, Version 2.0 available at http://www.wavelab.at/sources/USIT/) software toolkit in version 1.0.3. For the segmentation step the WAHET (Weighted Adaptive Hough and Ellipsopolar Transform) method is used. Figure 16.5 shows one example of an unrolled and normalized iris image for each dataset. For further details on the exact implementation of WAHET and the iris unrolling the interested reader is referred to [61]. The next step is the feature extraction using DenseSIFT, DMD and LBP. Afterwards, the features are dimensionality reduced using a GMM and then Fisher Vector encoding is applied before they are put into a linear SVM for classification. A fivefold cross validation is performed and the mean results of all five runs are used as final results shown below.

For the PRNU-based Sensor Identification (PSI) approach we decided to extract the PRNU from a central patch with varying sizes ranging from $64 \times 64$ up to $576 \times 576$ pixels because of the varying image size of the data sets. We furthermore evaluate the effect of applying the content attenuation PRNU enhancement *ELi*, described in Section 16.1.1, in contrast to not applying it. The configurations for the extraction and post-processing of the PRNU are: Extracted PRNU sizes (from $64 \times 64$ up to $576 \times 576$ pixels), denoising filters (Wavelet and BM3D), PRNU enhancements (ELi and NoEnh), and PRNU detectors (NCC and PCE). Due to the different image sizes of the datasets the number of sensors to discriminate decreases for an increasing PRNU size as shown by the value in parentheses next to the PNRU size in Table 16.8. For each run we selected 192 random images for each data set for the generation of the PRNU fingerprint ("training" set) and another 192 random images as the test set, without overlapping images between both sets. We compute the NCC and PCE correlation scores for all test images with all generated PRNU fingerprints, where the predicted sensor (or class) is determined by means of the highest (rank one) correlation score. The larger the size of the extracted PRNU, the less sensors could be used for evaluating the identification performance for the sensors. The experiment was repeated five times (fivefold cross validation), where the final result is the average of

Table 16.2 *Mean accuracies (mACC) and mean average*
*precisions (mAP) for DenseSIFT, DMD, and LBP*

| Method | DenseSIFT | DMD | LBP |
|--------|-----------|-----|-----|
| mAcc | 0.9838 | 0.9688 | 0.8715 |
| mAP | 0.9968 | 0.9878 | 0.9172 |

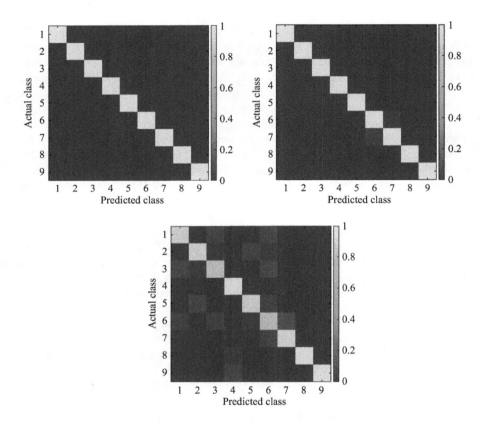

Figure 16.6 *Confusion matrix for DenseSIFT, DMD, and LBP*

all five runs. The described parameters have been chosen to make the results of both
identification/classification approaches as comparable as possible.

## 16.4 Experimental results

Table 16.2 summarizes the results of the ICT approach. It lists the mean accuracy
(mAcc) as well as the mean average precision (mAP) over all five runs. The accuracy

describes the number of correctly classified items (true positives + true negatives) over the number of total items per class calculated per class. The mAcc is just the mean over all single accuracies. The average precision (AP) describes the area under the precision/recall curve calculated per query/class. The mAP is the mean over all AP values. It can be clearly seen that the ICT approach works best using DenseSIFT features. Using DMD and LBP the recognition performance both in terms of the mACC and the mAP is still clearly over 90%. The results show that the ICT approach is able to determine the source of an unrolled iris texture image with a very high accuracy considering the nine iris datasets.

Figure 16.6 shows the confusion matrices for ICT. The numbers on the axes are corresponding to the class IDs in Table 16.1. Considering DenseSIFT it can be seen that for CASIA V2, ICE2005, MMU2, UBIRIS, UPOL, and UTIRIS all images are correctly classified as belonging to the actual dataset. Only some of the CASIA V3, CASIA V4, and IITD images are misclassified.

Figure 16.7 shows the average precision plots for ICT. Again it can be seen that classification works perfectly for CASIA V2, MMU2, UBIRIS, UPOL, ICE2005, and UTIRIS considering DenseSIFT. Considering DMD it still works perfectly for

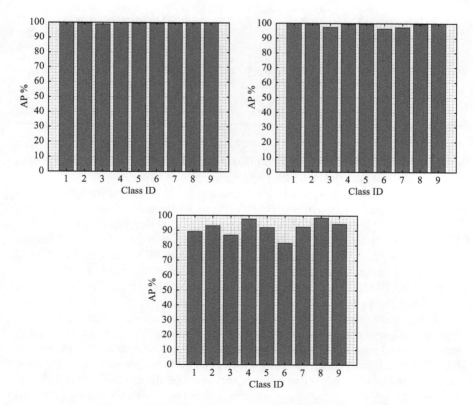

*Figure 16.7　Average precision for DenseSIFT, DMD, and LBP*

*Table 16.3   Mean accuracy (mAcc) and mean average precision (mAP) for all tested PRNU patch sizes and PRNU extraction configurations. The number in parentheses next to the PRNU size indicates the number of different sensors*

| | | Wavelet | | | | BM3D | | | |
| | | NoEnh | | ELi | | NoEnh | | ELi | |
| | **PRNU** Size | **NCC** | **PCE** | **NCC** | **PCE** | **NCC** | **PCE** | **NCC** | **PCE** |
|---|---|---|---|---|---|---|---|---|---|
| mAcc | 64 (9) | 0.4633 | 0.4587 | 0.4545 | 0.4456 | 0.5451 | 0.5397 | 0.5068 | 0.4845 |
| | 128 (9) | 0.7326 | 0.7300 | 0.7437 | 0.7380 | 0.7752 | 0.7696 | 0.7620 | 0.7554 |
| | 192 (8) | 0.9007 | 0.8932 | 0.9112 | 0.9008 | 0.9210 | 0.9279 | 0.9237 | 0.9197 |
| | 256 (6) | 0.9300 | 0.9358 | 0.9347 | 0.9326 | 0.9545 | 0.9507 | 0.9505 | 0.9446 |
| | 320 (5) | 0.9946 | 0.9963 | 0.9973 | 0.9988 | 0.9981 | 0.9994 | 0.9981 | 0.9983 |
| | 384 (5) | 0.9988 | 0.9987 | 0.9990 | 0.9990 | 0.9983 | 0.9981 | 0.9992 | 0.9992 |
| | 448 (5) | 0.9973 | 0.9988 | 0.9983 | 0.9992 | 0.9990 | 0.9998 | 0.9983 | 0.9971 |
| | 512 (2) | 0.9990 | 0.9974 | 0.9984 | 0.9932 | 0.9984 | 0.9990 | 0.9958 | 0.9943 |
| | 576 (2) | 0.9984 | 0.9979 | 0.9974 | 0.9995 | 0.9964 | 0.9974 | 0.9984 | 0.9964 |
| mAP | 64 (9) | 0.4184 | 0.4033 | 0.4325 | 0.4135 | 0.5227 | 0.5146 | 0.4925 | 0.4691 |
| | 128 (9) | 0.7209 | 0.7114 | 0.7438 | 0.7352 | 0.7674 | 0.7619 | 0.7675 | 0.7572 |
| | 192 (8) | 0.8832 | 0.8844 | 0.9166 | 0.9092 | 0.9271 | 0.9300 | 0.9287 | 0.9236 |
| | 256 (6) | 0.9264 | 0.9268 | 0.9389 | 0.9384 | 0.9576 | 0.9530 | 0.9549 | 0.9506 |
| | 320 (5) | 0.9934 | 0.9949 | 0.9977 | 0.9986 | 0.9985 | 0.9992 | 0.9989 | 0.9986 |
| | 384 (5) | 0.9989 | 0.9989 | 0.9992 | 0.9990 | 0.9988 | 0.9987 | 0.9994 | 0.9994 |
| | 448 (5) | 0.9976 | 0.9988 | 0.9985 | 0.9988 | 0.9991 | 0.9994 | 0.9986 | 0.9977 |
| | 512 (2) | 0.9991 | 0.9984 | 0.9987 | 0.9966 | 0.9990 | 0.9997 | 0.9977 | 0.9973 |
| | 576 (2) | 0.9995 | 0.9989 | 0.9990 | 0.9998 | 0.9982 | 0.9991 | 0.9994 | 0.9982 |

CASIA V2, ICE2005, UPOL, and UTIRIS but no longer for MMU2 and UBIRIS though still quite acceptably. LBP's performance is a bit worse.

Table 16.3 lists the PSI results which show that the PRNU size affects the identification performance most across all configurations. Reasonable mAcc and mAP rates can already be achieved with $192 \times 192$ pixel patches, while a patch size larger than $320 \times 320$ yields very good results for the identification of the different iris sensors through their PRNU fingerprint. Neither the choice of PRNU detector nor PRNU enhancement makes a big difference in this case, but better results can be achieved by choosing the BM3D denoising filter over the Wavelet filter for smaller PRNU sizes, as shown in Figure 16.8.

Next we are having a closer look at the results for the single classes or sensors. Figure 16.9 shows the confusion matrix for both denoising filters using a small patch size of $64 \times 64$ pixels, no content attenuation PRNU enhancement (NoEnh) and the NCC detector. It can be seen that the identification performance varies highly among the different classes, where class 9 shows very good results and the classes 1, 2, and 7 show very low identification performance independent of the denoising filter. All other classes show higher accuracies when the BM3D filter is used.

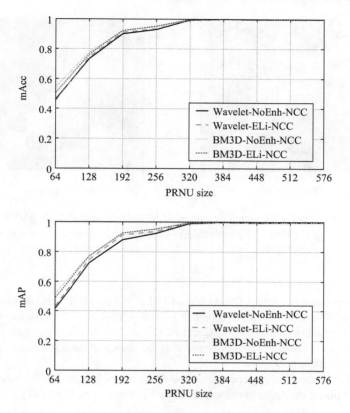

*Figure 16.8    Mean accuracy (mAcc) and mean average precision (mAP) for selected PRNU patch sizes and PRNU extraction configurations*

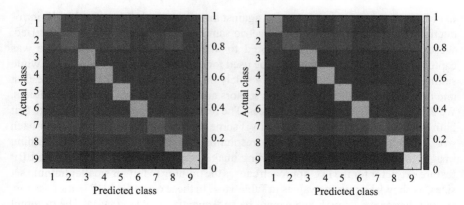

*Figure 16.9    Confusion matrices using the Wavelet (left) and BM3D (right) denoising filters for 64 × 64 pixels PRNU patch size, no content attenuation PRNU enhancement (NoEnh), and NCC detector*

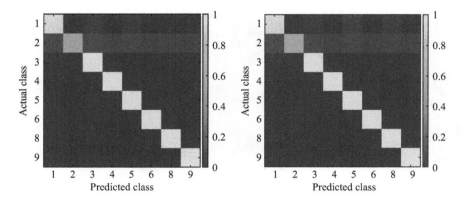

*Figure 16.10    Confusion matrices using the NCC (left) and PCE (right) detectors for 192 × 192 pixels PRNU patch size, ELi content attenuation PRNU enhancement, and BM3D denoising filter*

Having a look at a larger PNRU size of 192 × 192 pixels, BM3D denoising filter and ELi PRNU enhancement, as shown in Figure 16.10, reveals that the identification performance for both detectors, NCC and PCE, is practically identical with only slight differences for all classes. This indicates that the choice of detector is not critical for the overall performance.

The ITC approach in general and the PSI approach with PRNU sizes at least 512 × 512 pixels outperform the approach by El-Naggar and Ross [41].

## 16.5   Practical discussion

In order to secure a biometric system against insertion attacks (described in the introduction) the authenticity of the biometric samples, i.e., images, has to be verified. We examined two approaches tailored to identify the sensor an iris image was captured with. Both approaches can be used for existing biometric systems and while setting-up new ones because they rely solely on intrinsic image properties. The first one is based on the PRNU of the iris sensors and the second one on texture features. We examined different training set sizes. The results shown in Table 16.4 indicate that both achieve good classification results if some conditions are met. The ITC approach works well for a broad range of image resolutions as long as there are enough training images available, i.e., at least 10 training images should be available. Using the LBP feature extractor for the ITC approach causes the training to fail for smaller training set sizes, as shown by missing values in Table 16.4. In these cases the LBP feature vectors are not distinctive enough and cannot be soft-quantized by the GMM. The different image resolutions and iris sizes in the images among the datasets require varying unrolling parameters. Therefore, unrolling and normalization cause a separate level of interpolation for each dataset. Due to the texture classification nature of the ITC

*Table 16.4   Mean average precisions (mAP) for DenseSIFT, DMD, LBP, and PRNU with sizes $128 \times 128$, $256 \times 256$, and $512 \times 512$ for different training set sizes (TS size)*

| TS size | DenseSIFT | DMD | LBP | PRNU128 | PRNU256 | PRNU512 |
|---------|-----------|--------|--------|---------|---------|---------|
| 192 | 0.9937 | 0.9810 | 0.9116 | 0.8227 | 0.9540 | 0.9999 |
| 96 | 0.9919 | 0.9547 | 0.8143 | 0.7870 | 0.9348 | 0.9979 |
| 48 | 0.9833 | 0.9564 | 0.5038 | 0.7426 | 0.9069 | 0.9978 |
| 24 | 0.9668 | 0.9277 | – | 0.7073 | 0.8594 | 0.9937 |
| 12 | 0.9367 | 0.8805 | – | 0.6363 | 0.8003 | 0.9796 |
| 6 | 0.8766 | 0.8062 | – | 0.5244 | 0.7476 | 0.9565 |
| 3 | 0.7897 | 0.6921 | – | 0.4817 | 0.6905 | 0.9348 |
| 1 | 0.6320 | 0.5749 | – | 0.3297 | 0.5734 | 0.8623 |

approach the results could be biased by the interpolation artifacts that may improve the discriminative power of the datasets but not the sensors themselves. This effect could eventually be mitigated by using the ocular images as input in combination with different features like BSIF [62], which is designed to capture image characteristics similar to the PRNU.

The PSI approach works well for bigger PRNU sizes and also works for very few training images, cf. it even works with one single training image for the PRNU512 case. The PRNU is extracted directly from the ocular images as opposed to the unrolled iris texture in the ITC approach. Thus no additional bias is introduced and the discrimination relies solely on the sensors' characteristics.

In this work we only investigated different types of sensors, but not multiple sensors of the same model and manufacturer. As mentioned in the introduction the PRNU is able to distinguish between specific sensor instances of the same model, which is an advantage over the ITC approach in practical deployments since the attacker may have access to the same sensor model. It needs to be clarified whether the ITC approach is able to handle this kind of set-up as well.

Both approaches are suitable if it comes to securing a biometric system depending on the system's configuration. For biometric systems dealing with smaller images but with many training images available the ITC approach is favorable, for systems dealing with larger images but only very few training images available the PSI approach should be used. This is further illustrated in Figure 16.11.

If only few training images are available and the images are small, then a fusion of the two approaches could improve the results. The image size cannot be changed easily but it is easy to provide some more training images (just capture additional data with the sensor) and thus the ITC approach can be used again. For securing a biometric system at first the respective approach has to be trained (PRNU fingerprint generation for the PSI approach) using images of the biometric sensor(s). Every time a new biometric sample is captured the image is analyzed using the pretrained classifier which then tells if the image was captured by one of the biometric sensors

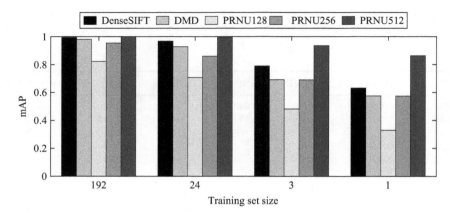

*Figure 16.11* *Exemplary mean average precision (mAP) scores for selected training set sizes*

it was trained with or not. In the latter case it is very likely that an insertion attack happened and the authentication process is aborted.

## 16.6  Conclusion

In this chapter we examined two passive approaches to secure an iris recognition system against insertion attacks by verifying the authenticity of the iris images. The first one, named PSI, is based on the photo response non-uniformity (PRNU) of image sensors and the second one, named ITC, exploits the texture information of unrolled iris images. The examination was performed using images from nine distinct iris databases or sensors, respectively.

The results show that both approaches perform well in identifying the correct sensor an iris image was captured with, though the performance of the PSI approach is dependent on the size of the extracted PRNU. The ITC approach worked well for all datasets. We furthermore examined the impact of the number of images available for the training of both approaches.

Each approach has its advantages and drawbacks depending on the configuration of the biometric system: The PSI approach gives better results if only a small number of high-resolution images is available, while the ITC approach needs a higher number of images to achieve an acceptable performance, but the advantage is they do not need to be of high resolution. In addition, the PSI approach is also suited to distinguish different sensors of the same model. This helps in detecting an injection of images from the same sensor model as deployed in the biometric system. If only a small number of low resolution images is available, a fusion of both approaches is likely to improve the overall performance.

# References

[1]   J. Daugman, "How iris recognition works," *International Conference on Image Processing*, vol. 1, pp. I-33–I-36, 2002.

[2]   N. Ratha, J. Connell, and R. Bolle, "Enhancing security and privacy in biometrics-based authentication systems," *IBM Systems Journal*, vol. 40, no. 3, pp. 614–634, 2001.

[3]   H. T. Sencar, M. Ramkumar, and A. N. Akansu, *Data Hiding Fundamentals and Applications. Content Security in Digital Multimedia*, San Diego, CA: Elsevier Academic Press, 2004.

[4]   C. Wu and C. Kuo, "Comparison of two speech content authentication approaches," in *Proceedings of SPIE, Security and Watermarking of Multimedia Contents IV*, vol. 4675, San Jose, CA, USA, 2002.

[5]   M. Schneider and S.-F. Chang, "A robust content based digital signature for image authentication," in *Proceedings of the IEEE International Conference on Image Processing (ICIP'96)*, Lausanne, Switzerland, 1996.

[6]   J. Tzeng, W.-L. Hwang, and I. Chern, "Enhancing image watermarking methods by second order statistics," in *Proceedings of the IEEE International Conference on Image Processing (ICIP'01)*, Thessaloniki, Greece, 2001.

[7]   C.-S. Lu and H.-Y. M. Liao, "Oblivious watermarking using generalized Gaussian," in *Proceedings of the Seventh International Conference on Fuzzy Theory and Technology*, Atlantic City, NJ, USA, 2000, pp. 260–263.

[8]   W.-K. Lin and N. Burgess, "Listless zerotree coding for color images," in *32nd Asilomar Conference on Signals, Systems and Computers*, Los Alamitos, CA, USA, 1998.

[9]   A. Uhl and Y. Höller, "Iris-sensor authentication using camera PRNU fingerprints," in *Proceedings of the Fifth IAPR/IEEE International Conference on Biometrics (ICB'12)*, New Delhi, India, 2012, pp. 1–8.

[10]  J. Hämmerle-Uhl, K. Raab, and A. Uhl, "Watermarking as a means to enhance biometric systems: A critical survey," in *Proceedings of the 2011 Information Hiding Conference (IH'11)*, Springer Lecture Notes in Computer Science, vol. 6958, Prague, Czech Republic, 2011, pp. 238–254.

[11]  M. M. Yeung and S. Pankanti, "Verification watermarks on fingerprint recognition and retrieval," in *Proceedings of the 11th SPIE Annual Symposium, Electronic Imaging'99, Security and Watermarking of Multimedia Contents*, vol. 3657, San Jose, CA, USA, 1999, pp. 66–78.

[12]  N. K. Ratha, J. H. Connell, and R. M. Bolle, "Secure data hiding in wavelet compressed fingerprint images," in *ACM Multimedia 2000*, Los Angeles, CA, USA, 2000.

[13]  N. K. Ratha, M. A. Figueroa-Villanueva, J. H. Connell, and R. M. Bolle, "A secure protocol for data hiding in compressed fingerprint images," in *ECCV Workshop BioAW*, Lecture Notes in Computer Science, vol. 3087, 2004, pp. 205–216.

[14]  N. Komninos and T. Dimitriou, "Protecting biometric templates with image watermarking techniques," in *ICB*, Lecture Notes in Computer Science, vol. 4642, 2007, pp. 114–123.

[15]    L. Li, C. S. Tong, and S. K. Choy, "Texture classification using refined histogram," *IEEE Transactions on Image Processing*, vol. 19, no. 5, pp. 1371–1378, 2010.

[16]    S. Ding, C. Li, and Z. Liu, "Protecting hidden transmission of biometrics using authentication watermarking," in *2010 WASE International Conference on Information Engineering (ICIE)*, vol. 2, 2010, pp. 105–108.

[17]    F. Ahmed and I. S. Moskowitz, "Composite signature based watermarking for fingerprint authentication," in *Proceedings of the Seventh Workshop on Multimedia and Security, MM&Sec '05*, New York, NY, USA, 2005, pp. 137–142.

[18]    J. Hämmerle-Uhl, K. Raab, and A. Uhl, "Experimental study on the impact of robust watermarking on iris recognition accuracy (best paper award, applications track)," in *Proceedings of the 25th ACM Symposium on Applied Computing*, 2010, pp. 1479–1484.

[19]    A. Lang and J. Dittmann, "Digital watermarking of biometric speech references: impact to the EER system performance," in *Electronic Imaging 2007*. International Society for Optics and Photonics, 2007, pp. 650513–650513-12.

[20]    M. R. Islam, M. Sayeed, and A. Samraj, "Biometric template protection using watermarking with hidden password encryption," in *Proceedings of International Symposium on Information Technology*, 2008, pp. 296–303.

[21]    B. Mahdian and S. Saic, "A bibliography on blind methods for identifying image forgery," *Image Communication*, vol. 25, no. 6, pp. 389–399, 2010.

[22]    T. Gloe and R. Böhme, "The Dresden image database for benchmarking digital image forensics," in *SAC 2010: Proceedings of the 2010 ACM Symposium on Applied Computing*, 2010, pp. 1584–1590.

[23]    M. K. Johnson and H. Farid, "Exposing digital forgeries in complex lighting environments," *IEEE Transactions on Information Forensics and Security*, vol. 2, no. 3, pp. 450–461, 2007.

[24]    K. Choi, E. Lam, and K. Wong, "Automatic source camera identification using the intrinsic lens radial distortion," *Optics Express*, vol. 14, no. 24, pp. 11551–11565, 2006.

[25]    Z. Geradts, J. Bijhold, M. Kieft, K. Kurosawa, K. Kuroki, and N. Saitoh, "Methods for identification of images acquired with digital cameras," in *Proceedings of SPIE, Enabling Technologies for Law Enforcement and Security*, vol. 4232, 2001, pp. 505–512.

[26]    M. Kharrazi, H. T. Sencar, and N. Memon, "Blind source camera identification," in *2004 International Conference on Image Processing, 2004. ICIP '04*, vol. 1, 2004, pp. 709–712.

[27]    S. Bayram, H. Sencar, N. Memon, and I. Avcibas, "Source camera identification based on CFA interpolation," in *Proceedings of the IEEE International Conference on Image Processing, ICIP '05*, vol. 2, Genoa, Italy, 2005, pp. 69–72.

[28]    A. De Rosa, A. Piva, M. Fontani, and M. Iuliani, "Investigating multimedia contents," in *2014 International Carnahan Conference on Security Technology (ICCST)*, 2014, pp. 1–6.

[29]   J. Fridrich, *Steganography in Digital Media: Principles, Algorithms, and Applications*, Cambridge: Cambridge University Press, 2009.

[30]   N. Bartlow, N. Kalka, B. Cukic, and A. Ross, "Identifying sensors from fingerprint images," in *IEEE Computer Society Conference on Computer Vision and Pattern Recognition Workshops, 2009. CVPR Workshops 2009*, 2009, pp. 78–84.

[31]   M. Chen, J. Fridrich, M. Goljan, and J. Lukas, "Determining image origin and integrity using sensor noise," *IEEE Transactions on Information Security and Forensics*, vol. 3, no. 1, pp. 74–90, 2008.

[32]   G. Chierchia, G. Poggi, C. Sansone, and L. Verdoliva, "A Bayesian-MRF approach for PRNU-based image forgery detection," *IEEE Transactions on Information Forensics and Security*, vol. 9, no. 4, pp. 554–567, 2014.

[33]   M. Goljan, J. Fridrich, and T. Filler, "Large scale test of sensor fingerprint camera identification," in *Proceedings of SPIE, Electronic Imaging, Security and Forensics of Multimedia Contents XI*, San Jose, CA, USA, 2009.

[34]   M. Goljan, J. Fridrich, and T. Filler, "Managing a large database of camera fingerprints," in *Proceedings of SPIE, Media Forensics and Security XII*, San Jose, CA, USA, 2010.

[35]   K. Rosenfeld and H. Sencar, "A study of the robustness of PRNU-based camera identification," in *Proceedings of SPIE, Media Forensics and Security XI*, vol. 7254, San Jose, CA, USA, 2009, pp. 72540M–725408M.

[36]   E. Alles, Z. Geradts, and C. Veenman, "Source camera identification for heavily jpeg compressed low resolution still images," *Journal of Forensic Sciences*, vol. 54, no. 3, pp. 628–638, 2009.

[37]   M. Goljan, J. Fridrich, and M. Chen, "Defending against fingerprint-copy attack in sensor-based camera identification," *IEEE Transactions on Information Security and Forensics*, vol. 6, no. 1, pp. 227–236, 2011.

[38]   N. Kalka, N. Bartlow, B. Cukic, and A. Ross, "A preliminary study on identifying sensors from iris images," in *The IEEE Conference on Computer Vision and Pattern Recognition (CVPR) Workshops*, 2015.

[39]   C.-T. Li, "Source camera identification using enhanced sensor pattern noise," *IEEE Transactions on Information Forensics and Security*, vol. 5, no. 2, pp. 280–287, 2010.

[40]   M. Goljan, J. Fridrich, and M. Chen, "Sensor noise camera identification: Countering counter-forensics," in *Proceedings of SPIE, Media Forensics and Security XII*, San Jose, CA, USA, 2010.

[41]   S. El-Naggar and A. Ross, "Which dataset is this iris image from?" in *2015 IEEE International Workshop on Information Forensics and Security (WIFS)*. Piscataway, NJ: IEEE, 2015, pp. 1–6.

[42]   J. Fridrich, "Digital image forensics," *IEEE Signal Processing Magazine*, vol. 26, no. 2, pp. 26–37, 2009.

[43]   J. Lukas, J. Fridrich, and M. Goljan, "Digital camera identification from sensor pattern noise," *IEEE Transactions on Information Forensics and Security*, vol. 1, no. 2, pp. 205–214, 2006.

[44]   K. Dabov, A. Foi, V. Katkovnik, and K. Egiazarian, "Image denoising with block-matching and 3D filtering," in *Electronic Imaging 2006*. International Society for Optics and Photonics, 2006, pp. 606414–606414-12.

[45]   A. Cortiana, V. Conotter, G. Boato, and F. D. Natale, "Performance comparison of denoising filters for source camera identification," in *Media Watermarking, Security, and Forensics XIII*, Proceedings of SPIE, vol. 7880, 2011, p. 788007.

[46]   J. Fridrich, "Sensor defects in digital image forensics," in *Digital Image Forensics: There Is More to a Picture Than Meets the Eye*, 2012, chapter 6, pp. 179–218.

[47]   D. G. Lowe, "Distinctive image features from scale-invariant keypoints," *International Journal of Computer Vision*, vol. 60, no. 2, pp. 91–110, 2004.

[48]   L. Fei-Fei and P. Perona, "A Bayesian hierarchical model for learning natural scene categories," in *Computer Vision and Pattern Recognition, 2005. CVPR 2005*, vol. 2. Piscataway, NJ: IEEE, 2005, pp. 524–531.

[49]   R. Mehta and K. Egiazarian, *Texture Classification Using Dense Micro-block Difference (DMD)*, ser. Lecture Notes in Computer Science. Berlin: Springer-Verlag, 2015, pp. 643–658.

[50]   T. Ojala, M. Pietikainen, and D. Harwood, "Performance evaluation of texture measures with classification based on Kullback discrimination of distributions," in *Proceedings of the 12th IAPR International Conference on Pattern Recognition*, vol. 1, 1994, pp. 582–585.

[51]   F. Perronnin, J. Sánchez, and T. Mensink, "Improving the Fisher kernel for large-scale image classification," in *European Conference on Computer Vision (ECCV10)*. Berlin: Springer, 2010, pp. 143–156.

[52]   M. Cimpoi, S. Maji, I. Kokkinos, S. Mohamed, and A. Vedaldi, "Describing textures in the wild," in *Proceedings of the IEEE Conference on Computer Vision and Pattern Recognition (CVPR'14)*, 2014, pp. 3606–3613.

[53]   F. Perronnin and C. Dance, "Fisher kernels on visual vocabularies for image categorization," in *2007 IEEE Conference on Computer Vision and Pattern Recognition*. Piscataway, NJ: IEEE, 2007, pp. 1–8.

[54]   N. L. of Pattern Recognition. CASIA Iris v4 database. http://biometrics. idealtest.org/.

[55]   P. J. Phillips, K. W. Bowyer, P. J. Flynn, X. Liu, and W. T. Scruggs, "The iris challenge evaluation 2005," in *Second IEEE International Conference on Biometrics: Theory, Applications and Systems, 2008. BTAS 2008*. Piscataway, NJ: IEEE, 2008, pp. 1–8.

[56]   A. Kumar and A. Passi, "Comparison and combination of iris matchers for reliable personal authentication," *Pattern Recognition*, vol. 43, no. 3, pp. 1016–1026, 2010.

[57]   Cteo, "Mmu2 iris image database," available at: http://pesona.mmu.edu. my/~ccteo/, 2008.

[58]   H. Proenca and L. Alexandre, "UBIRIS: a noisy iris image database," in *Image Analysis and Processing – ICIAP 2005*, Lecture Notes on Computer Science, vol. 3617. Cagliari, Italy: Springer-Verlag, 2005, pp. 970–977.

[59]  M. Dobes and L. Machala, "Upol iris image database, 2004," available at: http://www.phoenix.inf.upol.cz/iris, 2013.

[60]  M. Hosseini, B. Araabi, and H. Soltanian-Zadeh, "Pigment melanin: Pattern for iris recognition," *IEEE Transactions on Instrumentation and Measurement*, vol. 59, no. 4, pp. 792 –804, 2010.

[61]  C. Rathgeb, A. Uhl, and P. Wild, *Iris Recognition: From Segmentation to Template Security*, Advances in Information Security, 2013, vol. 59.

[62]  J. Kannala and E. Rahtu, "BSIF: binarized statistical image features," in *Proceedings of the 21st International Conference on Pattern Recognition, ICPR 2012*, Tsukuba, Japan, November 11–15, 2012, pp. 1363–1366.

*Chapter 17*

# Matching iris images against face images using a joint dictionary-based sparse representation scheme

*Raghavender Jillela and Arun Ross*

In this chapter, the problem of matching face against iris images using ocular information is considered. Face and iris images are typically acquired using different sensors: face recognition is predominantly conducted in the visible (VIS) spectrum while iris recognition is performed in the near-infrared (NIR) spectrum. Further, the subject-to-camera distance for face and iris recognition is substantially different. Due to these and other factors, the problem of matching face images against iris images is riddled with several challenges. To address this, we propose a novel matching algorithm based on Joint Dictionary-based Sparse Representation (JDSR) that exploits the use of *ocular information* available in both face and iris images. Experimental results on a database containing 1,358 images of 704 subjects indicate that the ocular region can provide better performance than the iris region in this challenging cross-modality matching scenario.

## 17.1 Introduction

In the biometric literature, the face and iris modalities are typically viewed to be two distinct modalities. This is because these modalities are obtained using different sensors operating in different spectral bands. Face images are usually acquired in the visible spectrum (VIS) while iris images are captured at close quarters in the near-infrared spectrum (NIR). Thus, the possibility of matching face images against their iris counterparts has not been adequately investigated in the literature. However, this problem is not easy to solve due to the following factors:

1. *Cross-modality*: Matching images corresponding to two different biometric modalities is a relatively less explored research topic.

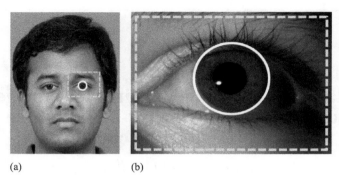

(a)                    (b)

*Figure 17.1    Sample images corresponding to the (a) face, and (b) iris modalities*
*acquired in VIS and NIR spectrum, respectively. The iris and ocular*
*regions have been highlighted using solid white circles and dashed*
*rectangles, respectively. Our goal is to match the face image in*
*(a) with the iris image in (b) using the iris and ocular region*
*information. Notice the variations in scale, image acquisition*
*spectrum, and the level of detail. The average number of pixels across*
*the irides in (a) and (b) are 35 and 110 pixels, respectively*

2.  *Cross-spectrum and cross-sensor*[1]: Variation in imaging spectral bands can
    impact the image formation model and highlight different features of the same
    object.
3.  *Cross-resolution*: Face and iris images typically tend to be of different spatial
    resolutions due to the variations in sensors used. Furthermore, the distance of
    the subject from the sensor and the geometric perspective of the image can be
    significantly different.

Owing to their common presence in both face and iris images, we propose the
usage of *iris* and *ocular*[2] *region* information to perform matching. A sample *RGB*
face and the corresponding *NIR* iris image are shown in Figure 17.1. Iris matching
involves using the highly discriminatory information offered by the rich texture of
the iris [1]. Ocular region matching is performed using information extracted from
the immediate vicinity of the eye [2].

Prior work in iris and ocular region matching has been limited to images acquired
from a single modality (i.e., using either iris or face images) [2–5]. Some researchers

---

[1]Note that cross-sensor does not always mean cross-spectrum. For example, cross-sensor iris recognition
can also refer to the task of matching iris images acquired using different sensors operating within the NIR
spectrum.
[2]Although we use the term *ocular* in this paper, the periocular information may also be available.

*Table 17.1* Publications studying the impact of various imaging factors on iris and ocular region matching

| Authors | Trait considered | Imaging factors considered | | | |
|---|---|---|---|---|---|
| | | Cross-resolution (or stand-off distance variation) | Cross-sensor | Cross-spectrum | Cross-modality |
| Jillela and Ross [6] | Iris | ✓ | ✗ | ✗ | ✗ |
| Ross *et al.* [7] | Iris | ✗ | ✗ | ✓ | ✗ |
| Bharadwaj *et al.* [9] | Ocular region | ✓ | ✗ | ✗ | ✗ |
| Connaughton *et al.* [8] | Iris | ✗ | ✓ | ✗ | ✗ |
| Xiao *et al.* [10] | Iris and ocular region | ✓ | ✓ | ✗ | ✗ |
| Tan and Kumar [11] | Iris and ocular region | ✓ | ✗ | ✗ | ✗ |
| Our work [12] | Iris and ocular region | ✓ | ✓ | ✓ | ✓ |

have studied the impact of various imaging factors (e.g., resolution [6], imaging spectrum [7], and sensors [8]) on iris and ocular region matching. Our work significantly differs from the existing work as we perform cross-modality matching. A list of major publications studying the impact of various imaging factors on iris and ocular region matching is provided in Table 17.1. The significance of our work can be clearly observed from the table.

There are several scenarios in which it may be necessary to match face images against iris images:

- *Matching legacy databases*: Given the growing interest in biometric recognition, it is possible to encounter situations when multiple databases may have to be merged. The biometric modalities available in the independent databases may not always be the same. An example of such a situation is illustrated in Figure 17.2. The process of reliably associating the identities between multiple, seemingly independent, databases can be complicated if: (a) the meta-data corresponding to the images in the individual databases are not comparable and (b) the organizations maintaining the databases do not allow complete meta-data sharing. In such cases, cross-modality matching of face and iris images can assist with the merging of identities if necessary.
- *Surveillance and law enforcement*: In law enforcement scenarios, when a face image acquired using a surveillance camera does not match with any images in a face database, it could be compared against images in an iris database. An example of such a situation is shown in Figure 17.3.
- *Privacy concerns*: In some applications, socio-cultural factors can impede reliable face recognition (e.g., subject wearing a scarf). In such cases, face recognition may not be viable and it may be necessary to match the input face with an iris image to elicit identity [2].

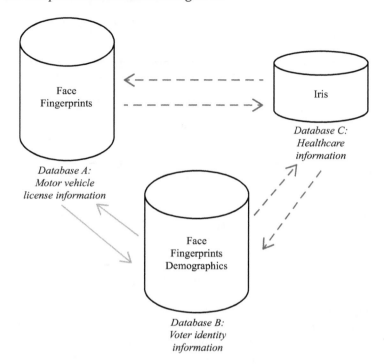

*Figure 17.2   In this example, matching the identities stored in Database C with those stored in other databases can be a challenging process*

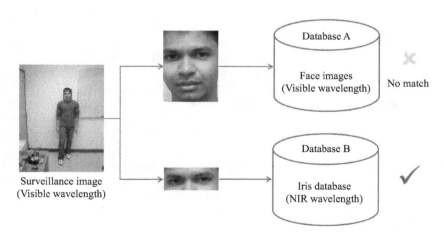

*Figure 17.3   An example scenario depicting the need for iris-to-face matching*

*Table 17.2   Specifications of the images in the database [13]*

| Modalities | Sensor used | Imaging spectrum | Image resolution (pixels) | Number of subjects | | Average iris radius (pixels) | Stand-off distance |
|---|---|---|---|---|---|---|---|
| | | | | Set 1 | Set 2 | | |
| Iris | Oki IrisPass M | NIR | 640 × 480 | 704 | 654 | 110 | 30–60 cm |
| Face | Olympus C-8080 wide zoom | VIS | 2,448 × 3,264 | 704 | 654 | 35 | 2 m |
| Ocular (cropped from face) | Olympus C-8080 wide zoom | VIS | 2,448 × 3,264 | 704 | 654 | 35 | 2 m |

## 17.2   Database

Face and iris images from the Biometric Collection of People (BioCoP) database [13] were used in this work. Both face and iris images were acquired in two different sessions (viz., *Set* 1 and *Set* 2). *Set* 1 and *Set* 2 contain images corresponding to 704 and 654 subjects, respectively. All 654 subjects of *Set* 2 are present in *Set* 1. Both sets contain 1 face, and 2 iris images (corresponding to the left and right sides) of a subject. Face images were acquired using a Olympus C-8080 wide zoom camera, operating in the visible spectrum with a resolution of $2,448 \times 3,264$ pixels. Iris images were acquired using an Oki IrisPass M sensor in the NIR spectrum, with a resolution of $640 \times 480$ pixels. The average radius of the iris in the face and iris images was observed to be 35 pixels and 110 pixels, respectively. The variation in iris radius and spatial resolution occurs due to different stand-off distances for the two modalities. The stand-off distances for face and iris images were maintained at 2 m and 30–60 cm, respectively. Ocular regions of size $225 \times 169$ pixels (approximately) were manually cropped from the face images.

A summary of the databases is provided in Table 17.2. Sample images of the face, cropped ocular region, and the iris are shown in Figure 17.4.

### 17.2.1   Challenges

The following factors make this database challenging:

- **Sensor and imaging wavelength**: The appearance of iris texture in the face and iris images is significantly different due to the use of different sensors and imaging wavelengths (see Figure 17.5).
- **Iris radius**: The stand-off distances and resolution of the face and iris sensors are different. This causes a significant difference in the radii of the irides of the face and iris images. Figure 17.5 illustrates this effect.
- **Viewing angle**: The face and iris sensors were placed at different heights from the ground level, causing perspective variations in the corresponding ocular regions.

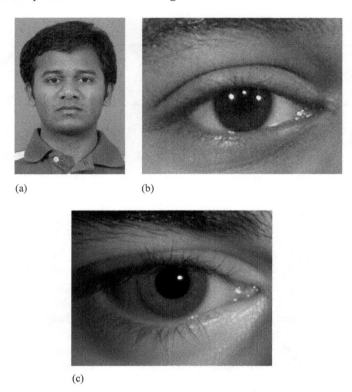

*Figure 17.4    Sample images from the BioCoP database showing the (a) face,*
*(b) cropped ocular region, and (c) iris of a subject. Subject's approval*
*to use the images for illustration purposes is on file*

Figure 17.6(a) and (b) illustrates the described effect. Notice that the folds between the upper eyelid and the eyebrow seen in (a) do not have the same appearance in (b).

- **Illumination**: Images obtained by the iris sensors were observed to exhibit significant illumination variations, as shown in Figure 17.7. Such variations render the task of iris segmentation and ocular feature extraction very challenging.
- **Occlusions**: A large number of images were observed to contain occlusions of the iris and ocular regions, caused by the eyelids, eyelashes, and the hair. Sample images showing such occlusions are provided in Figure 17.8.
- **Sensor-noise and non-uniformity in acquisition**: The Oki IrisPass M sensor depended on its in-built automatic face detection output for localizing and imaging the iris regions. Errors in this process resulted in non-uniform imaging. Furthermore, sensor noise was also observed in some images. Sample images of such cases are provided in Figure 17.9.

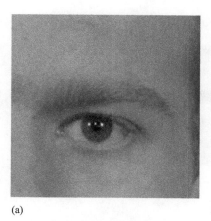

(a)

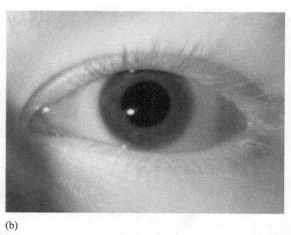

(b)

*Figure 17.5  Corresponding ocular regions from (a) a face image and (b) an iris image, acquired in the visible and NIR spectra, respectively. Note the variation in textural appearance of the iris across these two images*

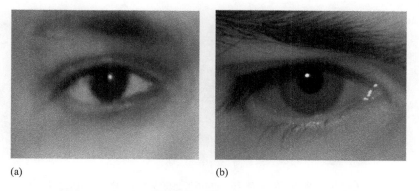

(a)                                        (b)

*Figure 17.6  Images showing variations in the viewing angle between corresponding ocular regions obtained from (a) a face image and (b) an iris image, respectively*

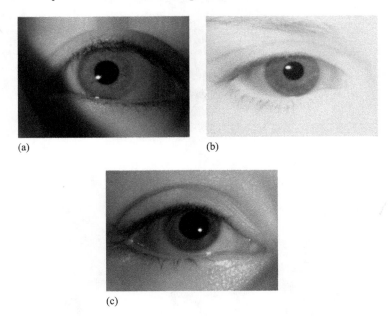

(a)     (b)

(c)

*Figure 17.7   Variations in illumination observed in three images acquired by the NIR iris sensor*

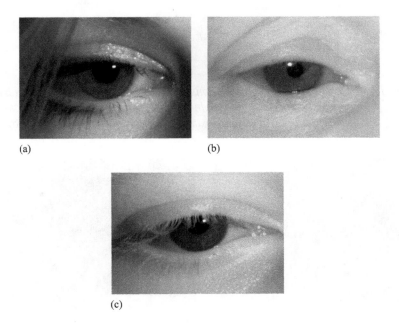

(a)     (b)

(c)

*Figure 17.8   Occlusions of the iris and ocular regions caused by (a) eyelids, (b) eyelashes, and (c) hair, as observed in the images acquired by the iris sensor*

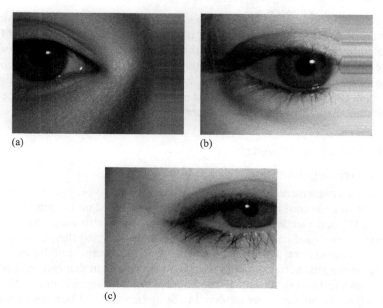

*Figure 17.9    Images depicting the sensor noise and non-uniformity in imaging*

## 17.3    Outline of experiments

Given the NIR iris and visible (VIS) ocular images, the following six different matching scenarios are possible:

1.  Iris matching – NIR iris images with NIR iris images
2.  Iris matching – VIS ocular images with VIS ocular images
3.  Iris matching – VIS ocular images with NIR iris images (cross-modality)
4.  Ocular region matching – NIR iris images with NIR iris images
5.  Ocular region matching – VIS ocular images with VIS ocular images
6.  Ocular region matching – VIS ocular images with NIR iris images (cross-modality)

## 17.4    Iris recognition

Two separate iris recognition algorithms were considered in this work:

1.  *Open-source*: Libor Masek's [14] implementation of Daugman's algorithm [1], and
2.  *Commercial*: VeriEye iris recognition system from Neurotechnology [15].

It was expected that VeriEye would provide better performance than Libor-Masek's implementation. This is because the latter is a rudimentary implementation of Daugman's algorithm. Utilizing both the algorithms allows for a comparison of the widely popular open-source implementation with one of the many available commercial systems. Compared to the commercial system, the open-source implementation allows more control on various factors that impact iris recognition performance (e.g., segmentation methods and feature template size).

## 17.4.1   Open source algorithm

### 17.4.1.1   Iris segmentation

Libor Masek's implementation utilizes Hough transforms to perform iris segmentation [16]. Two additional iris segmentation algorithms based on Integro-Differential Operators [1] and Geodesic Active Contours [17] were also used. All the three techniques were tested on a sample set of 100 iris images and their corresponding 100 ocular images selected from the BioCoP database. The purpose of this experiment is to choose the best performing iris segmentation algorithm that can be used with Libor Masek's feature extraction and matching scheme. The segmentation accuracies[3] of all the three techniques were observed to be around 76%.[4] The main reasons for poor segmentation performance were observed to be:

1.   non-uniform illumination and occlusions in NIR iris images, and
2.   low resolution and presence of dark colored irides in VIS ocular images.

The pupillary boundary in a dark colored VIS image is often difficult to distinguish, even for a human expert. Sample images showing correct and incorrect segmentation outputs obtained using the considered algorithms are shown in Figures 17.10 and 17.11, respectively.

The poor segmentation performances highlight the need for robust segmentation algorithms that can operate on both NIR and VIS images. As the focus of this work is on matching, and not on segmentation, iris regions were manually segmented for further analysis. This process helps in having a reasonably reliable ground truth, while minimizing the impact of incorrect segmentation on the recognition performance. As manual segmentation is a time-consuming process,[5] the open source algorithm was tested only on the sample set containing 100 subjects. Using 1 sample per subject does not generate genuine scores for intra-modality comparison. However, this experiment allows in observing the following aspects of iris recognition:

1.   *Imaging wavelength*: Boyce *et al.* [18] suggest that cross-spectral iris matching performance depends on the difference of imaging wavelengths considered. In this regard, iris regions extracted from the three separate channels of the VIS

---

[3] Segmentation accuracy $= \dfrac{\text{Number of correctly segmented images}}{\text{Number of input images provided}} \times 100$.

[4] Reported using the considered sample set containing 100 NIR iris, and 100 VIS ocular images.

[5] As the iris regions in VIS ocular images are of very small resolution.

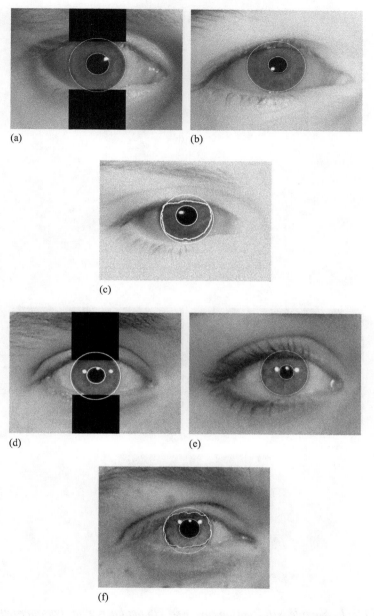

*Figure 17.10 Sample NIR iris (top row) and VIS ocular region (bottom row) images showing correct iris segmentation output obtained using: Integro-Differential Operator [(a) and (d)], Hough transform [(b) and (e)], and Geodesic Active Contours [(c) and (f)] based algorithms*

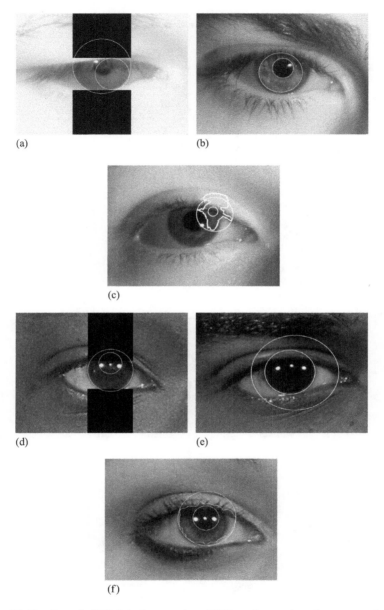

(a)                              (b)

(c)

(d)                              (e)

(f)

*Figure 17.11    Sample NIR iris (top row) and VIS ocular region (bottom row)
images showing incorrect iris segmentation output obtained using:
Integro-Differential Operator [(a) and (d)], Hough transform
[(b) and (e)], and Geodesic Active Contours [(c) and (f)] based
algorithms, respectively*

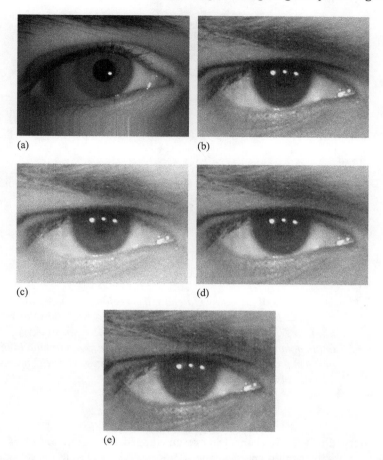

*Figure 17.12   (a) Sample NIR iris image. (b) Corresponding VIS ocular*
*region cropped from an RGB face image. (c), (d), and (e) are*
*R, G, and B channel images extracted from the VIS ocular image,*
*respectively*

ocular images, viz., R, G and B, were individually matched against those extracted from the NIR iris images. An NIR iris image along with the corresponding R, G, and B channel images extracted from a VIS ocular image is shown in Figure 17.12.

2.  *Resolution of the unwrapped iris*: Daugman's rubber sheet model unwraps the segmented iris into a rectangular entity of specific width and height [1]. Two different normalization resolutions were tested: $64 \times 360$ and $32 \times 180$. These resolutions were empirically chosen based on the pupillary and limbic radii observed in the NIR and VIS images.

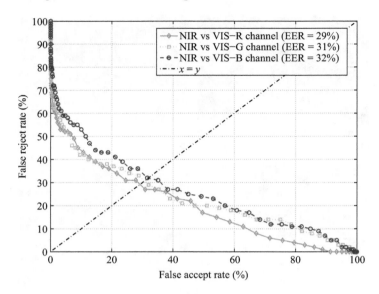

*Figure 17.13    ROC curves obtained by matching iris regions extracted from R, G, and B channels of the VIS ocular images with those extracted from the NIR iris images, using Libor Masek's open source implementation. Note that these curves correspond to matching performance obtained using a subset of images (100 NIR iris and 100 VIS ocular images)*

It was observed that the R channel image unwrapped to a resolution of $64 \times 360$ provided the best recognition performance (EER = 29%). The ROC curves obtained using the above matching considerations are shown in Figure 17.13. The low recognition performance, even with accurate segmentation on a small dataset, indicates the need for a better cross-spectral NIR–VIS iris matching algorithm.

## 17.4.2    Commercial algorithm

Owing to the poor performance of the open source implementation, a commercial iris recognition system, VeriEye [15] was used. Both *Set* 1 and *Set* 2 were combined to generate 1,358 VIS ocular and 1,358 NIR iris images from 704 subjects. Based on the performances observed in the previous experiment, only R channel images were considered. The ROC curves, along with the EERs obtained using VeriEye on the left-side images, are shown in Figure 17.14.

It was observed that 74, out of the 1,358 VIS ocular images, could not be processed by VeriEye. From the ROC curve, it can be observed that VeriEye provides good recognition performance only for the intra-spectral NIR–NIR iris matching.

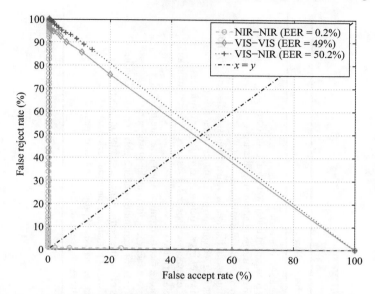

*Figure 17.14   ROC curves obtained using VeriEye to match irides from left-side NIR iris images and left-side VIS ocular images*

Iris matching performance using VeriEye on cross-spectral NIR–VIS and intra-spectral VIS–VIS images was very poor. The reasons for such poor performance could not be deduced as VeriEye does not provide the intermediate details of segmentation and matching.

## 17.5   Ocular recognition

Based on the results presented in Sections 17.4.1 and 17.4.2, it can be observed that the iris biometric does not result in good recognition performance when cross-spectral NIR–VIS images are used. Better performance could be expected if segmentation and matching schemes are significantly improved. However, this can be a very challenging task due to the presence of multiple non-ideal factors. Ross *et al.* [4] and Woodard *et al.* [3] suggest that the ocular region can provide better recognition performance under non-ideal conditions. In this regard, the following sections present various techniques used to perform cross-spectral NIR–VIS ocular region matching.

### 17.5.1   Baseline – local binary patterns

From Table 17.3, it can be observed that the three most popular techniques used for ocular image matching are: Gradient Orientation Histograms [29], Local Binary

*Table 17.3  List of major biometrics research publications focused on the ocular region*

| Year | Authors | Feature extraction | Visible or NIR | Major observation |
|---|---|---|---|---|
| 2009 | Park et al. [19] | SIFT, LBP, GOH | Visible | Serves as a soft biometric |
| 2010 | Woodard et al. [3] | LBP | NIR | Aids iris recognition under non-ideal conditions |
| 2010 | Bharadwaj et al. [9] | GIST, CLBP, SIFT | Visible | Alternative to iris recognition at a distance |
| 2010 | Merkow et al. [20] | LBP with SVM | Visible | Can help in gender classification |
| 2010 | Hollingsworth et al. [21] | Human expertise | NIR | Identifying useful features in ocular images, as discerned by humans |
| 2011 | Xu et al. [22] | WLBP | Visible | Improves face recognition performance under aging |
| 2011 | Boddeti et al. [5] | PDM | NIR | Comparison of iris and ocular recognition performances |
| 2011 | Dong and Woodard [23] | Manual segmentation with LDA and SVM | Both | Analysis of eyebrow shape based features for recognition and gender classification |
| 2011 | Park et al. [2] | SIFT, LBP, GOH | Visible | Can improve face recognition when faces are masked |
| 2012 | Jillela and Ross [24] | SIFT, LBP | Visible | Can improve face recognition under plastic surgery |
| 2012 | Ross et al. [4] | SIFT, LBP, PDM | NIR | Outperforms iris recognition under highly non-ideal conditions |
| 2012 | Padole and Proenca [25] | SIFT, LBP, GOH | Visible | Analysis of various performance degradation factors |
| 2012 | Hollingsworth et al. [26] | SIFT, LBP, GOH | Both | Comparison of human and machine performances |
| 2012 | Oh et al. [27] | Variations of PCA, LDA | Visible | Performance analysis of projection-based methods |
| 2013 | Hollingsworth et al. [28] | Active Shape Models | Visible | Automatic eyebrow segmentation |

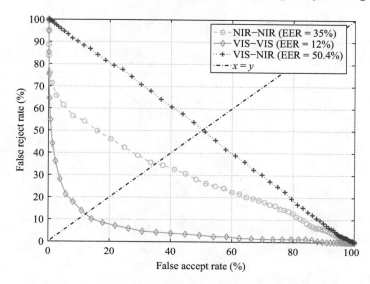

*Figure 17.15   ROC curves corresponding to the ocular region matching using LBP on left-side images*

Patterns [30], and Scale Invariant Feature Transform [31]. Based on an empirical evaluation on a sample dataset,[6] LBP was observed to provide better recognition performance on BioCoP images, in comparison with GOH and SIFT. Therefore, LBP was chosen as a baseline algorithm to perform ocular region recognition on the entire set of 1,308 VIS and 1,308 NIR images. Both the NIR and VIS images were resized to a fixed resolution of (225 × 169) pixels. Such resizing helps in having: (a) rough localization of the regions of interest, and (b) fixed size feature vectors. The EERs obtained using LBP on left-side images corresponding to (a) NIR–NIR, (b) VIS–VIS, and (c) VIS–NIR ocular region matching were observed to be 35%, 12%, and 50.4%, respectively. The corresponding ROC curves are provided in Figure 17.15.

From the results, it can be observed that the cross-spectral VIS–NIR ocular region recognition performance is no better than that of the iris biometric. Two main reasons for the low performance of LBP were observed to be: (a) appearance variations of the ocular regions, caused by different viewing angles of the sensors, and (b) reduced textural quality of the ocular regions in VIS images.

## 17.5.2   Normalized gradient correlation

Correlation-based approaches have been observed to provide better recognition performance when compared to histogram-based approaches (e.g., LBP, SIFT, and GOH) on non-ideal ocular images [4,5]. To test this observation, the Normalized Gradient

---

[6]100 NIR iris and 100 VIS ocular images.

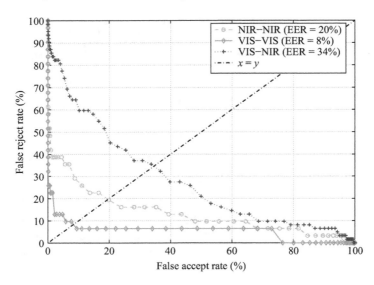

*Figure 17.16   ROC curves corresponding to ocular matching using the Normalized Gradient Correlation technique on left-side images*

Correlation (NGC) method proposed by Tzimiropoulos *et al.* [32] was used. NGC was initially proposed for image registration and alignment. In this work, the technique is modified to perform image matching by using the gradients within a considered region of interest. The advantages offered by NGC are that (a) it can well handle illumination variations and (b) it does not require any learning.

Normalized Gradient Correlation computation between two images is similar to that of the 2D normalized cross-correlation. The only difference is that it operates on the image gradients instead of the raw pixel intensity values. Given two images $I_1$ and $I_2$ of the same resolution, the normalized cross-correlation can be computed as

$$F^{-1}\left\{\frac{\widehat{I_1}\widehat{I_2^*}}{|\widehat{I_1}||\widehat{I_2^*}|}\right\} \tag{17.1}$$

where $\widehat{I_1} = F\{I_1\}, \widehat{I_2} = F\{I_2\}$, and $F$ denotes the Fourier transform operation. On the other hand, the Normalized Gradient Correlation can be computed as

$$F^{-1}\left\{\frac{\widehat{G_1}\widehat{G_2^*}}{|\widehat{G_1}||\widehat{G_2^*}|}\right\} \tag{17.2}$$

where $\widehat{G_1} = F\{G_1\}, \widehat{G_2} = F\{G_2\}$, and $G_1 = G_{1x} + G_{1y}, G_2 = G_{2x} + G_{2y}$. The terms $G_{ix}$ and $G_{iy}$ represent the gradients of the image $I$ in the $x$- and $y$-directions, respectively.

To perform ocular matching between two images using NGC, each image was first tessellated into 12 non-overlapping patches of equal size. The NGC value between the corresponding pairs of patches between NIR and VIS images was computed, yielding 12 different patch scores. The value obtained by the summation of all such patch scores was used as the final match score between an image pair. The ROC curves corresponding to NGC-based ocular matching are provided in Figure 17.16. The EER values corresponding to NIR–NIR, VIS–VIS, and VIS–NIR matching using left-side images were observed to be 20%, 8%, and 34%, respectively.

From the results, it can be observed that NGC provides better recognition performance for cross-spectral VIS–NIR image matching compared to (a) iris recognition, and (b) LBP-based ocular region recognition. The reason for such improved performance is that the process of comparing the patches on a one-to-one basis eliminates the need for having tight correspondences between the sampling points of histogram based methods. Furthermore, using the image gradients ensures that the edge information corresponding to the shapes within the image are accounted for. It has to be noticed that the NIR–NIR ocular matching performance is also improved by NGC.

## 17.6    Ocular matching using joint dictionary approach

Sparse representation-based approaches for image matching have been gaining significant attention in the recent past. Such approaches have been successfully applied in biometrics for face [33], iris [34], and ear [35] recognition. Sparse coding approximates a given image by a linear combination of a few atoms from a dictionary learned from a training set of images. Sparse approaches allow encoding of images into sparse vectors even under various challenging variations. Such methods have also been used in many image processing problems such as denoising [36], restoration [37], and super-resolution [38].

### 17.6.1   Sparse representation framework

The basic framework for sparse representation based approaches for pattern classification mainly depends on: (a) the dictionary formed using the training samples, and (b) the conditions used for obtaining the sparse representation of a given test sample. Consider a training dataset that contains $k$ image samples corresponding to each of $n$ different subjects (i.e., classes). A data matrix, $A$, can be obtained by concatenating all the given training images as column vectors as:

$$A = [I_{1,1}, I_{1,2}, \ldots, I_{1,k}, \ldots I_{n,1}, I_{n,2}, \ldots, I_{n,k}], \tag{17.3}$$

where $I_{i,j}$ denotes the $i$th image sample ($i = 1, 2, \ldots, k$) of the $j$th subject ($j = 1, 2, \ldots, n$), represented as a column vector. Assume that a sufficient number of training images corresponding to each class are available. A new test image, $y$, can then be represented as a linear representation of the data matrix entries. This process can be mathematically represented as:

$$y = \alpha_{1,1} I_{1,1} + \alpha_{1,2} I_{1,2} + \cdots + \alpha_{n,k} I_{n,k}, \tag{17.4}$$

where $\alpha_{i,j}$ represents a scalar coefficient corresponding to $i$th image of the $j$th subject. The above equation can be summarized as:

$$y = Ax, \tag{17.5}$$

where $x = [\alpha_{1,1}, \alpha_{1,2}, \ldots, \alpha_{n,k}]$.

In an identification scenario, the identity of $y$ can be determined by solving the following minimization problem:

$$\hat{x} = \arg\min \|x\|_1 \text{ subject to } Ax = y. \tag{17.6}$$

The coefficient vector, $\hat{x}$, typically contains non-zero entries that correspond to the identity of the test sample, and zeros everywhere else. The generic structure of $\hat{x}$ can therefore be given as:

$$\hat{x} = [0, \ldots, 0, \alpha_{p,1}, \alpha_{p,2}, \ldots, \alpha_{p,n}, 0, \ldots, 0], \tag{17.7}$$

where $p$ corresponds to the true identity of $y$. In the presence of noise, a stable solution can be determined by rewriting (17.6) as:

$$\hat{x} = \arg\min \|x\|_1 \text{ subject to } \|Ax - y\|_2 \leq \epsilon, \tag{17.8}$$

where $\epsilon$ represents a desired threshold.

The above framework was modified by Guo *et al.* [39] to perform *verification* using face images. Given a pair of face images $I_p$ and $I_q$, their sparse representation vectors, $\hat{x}_p$ and $\hat{x}_q$, are first computed using (17.8). The Euclidean distance between $\hat{x}_p$ and $\hat{x}_q$ is then used to determine the similarity between the two images.

It has to be noted that the data matrix $A$ is typically referred as an *overcomplete dictionary* whose base elements are the training images themselves. This leads to a large dimensionality of $A$, resulting in expensive computations. A large number of algorithms have been proposed for learning a compact dictionary while ensuring sparsity [40,41]. One such method [42] to determine a compact dictionary $D$, from a given training data matrix $A$ involves the following equation:

$$D = \arg\min_{D,Z} \|A - DZ\|_2 + \lambda \|Z\|_1, \tag{17.9}$$

where $Z$ and $\lambda$ represent the sparse coefficient matrix and the regularization parameter, respectively.

## 17.6.2  Joint dictionary approach

It has to be noted that the above described sparse representation framework may not be directly applicable to the current VIS–NIR ocular image matching problem. A vast majority of the existing approaches generate a single dictionary using the training images. To perform VIS–NIR ocular image matching, however, two different dictionaries are required. This is because the sparse representations for even the genuine pairs of VIS and NIR images can be significantly different due to variations in image acquisition.

Consider two different dictionaries $D_{NIR}$ and $D_{VIS}$ generated using NIR only, and VIS only training images, respectively. Let $I_{NIR}$ and $I_{VIS}$ represent a pair of NIR and VIS ocular images that have to be matched. The sparse representation vector of $I_{VIS}$, represented by $\hat{x}_{VIS}$, can be computed as:

$$\hat{x}_{VIS} = \arg\min \|x_{VIS}\|_1 \text{ subject to } \|D_{VIS}x_{VIS} - I_{VIS}\|_2 \leq \epsilon_{VIS}. \tag{17.10}$$

Similarly, the corresponding sparse representation vector of $I_{NIR}$, denoted by $\hat{x}_{NIR}$, can be computed as:

$$\hat{x}_{NIR} = \arg\min \|x_{NIR}\|_1 \text{ subject to } \|D_{NIR}x_{NIR} - I_{NIR}\|_2 \leq \epsilon_{NIR}. \tag{17.11}$$

The similarity between $\hat{x}_{NIR}$ and $\hat{x}_{VIS}$ cannot be directly used as a measure of the similarity between the images due to the differences in $D_{NIR}$ and $D_{VIS}$. If the relation between NIR and VIS images could be modeled, $D_{NIR}$ and $D_{VIS}$ could then be related to each other. However, such a modeling is very difficult due to a multitude of factors that cause variations within NIR and VIS images. This problem can be mitigated by combining the dictionaries $D_{NIR}$ and $D_{VIS}$ by a joint dictionary training approach. Such an approach ensures that $\hat{x}_{NIR}$ and $\hat{x}_{VIS}$ have similar non-zero coefficients if $I_{NIR}$ and $I_{VIS}$ correspond to the same subject.

### 17.6.3 Dictionary learning and matching

Consider a data matrix, $A_{VIS}$, generated from a set of VIS images, using (17.3):

$$A_{VIS} = \left[I_{1,1}^{VIS}, I_{1,2}^{VIS}, \ldots, I_{1,k}^{VIS}, \ldots I_{n,1}^{VIS}, I_{n,2}^{VIS}, \ldots, I_{n,k}^{VIS}\right]. \tag{17.12}$$

Let the corresponding NIR images be used to generate $A_{NIR}$:

$$A_{NIR} = \left[I_{1,1}^{NIR}, I_{1,2}^{NIR}, \ldots, I_{1,k}^{NIR}, \ldots I_{n,1}^{NIR}, I_{n,2}^{NIR}, \ldots, I_{n,k}^{NIR}\right]. \tag{17.13}$$

A number of approaches have been proposed for effective dictionary learning [40]. The formulation used in this work is inspired by the joint dictionary learning approach proposed by Yang *et al.* [43] for image super-resolution. The *independent* compact dictionaries for VIS and NIR images, $D_{VIS}$ and $D_{NIR}$, can be determined by:

$$D_{VIS} = \arg\min_{D_{VIS},Z} \|A_{VIS} - D_{VIS} * Z_{VIS}\|_2 + \lambda_{VIS}\|Z_{VIS}\|_1 \tag{17.14}$$

and

$$D_{NIR} = \arg\min_{D_{NIR},Z} \|A_{NIR} - D_{NIR} * Z_{VIS}\|_2 + \lambda_{NIR}\|Z_{NIR}\|_1, \tag{17.15}$$

where $Z$ and $\lambda$ represent the sparse coefficient matrix and regularization parameter, respectively, for the considered set of VIS or NIR test images.

The goal here is to learn a *joint* dictionary such that the sparse representation of an NIR test image will be similar to that of its corresponding VIS image of the same subject. Therefore, (17.14) and (17.15) can be combined as:

$$\arg\min_{D_{NIR},D_{VIS},Z} \|A_{VIS} - D_{VIS} * Z\|_2 + \|A_{NIR} - D_{NIR} * Z\|_2 + \lambda\|Z\|_1. \tag{17.16}$$

The above equation could be rewritten as:

$$\arg\min_{D_{NIR},D_{VIS},Z} \|A_{joint} - D_{joint} * Z\|_2 + \lambda\|Z\|_1, \tag{17.17}$$

where $A_{joint} = \begin{bmatrix} A_{VIS} \\ A_{NIR} \end{bmatrix}$ and $D_{joint} = \begin{bmatrix} D_{VIS} \\ D_{NIR} \end{bmatrix}$.

Efficiently solving the above formulation using numerical methods is a challenge by itself. To this end, multiple solutions have been proposed in the machine learning domain [40]. In this work, the approach suggested by [42] is used. Equation (17.17) is considered to be non-convex in both $D$ and $Z$ collectively, but is convex in one of them if the other is fixed. Therefore, the optimization is performed in an alternate manner over $D_{joint}$ and $Z$. The optimization algorithm is outlined in Algorithm 1. MATLAB® packages provided by [42,43] were used for solving the algorithm. A variation of the joint dictionary approach has been used by Shekhar *et al.* [44]. However, such techniques have been used for identification and not for verification.

The proposed VIS–NIR ocular image matching technique is outlined in Algorithm 2. Thirty percent of the database was used for training and the remaining

---

### Algorithm 1: Joint Dictionary Learning

*Step 1*
  Use a Gaussian random matrix to initialize $D_{joint}$
*Step 2*
  With $D_{joint}$ fixed, update $Z$ by solving the following formulation:
  $$Z = \arg\min_Z \|A_{joint} - D_{joint} * Z\|_2 + \lambda\|Z\|_1$$
*Step 3*
  With $Z$ fixed, update $D_{joint}$ by:
  $$D_{joint} = \arg\min_{D_{joint}} \|A_{joint} - D_{joint} * Z\|_2 \text{ such that } \|D_{joint}\|_2 \leq 1$$
*Step 4*
  Iterate between steps 2 and 3 until convergence.
*Final Output*: $D_{joint}$

---

### Algorithm 2: Proposed VIS–NIR Ocular Image Matching Approach

*Training*
  1. Input: VIS and corresponding NIR training image pairs
  2. Obtain $D_{joint} = [D_{VIS}D_{NIR}]$ using Algorithm 1.
*Testing*
  1. Input: Given VIS and NIR test images, $I_{VIS}$ and $I_{NIR}$
  2. Compute the sparse representation vectors $\hat{x}_{VIS}$ and $\hat{x}_{NIR}$ (use (17.10) and (17.11))
  3. Compute the Euclidean distance, $d$, between $\hat{x}_{VIS}$ and $\hat{x}_{NIR}$
  4. Determine a vector $K$ whose entries satisfy the condition:
    $\{\hat{x}_{VIS}(k) > 0 \text{ and } \hat{x}_{NIR}(k) > 0\}$ or $\{\hat{x}_{VIS}(k) < 0 \text{ and } \hat{x}_{NIR}(k) < 0\}$
  5. Match score between $I_{VIS}$ and $I_{NIR}$ is considered as $d/size(K)$
*Final Output*: $d/size(K)$

70% was used for testing (disjoint subjects). This results in considering 407 and 951 images for training and testing, respectively. The obtained match scores are used as similarity measures between the given images. The ROC curves obtained using the proposed joint dictionary based sparse representation approach on the left-side images are provided in Figure 17.17. The EERs obtained using all the ocular matching techniques considered in this work are listed in Table 17.4. From the results it can be noticed that the proposed joint dictionary based sparse representation approach improves the recognition performance in all the three matching scenarios (i.e., NIR–NIR, VIS–VIS, and VIS–NIR).

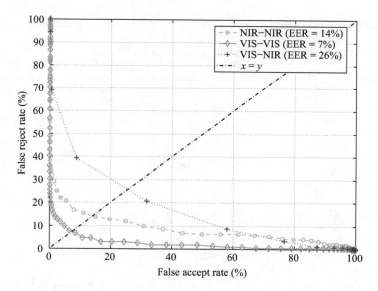

*Figure 17.17   ROC curves corresponding to ocular matching using the joint dictionary based sparse representation approach on left-side images*

*Table 17.4   Equal Error Rates obtained using left-side images of the considered BioCoP database*

|  | NIR–NIR (%) | VIS–VIS (%) | VIS–NIR (%) |
| --- | --- | --- | --- |
| Iris Recognition – VeriEye | 0.2 | 49 | 50.2 |
| Ocular Recognition – Local Binary Patterns (LBP) | 35 | 12 | 50.4 |
| Ocular Recognition – Normalized Gradient Correlation (NGC) | 20 | 8 | 34 |
| Ocular Recognition – Joint Dictionary based Sparse Representation (JDSR) approach | **14** | **7** | **26** |

*Table 17.5   Equal Error Rates corresponding to different values of* λ

| λ value | EERs for VIS–NIR matching (%) |
|---------|-------------------------------|
| 0.01    | 36.5                          |
| 0.05    | 33.1                          |
| 0.09    | 26.0                          |
| 0.15    | 36.2                          |

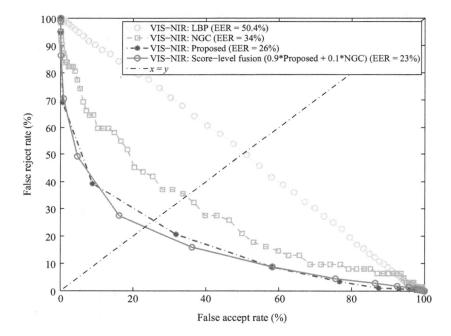

*Figure 17.18   ROC curves for the cross-spectral ocular matching using (a) LBP, (b) NGC, (c) proposed joint dictionary based sparse representation approach, and (d) weighted score-level fusion of all the considered techniques*

## 17.7   Computational details

The value of λ was set to 0.09 in this work. This was done based on observing the recognition accuracies corresponding to different values of λ. The value corresponding to minimal EER was chosen for experiments. It has to be noted that other methods exist to choose the value of λ (e.g., based on the objective function, based on the sparsity of the vector, etc.). However, since the focus of this work is on recognition performance, $\lambda$ is chosen empirically. The EERs obtained for various values of $\lambda$ are listed in Table 17.5. The time required for generating the sparse representation of a given image was observed to be 0.8 seconds using an Intel Core i7 processor with a 3.4 GHz processor and 8 GB RAM.

## 17.8 Score-level fusion

To study the impact of score-level fusion, match scores obtained by all the three considered ocular matching techniques were combined using a simple sum rule. LBP and NGC scores that correspond only to those image pairs used in the testing phase of the joint dictionary-based sparse approach were fused. The weights for fusion were determined empirically with an objective of minimizing the EER. The ROC curves obtained using the individual techniques and by the score-level fusion for VIS–NIR ocular matching are shown in Figure 17.18. From the results, it can be observed that score-level fusion enhances the recognition performance only by a small margin (EER = 23%). This is because of the low recognition performances offered by LBP and NGC techniques.

## 17.9 Summary

The problem of matching ocular regions in *RGB* face images and *NIR* iris images is studied. The variations in modalities, wavelengths, resolutions and sensors render this problem very challenging. A sparse representation-based approach which generates a joint dictionary from corresponding pairs of ocular regions in NIR and VIS images is proposed. The proposed technique is observed to outperform some of the well-known ocular matching techniques. Additionally, this work highlights the potential of ocular region in non-ideal conditions when iris information may not be reliable. Future work would include investigating the robustness of the proposed approach when accurate localization of the ocular regions is not possible. Also, the current work does not account for geometric deformations that can occur in the ocular region. A robust ocular matching model that can simultaneously handle photometric and geometric variations has to be developed.

## References

[1] J. Daugman, "How iris recognition works," *IEEE Transactions on Circuits and Systems for Video Technology*, vol. 14, no. 1, pp. 21–30, Jan. 2004.

[2] U. Park, R. Jillela, A. Ross, and A. K. Jain, "Periocular biometrics in the visible spectrum," *IEEE Transactions on Information Forensics and Security*, vol. 6, no. 1, pp. 96–106, Mar. 2011.

[3]    D. Woodard, S. Pundlik, P. Miller, R. Jillela, and A. Ross, "On the fusion of periocular and iris biometrics in non-ideal imagery," in *20th International Conference on Pattern Recognition*. Piscataway, NJ: IEEE, 2010, pp. 201–204.

[4]    A. Ross, R. Jillela, J. Smereka, *et al.*, "Matching highly non-ideal ocular images: An information fusion approach," *Proceedings of the Fifth IAPR International Conference on Biometrics*, Mar. 2012.

[5]    V. Boddeti, J. Smereka, and B. Vijaya Kumar, "A comparative evaluation of iris and ocular recognition methods on challenging ocular images," in *International Joint Conference on Biometrics*, Oct. 2011.

[6]    R. Jillela, A. Ross, and P. Flynn, "Information fusion in low-resolution iris videos using principal components transform," in *IEEE Workshop on Applications of Computer Vision*. Piscataway, NJ: IEEE, 2011, pp. 262–269.

[7]    A. Ross, R. Pasula, and L. Hornak, "Exploring multispectral iris recognition beyond 900 nm," in *IEEE Third International Conference on Biometrics: Theory, Applications, and Systems (BTAS)*, Sep. 2009, pp. 1–8.

[8]    R. Connaughton, A. Sgroi, K. Bowyer, and P. Flynn, "A cross-sensor evaluation of three commercial iris cameras for iris biometrics," in *2011 IEEE Computer Society Conference on Computer Vision and Pattern Recognition Workshops (CVPRW)*. Piscataway, NJ: IEEE, 2011, pp. 90–97.

[9]    S. Bharadwaj, H. Bhatt, M. Vatsa, and R. Singh, "Periocular biometrics: when iris recognition fails," in *Biometrics: Theory, Applications and Systems*, Sep. 2010.

[10]   L. Xiao, Z. Sun, and T. Tan, "Fusion of iris and periocular biometrics for cross-sensor identification," in *Biometric Recognition*. Berlin: Springer, 2012, pp. 202–209.

[11]   C. Tan and A. Kumar, "Towards online iris and periocular recognition under relaxed imaging constraints," in *IEEE Transactions on Image Processing*. Piscataway, NJ: IEEE, 2013.

[12]   R. Jillela and A. Ross, "Matching face against iris images using periocular information," in *IEEE International Conference on Image Processing (ICIP)*, Oct. 2014, pp. 1–5.

[13]   Federal Bureau of Investigation (FBI) and West Virginia University (WVU), "Biometric Collection of People (BioCoP) database," Proprietary multi-modal biometric database, 2012.

[14]   L. Masek and P. Kovesi, "MATLAB source code for a biometric identification system based on iris patterns," 2003.

[15]   Neurotechnology, "VeriEye Iris Recognition Software," http://www.neurotechnology.com/verieye.html/, 2009.

[16]   R. Wildes, "Iris recognition: an emerging biometric technology," *Proceedings of the IEEE*, vol. 85, no. 9, pp. 1348–1363, Sep. 1997.

[17]   S. Shah and A. Ross, "Iris segmentation using geodesic active contours," *IEEE Transactions on Information Forensics and Security*, vol. 4, no. 4, pp. 824–836, Dec. 2009.

[18]   C. Boyce, A. Ross, M. Monaco, L. Hornak, and X. Li, "Multispectral iris analysis: a preliminary study," in *Computer Vision and Pattern Recognition Workshop*, Jun. 2006.

[19]  U. Park, A. Ross, and A. Jain, "Periocular biometrics in the visible spectrum: a feasibility study," in *IEEE Third International Conference on Biometrics: Theory, Applications, and Systems*. Piscataway, NJ: IEEE, 2009, pp. 1–6.

[20]  J. Merkow, B. Jou, and M. Savvides, "An exploration of gender identification using only the periocular region," in *Fourth IEEE International Conference on Biometrics: Theory Applications and Systems (BTAS)*, Sep. 2010, pp. 1–5.

[21]  K. Hollingsworth, K. Bowyer, and P. Flynn, "Identifying useful features for recognition in near-infrared periocular images," in *Fourth IEEE International Conference on Biometrics: Theory Applications and Systems (BTAS)*, Sep. 2010, pp. 1–8.

[22]  F. Juefei-Xu, K. Luu, M. Savvides, T. Bui, and C. Suen, "Investigating age invariant face recognition based on periocular biometrics," in *International Joint Conference on Biometrics*. Piscataway, NJ: IEEE, 2011, pp. 1–7.

[23]  Y. Dong and D. Woodard, "Eyebrow shape-based features for biometric recognition and gender classification: a feasibility study," in *International Joint Conference on Biometrics*. Piscataway, NJ: IEEE, 2011, pp. 1–8.

[24]  R. Jillela and A. Ross, "Mitigating effects of plastic surgery: fusing face and ocular biometrics," in *International Conference on Biometrics: Theory, Applications and Systems*. Piscataway, NJ: IEEE, 2012, pp. 402–411.

[25]  C. Padole and H. Proenca, "Periocular recognition: analysis of performance degradation factors," in *IAPR International Conference on Biometrics*. Piscataway, NJ: IEEE, 2012, pp. 439–445.

[26]  K. Hollingsworth, S. Darnell, P. Miller, D. Woodard, K. Bowyer, and P. Flynn, "Human and machine performance on periocular biometrics under near-infrared light and visible light," *IEEE Transactions on Information Forensics and Security*, vol. 7, no. 2, pp. 588–601, Apr. 2012.

[27]  B. Oh, K. Oh, and K. Toh, "On projection-based methods for periocular identity verification," in *IEEE Conference on Industrial Electronics and Applications*, 2012, pp. 871–876.

[28]  K. Hollingsworth, S. Clark, J. Thompson, P. Flynn, and K. Bowyer, "Eyebrow segmentation using active shape models," in *SPIE Defense, Security, and Sensing*. International Society for Optics and Photonics, 2013, pp. 871208–871208-8.

[29]  N. Dalal and B. Triggs, "Histograms of oriented gradients for human detection," in *IEEE Computer Society Conference on Computer Vision and Pattern Recognition*. Piscataway, NJ: IEEE, 2005, vol. 1, pp. 886–893.

[30]  M. Pietikamen, "Image analysis with local binary patterns," *Image Analysis, Lecture Notes in Computer Science*, vol. 3540, pp. 115–118, 2005.

[31]  D. Lowe, "Distinctive image features from scale-invariant key points," *International Journal of Computer Vision*, vol. 60, no. 2, pp. 91–110, 2004.

[32]  G. Tzimiropoulos, V. Argyriou, S. Zafeiriou, and T. Stathaki, "Robust FFT-based scale invariant image registration with image gradients," *IEEE Transactions on Pattern Analysis and Machine Intelligence*, vol. 32, no. 10, pp. 1899–1906, 2010.

[33]   J. Wright, A. Yang, A. Ganesh, S. Sastry, and Y. Ma, "Robust face recognition via sparse representation," *IEEE Transactions on Pattern Analysis and Machine Intelligence*, vol. 31, no. 2, pp. 210–227, 2009.

[34]   J. Pillai, V. Patel, R. Chellappa, and N. Ratha, "Secure and robust iris recognition using random projections and sparse representations," *IEEE Transactions on Pattern Analysis and Machine Intelligence*, vol. 33, no. 9, pp. 1877–1893, 2011.

[35]   A. Kumar and T. Chan, "Robust ear identification using sparse representation of local texture descriptors," *Pattern Recognition*, vol. 46, no. 1, pp. 73–85, Jan. 2013.

[36]   M. Elad and M. Aharon, "Image denoising via sparse and redundant representations over learned dictionaries," *IEEE Transactions on Image Processing*, vol. 15, no. 12, pp. 3736–3745, 2006.

[37]   J. Mairal, G. Sapiro, and M. Elad, "Learning multiscale sparse representations for image and video restoration," Tech. Rep., DTIC Document, 2007.

[38]   J. Yang, J. Wright, T. Huang, and Y. Ma, "Image super-resolution as sparse representation of raw image patches," in *IEEE Conference on Computer Vision and Pattern Recognition*. Piscataway, NJ: IEEE, 2008, pp. 1–8.

[39]   H. Guo, R. Wang, J. Choi, and L. Davis, "Face verification using sparse representations," in *IEEE Computer Society Conference on Computer Vision and Pattern Recognition Workshops*. Piscataway, NJ: IEEE, 2012, pp. 37–44.

[40]   I. Tosic and P. Frossard, "Dictionary learning," *IEEE Signal Processing Magazine*, vol. 28, no. 2, pp. 27–38, 2011.

[41]   J. Mairal, F. Bach, J. Ponce, and G. Sapiro, "Online dictionary learning for sparse coding," in *Proceedings of the 26th Annual International Conference on Machine Learning*. New York, NY: ACM, 2009, pp. 689–696.

[42]   H. Lee, A. Battle, R. Raina, and A. Ng, "Efficient sparse coding algorithms," in *Advances in Neural Information Processing Systems*, 2006, pp. 801–808.

[43]   J. Yang, J. Wright, T. Huang, and Y. Ma, "Image super-resolution via sparse representation," *IEEE Transactions on Image Processing*, vol. 19, no. 11, pp. 2861–2873, 2010.

[44]   S. Shekhar, V. Patel, N. Nasrabadi, and R. Chellappa, "Joint sparsity based robust multimodal biometrics recognition," in *European Conference on Computer Vision (ECCV) Workshop*. Berlin: Springer, 2012, pp. 365–374.

*Part VI*

**Future Trends**

*Chapter 18*

# Iris biometrics for embedded systems

*Judith Liu-Jimenez and Raul Sanchez-Reillo*

Embedded systems are becoming widespread nowadays. We can find these systems in many applications, as part of bigger systems but as well as standalone systems. Designed for the increasing necessity of access control, many implementations of iris biometrics systems are being carried out in embedded systems. This chapter is focused on such developments, and therefore is structured as follows: first an introduction to embedded systems is made, to later describe some of the most common architectures of these systems and the design alternatives. After that, considering the particular case of an iris biometric system, detailed requirements are described. These requirements are divided into functionality and security requirements. Later on, state-of-the-art implementations according to different design alternatives are presented. The chapter will come to a set of conclusions at its end.

## 18.1 Introduction to embedded systems

An exact definition of an embedded system is rather complex due to the large variety of them and the great difference in their designs and shapes. An embedded system is a generalized term for many systems which satisfy most of the following requirements [1,2]:

- An embedded system is always designed for a specific purpose, although in some cases it can be used for other tasks that it has not been designed for, but then, the performance is significantly reduced or in some cases null.
- Due to their performance restrictions, an embedded system is typically small in dimensions. The size of an electronic device is highly influenced by the number of peripherals and boards required to perform the task it is designed for.
- The cost of these systems is lower than that of a general purpose machine, the cost has as a direct influence on the optimized design for the specific task.
- These systems usually make use of ROM (Read Only Memory) memories to store their programs and not hard disks or any other big storage systems. ROM memories reduce the storage capability, and therefore, most of the systems use

Real-Time Operating Systems or an embedded code called firmware, where its
size is significantly smaller than general purpose operating systems.
- Due to the application, most of these systems work under real-time constraints,
  such applications range from time sensitive to time critical.
- Depending on the functionality, power restrictions can be applied, i.e., use of
  batteries or power supply from a separate system.
- They may also form part of a larger system. Thus, an embedded system may be
  a stand-alone device or a co-processor.

The term embedded system is quite broad, so is its deployment. The large deployment
of these systems has been motivated by the reduction in the hardware area required to
implement complex systems: The increase in the number of transistors per hardware
area has helped to develop more complex systems within a reduced space which are
more specialized in performing determined tasks. For this reason, it is possible to
find large quantities of embedded systems on the market and this number is con-
stantly increasing as new embedded systems are available which provide increased
performance within a smaller space.

## 18.2   Design strategies for developing embedded systems

### 18.2.1   General architecture of an embedded system

It is difficult to present a general architecture for an embedded system, this is because
each system is designed for a specific purpose. Therefore, the architecture of each
system is unique and optimized for the specific task to be performed [2,3]. However,
a general architecture under which most of these systems can be classified is shown
in Figure 18.1.

As Figure 18.1 shows, most embedded systems are formed using a central processor unit, one or more peripherals and several memories which are required for the
different tasks. All these elements are connected using two main paths or buses, one
which controls the processes and the second the data transferral. The protocols used
by these buses depend on each device and the peripherals connected to them.

The main processor may be a microprocessor (or a microcontroller), a specific
chip, an FPGA (Field Programmable Gate Array), etc. These devices are easily con-
figurable and are capable of performing almost any program. Generally, this main
processor is in charge of controlling the overall performance of the system, although
when necessary, it can also perform other tasks. For this reason, it is usually connected
to both busses so that it can access peripherals and memories depending on the task
required.

Peripherals are only able to perform a single function. Among all the possible
peripherals, here we highlight two different types because of their relevance in the
system:

- Input/output peripherals: these are used to interact with the outer world and
  therefore, are in charge of acquiring data for the system or receiving orders from
  a user or other devices.

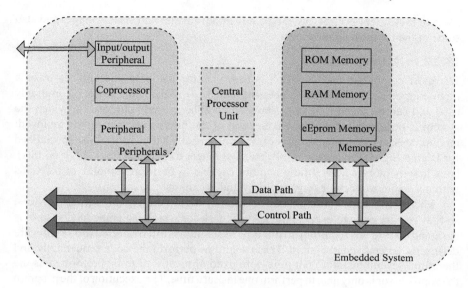

*Figure 18.1   General architecture for an embedded system*

- The so-called co-processors: a co-processor is a dedicated processor which is designed to perform a single function faster than any traditional microprocessor. In an embedded system, one or several coprocessors are developed according to the system requirements.

Another important block of the embedded system is the memory block. Commonly used memories are listed as follows:

- Memory for storing temporary data: RAM (Random Access Memories) memories.
- Memory for storing fixed non-volatile data such as programs: ROM memories.
- EEPROM memory (Electrical Erasable Programmable Read-Only Memory) or flash memory may be required to store data which is required to be kept even when the device is switched off.

Memories should be able to be accessed by the peripherals and central processor unit. Precisely, such circumstance leads to one of the major problems associated with this type of system, memory access may become a bottleneck if several components try to access the information simultaneously. In this case, a memory arbiter is required to control the access.

## 18.2.2   Architecture alternatives

In designing embedded systems, several architectures may be used [2]; this depends on the central element and the scheme to be followed. In this section, we will introduce some of these architectures, pointing out their advantages and disadvantages and also some of their typical applications.

### 18.2.2.1  Microprocessor

The most commonly used solution is based on the use of a microprocessor or a microcontroller as the central processing unit [3,4]. This architecture is commonly used and can be found in several applications. The specific purpose for which the system is designed, influences the election of the microprocessor to be employed: microprocessors used to control processes are generally 8 bits, while those specialized for Digital Signal Processing (DSP) manage longer data words (e.g., 32 or 64 bits), have low memory access timing and are capable to perform complex instructions within a single work cycle (e.g., multiply and accumulate operation).

Microprocessors or microcontrollers are, by definition, general purpose devices, so they require code to execute the specific task for which the embedded system is designed. Due to this requirement this type of solution is often referred to as a software design for an embedded system. These solutions perform the code sequentially and therefore, several tasks are not time-optimized or must wait to be performed as the microprocessor is only able to perform one task at a time. The execution of the program can be divided in two different types:

- Those where the program is executed into an operating system, such as Windows Mobile, Windows Phone, IOS, Android, Linux Embedded or a Real Time Operating System (RTOS). These solutions have the capability for loading and developing new programs where no knowledge of the internal configuration of the system is required. The main disadvantage in using these operating systems is related to the storage capacity and the high performance required to make the system work efficiently. In order to overcome these two disadvantages, the total cost of the system is sometimes increased by using hardware with much more advanced features than the ones really needed.
- Those where no Operating System is used and, therefore, the task to be performed is implemented directly on the microprocessor. This case is typically referred to as a firmware solution. In this case, improved use of the microprocessor is achieved, as the code is more efficiently optimized for the platform and task. Firmware code is usually carried out using high or medium level programming languages, such as C language; however, if a more optimized design is desired or the program storage capability is extremely reduced, assembly code may be needed. Therefore, using firmware requires specialized designers who are capable of programming microprocessors with high level of resource optimization.

The main advantages associated with this architecture are those related to the widely extended use of these types of systems. As a result, many efficient drivers and property intellectual processors cores are easily found where manufacturers also provide several libraries required to perform commonly used functions. These libraries, which are provided by manufacturers, are highly optimized for the processor to be used. However, in many cases, if they are used along with other processors, a poorer performance than expected is provoked.

Another important advantage pertaining to these systems is related to the potential upgrades. When designing these systems, the option to upgrade the firmware is generally available. These architectures are quite flexible, and at the same time, are relatively easy to work with, this due to all the facilities manufacturers provide. All these advantages have prompted the widespread use of this type of architecture. The main advantage then is the reduction of the time needed to develop the solution. On the other hand, the two main disadvantages are the cost when supplying a large number of units, and the lower performance (in particular considering execution time) compared with other architectures.

### 18.2.2.2   Full custom circuits

When massive production and high performance is desired, full custom circuits are the best option. As indicated by its name, a full custom circuit is a hardware solution created to perform a specific task, such as measuring a range of data or creating signals according to the system input. Due to this, the design should be carried out by a specialized designer with experience, resulting in the development of a specialized solution.

The main problem with this type of solutions is related to their fixed costs: designing a solution of this type requires not only an experienced designer, but also specific developing tools, longer design time and more complex facilities when compared to other more straightforward architectures.

However, once designed, these systems are cheaper than other solutions with respect to the manufacturing process. The investment can be recovered by massive production, as the cost per unit is highly reduced when compared to microprocessor architectures. At the same time, the hardware area required for these systems is smaller than other solutions which are used to perform the same task, this makes this solution suitable for small devices and helps to reduce the cost per unit.

The upgrading possibility is variable and depends on the hardware developed, but in most cases, it is not possible as the hardware may not be rebuilt. Therefore, these solutions are considered to be closed designs.

Finally, one major advantage of this type of solution is the time reduction in its execution. This reduction is a direct consequence of using dedicated hardware, as concurrent processes can be performed simultaneously.

Because of the above-mentioned disadvantages, full custom solutions are currently not much used, being only recommended in embedded systems with massive production, which usually means that they can be integrated in other more complex solutions and, therefore, commercialized for different purposes.

### 18.2.2.3   Combining solutions

Hardware/software architectures [5,6], also known as co-design architectures, are a half way solution when considering the previously discussed alternatives. These architectures are characterized by the use of both possible solutions in order to obtain the benefits from both of them. They use full custom solutions to perform some tasks and a microprocessor to perform other tasks. By using this combination, the inherent advantages of both systems are obtained: such as reduced time, reduced area, improved performance and also low power consumption (see Figure 18.2).

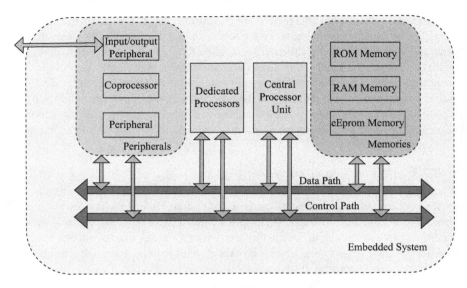

*Figure 18.2    Architecture for a hardware/software co-designed embedded system*

However several disadvantages are also inherited: the designer in this case, should be familiar with the technologies involved in both alternatives; the expected design time is expected to be greater than when considering only the use of a microprocessor. In spite of this disadvantage, such combinations are commonly used and are being recommended for many embedded systems.

This solution is directly related to the co-processor approach mentioned in previous sections. When the coprocessor design is performed exclusively for the system and using a dedicated hardware, while the central processor unit if formed by a microprocessor.

Among all the potential architectures, we would like to highlight the differences between a conventional hardware/software solution and System on Chip (SoC) solutions. In the first case, the hardware is implemented on a dedicated chip and used to perform the task, the microprocessor and the peripheral are located on different chips. However, due to the transistor integration capacity, many chips which contain both a microprocessor and dedicated hardware are becoming increasingly popular. These chips are a half way solution between the conventional microprocessors and the full custom chip solutions. These are commonly referred to as System on Chip (SoC) solutions.

So using SoC solutions gets the benefits of being faster to develop as tools are provided by manufacturers; update mechanisms are possible (full custom solutions are generally closed solutions and no update is allowed); and they are faster in execution and present lower power consumption than microprocessors solutions alone. On the other hand, they do have the complementary disadvantages when compared to the other architectures.

## 18.3 Biometric requirements for embedded systems

An embedded system for biometrics will have as the main purpose to perform either an identification or most probably a verification. These two processes and the data managed during them will have to fill some requirements, which may be quite restrictive under the constraints of an embedded system. In this section such general requirements will be described, under a modality-independent point of view. To illustrate such requirements the extreme case of an embedded system used as a citizen ID Token will be used (e.g., an on-card comparison smartcard).

### 18.3.1 *Functional requirements*

A biometric embedded system has to cover a minimum set of functional requirements [7–9]. The most important functional requirements to consider are the following:

- Portability
  An obvious requirement is related to the size of the device. An ID token should be portable, allowing the user to carry it comfortably. Its size should be as small as possible and should also be lightweight. Commonly used ID tokens exist with different shapes and sizes, from the cards we use every day to tiny chips situated under the skin of a dog.
- User friendly
  Since ID tokens may be used many times on a daily basis to gain access to a particular system, the interaction with the user and the terminal should be as user friendly as possible to prevent any uncomfortable feeling that may result and which may lead to technology rejection.
- Real-time solution
  User access to a system should be as fast as possible. In many cases, it is not permissible to have users queuing up, awaiting identification. This scenario is commonly seen at airports and border controls, where time-consuming identification procedures retain users excessively at the control point.
- Power Consumption
  The power consumption of these devices should be minimal, as the addition of batteries increases the size and weight of the device making it more inconvenient for the user to carry.
- Storage requirement
  The token should be able to store all user data that is required, and also if used, cryptographic keys. For this reason, an EEPROM memory is used as it is important that the information is not erased when the devices is un-powered. The size of this memory depends on several factors: if the device is to contain a cryptographic key, the size of the biometric template, etc. When using Iris Biometrics, the storage capacity required for the complete set of biometric information is less than 2 kB, which is much lower than in other biometric modalities, such as fingerprint correlation algorithms, where storing an image as a template is required.

- Processing capabilities
  The processes designed to be executed by the Token determine its processing capabilities. When using cryptography the token should be able to cipher and decipher the information exchanged. When using Biometrics as an on-token comparison device, the comparison process should be performed within it. This generally signifies very different requirements, i.e., the use of simple algorithms such as the Hamming Distance, or complex processes such as Gaussian models or neural networks [10].

## 18.3.2   Security requirements

As well as the functional requirements, the ID tokens should satisfy several security requirements as the information they store is highly sensitive (i.e., the biometric data of the user is considered personal data, and therefore sensitive). These requirements are related to three main topics [7,11]:

- Communications security
- Physical security
- Electronic hardening

### 18.3.2.1   Communications security

When the system requires high security levels due to potential forgery risks of the information stored, or due to the sensitivity of the services it provides, the system should follow the PKI (Public Key Infrastructure) scheme. Apart from PKI there is also symmetric cryptography, where the communication confidentiality is based on just one cryptographic key which is known by both the user and the service provider (symmetric key). PKI secrecy is based on two keys per user and/or token:

- The public key, as its name points out, is public and can be known by everybody.
- The private key is only known by the user, where this user is in charge of storing and maintaining its secrecy.

Both keys are mathematically dependant, although it is important to point out that just one of these is known and the other one is widely known, and due to their mathematical relation, the private one is not deducible. This is commonly referred to as asymmetric cryptography.

Tokens cipher the information they send using the private key, which can only be deciphered using the corresponding public key. Therefore, both the service provider of the transmitted information and the user's public key may receive this information at the same time, confirming the transmitter identity, i.e., the user. Also the vice versa situation arises where the terminal can provide information to the Token, encrypting it by means of the public key which can only be deciphered by the Token with its private key.

With the key provided once the recognition is performed, the token encrypts all further communications. The algorithms used for encryption varies from one token to another. Among these algorithms, the following two are currently the most recommended for use:

- RSA (Rivest, Shamir and Adleman): Asymmetric. The length of the key is quite long and requires both a public key and a private key; for this reason, this algorithm is widely used in several National ID's such as is the case with the Spanish system.
- AES (Advanced Encryption Standard): Symmetric. This algorithm is being used as a standard by the American government to encrypt their national identification cards or passports. There have been several attempts to decipher codes encrypted by this algorithm without any success up to now.

### 18.3.2.2 Physical requirements

Even when protecting the communications link with the terminal, the token is still a physical device that can be subject to attacks, such as reverse engineering [12,13] or inspection using Focus Ion Beam systems. The main intention of these attacks is to access physical parts of the token to read information and whatever processes are performed.

Physical countermeasures should be considered when creating a token to protect it from such attacks. Some good practices are:

- Make physical intrusion obvious, the external casing needs to be breached to access the internal components: any potential attack of this kind ideally should leave evidence of it, such as scratches or pry marks.
- Make internal tampering externally visible: several tokens provide different means of showing the presence of internal intrusion such as changes in the colour of a tamper-proof window, zeroing the EEPROMs or tripping a "dead-man's switch".
- Encapsulate internal components in epoxy. If the intruder manages to access the token, the internal components should be protected using an additional encapsulation. Epoxy resin increases the difficulty of access to these components, where removing the epoxy heat is required that would most likely damage the encapsulated electronic components.
- Use glue with a high melting point: glue with a high melting point is similar to using epoxy in the encapsulation.
- Obscure part numbers: if the chips are recognizable, possible reconstruction of the board can be carried out.
- Restrict access to the EEPROMs: EEPROMs store non-erasable data, and thus, personal information or cryptographic keys. Therefore, the access to these memories should be as limited as possible to avoid potential intrusions where this data can be read and later replicated.
- Remove or deactivate future expansion points: expansion points are commonly used in hardware for later modifications, debugging or expansion. These may be access points to data and processes stored within the system, and therefore, should be removed.
- Reduce electro-magnetic frequency (EMF) emissions: the information transmitted between the token and the terminal can be deduced from the time spacing between signals, and the electronic-magnetic pulses emitted.

- Use good board layout and system design: Distributing the connections between the components on different layers increases the difficulty of possible eavesdropping of the processes that are taking place within the token, and therefore avoids replication of the processes.

### 18.3.2.3    Electronic hardening

The electronic countermeasures that a token should consider are related to potential electronic attacks such as firmware modification or other potential changes in the electronic parts:

- Protection of flashable firmware: potential changes of the firmware may lead to the key stored in the system being released. If the firmware is changed, the identification process can be avoided by introducing a firmware which always leads to a positive identification. For this reason:
   o Firmware should be encrypted and decrypted by the device before updating it. Additionally before accepting this firmware, both the token and terminal should recognize the rejection of firmware from other devices or intruders.
   o The updated firmware needs to be signed digitally.
   o The firmware should be compiled-optimized before being released, and all debugging and symbol information should be removed.
- Integrity of communications with the terminal: The communication between the terminal and the token should be encrypted before being sent. The data should also be signed and time-stamped to prevent an attacker from recording a transaction by playing it back later. To accomplish this, a shared secret is required. A shared secret can be transmitted in the clear, however this may be intercepted by tapping the communications channel.
- Protection from external injections of spurious data: in the I/O (Input/Output) points of the system spurious data can be introduced to compromise the system or impede its use. Different attacks related to I/O should be considered:
   o Buffer overflows: when receiving data, this data is usually stored in a buffer. If the amount of data received is superior to the buffer capacity, information collapse or overwriting is possible.
   o Use of undocumented command sets and functionality: The communication between the terminal and the token is done using a number of burst bits. These bits should be interpreted by the token as commands. When the bits do not correspond to any specific command, the token should be able to realize this and not perform any actions nor interpret it as a possible attack, denying access to the information and the services provided.
   o Inappropriately structured data elements.
   o Improper failure states: when a token goes into a possible error state, a method should exist to recover from such an error. If not, potential denial of use may occur or the functionality of the device may be altered.

The embedded nature of these devices conforms to its portability and single task operation of storing identification data and in some cases performing this identification. However, and as presented, other considerations should also be taken into

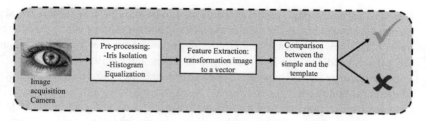

*Figure 18.3    General schema of iris biometrics*

consideration; these are the operational environment, security of the information contained within the device and the security of the device itself.

As can be noticed from this section, several different facts should be considered when developing ID embedded systems: they should be portable and are only designed to perform a single task: store ID data and to perform identification. However, not only these considerations should be taken into account. An ID embedded system, due to the functionality it provides and the environment it works in, should also be able to secure the information it stores and its processing, both physically and electronically.

## 18.4    Iris biometrics analysis for embedded systems

After covering the requirements needed when embedding a biometric solution, it is time to particularize them in the case of an iris recognition system. This section will cover such particularization (see Figure 18.3).

### 18.4.1    Data capture

Firstly, an image of the iris shall be captured by a sensor, this capture is done using a camera which typically works in the Near Infrared range. The image captured should fulfil the quality requirements provided in ISO/IEC 29794-6 [14], and if the image captured is to be transferred to another system, then it should use the ISO/IEC 19794-6 [15] data format. As a summary, a subset of the requirements to fulfil is the following:

- The data transmitted shall contain one or more representations of the eyes of the same user.
- Each representation may be an un-cropped image with the iris far enough of the image borders, or it can be a cropped version of such image, either with 640×480 pixels, or just with the iris centred with a horizontal margin of 0,6R in each side, and a vertical margin of 0,2R up and down, being R the iris radius. In this latter case, the image may even have masked the information on eyelids.
- Usable iris area: the fraction of iris portion of the image which is not occluded by eye lids, eyelashes or specular reflections. At least 70% of the iris should be visible.

- Iris–sclera and iris–pupil contrast: which should be enough to be used in iris segmentation algorithms. In the case of iris–sclera contrast values over 5 are required, meanwhile, between iris and pupil the values required are over 30. If these values are not achieved, the standard recommends increasing the illumination.
- Pupil boundary circularity which is measured using the Fourier series expansion of the pupil boundary.
- Grey scale utilization: a useful iris image should have a dynamic range of 256 grey levels, allocating at least 8 bits with a minimum of 6 bits of useful information.
- Iris radius: which should be at least 80 pixels.
- Pupil dilation: which is measured by the pupil-iris ratio, and shall be between 20 and 70. This value is referred to the pupil constriction or dilation.
- Iris pupil concentricity: it measures the distance between the approximated iris circle and the approximation of the pupil circle. It shall be 90 or more.
- Sharpness: this is an important factor in embedded system, as it depends on the direction of the focus system of the acquisition camera. According to the standard it has to be measured as a function of the power spectrum after filtering with a Laplacian of Gaussian operator.
- Motion blur: measures the degree of distortion in the image due to motion. Also becomes important for embedded system which can have the camera not fixed in a position, or can be part of a portable system such as smartphone, and therefore, motion blur can happens easily.
- Illumination similarity: which refers to a different illumination sources for different images which are going to be compared.
- Spatial sampling rate: requirement which influences in the resolution of the camera. It shall be equal or bigger than 15.7 pixels per millimetre.
- Optical distortion: an optical distortion in the acquisition sensor will lead to spherical or chromatic aberrations, astigmatism, etc. Theses distortions should be avoided.
- Pixel aspect ratio: ratio between the width and the height of rectangular pixels. Values of 0.99 and 1.01 are the one suggested in the standard.
- Sensor signal-to-noise ratio: related to the information captured by the camera and the noise also captured. These values shall be larger or equal to 36 dB.

The data capture module becomes one of the major bottlenecks for implementing iris biometrics in embedded systems. Satisfying the ISO/IEC 29794-6 and 19794-6 requirements in addition to the embedded systems requirements mentioned in previous section lead to a trade-off which should be considered:

- External Illumination, cameras with a large depth of focus for capturing quality images increases the hardware required, and therefore, portability can be affected.
- Interacting with such a system using a portable device can require the user to provide several images and several tries to satisfy the quality measurements. Embedding complex focus systems to acquire quality images requires a considerable hardware area, which in some cases can be difficult or not affordable. Other possible solution for this acquisition is to provide user instructions to provide this quality image. For such, user-friendly interfaces, comfortable systems,

feedback, etc. should be considered in the design. This solution in spite of all this features will probably require the user to interact with the sensor some times.

## 18.4.2   Preprocessing

After capturing an image of the eye, a **pre-processing** block takes place to perform some processes necessary for the following blocks. Among these processes we can find:

- Locating the iris in the image by segmentation, detecting both iris boundaries (iris-sclera and iris-pupil).
- Those referred to improve image quality, such as image equalization.
- Other processes which reduce the computational load of the system or make it easier to follow block tasks, such as image resizing or image transformations.

Generally this block is the most costly computationally speaking. Most of the processes of this block relays in examining the image either recursively or managing images, transformation of them, etc. All these processes are based on managing big amount of data, therefore, this data should be read from memory, manage, etc. This introduces delays in time.

Taking into consideration, for example, John Daugman's integro-differential operator [16], the pre-processing block looks for the iris in the image and separate it from rest of the eye parts. The strategy proposed is based on coarse-to-fine search. For obtaining the maximum of that operator, several different points are used as centre points, different radii values, etc. and with those values, data accesses are made to obtain the intensity value at such point. Therefore, this process implies multiple memory accesses, which should be done sequentially for two reasons:

- Memory access is sequential.
- As a coarse-to-fine search is proposed, results from the coarse search will determine the area where to perform the fine search.

Because of this sequential search, the time required for this process is relatively high when comparing with other modules.

Other quite common approach is considering the iris as a circular corona use the Hough transform for detecting circles [17–20]. Again this approach requires a lot of computational time, as it is based on a recursive schema.

Active contours approaches, etc. are based on, again, coarse-to-fine search [21], and several iterations should be performed.

As can be observed, most of the iris segmentation algorithms are based on continuous accesses to memories. This implies high processing time as well as an important use of intermediate memories. Intermediate memories should be used in this module to store temporary data. As so, it is not necessary to use ROM memories which access is slower than RAM memories. Other possible design option instead of using dedicated memory chips is to use the flip flops contained in an FPGA (if this is the dedicated hardware solution used). This option is faster than using a dedicated chip, however, as at this point, images are being managed, and the amount of memory required can be enough to increase significantly the hardware area and affects portability.

Data managed in the pre-processing cannot be considering as data to be protected, as it is just segmentation or slights modifications of the iris image.

### 18.4.3    Feature extraction

Feature extraction consists in transforming the iris into a number of features, denoted as feature vector or iris code, which represents the iris under study. This transformation is the most characteristic part of the algorithm as it determines the performance of the rest of them. Different algorithms are presented, but all of them try to represent the iris structure of ridges or valleys into a measurable vector. The feature extraction never tries to detect colour or size of the iris, as these two features change along human life.

One first step that is usually performed is a normalization of the iris image to overcome problems due to different iris sizes. The iris size depends on the pupil dilation or other factors, such as eye lashes, etc. which can occlude partially the iris. For compensating such factors, a normalization is made. This normalization can be considered as reorganization of the information, and for that several memory accesses are performed following one order. In addition to the sequential access to memories, this access is performed using complex operations with float point operations, such trigonometric functions. This also establishes an important point to consider in implementing one module in one or another platform: the data accuracy.

Dedicated hardware for providing advantages in speed works usually in fixed point operations with or without decimals, this implies a potential recursive loss of accuracy depending on the iterations, etc.

Software, especially if using operating systems, can support floating point operations or fixed point operations. But on the other hand they require more time to compute a single operation.

Regarding the feature extraction itself, many approaches can be found. In some cases, such as [22] some operations are performed recursively, the output of on step is the input of the following.

When opting for one or another potential platform to implement it, parallelism should be considered. In opposite to what usually happens in the pre-processing block, in the feature extraction in some cases, it is possible to parallelize some processes and use either different platforms or just dedicated hardware and accelerate the process.

### 18.4.4    Comparison

The code obtained by the feature extraction should represent the iris and be different enough from one user to another. This differentiation from the sample to any previous data stored, i.e., template, is performed by the comparison algorithm. This algorithm calculates the distance between the sample and the template, and depending on these results determines if the sample belongs or not to the user whose template is being used in the comparison.

Usage of one or another comparison algorithm is dependent on the nature of the feature vectors. In iris many implementations lead to binary feature vectors. The most common and with best results algorithm is the Hamming distance. This distance can be easily computed on dedicated hardware using different approaches [23], determined for the access to memories either intermediate or storage memories.

Comparison block is the block which requires more secure countermeasures to protect the data managed. In this block two main data are used: the feature vector obtained from the sample and the template vector. These vectors can be obtained by MITM, eavesdropping attack for latter replicate them or replicate slight modifications (hill climbing) in the comparison module. Therefore, special countermeasures should be considered in this module and the memories it uses in order to prevent both security and privacy attacks.

## 18.5   Existing implementations

Considering the different strategies and the requirements that an Iris biometric system has, several embedded proposals can be found in the literature. Proposals for different platforms can be found and some results can be extrapolated. In this section we will describe some of these state-of-the-art implementations.

It is important to note that in the case of implementations not related to smartphones, the input sensor is mostly not considered in the design, although there are some authors that have decided to use a standard camera (i.e., not an iris camera) instead of images coming from a pre-captured database. In the case of smartphone implementations, there are two main approaches: one using the already embedded standard camera and another one, connecting to an external camera built up for a future integration in a mobile device, or adapting the lighting conditions. But in few words, there are not improvements and/or recommendations for embedding the capture process, as most publications focus on the signal processing and comparison parts.

As can be seen from previous sections, modifications in the algorithm sometimes have to be done. Most of those modifications are either algorithm simplifications or just avoiding to use some operations, such as square root, divisions, etc. These limitations become necessary for satisfying specific restrictions which are imposed by design, such as the reduction of processing time. Another important reason to modify or expect different results between conventional software implementations and embedded system implementations is the precision of the data processed. Embedded systems due to their reduced functionality do not use data lengths as PCs do. PCs may work with 64 bits, bigger ALUs, data buses, memories, etc. Embedded systems typically use a lower data length, although recent advances in microcontrollers allow the use of 32-bit architectures in most systems. Moreover, in many cases, embedded systems do not work with floating point operations, which implies in many cases small deviations from the performance results obtained.

Therefore, designers have to modify the algorithm to fit the platform necessities. An example of it can be found in [24], where authors present a modified Iris Biometrics with reduced operations to be implemented in a hardware coprocessor.

## 18.5.1    Custom hardware-FPGA

Most of the state of the art in implementations using dedicated hardware for Iris recognition is based on FPGAs. FPGAs are slower than ASIC (Application Specific Integrated Circuits), however, the commodity and the development time makes them suitable for prototype purposes. Additionally, FPGAs can be configured using a PC, so no extra hardware is necessary.

Hematian *et al.* in [25] present a proposal where the iris locator and the iris unwrapper are performed in an FPGA. In their proposal they obtain a system which leads to a zero delay in those processes. This is done by including in the FPGA the camera controller and the SRAM controller decreasing the memory access time.

Considering existing algorithms, several proposals have been made in literature for implementing partially or totally all the processes of an iris biometric system: [26–29]. Different FPGAs and different algorithms have been used, but in all cases the conclusions are:

Implementing any process in the FPGA, thanks to the hardware speed and the possibility of parallelizing processes, decreases the computational time.

For implementing such solutions, an expertise from the designer is required, both in the algorithm itself, but also in the specific architecture used in order to take advantages of it.

The power consumed in much lower than using a PC.

## 18.5.2    Software implementations

### 18.5.2.1    General purpose microprocessors

In many cases, the effort required to make an implementation on dedicated hardware. Time needed for such implementations is most of the times not affordable, so faster implementations are needed and this is where microprocessor-based designs get their importance. Additionally, microprocessors provide further advantages such as:

Code can be updated easily and in convenient manners.

Due to their general purpose functionality, they provide the possibility of not only performing the biometric recognition process, but also other functionalities.

A microprocessor can include an operating system, which will provide transparency for the developers regarding the management of some peripherals, such memory, input/output, etc.

The main disadvantage is that a microprocessor is not as fast as dedicated hardware.

Several cases can be found in the literature: in [30], an implementation based on an ARM7 is detailed. This proposal is done on the microprocessor directly, i.e., managing the peripherals from the source code. However, in [31] an ARM9-based proposal is made, embedding a Linux operating system on the microprocessor, for later implementing the algorithm.

### 18.5.2.2    DSP

Another different approach will be using a Digital Signal Processor (DSP) instead of a general-purpose microprocessor. A DSP is a specific device optimized for performing digital signal processing operations such as filtering, thanks to its architecture which includes specific busses and memory organization, together with the addition of MACs (multipliers-accumulators) in the computational unit. In [32,33] two implementations are shown using DSP for Iris recognition. Compared to general-purpose microprocessor, an important time reduction is achieved using DSPs, although the possibility of including additional functionality is reduced. In a real system, a DSP will act as a co-processor of a general purpose platform, being the DSP devoted to the biometric recognition process.

### 18.5.2.3   Co-design

As can be seen, using dedicated hardware- or software-based solutions provide advantages which should be considered in embedded systems. But also, each of the alternatives carry some disadvantages. Combining both architectures can provide the best of both worlds into a single implementation:

Dedicated hardware can be used as the implementation platform for those processes which requires more time, are suitable of being parallelized.

Software-based approaches can provide an important tool for controlling peripherals or just the overall functionality and coordinating the tasks performed by the dedicated hardware.

The idea of a co-design approach is simple, but it also presents some challenges. The main challenge is how to partition all processes involved, i.e., to decide which process should be implemented in hardware and which ones in software. The problem of partitioning is not trivial and is widely studied under the generic title of "hardware-software co-design".

Grabowski *et al.* presented in [34] an architecture based on two FPGAs and two DSPs. The different processes of the algorithm were divided either on the FPGA or DSP, establishing a communication channel among different devices. The election of one platform or another was made manually and attending to parallelism motives.

In [35] Lopez *et al.* followed a similar idea for partitioning, but they implemented processes both in hardware and in software to determine the time reduction achieved and later on to take a decision on the partitioning. Also they compared their results with an ARM9 implementation.

Finally, a further partitioning implementation is described in [36] where authors propose a partitioning method which considers time, hardware area, security and power consumed considering not only processes but also communication of data among them.

### 18.5.2.4   Mobile devices integrations

But although some work has been done with other platforms the most used platform nowadays is smartphones. Smartphones can be considered as embedded systems in the sense that they follow the architecture mentioned before, and fulfil the requirements mentioned. However, the market has impulse their development to almost make them PCs. The interest on these platforms lie on the increased use of these devices for social communications, internet banking, and other applications. In many of them, the necessity of recognizing the user univocally and comfortable has led to the necessity of developing iris systems on them.

The interest in this topic is so high that some competitions have appeared, such as MICHE [37] or MIR2016 [38]. In both cases, the main objective is to test algorithms considering images obtained in mobile devices. For such, databases have been acquired using different smartphones.

As close embedded systems smartphones are limited in some facts, as well as the possibility of increasing their hardware is quite limited. The main concern and problem is related to the acquisition camera. Cameras in smartphones work on visible range and the focus system is electronic, therefore, problems related to not suitable images are quite common. However, everyday new and improved smartphones appear in the market, and more sophisticated cameras are included in them.

For solving these gaps, some solutions can be found in the market. These solutions are based on introducing an additional hardware module in the mobile device. This module provides an external illumination [39], a mirror to help the user to find the best position for providing a good iris image and in some cases a camera [40,41].

Other possible approach is trying to improve the image obtained by regular cameras by software means. This solution is proposed in [42,43]. Or by using the camera to obtain a face image, and from that image segment the periocular part and from there isolate the iris [44]. By this approach, not only iris biometrics is used, but also periocular and face recognition, so a multimodal solution is proposed.

An extended study about the free constraints these systems have to deal with can be found at [45].

## 18.6   Conclusions

As with many biometric modalities, the interest in implementing iris recognition in an embedded system is growing as new services and platforms appear in the market. Iris recognition presents, as an advantage, one of the simplest and fastest comparison algorithms, which makes it perfect for a whole range of platforms. Unfortunately, the acquisition module as well as the pre-processing stages is extremely complicated. These two inconveniences open a whole range of R&D activities. Pre-processing and feature extraction algorithms require adapted processing capabilities, which can be solved using different strategies: from general purpose microprocessor approaches, till the design and manufacturing of ASICs. Several references can be found covering the different alternatives, as have been mentioned along this chapter. Depending on the target application and the time allowed for the development of the solution, the designer will have to decide on which approach to consider.

But the acquisition process is still an open issue, as there is a need of dedicated cameras with specific focus capabilities, adaptable to different scenario conditions. If this challenge is added to the current trend of migrating this technology into mobile devices, the number of scientific works is expected to increase in the near future. This shows an interesting field within this biometric modality.

# References

[1] Berger A. S., *Embedded systems Design: An Introduction to Processes, Tools & Techniques*. Ed. CMP Books, 2001.

[2] Noegaard T., *Embedded System Architecture: A Comprehensive Guide for Engineers and Programmers (Embedded Technology)*. Ed. Newness, 2005.

[3] Gupta R. K., Embedded processors. *Project Report for ICS 212*, March 2000.

[4] Jerraya A., Wolf W., *Multiprocessor Systems-On-Chips (Systems on Silicon)*. Ed. Morgan Kaufmann, 2004.

[5] De Micheli G., Gupta R. K., "Hardware/software co-design," *Proceedings of the IEEE*, 1997, Vol. 85(3), pp. 349–365.

[6] Ernst R., "Codesign of embedded systems: status and trends," *IEEE Design & Test of Computers*, 1998, Vol. 15(2), pp. 45–54.

[7] Reid P., *Biometrics for Network Security*. Prentice-Hall, 2003.

[8] Sanchez-Reillo R., Liu-Jimenez J., and Entrena L., "Architectures for biometric match-on-token solutions," *ECCV Workshop BioAW*, Prague, Czech Republic, 2004, pp. 195–204.

[9] O'Gorman L., "Comparing passwords, tokens and biometrics for user authentication," *Proceedings of the IEEE*, 2003, Vol. 91(12), pp. 2021–2040.

[10] Sanchez-Reillo R., *"Mecanismos de Autenticacion Biometrica mediante tarjeta inteligente,"* PhD thesis, Polytechnic University of Madrid, 2000.

[11] Burr W. E., Dodson D. F., and Polk W. T., *"Electronic authentication guideline – information security,"* Technical report, Nist National Institute of Standards and Technology, 2006.

[12] Kingpin, "Attacks on and countermeasures for USB hardware token devices," *Proceedings of the Fifth Nordic Workshop on Secure IT Systems*, 2000.

[13] Oblivion B. and Kingpin, *Secure Hardware Design*, 2000. Black Hat Briefings.

[14] ISO/IEC 29794-6 "Information technology – biometric sample quality – part 6: iris image," JTC1: SC37, International Standard Edition, 2014.

[15] ISO/IEC 19794-6 "Information Technology – Biometric Data Interchange Formats – Part 6: Iris image," JTC1: SC37, international standard edition, 2014.

[16] Daugman J. G., "High confidence visual recognition of persons by a test of statistical independence," *IEEE Transactions on Pattern Analysis and Machine Intelligence*, November 1993, Vol. 15(11), pp. 1148–1161.

[17] Wildes R. P., "Iris recognition: an emerging biometric technology," *Proceedings of IEEE*, 1997, Vol. 85(9), pp. 1348–1363.

[18] Li Ma, Tieniu Tan, Yunhong Wang, and Dexin Zhang. "Efficient iris recognition based characterizing key local variations," *IEEE Transactions on Image Processing*, 2004, Vol. 13 (16), pp. 739–750.

[19] Li Ma, Yunhong Wang, and Tieniu Tan. "Iris recognition using circular symmetric filters," *Proceedings of the 16th International Conference on Pattern Recognition*, Quebec, Canada, 2002, Vol. 2, pp. 414–417.

[20]    Masek L., "Recognition of Human Iris Patterns for Biometric Identification," Master's thesis, School of Computer Science and Software Engineering, University of Western Australia, 2003.
[21]    Daugman J., "New methods in iris recognition," *IEEE Transactions on Systems, Man and Cybernetics – Part B: Cybernetics*, 2007, Vol. 37(5), pp. 1167–1175.
[22]    Sanchez-Avila C., Sanchez-Reillo R., "Two different approaches for iris recognition using Gabor filters and multiscale zero-crossing," *Pattern Recognition*, 2005. Vol. 38(2), pp. 231–240.
[23]    Liu-Jimenez J., Sanchez-Reillo R., Sanchez-Avila C., Entrena L., "Iris biometrics verifiers for low cost identification tokens," *Proceedings of the XIX International Conference on Design of Circuits and Integrated Systems*, Bordeaux, France, 2004.
[24]    Grabowski K., Sankowski W., Napieralska M., and Zubert M., "Iris recognition algorithm optimized for hardware implementation," *Proceedings of the 2006 IEEE Symposium on Computational Intelligence and Bioinformatics and Computational Biology*, Toronto, Canada, 2006, pp. 1–5.
[25]    Hematian A., Chuprat S, Abdul Manaf A., Yazdani S., and Parsazadeh N., "Real-time FPGA-based human iris recognition embedded system: zero-delay human iris feature extraction," *Proceedings of the Ninth International Conference on Computing and Information Technology (IC2IT2013)*, Khulna, Bangladesh 2013, pp. 195–204.
[26]    Rakvic R. N., Ulis B. J, Broussard R. P., Ives R. W., and Steiner N., "Parallelizing Iris Recognition," *IEEE Transactions on Information Forensics and Security*, 2009, Vol. 4 (4), pp. 812–823.
[27]    Hentati R., Bousselmi M., and Abid M., "An embedded system for iris recognition," *Proceedings of Fifth International Conference on Design & Technology of Integrated Systems in Nanoscale Era*, Hammamet, Tunisia, 2010, pp. 1–5.
[28]    Hentati R., Abid M., and Dorizzi B., "Software implementation of the OSIRIS iris recognition algorithm in FPGA," *Proceedings of International Conference of Microelectronics ICM 2011*, Hammamet, Tunisia, 2011, pp. 1–5.
[29]    Liu-Jimenez J., Sanchez-Reillo R., and Fernandez-Saavedra B., "Iris biometrics for embedded systems," *IEEE Transactions on Very Large Scale Integration (VLSI) Systems,* 2011, Vol. 19 (2), pp. 274–282.
[30]    Patil A. S. and Rajbhoj S. M., "Embedded systems for Iris recognition," *International Journal of Electronics and Computer Science Engineering*, 2015, Vol. 4(3), pp. 201–205.
[31]    Wang Y., He Y., Hou Y., and Ting Liu, "Design method of ARM based embedded iris recognition system," *Proceedings of International Symposium on Photoelectronic Detection and Imaging 2007: Related Technologies and Applications*, 2008, SPIE 6625.
[32]    Zhao X. and Xie M., "A practical design of iris recognition system based on DSP," *Proceedings of International Conference on Intelligent Human–Machine Systems and Cybernetics, 2009.* IHMSC'09, Hangzhou, Zhejiang, China, 2009, pp. 66–70.

[33]   http://www.ti.com/solution/iris_biometrics, accessed May 31, 2016.

[34]   Grabowski K. and Napieralski A., "Hardware architecture optimized for iris recognition," *IEEE Transactions on Circuits and Systems for Video Technology*, 2011, Vol. 21(9), pp. 1293–1303.

[35]   Lopez M., Daugman J., and Canto E., "Hardware-software co-design of an iris recognition algorithm," *IET Information Security*, 2011, Vol. 5 (1), pp. 60–68.

[36]   Liu-Jimenez J., Sanchez-Reillo R., Mengibar-Pozo L., and Miguel-Hurtado O., "Optimisation of biometric ID tokens by using hardware/software co-design," *IET Biometrics* 2012, Vol. 1(3), pp. 168–177.

[37]   De Marsico M., Nappi M., Riccio D., and Wechsler H., "Mobile iris challenge evaluation (MICHE)-I, biometric iris dataset and protocols," *Pattern Recognition Letters* (Elsevier), 2015, Vol. 57, pp. 17–23.

[38]   MIR2016 Competition:   http://biometrics.idealtest.org/2016/MIR2016.jsp, accessed June 15, 2016.

[39]   Lu H., Chatwin C. R., and Young R. C. D., "Iris recognition on low computational power mobile devices," in: *Biometrics – Unique and Diverse Applications in Nature, Science, and Technology*, M. Albert (Ed.), InTech, 2011, pp. 107–128.

[40]   http://www.iritech.com/products/solutions/iris-mobile-solution,   accessed June 20, 2016.

[41]   http://www.techshinobiometrics.com/products/iris-recognition-products/iris-module-for-mobile-devices/, accessed June 20, 2016.

[42]   Kang J.-S., "Mobile iris recognition systems: an emerging biometric technology," *Procedia Computer Science*, 2010, Vol. 1 (1), pp. 475–484.

[43]   Kang B. J. and Park K. R., "A new multi-unit iris authentication based on quality assessment and score level fusion for mobile phones," *Machine Vision and Applications*, June 2010, Vol. 21 (4), pp. 541–553.

[44]   Raja K. B., Raghavendra R., Stokkenes M., and Busch C., "Multimodal authentication system for smartphones using face, iris and periocular," in *IAPR International Conference on Biometrics (ICB-2015)*, 2015, Phuket, Thailand.

[45]   Shejin T., "Contributions to practical iris biometrics on smartphones." PhD thesis, College of Engineering and Informatics, National University of Ireland, Galway, 2015.

*Chapter 19*

# Mobile iris recognition

*Akira Yonenaga and Takashi Shinzaki*

## 19.1 Background

An environment is now a place that can offer services utilizing information and communication technology (ICT) in various scenes of daily life, and a wide range of operations and commercial transactions are becoming cloud-based. In this situation, biometric authentication is becoming widespread as a reliable and simple means of user authentication.

Fujitsu started providing biometric authentication devices for PCs in 1999. Subsequently, we have worked on the development of biometric authentication technologies for notebook PCs and smartphones, pursuing convenience as well as security. This paper presents Fujitsu's activities related to biometric authentication technologies, centering on the successful integration of iris authentication in a smartphone for the first time in the world.

## 19.2 Mobile iris authentication

Easy use of iris authentication by smartphone users requires that users be able to hold the smartphone in a natural way during authentication. The distance from the eyes to the sensor of the smartphone is an important factor in this regard. Conventionally, the smartphone needed to be brought near the face in order to achieve sufficient authentication accuracy. To solve this problem, a high-resolution infrared camera and high-power infrared LED were developed, allowing authentication with a practical eye-sensor distance range of 25–35 cm.

Moreover, for convenient use of iris authentication by users, we also developed reliable registration technology for dependable iris image capture, and split-second authentication technology for instant authentication. These technologies make possible the optimal design of the mounting position, spacing, and angle of the infrared camera and infrared LED within the restricted space of smartphones to ensure reliable image capture of the iris (Figure 19.1).

Furthermore, as this would be the first time for users to perform iris authentication on a smartphone, it was important that the operation procedure be clearly relayed. As it is difficult for users to visualize a procedure based on words alone, we decided

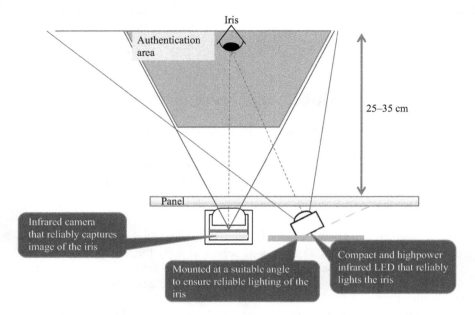

*Figure 19.1   Authentication area using infrared camera and infrared LED*

to show the actual procedure through an animation. To further ensure easy and accurate registration, we decided to provide users with illustrations clearly showing the recommended approach (Figure 19.2).

The arrows NX F-04G smartphone for NTT DOCOMO was the first smartphone in the world to feature an iris authentication function that unlocks the screen through iris authentication that uses the above-described technologies (Figure 19.3) It is also implemented password-less login to online service through the above-mentioned FIDO UAF 1.0 compliance. The combination of the iris authentication and the FIDO UAF 1.0 has realized the security and the ease of use for user authentication on a smartphone.

Since then, the arrows NX F-02H for NTT DOCOMO has been released as a smartphone with the iris authentication. In addition, as the world's first iris authentication equipped tablet, the arrows Tab F-04H for NTT DOCOMO has been shipped since May 2016.

After the arrows NX F-04G, other smart phone vendors also started to present smart phones with built-in iris authentication as a product.

The iris authentication function in smart phones will spread as an important function to improve the user convenience.

## 19.3   Technologies for the future

An iris authentication technology using a visible light camera is a technology for the future. So far, an iris cameras embedded in smart phone products use the near

(1) Display of easy-to-understand animation

Perform registration indoors

Better to perform registration
with glasses removed

Open your eyes wide

(2)To-the-point and easy-to-understand explanations

*Figure 19.2   Iris authentication guidance for smartphone users*

infrared light illumination. On the other hand, the research of the iris authentication
using visible/infrared camera is progressing as an academic research [1]. This type of
technology will integrate in-camera and iris camera into one. Moreover, a research
for the iris image acquisition device for constraint-free in mobile use is progressing
as an academic research [2].

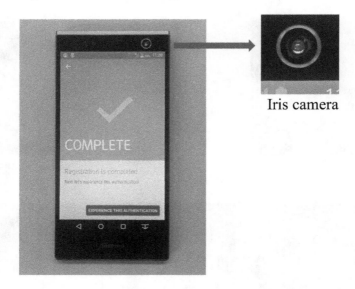

Iris camera

*Figure 19.3    Arrows NX F-04G smartphone (NTT DOCOMO)*

These kinds of technologies will contribute for deployment of an iris authentication in a mobile use.

## 19.4    Summary

We have successfully downsized iris authentication to a level that allows its use in smartphones by using an infrared camera and infrared LED.

Fujitsu aims to further evolve biometric authentication technologies to allow their everyday use by everyone, from children to the elderly. We will proceed with research and development work on biometric authentication technology to make our lives safer, more secure, and more convenient.

## References

[1]    S. Thavalengal, I. Andorko, A. Drimbarean, P. Bigioi, and P. Corcoran. "Proof-of-concept and evaluation of a dual function visible/NIR camera for iris authentication in smartphones" *IEEE Transactions on Consumer Electronics*, vol. 61, no. 2, pp. 137–143, 2015.

[2]    S. Thavalengal, P. Bigioi, and P. Corcoran, "Iris authentication in handheld devices – considerations for constraint-free acquisition" *IEEE Transactions on Consumer Electronics*, vol. 61, no. 2, pp. 245–253, May 2015.

*Chapter 20*

# Future trends in iris recognition: an industry perspective

*Daniel Hartung, Ji-Young Lim, Sven Utcke,*
*and Günther Mull*

Since 1994 iris recognition was established as a state-of-the-art technology in the field of biometric recognition, and after central intellectual property rights expired in 2011 it was established as a reliable alternative to fingerprint and face recognition-based systems. Its relevance is further backed by the integration in the currently largest biometric project (UID [1]) as one of the main biometric modalities.

In this chapter we discuss the current state, pending challenges, and upcoming trends of iris recognition from an industry perspective.

## 20.1 Customer requirements

Industrial research and development is driven by customer requirements. Here we highlight the most relevant aspects, namely robustness, performance, and usability to motivate the following sections.

### 20.1.1 Robustness

Customer requirements with regards to robustness are usually quite simple: iris recognition should be resilient to factors such as the capturing device and its resolution, the quality of the data, or demographic factors.

Some of these requirements can be met simply through the choice of adequate technology—the reason why all commercially obtainable iris cameras work in the Near Infrared (NIR) as recommended by [2] (see Section 20.3.1.1). This will automatically alleviate the problems with low contrast iris patterns in dark brown eyes (Figure 20.1), which are dominant in humans and virtually the only eye colour in many parts of the world (the influence of other demographic factors will be discussed in more detail in Section 20.2.1).

Other factors affecting the robustness, unfortunately, are not so easily addressed. A study by the International Biometric Group [3] suggests that even among high quality sensors the performance decreases considerably when the gallery- and probe-images come from different camera models rather than the same camera. This means

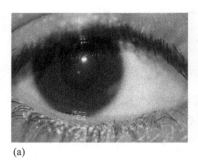 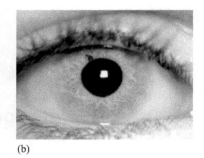

(a)                                                (b)

*Figure 20.1   Irises that show little or no texture in the visible light range
(a) show distinct markings in the NIR (b). Images are taken from the
UTIRIS DB [4]*

that while it is in fact possible to use different iris camera models, or even change camera models during the lifetime of a system, a degradation of the system performance should be expected.

Similarly, the image quality, e.g., off-angle views of the iris, motion blur, and defocus, can considerably influence and in fact dominate the system performance, as demonstrated in [5–7]. Daugman showed that while the distribution of similarity scores for non-mated comparisons is stable and essentially independent of the sample population and camera used, the distribution of similarity scores for mated comparisons can be highly dependent on camera quality and imaging conditions. This can result in vastly different accuracies even over the same population and using the same algorithm. Robustness therefore is, to a very high degree, a direct measure of the quality of the available images and, ultimately, the iris camera and even the exact wavelength used [7]. A good compromise is reported at around 800–850 nm [8].

Figure 20.2 shows the effect when comparing histograms for iris data sets from different capture devices. While the non-mated (impostor) distribution is essentially the same for both databases, the mated (genuine) distribution is considerably different. Figure 20.2(a) indicates higher fidelity for the samples than the ones in Figure 20.2(b).

## 20.1.2   Performance

In this section we adhere to the ISO/IEC biometric performance testing and reporting standard [9] for discussing the capabilities of iris-based biometric systems.

### 20.1.2.1   Accuracy

In combination with the individually unique iris texture and its apparent stability in the feature sets, iris recognition has gained popularity as a robust and reliable biometric technology. Daugman states in [6] that the dimensionless as well as fixed-length compact iris feature representation enables real-time decisions about personal identity with high confidence. This allows even a very large database to be searched

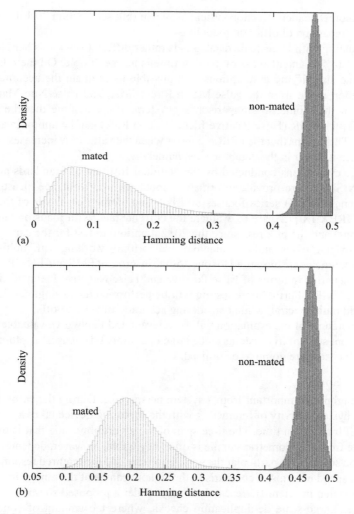

*Figure 20.2   Differences in the distributions for mated comparisons depending
on image capturing device: better quality samples in distribution
(a) than in (b)*

exhaustively (i.e., one-to-many, 'identification' mode) without making significant
errors.

The extent of iris usage in practice is very large: by June 2016 the number of
data subjects enrolled in the Aadhaar project exceeded 1 Billion, with each new
enrolee's iris pair being compared to all others for de-duplication [1]. Also, recent
publications [10,11] show that iris recognition is regarded as the most reliable and

accurate biometric identification system available (but see Section 20.1.4 for a more detailed comparison of different modalities).

Evaluating with large-scale databases is rather difficult since it is hard to obtain such big data sets and also the evaluation time is not negligible. On the other hand, under some simplifying assumptions it is possible to estimate the accuracy in the identification mode from the False Match Rate (FMR) and False Non Match Rate (FNMR) of the biometric comparison subsystem. If we assume that the FMR is very small, the FPIR (False Positive Identification Rate) can be approximated with $N \times$ FMR. This means that the FPIR increases when the value of $N$ increases and often the increase is linear in the database size, namely $N$.

Public evaluations conducted by the National Institute of Standards and Technology (NIST) aim to provide information about such performance measures. The Iris Exchange (IREX) series focuses on iris-based biometrics. Goals of their latest iteration, IREX IX [12], are to evaluate the current state (with focus on large-scale applications) and the progress since the 2012 iteration, to test for the capabilities to handle 'difficult' cases, and to evaluate new algorithms working with multi-spectral images. Accuracies are measured for one-to-one in terms of FMR and FNMR, and one-to-many scenarios in terms of False Positive and False Negative Identification rates (FPIR and FNIR). Furthermore testing will be performed using single or dual-eyes, in NIR and multispectral, while considering accuracy-time trade-offs.

The outcome of the evaluation will be relevant and likely give valuable insights in the current state of iris systems since large operational databases[1] captured with a variety of sensors are going to be utilised.

### 20.1.2.2   Speed

High throughput is important from a system perspective. During the enrolment process the physical activity of interacting with the capturing device takes a majority of the overall transaction time. The time span of the algorithmic side that is extracting a template from the biometric sample is often negligible. However, depending on the application scenario, in particular the population size and the required response times, algorithm speed in terms of comparison throughput can be of the utmost importance. Often neglected in scientific research, it may render a proposed solution practically unfeasible. Large-scale de-duplication checks, where the amount of comparisons grows quadratic with the total number of templates available, require high speed and high accuracy comparison algorithms. Similarly response times for an identification scenario can be critical, where a single template is compared against a large background data set.

One of the strong points of common algorithms is the possibility to transform the iris image and texture information into a binary representation of fixed length and structure. In this structure, rotations can be compensated by simply shifting the feature set in one dimension. While a certain degree of rotational tolerance is important, larger tolerances lead to decreasing accuracies due to the increasing number of possibilities for false matches. It is noteworthy that with increased rotational tolerance

---

[1]Multi-spectral data set contains 220,000 iris samples, operational data sets OPSII/III with several million respectively 700,000 iris samples.

the comparison speed is negatively influenced. In our experience, a tolerance of about ±10–15° is recommendable depending on the system set-up.

The simple structure and the binary form of the feature sets help to achieve high comparison speed: if classic fractional Hamming Distance-based comparison algorithms are used, manifold speedup times can be achieved with implementations based on CPU intrinsics or GPU parallelisation compared to classic implementations. Our technology achieves millions of comparisons per CPU and second.

## 20.1.3 Usability

This requirement is often less considered when designing biometric systems. However, the interfaces and ergonomics have a major influence on the overall user acceptance and on the performance in terms of throughput and recognition accuracy.

There are two different perspectives on usability. On the one hand, usability can be approached in the context of ergonomics of human–system interaction as defined in ISO 9241-11 [13] as 'The extent to which a product can be used by specified users to achieve specific goals with effectiveness, efficiency, and satisfaction in a specified context of use'. This user-centric approach is independent of the application scenario and it is widely applied in *usability testing*.

On the other hand, usability can be seen from a *human computer interaction* perspective. The main purpose of this methodology is to improve certain goals. In the context of biometric systems the influence of usability on the system performance, such as the throughput or the biometric accuracy, is the major concern. The Human-Biometric-Sensor Interaction (HBSI) usability model [14] follows this approach and was introduced to investigate the thoroughness of traditional performance evaluation metrics. It focuses on real time recommendations and user feedback during the inter-action with a biometric system. Less widely applied, it aims in particular to reduce the cases of failed enrolments due to operating errors.

A new work item proposal at the ISO SC37 level aims at unifying the two different approaches into a single usability performance methodology.

Most recently, at IBPC 2016, several talks focused on iris usability related topics. Sirotin [11] reported on the usability and user perceptions of self-service biometric technologies. In this work, three different capturing procedures for face and iris systems are investigated in an immigration scenario: (1) the user is responsible for correct placement inside a narrow capture volume, (2) the user stands still and the device locates and captures the region of interest within a larger capture volume, and (3) the operator adjusts a narrow capture volume device. Then, three measurements were evaluated, namely the rate of Failures To Acquire (FTA), where no image could be captured, the transaction time, and user satisfaction. In conclusion, scenario (1) led to low customer satisfaction, large FTA rates, and high transaction times whereas the other scenarios performed significantly better. (2) and (3) performed on par with a ten-print fingerprint system in terms of user satisfaction but with faster transaction times. Further investigations were performed to identify the major usability issues leading to slow transactions. It was revealed that positioning is the main issue in scenario (1), gaze and interference (e.g., eyelashes, eye lids, glasses) were issues in (2) and (3). Consequently they recommend a system layout that allows for still standing users without the need to correctly positioning themselves in front of the camera, and furthermore a design with reduced need for a long steady gaze.

Another real-life study presented by Bachenheimer [10] showed that when measuring the FTA rates, iris recognition systems performed better than fingerprint or face systems for biometric data subjects that were four years or older. The study could not clarify if the failures were caused by the inherent properties of the biometric trait, the capability of the capturing system, the usability, or other unknown factors. However, in practice the source of such errors is often negligible if customer requirements are met.

Customers require high throughput rates for the enrolment and the authentication phases to minimise costs and maximise user satisfaction, on the other hand the quality of the biometric samples has to be sufficient. Easy-to-use interfaces for data subjects as well as operators help to achieve these goals. Detailed information about usability in biometric recognition systems can be found in [15].

### *20.1.4   Choosing a modality*

Regarding the before-mentioned requirements, the most prominent alternatives when choosing a biometric modality are fingerprint and face. However, a general recommendation for one or the other is difficult and it depends on many factors. Oftentimes it comes down to keeping legacy data usable: a clear advantage for fingerprint. Besides the obvious difference in available data sources (e.g., ten fingerprints, one face, two irises), there are interesting other aspects.

Due to our background with fingerprint systems, we want to state our experience about performance differences: generally the accuracy of a system is growing with the availability of distinct biometric samples from a data subject. Regular fingerprint capturing set-ups make therefore use of one, two, four, eight, or ten fingers for each enrolee. Figure 20.3 shows an estimation of the distinguishing power of these set-ups

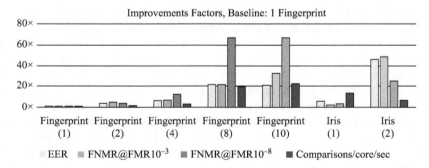

*Figure 20.3    Improvement factors in terms of accuracy (equal error rates (EER), false nonmatch rates (FNMR) at specific false match rates (FMR)) and comparison speed of multiple fingerprint and multiple iris systems compared to a single fingerprint system. The number in brackets denotes the number of distinct biometric sources for each enrolee*

compared to one and two iris-based systems. The comparison speed improvements are given as well. Note that the baseline is a single fingerprint print system (factor $1\times$) and that despite adding more data with growing number of fingers, DERMALOG's technology is able to speed up the comparison significantly: the ten-print system set-up offers highest comparison speed ($23\times$ improvement) and highest accuracy for identification scenarios ($68\times$ improvement for FNMR@FMR=$10^{-8}$), whereas the accuracy of a two iris system set-up is highest for verification scenarios ($49\times$ improvement for FNMR@FMR=$10^{-3}$).

## 20.2 Development process

In the scientific literature, the challenges met during the development process are often merely skimmed over, or not mentioned at all. Here, we describe some of the pitfalls and workflows faced during the development process.

### 20.2.1 Bias and variability

As already discussed in Section 20.1.1, different iris cameras will produce at least subtly varying images. According to [16] such variations will be due to the sensor used—CMOS versus CCD, with the latter being somewhat less susceptible to noise, but also different pixel density and spacing, pixel–pixel crosstalk, sensor spectral sensitivity, photo response uniformity—the lens used—whether it is optimised for NIR, its overall quality with regards to image sharpness, distortion level and uniformity of brightness—the illumination used—wavelength, uniformity, light level. For cheaper or more slovenly assembled cameras this can lead to a high in-class variability even if all images were taken with the same camera, resulting in poor performance and high FNMR [5] (see Figure 20.2 for comparison). However, even for high quality cameras all those different design decision will generally result in an increase in in-class variability when compared to the use of a single camera, and will subsequently result in lower accuracy [3].

In addition to those technical factors which can be controlled at least theoretically, we also have to consider the variability within the population using this system. Eyeglasses, contact lenses, and even makeup (false eyelashes or heavy mascara) can all influence recognition performance. In [17] the authors have shown that both eyeglasses as well as contact lenses decrease the iris image quality scores defined in [2]. Also, according to our experience, at least glasses will also directly influence segmentation results, leading to a slightly increased Failure to Enrol (FTE) rate and decreased mated similarity comparison scores (mostly due to reflections or dirt on the glasses). How different eye-colours will affect comparison scores is not known to the authors, but [17] reported no differences in image quality scores. Figure 20.4 depicts some typical problem cases.

Easily visible characteristics such as glasses or makeup, however, are not the only factors influencing recognition rates. A phenomenon has been documented in [18]

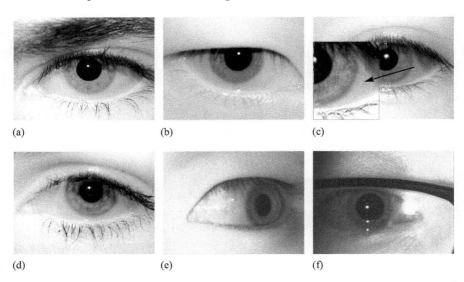

(a)                        (b)                        (c)

(d)                        (e)                        (f)

*Figure 20.4    Various challenging iris sample images: (a) shadows due to thick eye
lashes, (b) Asian eye with lashes interfering iris texture, (c) additional
circular structures from soft contact lens outside iris, (d) extensive eye
lash interference due to mascara, (e) off-axis gaze angle, (f) irregular
reflections from glasses. Images (a)–(d) are taken from
ND-IRIS-0405 [19] and (e) and (f) from CSIR2015 [20]*

and could be reproduced in our own research work, where a pronounced difference in
performance between left and right eyes is measured. Left eyes on average creating
higher mated similarity scores than the right eyes. The exact mechanism behind
this is, however, still unclear, except for noting that it is related to eye dominance.
For subjects with a right-dominant eye there is no statistically relevant difference
in iris recognition accuracy between the two eyes, while for subjects with a left-
dominant eye iris recognition performance is statistically significantly better for the
left iris.

## 20.2.2    Parameter optimisation

Probably the greatest challenge in iris recognition is to be robust across multiple
imaging devices. The algorithms need to avoid overfitting while still reaching excellent accuracies without specialised parameterisations. This proves to be a non-trivial
task and takes up a significant amount of the industrial development process. While
approaching the challenge scientifically, it still is a game of 'whack-a-mole' (once
you straightened out one problem, others cop up) due to the different nature of the

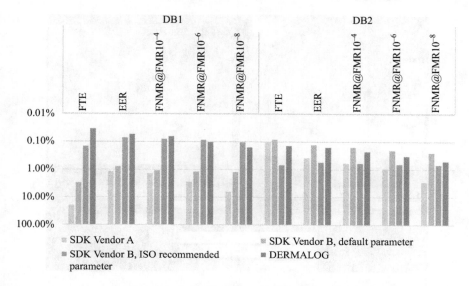

*Figure 20.5   Accuracy comparison of commercially available SDKs (latest releases May 2016, higher bars correspond to better performance) on two different data sets DB1 [21] and DB2 (subset from LG4000 iris sensor) [22]. The accuracy is measured in terms of equal error rate (EER), false nomatch rates (FNMR) at specific false match rates (FMR), and the failure to enrol rate (FTE). Error rates are given in percent. DERMALOG's solution shows consistently low error rates without the need for adapting parameters*

data sources. Internally we utilise several open and commercially available iris data sets for training, testing, and evaluation.

Some of the commercially available SDKs show a strong reliance on finding specific parameters for specific input data to reach consistently low error rates. This dependency in real application scenarios is often critical since the data is not available during algorithm development and tuning. In Figure 20.5, it can be seen that the accuracy for three different commercially solutions (Vendor A, B, and from DERMALOG) on two data sets. The data sets DB1 [21] and DB2 (subset from LG4000 iris sensor) [22] consist of about 30,000 iris images each resulting in about 450 Million symmetric comparisons.

When comparing the SDKs, the before-mentioned phenomenon can be observed in the behaviour of SDK B: on DB2 SDK B outperforms the others with default parameters, however it performed significantly worse on DB1. SDK A causes an unacceptable high failure to enrol rate (19.7%) on DB1, which renders it unusable. DERMALOG's solution performs consistent and proves its robustness with the standard

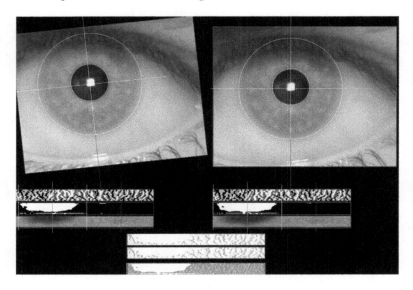

*Figure 20.6    A mated comparison with rotated images shown in the comparison analyse tool: best rotation at 6° with a fractional Hamming Distance of 0.1*

parameterisation over different data sets with lowest error rates and FTE on DB1 (until $FNMR@FMR = 10^{-6}$).

## 20.2.3    Evaluation and tools

In the development process of iris recognition technology, we use several tools, in particular a benchmarking engine and a comparison analyser. The main purpose of using such tools is to accelerate the development process, to make benchmark evaluation repeatable, reproducible, and transparent.

The benchmark engine enables us to benchmark different modalities and technologies with a bundle of various database and parameter sets. Benchmarking a technology with different data sets makes it possible to evaluate its robustness. And it allows for comparing the accuracy of the different technologies and parameter sets on the same data sets. We systematically analyse and compare the results of a benchmark in the form of DET / ROC curves, cumulative histograms (linear or logarithmic), or CMC curves.

The comparison analyser is designed to manually check single comparisons (see Figure 20.6). It decodes the extracted iris templates and shows the matching details. In this way it is possible to identify problem classes in lowest mated and highest nonmated comparisons. For a given database, we could reliably find labelling errors with the before-mentioned selection of scores (or conceal attacks as in Fig 20.7). Since the lower scores of mated comparisons are usually caused by segmentation errors, we can collect problem cases even for large databases.

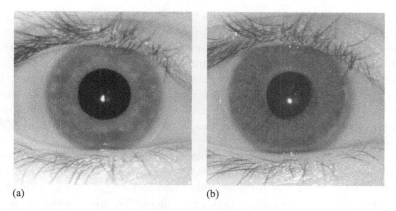

(a)                                                     (b)

*Figure 20.7    Cropped mated sample images from the ND-CrossSensor-Iris-2013*
*data set [22], (a) image 05845d54 with regular contact lens,*
*(b) image 05845d257 with patterned contact lens (covering the iris*
*texture and parts of the pupil) showing a conceal attack*

## 20.3    Future trends and challenges

As reflected by the structure of the book, there are still many interesting research
lines and challenges in iris recognition. Here we are going to discuss relevant trends
from an industrial perspective. In particular we give an overview of standardisation
activities, presentation attack detection (aka liveness detection), as well as algorithm-
related trends. Furthermore sensor-related trends, such as multimodality, visible light
iris recognition, and mobile biometrics are introduced.

### 20.3.1    Sensor-related trends

#### 20.3.1.1    Capturing technology

Iris cameras come in a variety of physical form-factors: from hand-held binocular-like
devices over tablet-like hand-held and wall-mounted devices up to automated e-gate
solutions. The latter type usually offers stop-and-stare logic capturing iris images
from as far away as 3 m.

However, all of these cameras use essentially the same technology: active illumination in the near infrared (NIR) and sensors with daylight cut-off filters capturing
images of the iris with sufficient resolution, usually at least $R = 80$ pixel as recommended by [23] (compare Section 20.3.4). Especially the e-gate type iris cameras,
however, tend to capture irises at lower resolutions due to technical constraints. The
artificial NIR illumination has to be mounted in such a way that no reflections will
obscure the iris. This requires either a nearly confocal light source, resulting in a
reflection directly on the pupil, where it does not interfere with the iris-image, or
illumination from the side at nearly right angles, leaving the reflection well outside
the iris on the sclera.

The main reason for this reliance on NIR is the low contrast of iris structures in dark brown eyes when viewed under visible light (see Figure 20.1(a)). In the NIR the same eye shows rather pronounced markings, see Figure 20.1(b), while blue eyes usually exhibit marked striations in both, the visible light range as well as in the NIR.

Another factor common to most cameras is the image format. While many cameras incorporate some logic and early pre-processing of the image, most are capable of outputting a VGA-sized (640 × 480 pixel) image, often with the pupil already centred. Many cameras can even output a pre-segmented image according to ISO/IEC 19794-6, compare Section 20.3.4.

However, the technical challenges do not stop there. Most off-the-shelf components today are geared towards the visible spectrum and colour imaging. As such they generally feature infrared cut-off filters rather than the daylight cut-off filters needed, RGB-Bayer filters and anti-aliasing (low-pass) filters, all of which are unsuited for an iris camera. Additionally, the current trend goes towards higher pixel densities due to the reduction of sensor surfaces and increase of sensor resolution. This leads to ever more problems with noise due to fewer and fewer photons being captured by each individual pixel, as well as problems with diffraction, see Section 20.3.1.2.

In the same vein, standard commercial lenses are usually designed for the visible light range and generally perform poorly in the infrared. This starts with the lens coating for reducing reflection and light loss. Each glass surface absorbs about 4%–5% of near infrared light, typically a lens will have between 10 and 20 of those. Here a coating optimised for the NIR can reduce those losses to about 0.5%–1% each. Another problem is caused by diffraction, the so-called 'Airy disk'. This describes the best focused spot of light that even a perfect lens can make at a specific aperture. Its diameter is directly proportional to the wavelength and therefore an image in the infrared can never be as sharp as one taken in visible light. More on the problems of sensor design can be found in the next section.

## 20.3.1.2    Raising the capture distance

As mentioned in Section 20.3.1.1, iris sensors come in several different form factors. However, until quite recently they all shared a common feature: they required the active cooperation of the data subject. This might come in the form of a binocular-like device raised to your eyes or, in some cases, requiring the user to bend over the ocular as if looking through a microscope: raising concerns about hygiene as well as causing discomfort of looking into a 'dark tunnel'. Or they might come in the form of handheld- or wall-mounted devices that either have to be positioned at just the right distance and angle in front of a user's eyes or, possibly worse, require the users to bend over and position their eyes at just the right distance and angle in front of the device.

While all these cameras can produce qualitatively good images and are usually convenient for experienced users, they can provide considerable difficulties for the first time user and are therefore particularly unsuited for unattended scenarios, and for application areas such as border control. While many such systems are in fact in use at international borders (e.g., in the United Arab Emirates, at Schiphol airport in Amsterdam, and in several Canadian and US Airports as part of the CANPASS or

NEXUS program) others such as the British customs have dropped their intent due to difficulties of use [24]. Moreover, wheelchair bound users are usually not able to use these systems at all, as at least the wall-mounted units require a minimum size of, e.g., 150 cm in the case of Schiphol airport. Compare Section 20.1.3 for more information about the influence of ergonomics on both user satisfaction as well as recognition results.

Those constraints have given rise to a new generation of iris-cameras, so-called 'stop-and-stare' type scanners usually integrated into e-gates. Here, users are only required to stop at a line and stare at a screen for around one second while iris images are taken from a distance of between 1 and 3 m. These systems usually also work over a much larger height-range (typically for heights between 1 and 2 m) and might therefore also be usable for users with disabilities. The most current crop of these systems do not even require you to stop, but can capture images of subjects moving at normal walking speed.

This marks the beginning, and current research is focusing on several extensions to existing systems:

1. Iris at *large* distances: rather than taking images of irises at distances from 1 m to 3 m and with the subject's (implied) consent and cooperation, for many security applications it would be beneficial to take images from much greater distances, and current research has reached distances of more than 10 m [25–27].
2. Visible light irises: so far, iris capture requires special equipment working in the NIR. Should it be possible to use the visible light spectrum instead, use of iris recognition could be greatly expanded, from mobile devices on the one side (compare Section 20.3.1.5) up to...
3. ...Ubiquitous (or latent) iris: with more and more cameras, with increasing resolutions for every new generation, surrounding us in everyday life, it may only be a question of time if and when the first applications will develop both for law enforcement as well as other tasks such as personalised services.

However, quite a few technical obstacles would need to be overcome in order to extend the possible capture range far beyond the now customary 1–3 m. The dilemma faced by industry and researchers alike is that, with increased capture distance, either lenses with longer focal length or sensors with (considerably) more pixels are needed (as the number of pixels is proportional to the square of the distance covered). Unfortunately, both approaches come with their own set of limitations.

Using longer lenses results in an ever decreasing depth of field (inverse proportional to the focal length). This could in theory be offset by using higher f-stops, but this then would result in less light reaching the sensor, which could in turn be offset by using larger sensors (i.e., sensors with larger individual bin-size capable of capturing more photons—but leading in turn to larger focal length) or longer exposure times. Increasing the amount of available light is usually not an option due to eye safety considerations, and in any case poses its own difficulties due to the amount of light being proportional to the inverse square of the distance.

However, longer exposure times usually are not an option either, as short exposure times are needed to counter motion blur due to the object moving and, even for a cooperative subject standing still, due to saccadic eye movement. And using longer lenses means a severely restricted field of view, requiring pan-tilt-zoom (PTZ) heads to position the camera in the direction of the iris as well as a second, larger field of view camera to direct the first. But these, in turn, would probably be susceptible to vibrations, again requiring short exposure times.

Using sensors with more pixels is no easy option either. Apart from the fact that due to the quadratic growth of the number of pixels needed one quickly ends up in the 100th or 1,000th of megapixel, far surpassing any currently existing sensor, we would face essentially the same problems as before, with the exception of those introduced by the PTZ. Additionally the higher number of pixels would lead to smaller individual pixel sizes, confounding the problems with sensor noise, illumination, and (due to the smaller allowable circle of confusion) depth of field.

Illumination, finally, comes with its own sets of problems. The inverse-square law basically determines that the amount of illumination needed grows quadratic with the distance. This, however, poses considerable eye-safety concerns, as the same radiant flux that leads to a perfectly save irradiance at a distance of 10 m will lead to a $100\times$ higher irradiance at 1 m distance. This problem is exacerbated by the fact that the human eye lacks any protective counter measures against the invisible NIR illumination. Only a light source emitting completely parallel rays such as through a telecentric lens would not have that problem, but would require a bulkier set-up and a PTZ for the illumination too. Note that IR lasers, while originally affording parallel rays, would not offer a solution as the illuminated area is too small without prior beam widening.

And even if we solved all the aforementioned problems for NIR-based cameras, this would still require the costly installation of custom cameras, as current NIR-based algorithms cannot work with any of the currently already deployed cameras working in visible light, be it surveillance cameras or cameras in mobile phones. This maybe the main reason for the interest in algorithm working in the visible light spectrum, as discussed in Section 20.3.1.4, and in particular periocular-based approaches as described in Section 20.3.1.5.

### 20.3.1.3   Mobile iris biometrics

Mobile biometrics—and in our case mobile iris biometrics—come in two distinct flavours. One is the use of mobile solutions in the context of law enforcement or official security applications; the other is the use for authentication to log into mobile phones, tablets, and laptops—but also, at least conceptually, for authentication in mobile e-pay solutions.

Since the beginning of this decade law enforcement agencies all over the United States have been equipped with handheld biometric devices which attach to smart phones and provide iris, face and fingerprint recognition. And in mid-2015, the Kenya Ministry of Education started using iris-cameras connected to an android phone or tablet in order to track student attendance in school busses or in class. The system can hold up to 500 identities (expandable to 5,000 identities) locally, no communication with other devices is needed or used.

As different as the mentioned cases might sound, what both share is the use of dedicated hardware connected to a mobile phone, rather than use of the mobile phone's hardware itself.

On first glance this seems to be different from the second flavour, using iris recognition to log into mobile phones. Microsoft's Lumia 950 and 950 LX prominently feature this technology, as did the mid-2015 Fujitsu NX F-04G before and the Fujitsu NX F-02H or the withdrawn Samsung Galaxy Note 7 since. However, despite their 'normal' appearance, all these phones internally use a dedicated iris camera and NIR illumination and therefore essentially the same technology common in 'standard' iris recognition. And despite a growing interest in algorithms capable of dealing with images from the visible light spectrum (compare Section 20.3.1.4) it stands to reason that at least for the foreseeable future mobile iris biometrics will rely on dedicated hardware.

### 20.3.1.4 From NIR to visible spectrum

One of the new trends of iris recognition technologies is to overcome the necessity for NIR light and to extend its application area into different wavelength. This trend, while driven from mobile phone and security applications, can easily be observed in the scientific community too (see also Sections 20.3.1.1 and 20.3.1.3).

Multi-purpose sensors, as in the average mobile phone, are designed to capture the visible light spectrum and to block the NIR spectrum that is commonly used in iris recognition systems. Research line focuses on making use of such visible light iris images, despite potential issues regarding dark iris pigmentation and reflections.

Most iris cameras capture light in the 700–850 nm range of the electromagnetic spectrum. Since the transmission, absorption, and reflection vary within biological iris components (i.e., the pupillary sphincter muscle and dilator muscle), some researches ([28–30]) propose to incorporate different portions of the electromagnetic spectrum for representing certain physical characteristics of the epigenetic iris pattern. The importance of multi-spectral iris analysis can be assessed by the IREX IX [12] evaluation plan (compare Section 20.1.2). IREX IX aims to test how well recognition algorithms can segment and compare iris samples captured over wavelengths ranging from 400 nm to the infrared. Cross-spectral comparisons will also be tested. Additionally, the evaluation will cover the ability of comparison algorithms to handle colour-captured visible-spectrum images with possible applications in forensic iris recognition.

Boyce [28,29] concentrated on the imagery of visible (400–700 nm) and the near IR (700–1,000 nm) ranges of iridial light reflection. An assessment involving various eye colours across these ranges is performed. In particular, the role of information represented in individual spectral channels/wavelengths (i.e., IR, Red, Green, and Blue) on the comparison performance of iris recognition is studied. This work addressed that the use of multispectral information has the potential to enhance the segmentation and enhancement procedures thereby improving the performance of iris recognition systems.

Recently, Wild *et al.* [30] studied the impact of score-level fusion and cross-spectral performance in multispectral iris recognition systems using four feature extraction techniques. Namely LG (Log Gabor), QSW (Wavelet-based), DCT (Discrete Cosine Transform), and 2DG (2D Gabor) were tested on the public UTIRIS multispectral iris dataset [4]. The results indicated that features can be rather susceptible to spectral channels (such as DCT delivering best NIR but worst RGB performance). Cross-spectral performance turned out to be highly challenging (EERs > 33% for comparing NIR against RGB channels across feature types), confirming increases in EERs as the difference in wavelength increases, but further also indicating a more pronounced degradation for Green versus Blue compared with Red versus Green intra-colour cross-spectral application.

Regarding iris recognition technologies based on the visible spectrum, we cannot identify any established standard technologies and scientific approaches yet. We are following the developments with a great deal of interest.

### 20.3.1.5  Multi-modality and periocular

Iris-based biometric systems are well established and commercially available. In marked contrast, research on periocular based recognition systems is still young. From previous works on forensic face biometrics it is however known that the eye area contains rich discriminative features even in low quality images from CCTV cameras [31].

Potentially holistic eye recognition systems can overcome current iris recognition systems in terms of robustness and accuracy while avoiding known issues. If quality control fails during enrolment or we are dealing with legacy or forensic data, periocular information may be used to further lower the rate of biometric samples that cannot be processed, or to avoid interference issues. In fact, disruptive factors for iris recognition such as eye lids, eye lashes, and other facial hair covering the essential iris texture, could be used as discriminating features. Additionally features such as the eye shape, skin texture, and information from the sclera area are potentially extractable from common iris images. Advantage of this approach would be the possibility to augment iris systems that are based on the ISO 19794-6 standard [2] without modifications of the capture devices.

Another approach focuses on visible light images where textures from dark irises can hardly be captured [32–34]. Hollingsworth *et al.* [35] studied the usefulness of different features from the periocular region for human experts to distinguish subjects. From most useful to least useful the list reads as follows: eye lashes, tear duct, eye shape, eye lid, eye brows, eye corner, and skin texture. When designing algorithms the feature list should be considered, but it remains unclear if over longer periods of the time features remain stable enough. Interested readers can refer to [36–42] for further works on periocular biometrics. Elsewhere in this book, Smereka *et al.* were presenting their work on the best periocular regions for biometric recognition.

In conclusion, following the trend towards multi-modal systems, periocular information will be used for augmenting common NIR iris based systems to increase the recognition accuracy and to overcome some limitations. On the other hand, periocular information, also available in the visible spectrum, can augment face recognition systems. The information can prove valuable in particular when parts of the face are covered. Since there are no accepted standard approaches, as available for iris recognition systems, research effort has to be invested to bring periocular into products. Currently DERMALOG is supported by ZIM (Zentrales Innovationsprogramm Mittelstand) to follow this goal.

## 20.3.2 Presentation attack detection

As mentioned in Section 20.3.4, the importance of presentation attack detection (PAD) is reflected by recent standardisation activities [43]. Scope of the standard is presentation attacks at the sensor level. Following the nomenclature two types of attackers are differentiated: biometric imposters (who intend to be recognised as another individual) and biometric concealers (who seek to conceal their biometric characteristics).

In the context of iris recognition, one obvious conceal presentation attack is the use of patterned contact lenses worn on top of the eye ball. Classified as partial artificial attack instrument, these contact lenses are challenging to detect and can be used to conceal the biometric characteristics. Two mated images in Figure 20.7 from the ND-CrossSensor-Iris-2013 data set [22] indicate the difficulty to reach high similarity scores when the texture information is overlaid with a patterned contact lens.

The standard [43] gives a framework for PAD methods and defines two categories: (1) through data capture subsystem and (2) through system-level monitoring. Iris-based PAD systems could make use of challenge-response liveness detection—for instance with changing visible light illumination the pupil size is expected to change involuntarily. A voluntary challenge could be a cue to close a certain eye. Examples for non-challenge liveness detection possibilities are the Hippus motion (rhythmic and regular movements of the iris muscles) and multispectral absorption properties of the eye area.

The first work on iris and face presentation attack detection shows promising results, in [44,45] an Average Classification Error Rate (ACER) of 0% is reached for the ATVS fake iris database [44]. Another database of 3 300 visible spectrum iris artefacts is presented in [46] together with an analysis of print and electronic display attacks. A visible light iris database gathered with smart phone cameras is presented in [47], their PAD mechanism based on Eulerian Video Magnification (EVM) on replayed iris presentation reaches an ACER of 0% as well. In [48] the Hippus motion is investigated. An experiment is set up and a database is gathered on which their proposed algorithm, based on SVM (Support Vector Machine) decisions on features describing the Kohn and Clynes pupil dynamics model [49], separates presentation attacks from normal presentations perfectly.

The issue of conceal attacks using contact lenses is approached in the literature as well. Databases such as the ND Contact Lens Database 2013[2] or the IIT-Delhi Contact Lens Iris Database[3] are utilised extensively in current research works [50,51]. Other

---

[2] https://sites.google.com/a/nd.edu/public-cvrl/data-sets, June 2016.
[3] https://research.iiitd.edu.in/groups/iab/irisdatabases.html, June 2016.

publications use stereo imaging in the visible light to make use of the three dimensional structure of the eyeball to detect presentation attacks [52].

From a recent approach [53], we could find an investigation of iris spoofing detection based on deep representations (see Section 20.3.3 for more on deep learning). This method dealt with iris spoofing printed attacks and some experimental datasets using cosmetic contact lenses.

PAD is going to attract further attention when the relevant ISO/IEC standards are finalised and customers are requesting compliance.

## 20.3.3    Deep learning

While the classical approach to iris detection, segmentation, and coding has been described in the first chapter, there exist alternative approaches. Among those, representation learning based on deep architectures has attracted a great deal of attention not only in the image processing and computer vision community but also in the biometrics community since learning representations of the data make it easier to extract useful information when building classifiers or other predictors.

Besides earlier neural network research work inspired by the architectural depth of brain, a breakthrough happened in 2006 by proposing learning algorithms such as Restricted Boltzmann Machine and auto encoders, apparently exploiting the same principle: guiding the training of intermediate levels of representation using unsupervised learning, which can be performed locally at each level (Bengio [54]).

In the field of face recognition, for example, a deep-learning based technology using a Convolutional Neural Network (CNN) is widely used. Extensive researches have been made to reach a better performance and indeed they achieve significant improvement in many applications. The success of such methods is due to (1) the recent technological improvements for data acquisition, storage, and processing, (2) mastering of recognition and classification tasks superior to classically feature-based methods, (3) availability of large quantities of training data, and (4) open source communities that enable sharing the know-how and trained nets information. One drawback, however, of learning-based methods is the need to label ground truth for the training data, which is laborious and time-consuming.

In the field of iris recognition, deep learning-based methods are not yet established. As described in Section 20.3.2, recent work describes presentation attack detection methods based on deep representations [53]. Otherwise, iris recognition methods currently are mainly based on conventional feature-based methods.

Motivated by the success from the scientific field, we had carried out some feasibility studies of deep learning-based methods—besides others on iris masking and on discrimination of left and right eyes (see Figure 20.8):

**Masking:** our experience shows that the accuracy of iris recognition varies strongly with the quality of the iris mask. We compared purely learned representations vs. hand-crafted features with learned classifiers. The accuracies are comparable; however, purely learning-based approaches are more sensitive to the amount and quality of the training data.

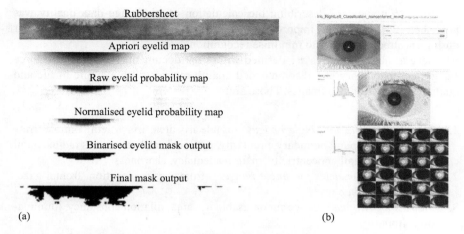

Figure 20.8   *Potential applications of deep learning in iris recognition:*
              *(a) iris masking and (b) left- and right-eye discrimination*

**Left- and right-eye discrimination:**  based on the Caffe,[4] and DIGITS[5] frameworks,
the trained CNNs reach high classification accuracy.

**Comparison:** Promising results could be generated from Siamese comparison
networks for iris feature sets.

It remains to be seen if deep learning based methods can also help to further
improve iris recognition algorithms or if the traditional approaches can keep ahead.

## 20.3.4  Standards

The main goals of standardisation activities are to guarantee interoperable systems
and to avoid vendor lock-in. On the international level, the ISO/IEC SC37 deals with
the standardisation of biometric systems. Here we introduce the current state relevant
to iris biometrics and we give some future directions.

The importance of the iris modality is reflected by the adoption of the data interchange format in 2005 and its revised version from 2011 [2]. The standard defines
an image based data format. For the capturing process the recommended range of
wavelengths is in the NIR (700–900 nm). Inspired by the first Interoperable Iris
Exchange report by NIST [55], the compression formats were changed from JPEG
to JPEG2000 and lossless PNG. Additionally, optionally considerable data reduction
can be achieved by utilising segmentation information to crop and mask-out every-
thing (with predefined grey-values for lids, sclera, and pupil) except for the relevant
iris texture information.

---

[4] http://caffe.berkeleyvision.org/tutorial/, June 2016.
[5] https://developer.nvidia.com/digits, June 2016.

In 2015 a standard describing the calculation of iris image data quality was published [23]. This is an important aspect of any biometric system, in particular during enrolment, in order to minimise recognition errors.

Several quality metrics are defined within the document, separated into three categories: mandatory and recommended quality metrics from a single image and quality metrics from two images. Those are:

1. *Required metrics for single images*—usable iris area, iris–sclera contrast, iris–pupil contrast, pupil boundary circularity, grey scale utilisation, iris radius, pupil dilation, iris pupil concentricity, margin adequacy, sharpness.
2. *Recommended metrics for single images*—frontal gaze-elevation, frontal gaze-azimuth, motion blur.
3. *Metrics for two images*—common usable iris area, dilation constancy, illumination similarity.

Also worth mentioning is the ISO/IEC biometric performance testing and reporting standard [9] that defines a common basis for comparing the performance of biometric systems. Its nomenclature and protocols are widely adopted. They were also utilised here to create comparisons in Section 20.1.2.1. Currently a multipart standard on presentation attack detection (PAD), also referred to as spoofing and liveness detection, is being developed. The framework part is already published under [43]. PAD is going to be one of the key technologies, especially considering a growing interest in unattended solutions.

In the continuation of this section we are describing some limitations of the current iris-related standards and we provide directions for further developments.

Despite the importance of the ISO iris quality estimation standard [2], it has some limitations that may affect its usefulness. A general development in biometric quality estimation, as seen within the NFIQ 2.0 project, leads to estimating the *utility* (defined in [56]) of a biometric sample as its quality, ultimately defining quality as an estimation of its influence on biometric accuracy. In contrast to focusing on human readable hand-crafted features, as in the current iris sample quality standard, learning-based combinations of more complex features are state of the art in accurate estimation. The current standard lacks some clarity in the aggregation of a final quality scalar: absolute values for the coefficients combining the various metrics are missing, instead normative instructions for optimisations are given.

Besides the general approach of how to define quality, one major issue of the selected metrics is their dependence on a correct segmentation. Out of the ten mandatory single image quality metrics, solely the *Sharpness* and the *Grey scale utilisation* metrics are segmentation-independent. Images that cannot be properly segmented therefore produce unreliable values for the other eight metrics. Furthermore, despite describing standard image processing procedures, the *Sharpness* metric algorithm has an intellectual property (i.e., protected by patents) issue. A possible next iteration could even further improve the standard regarding these points.

## 20.4 Summary

Iris is a well-established biometric modality; next to fingerprint and face recognition it is reliable, mostly intellectual property free, and thus interesting for commercial use. Current developments open further possibilities for scientific research and for significant improvements: as reflected by the book chapters, multi-modal systems making use of periocular data, new sensors and mobile biometrics, presentation attack, and contact lens detection as well as privacy preserving technologies are among the most promising directions. Furthermore with the latest IREX benchmark [12], capabilities of multi-spectral iris recognition are going to be pushed forward.

## Acknowledgement

Currently DERMALOG is supported by ZIM (Zentrales Innovationsprogramm Mittelstand)[6] conducting the project BIPED (Biometric Iris PEriocular Development) to approach and combine iris and periocular into an holistic eye recognition system.

## References

[1]    Government of India. *Aadhaar—Unique Identification*, 2015 (Retrieved 2016-06-16). https://uidai.gov.in/library/references.html.

[2]    ISO/IEC JTC 1/SC 37. Information technology—biometric data interchange formats—part 6: Iris image data. ISO 19794-6:2011, International Organization for Standardization, Geneva, Switzerland, October 2011.

[3]    International Biometric Group. Independent testing of iris recognition technology (ITIRT)—final report. Technical report, International Biometric Group, May 2005.

[4]    M. S. Hosseini, B. N. Araabi, and H. Soltanian-Zadeh. "Pigment melanin: Pattern for iris recognition," *IEEE Transactions on Instrumentation and Measurement*, 59(4):792–804, April 2010.

[5]    J. Daugman. "The importance of being random: Statistical principles of iris recognition," *Pattern Recognition*, 36(2):279–291, February 2003.

[6]    J. Daugman. "How iris recognition works," *IEEE Transactions on Circuits and Systems for Video Technology*, 14(1):21–30, January 2004. Invited Paper.

[7]    J. Daugman. Results from 200 billion iris cross-comparisons. Technical Report UCAM-CL-TR-635, University of Cambridge, Computer Laboratory, June 2005.

[8]    H. T. Ngo, R. W. Ives, J. R. Matey, J. Dormo, M. Rhoads, and D. Choi. "Design and implementation of a multispectral iris capture system," in *43rd Asilomar*

[6]http://www.zim-bmwi.de/, June 2016.

*Conference on Signals, Systems and Computers*, Asilomar'09, pp. 380–384, Piscataway, NJ, USA, November 2009. IEEE.

[9] ISO/IEC JTC 1/SC 37. Information technology—biometric performance testing and reporting—part 1: Principles and framework. ISO 19795-1:2006, International Organization for Standardization, Geneva, Switzerland, April 2006.

[10] D. Bachenheimer. "Use of biometrics in migration: UN high commission on refugees," International Biometric Performance Conference, http://www.nist.gov/itl/iad/ig/ibpc2016_presentations.cfm, 2016.

[11] Y. Sirotin. "Usability and user perceptions of self-service biometric technologies," International Biometric Performance Conference, http://www.nist.gov/itl/iad/ig/ibpc2016_presentations.cfm, 2016.

[12] G. W. Quinn, P. Grother, and J. Matey. "IREX IX: Multi-spectral iris evaluation—concept, evaluation, and API specification," Technical Report Version 1.1, National Institute of Standards and Technology, June 2016.

[13] ISO/TC 159/SC 4. Ergonomics of human-system interaction—part 11: Usability: Definitions and concepts. ISO 9241-11, International Organization for Standardization, Geneva, Switzerland, 2015. under development.

[14] E. P. Kukula. *Design and evaluation of the human-biometric sensor interaction method*. PhD thesis, Purdue University, West Lafayette, IN, USA, August 2008. AAI3337302.

[15] R. Blanco Gonzalo. *Usability in biometric recognition systems*. PhD thesis, Universidad Carlos III de Madrid, February 2016.

[16] P. D. Wasserman. "Digital image quality for iris recognition," in *Biometric Quality Workshop I*, p. 35, USA, March 2006. National Institute of Standards and Technology, NIST.

[17] L. Wang. "Iris verification and ANOVA for iris image quality," *Journal of Automation and Control*, 2(1):33–38, March 2014.

[18] A. Sgroi, K. W. Bowyer, and P. Flynn. "Effects of dominance and laterality on iris recognition," in *Computer Vision and Pattern Recognition Workshop (CVPRW)*, pp. 52–58. IEEE, Piscataway, NJ, June 2012.

[19] P. J. Phillips, W. T. Scruggs, A. J. O'Toole, *et al.*, "FRVT 2006 and ICE 2006 large-scale experimental results," *IEEE Transactions on Pattern Analysis and Machine Intelligence*, 32(5):1–1, May 2010.

[20] CASIA (Chinese Academy of Sciences' Institute of Automation). CSIR2015 database, used for 2015 ICB competition on cross-sensor iris recognition. http://biometrics.idealtest.org/2015/csir2015.jsp, 2015.

[21] University Bath. Iris DB1600 image database. http://www.smartsensors.co.uk/es/wp-content/uploads/2010/10/SSL-price-list-July-2011.pdf, 2014.

[22] University of Notre Dame du Lac (UND) CVRL Lab. ND-CrossSensor-Iris-2013. https://sites.google.com/a/nd.edu/public-cvrl/data-sets, 2013.

[23] ISO/IEC JTC 1/SC 37. Information technology—biometric sample quality—part 6: Iris image data. ISO 29794-6:2015, International Organization for Standardization, Geneva, Switzerland, July 2015.

[24] J. Tozer. £9 million down the drain as airports scrap iris passport scanners which were meant to speed up queues...because they are slower than manual checks. *Mail Online*, February 2012. http://www.dailymail.co.uk/news/article-2102076/Millions-drain-airports-SCRAP-iris-passport-scanners.html.

[25] R. Meyer. "Long-range iris scanning is here," *The Atlantic*, May 2015. http://www.theatlantic.com/technology/archive/2015/05/long-range-iris-scanning-is-here/393065/.

[26] S. Venugopalan, U. Prasad, K. Harun, *et al.*, "Long range iris acquisition system for stationary and mobile subjects," in *International Joint Conference on Biometrics (IJCB)*, pp. 1–8. IEEE, Piscataway, NJ, October 2011.

[27] J. A. De Villar, R. W. Ives, and J. R. Matey. "Design and implementation of a long range iris recognition system," in *44th Asilomar Conference on Signals, Systems and Computers*, pp. 1770–1773. IEEE, Piscataway, NJ, November 2010.

[28] C. K. Boyce. *Multispectral iris analysis: Techniques and evaluation*. Master's thesis, College of Engineering and Mineral Resources at West Virginia University, Lane Department of Computer Science and Electrical Engineering, Morgantown, West Virginia, USA, 2006.

[29] C. K. Boyce, A. Ross, M. Monaco, L. Hornak, and X. Li. "Multispectral iris analysis: A preliminary study," in *Computer Vision and Pattern Recognition Workshop (CVPRW)*. IEEE, June 2006.

[30] P. Wild, P. Radu, and J. Ferryman. "On fusion for multispectral iris recognition," in *Eighth IAPR International Conference on Biometrics (ICB)*, pp. 31–37. IEEE, Piscataway, NJ, May 2015.

[31] T. Ali, P. Tom, J. Fierrez, R. Vera-Rodriguez, L. J. Spreeuwers, and R. N. J. Veldhuis. "A study of identification performance of facial regions from CCTV images," in *5th International Workshop on Computational Forensics (IWCF)*, pp. 1–9, Madrid, Spain, November 2012. Autonomous University of Madrid, Spain.

[32] S. Bharadwaj, H. S. Bhatt, M. Vatsa, and R. Singh. "Periocular biometrics: When iris recognition fails," in *Fourth International Conference on Biometrics: Theory, Applications, and Systems (BTAS)*, pp. 1–6. IEEE, Piscataway, NJ, September 2010.

[33] U. Park, R. R. Jillela, A. Ross, and A. K. Jain. "Periocular biometrics in the visible spectrum," *IEEE Transactions on Information Forensics and Security*, 6(1):96–106, March 2011.

[34] U. Park, A. Ross, and A. K. Jain. "Periocular biometrics in the visible spectrum: A feasibility study," in *Third International Conference on Biometrics: Theory, Applications, and Systems (BTAS)*, pp. 1–6. IEEE, Piscataway, NJ, September 2009.

[35] K. P. Hollingsworth, K. W. Bowyer, and P. J. Flynn. "Useful features for human verification in near-infrared periocular images," *Image and Vision Computing*, 29(11):707–715, October 2011.

[36] K. P. Hollingsworth, S. S. Darnell, P. E. Miller, D. L. Woodard, K. W. Bowyer, and P. J. Flynn. "Human and machine performance on periocular biometrics

under near-infrared light and visible light," *IEEE Transactions on Information Forensics and Security*, 7(2):588–601, April 2012.

[37]    R. R. Jillela, A. A. Ross, V. N. Boddeti, *et al.* "Iris segmentation for challenging periocular images," in *Handbook of Iris Recognition*, pp. 281–308. Springer, Berlin, January 2013.

[38]    J. R. Lyle, P. E. Miller, S. J. Pundlik, and D. L. Woodard. "Soft biometric classification using periocular region features," in *Fourth International Conference on Biometrics: Theory, Applications, and Systems (BTAS)*, pp. 1–7. IEEE, Piscataway, NJ, September 2010.

[39]    P. F. G. Mary, P. S. K. Paul, and J. Dheeba. "Human identification using periocular biometrics," *International Journal of Science, Engineering and Technology Research (IJSETR)*, 2, May 2013.

[40]    P. E. Miller, J. R. Lyle, S. J. Pundlik, and D. L. Woodard. "Performance evaluation of local appearance based periocular recognition," in *Fourth International Conference on Biometrics: Theory, Applications, and Systems (BTAS)*, pp. 1–6. IEEE, Piscataway, NJ, September 2010.

[41]    A. Ross, R. Jillela, J. M. Smereka, *et al.*, "Matching highly non-ideal ocular images: An information fusion approach," in *Fifth IAPR International Conference on Biometrics (ICB)*, pp. 446–453. IEEE, Piscataway, NJ, March 2012.

[42]    M. Uzair, A. Mahmood, A. Mian, and C. McDonald. "Periocular biometric recognition using image sets," in *Workshop on Applications of Computer Vision (WACV)*, pp. 246–251. IEEE, Piscataway, NJ, January 2013.

[43]    ISO/IEC JTC 1/SC 37. Information technology—biometric presentation attack detection —part 1: Framework. ISO 30107-1:2016, International Organization for Standardization, Geneva, Switzerland, January 2016.

[44]    J. Galbally, J. Ortiz-Lopez, J. Fierrez, and J. Ortega-Garcia. "Iris liveness detection based on quality related features," in *Fifth IAPR International Conference on Biometrics (ICB)*, pp. 271–276. IEEE, Piscataway, NJ, March 2012.

[45]    R. Raghavendra and C. Busch. "Presentation attack detection algorithm for face and iris biometrics," in *22nd European Signal Processing Conference (EUSIPCO)*, pp. 1387–1391. IEEE, Piscataway, NJ, September 2014.

[46]    R. Raghavendra and C. Busch. "Robust scheme for iris presentation attack detection using multiscale binarized statistical image features," *IEEE Transactions on Information Forensics and Security*, 10(4):703–715, April 2015.

[47]    K. B. Raja, R. Raghavendra, and C. Busch. "Video presentation attack detection in visible spectrum iris recognition using magnified phase information," *IEEE Transactions on Information Forensics and Security*, 10(10):2048–2056, October 2015.

[48]    A. Czajka. "Pupil dynamics for iris liveness detection," *IEEE Transactions on Information Forensics and Security*, 10(4):726–735, April 2015.

[49]    M. Kohn and M. Clynes. "Color dynamics of the pupil," *Annals of the New York Academy of Sciences*, 156(2):931–950, April 1969.

[50]    D. Gragnaniello, G. Poggi, C. Sansone, and L. Verdoliva. "Contact lens detection and classification in iris images through scale invariant descriptor," in *10th*

International Conference on Signal-Image Technology and Internet-Based Systems (SITIS), pp. 560–565. IEEE, Piscataway, NJ, November 2014.

[51] D. Yadav, N. Kohli, J. S. Doyle, R. Singh, M. Vatsa, and K. W. Bowyer. "Unraveling the effect of textured contact lenses on iris recognition," *IEEE Transactions on Information Forensics and Security*, 9(5):851–862, May 2014.

[52] K. Hughes and K. W. Bowyer. "Detection of contact-lens-based iris biometric spoofs using stereo imaging," in *46th Hawaii International Conference on System Sciences (HICSS)*, pp. 1763–1772. IEEE, Piscataway, NJ, January 2013.

[53] D. Menotti, G. Chiachia, A. da Silva Pinto, *et al.*, "Deep representations for iris, face, and fingerprint spoofing detection," *IEEE Transactions on Information Forensics and Security*, 10(4):864–879, April 2015.

[54] Y. Bengio. *Learning Deep Architectures for AI*. NOW, the essence of knowledge, now Publishers Inc., PO Box 179, 2600 AD Delft, The Netherlands, November 2009.

[55] P. Grother, E. Tabassi, G. W. Quinn, and W. Salamon. IREX interoperable iris exchange I: Performance of iris recognition algorithms on standard images. Technical Report NIST Interagency Report 7629, National Institute of Standards and Technology, September 2009.

[56] ISO/IEC JTC 1/SC 37. Information technology—biometric sample quality—part 1: Framework. ISO 29794-1:2009, International Organization for Standardization, Geneva, Switzerland, August 2009.

# Index